METROPOLITAN PROBLEMS

METROPOLITAN PROBLEMS

International Perspectives: A search for comprehensive solutions

Edited by

SIMON R. MILES

Routledge
Taylor & Francis Group

LONDON AND NEW YORK

First published in 1970

This edition published in 2007
Routledge
2 Park Square, Milton Park, Abingdon, Oxon, OX14 4RN

Routledge is an imprint of Taylor & Francis Group, an informa business

Transferred to Digital Printing 2007

© 1970 Methuen publications

The publishers have made every effort to contact authors and copyright
holders of the works reprinted in the *The City* series. This has not been
possible in every case, however, and we would welcome correspondence
from those individuals or organisations we have been unable to trace.

These reprints are taken from original copies of each book. In many cases
the condition of these originals is not perfect. The publisher has gone to
great lengths to ensure the quality of these reprints, but wishes to point out
that certain characteristics of the original copies will, of necessity, be
apparent in reprints thereof.

British Library Cataloguing in Publication Data
A CIP catalogue record for this book
is available from the British Library

Metropolitan Problems
ISBN10: 0-415-41834-8 (volume)
ISBN10: 0-415-41930-1 (subset)
ISBN10: 0-415-41318-4 (set)

ISBN13: 978-0-415-41834-8 (volume)
ISBN13: 978-0-415-41930-7 (subset)
ISBN13: 978-0-415-41318-3 (set)

Routledge Library Editions: The City

Metropolitan Problems

International perspectives

A search for comprehensive solutions

Edited by
Simon R. Miles

INTERMET
Metropolitan Studies
Series

Toronto · London · Sydney
Wellington

Methuen

Library of Congress Catalog Card Number 75-94381

ISBN 0-485-90290-7

Designed by Hiller Rinaldo

1 2 3 4 5   74 73 72 71 70

Contents

v

Foreword

Much has been written, and even more said, about the potential of concerted action by the public and private sectors in attempts to evolve solutions to the problems facing major metropolitan areas throughout the world. Undaunted by the fact that the record of accomplishment in this area was minimal, the Bureau of Municipal Research made an irreversible commitment when, in 1966, it decided to sponsor and mount the Centennial Study and Training Programme on Metropolitan Problems.

Believing that such a multi-faceted subject required an equally integrated and embracive approach, we selected eleven infrastructural and service fields, commissioned study papers on each, assisted in the formation of multi-interest study groups in forty participating metropolitan areas in thirty countries, held a preliminary meeting of group leaders, and then conducted the Seminar-Conference proper at York University, in Toronto, in August of 1967. Because the project was Canadian-sponsored, we were able to enlist the participation of metropolitan areas representative of developed and developing technologies and of a full spectrum of political ideologies. That we were successful is testimony to the virtue of immodesty in concept.

It is always difficult to credit those

individuals and institutions whose contributions led to the success of a complex undertaking. Perhaps the most prudent course would involve an exhaustive listing of these acknowledgements, as is attempted by the editor. Another would be to mention only those persons who were absolutely essential to the venture. In the case of the Centennial Study and Training Programme on Metropolitan Problems, three individuals meet this demanding criterion – Donald M. Deacon, Programme Chairman; F. Warren Hurst, Bureau President; and Simon R. Miles, Assistant Director of the Bureau, who was seconded as Programme Associate Director, and who has since become Executive Director of The International Association for Metropolitan Research and Development (INTERMET). It is revealing that the first two of these key persons are from the leadership of the private sector of Metropolitan Toronto.

Of all the benefits and rewards that sponsorship of this project has bestowed upon the Bureau of Municipal Research, none can equal the fact that formation of a permanent international association was requested by the participants. By building upon the invaluable network of concerned and knowledgeable persons who participated in the initial phases of this project, an even more significant contribution to the solution of world metropolitan problems may be made with the emergence of that permanent organization – INTERMET – in 1968. Failure will not be the result of lack of effort.

Dominic DelGuidice
Toronto, 1970

Acknowledgements

This book is based upon the Centennial Study and Training Programme on Metropolitan Problems. The very nature of the Programme, the objectives of which are described in the Introduction, implies the active involvement of many persons from varied institutions in different parts of the world. The scale of the Programme, its duration and total cost have necessarily called for the support of many more individuals and institutions. Although this book is only one of the many products of this Programme, it provides the only opportunity to acknowledge the help and co-operation so generously provided. As the Associate Director of the Programme, and being responsible for its day-to-day operations, I would not wish to miss this opportunity to express my personal thanks to all those individuals who played such an important part in its development.

I arrived in Canada after the idea had already been born out of discussions between John Pearson of the Department of Municipal Affairs of Ontario, and Michael Goldrick, who was then Executive Director of the Bureau of Municipal Research. The idea then took shape through the beatings and gentle manipulations received at the hands of then younger notables, such as Martin Meyerson and Robert C. Wood, at the M.I.T.-Harvard Joint Center for

Urban Studies, and others, such as Emil Sady and Ernest Weissmann, at the United Nations in New York. Learned visitors to Toronto, and I recall Jean Gottman as one of them, were prevailed upon for further comment.

The Bureau staff has always been small and it was not until the Autumn of 1965 that we had a proposal that seemed to be viable. However, at this point Warren Hurst, the new President of the Bureau, invited Donald Deacon to become Chairman of the Centennial Programme. The leadership of these two outstanding members of the business community was lent further administrative and technical support when Dominic DelGuidice joined the Bureau staff as Executive Director and we acquired the consultative expertise of Meyer Brownstone, as Executive Director of the Programme, and David Smith, as Conference Consultant. The appreciation of this entire staff must be extended to the local Advisory Committee members: Hans Blumenfeld, Eli Comay, J. Stefan Dupré, James Gillies, George O. Grant, Macklin Hancock, Harold Kaplan, William Laird, William J. McCordic, John O. E. Pearson and Albert Rose.

In establishing study groups around the world we benefited tremendously from the contacts provided by the United Nations Public Administration Division, and shortly thereafter the Programme was formally recognized as being sponsored with the cooperation of the Division. I owe much to the support offered us by the Division and, in particular, Emil Sady. Similarly, invaluable assistance was provided by the International Union of Local Authorities and the Toronto Dominion Bank in establishing contacts in places remote to the Bureau.

Shortly after the forty study groups were established, the group leaders became involved in the final planning stages of the Programme. Thanks are due to the Government of France for making available the Abbaye de Royaumont for a meeting of the group leaders. The group leaders were the key force in this Programme of self-help and cooperation. Without them there would have been no exchange of ideas for the further development of a common understanding of very similar problems and the pursuit of solutions to these problems. Nor would there have been the rapport that developed among the groups and the *esprit de corps* that emerged throughout the entire network. In some cases these group leaders were supported by excellent secretaries, in others the leader assumed all the responsibility. In a few cases the leadership changed.

Thus, while wishing to extend my general gratitude to all members of all groups, in particular I should like to thank the following: Michael P. Goutos, Athens; Chamnan Yuvapurna, Bangkok; Mihailo Radovanovic, Belgrade; Istvan Bartos, Budapest; Carlos Mouchet, Buenos Aires; Mohsen Idris and M. M. El-Hafnawi, Cairo; Jehangir Kabir and M. G. Kutty, Calcutta; Peter B. Spivak, Detroit; H. B. Fraser and R. E. McClary, Edmonton; George P. Richardson and R. E. Nicoll, Glasgow; Erika Spiegel and Rudolf Hillebrecht, Hanover; O. Bateye and Adebayo Adedeji, Ibadan; Kemal Ahmet Aru, Istanbul; Kenneth Roome, E. M. Penrose and Patrick R. B. Lewis, Johannesburg; Chauhdri Shah Nawaz and Khalid Shibli, Karachi; Valentin Kamensky, Leningrad; Eduardo Dibos and Alfredo Saravia T., Lima; Sir William Hart, W. L. Abernethy and A. W. Peterson, London; Luis Sanches Agesta and Amado de la Cruz, Madrid; Carlos Ramos, Aprodicio Laquian and L. A. Viloria, Manila; Gustavo Martinez Cabanas, Mexico City; Miro Allione, Bernardo Secchi and Cesare Macchi-Cassia, Milan; Lucien Saulnier and Georges Longval, Montreal; Alexey Kudriyavsev, Moscow; Takeo Saito and Tadao Kameda, Nagoya; Arturo Polese and Luciana de Rosa, Naples; Henri van der Weidje, Netherlands; John P. Keith and Pearl Hack, New York; Michel Piquard, Pierre Viot, Paul Delouvrier and Maurice Doublet, Paris; William Rafsky and Dale Taylor, Philadelphia; Jan Formanek and Frantisek Budsky, Prague; Diogo Lordello de Mello, Rio de Janeiro; Victor Jones, Willis Hawley and Stanley McCaffrey, San Francisco; Il Such Cha and Chung-Hyun Ro, Seoul; Ruth Atkins and George Clarke, Sydney; Katsuhiko Onogi and Masamichi Royama, Tokyo; William Allen and Dominic Del-Guidice, Toronto; Hilda Symonds and H. Peter Oberlander, Vancouver; Zygmunt Skibniewski and Jerzy Przezdiecki, Warsaw; Moya Tweedie and Elswood Bole, Winnipeg.

These study groups produced a wealth of material, translated in English or French, upon which I have been able to draw in putting together this book. In this regard I cannot extend thanks enough to the authors of the original study papers; first for their willingness to subject themselves to extensive commentary and criticism, secondly for providing leadership in the seminars at the conference, and thirdly, for their cooperation in the revision of their studies in the final stages of the Programme. Again, I should like to thank the International Union of Local Authorities for permitting us to use the paper by Professor Hanson that was based on a paper prepared initially for an I.U.L.A. conference in Bangkok in 1966.

The Seminar-Conference, in Toronto, August, 1967, again made great demands upon many individuals, and at this juncture I should like to thank Ed Annis and his administrative staff at York University, Geoff Milburn of Consumers' Gas Company, Jim Main, of the Ontario Government, Edgar Norris of the City of Toronto, Stephen Zacks and Dick Dodds as additional recruits to the Secretariat staff, Michael Goldrick and his team of rapporteurs, and the many Toronto families who welcomed the visitors to their homes.

Financial support for the Programme was broad-based, including all levels of government, foundations and the business community. Of especial significance to the Programme were the contributions of the Province of Ontario and the Ford Foundation. These two institutions quickly recognized the merits of the proposal and their positive response enabled us to launch the Programme without further delay. Support then followed from A. E. Ames Company Limited; Atkinson Charitable Trust; Borough of North York; Canadian Bankers Association; Consumers' Gas Company; Dominion Securities Corporation Limited; Donner Canadian Foundation; T. Eaton Company; Government of Canada (through the Central Mortgage and Housing Corporation, the Department of Health and Welfare, the Department of Transport, and the National Capital Commission), Imperial Oil Company of Canada; Maclean-Hunter Publishing Company; the Province of Manitoba; The Municipality of Metropolitan Toronto; the Province of Quebec; Robert Simpson Company Limited; Alfred P. Sloan Foundation; Steel Company of Canada; Tippet Foundation; Toronto Harbour Commission; and Wood Gundy Securities Limited. Non-cash contributions in various forms were made available by Air Canada; Central Mortgage and Housing Corporation; the City of Toronto; Consumers' Gas Company; Gestetner (Canada) Limited; Guild Inn; I.B.M. Company Limited; Mission

Press; Olivetti-Underwood Limited; Ontario Hydro; Royal Trust Company; Robert's Gallery; St. John United Church and Parkview United Church, Stratford; Southam Press Limited; Swissair; Toronto Dominion Bank; Toronto Harbour Commission; Toronto Study Group; the Toronto Transit Commission and Xerox of Canada Limited.

There is one group of people whom I find difficult to place in the above listing, for their help has been of especial significance to me not only in the writing of this book but throughout the Programme. I refer to my secretaries. Thinking back, there seems to have been an alarming number of them, but at least they managed to retire before the going became too much for them! In particular I must thank Mrs. Ute Wright and Patricia Prescott-Clarke for the resilience they showed under pressure, Thelma Marco-Bilbao and Angela Arkell for their contributions at the later stages in the Programme, and Mrs. Geraldine Sharpe for typing much of the final manuscript. To Penny Dutton I owe thanks for having successfully taken on the arduous task of indexing and proofing the manuscript.

To Meyer Brownstone and Aprodicio Laquian I owe special thanks for reviewing parts of the manuscript, and to the staff of Methuen Publications I owe much for their patience. Any errors of interpretation, I regret to say, are mine.

S.R.M.
Toronto, 1970

Contributors

TIBOR BAKÁCS is Head of Medical Studies at the Faculty of Medicine in Budapest and Director-in-Chief of the Hungarian Institute of Public Health. He has published close to a hundred scientific papers and two university textbooks—*Hygiene* and *A Study of General and Particular Infectious Diseases*. He has also served as a lecturer at, and advisor on, international conventions sponsored by the World Health Organization.

J. STEFAN DUPRÉ is Chairman of the Department of Political Economy at the University of Toronto and a member of the National Research Council of Canada. An expert in the study of intergovernmental relations, he has published articles on the role of the federal government in support of research in Canadian universities and on the subject of intergovernmental finance in the Province of Ontario. He has also been a contributing author for several books, including *Area and Power* (edited by A. Maass).

RICHARD L. FORSTALL, presently editor of the *International Atlas* for Rand McNally and Company, has also been responsible for editing the *Commercial Atlas*. This publication is probably the most comprehensive reference available on the location, population and transportation services of American localities. He is a co-author of *The World's Metropolitan Areas* and has published articles in *The Municipal Yearbook*.

JULIUSZ GORYNSKI is Director of the Postgraduate Training Department for Regional Planners at the Central School of Planning and Statistics in Warsaw, Poland. He has written books on housing standards, urbanization, town planning and architecture in addition to approximately two hundred pamphlets, articles and reports.

VICTOR JONES is Professor of Administrative Science at the University of California, Berkeley. Although he has held this post since 1955, he has also served as visiting Assistant Professor at Yale University and, in 1967, Senior Resident Fellow at the Washington Center of Metropolitan Studies. His publications include *Local Government Organization in Metropolitan Areas*.

A. H. HANSON is Professor of Politics, University of Leeds, and has worked on problems of public enterprise and economic planning with special reference to developing countries for many years. He has taken part in missions under the auspices of the United Nations and Organization for Economic and Cultural Development.

URSULA (LADY) HICKS is a Fellow of Linacre College, Oxford University, and an Honorary Fellow of the Institute of Social Studies, The Hague. Books she has written include *Development Finance: Planning and Control* (1965) and *Public Finance* (1955, revised 1968). She has been published widely in journals as well.

JOHN F. KAIN is Assistant Professor of Economics, Harvard University, and is a member of the M.I.T.-Harvard Joint Center for Urban Studies. He is also a consultant to the Rand Corporation and to the U.S. Department of Housing and Urban Development. His published work includes *The Urban Transportation Problem* (with John R. Meyer and Martin Wohl).

SIMON R. MILES served as Associate Director of the Centennial Study and Training Programme on Metropolitan Problems and is now Executive Director of INTERMET. He also lectures at York University, in Toronto, and is the author of several papers on metropolitan area development.

HUGH PHILP is Professor of Education and Director of the School of Education at Macquarie University, New South Wales, Australia. He is a former director of UNESCO's Institute for Child Study in Bangkok (1959-1964) and Chief of their Division of Educational Studies and Research (1964-1966).

DR. ALBERT ROSE is Director of the School of Social Work, University of Toronto, and Chairman of the Consultative Committee on Housing Policies for the City of Toronto. He is engaged in a wide range of community activities and has published a number of books and articles on social problems, urban development and planning. In 1967 he was awarded the Canadian Centennial Medal for his outstanding contributions in the field of social welfare.

ZYGMUNT RYBICKI has been Rector of the University of Warsaw since February, 1969. He is president of the Polish Association for the Popularization of Science and Culture and a member of the International Political Science Association and other international associations. His publications include *The Citizen and the State, Provincial Economy Administration*, and a number of scientific articles.

HIDEHIKO SAZANAMI is Chief of the Urban Facilities Research Group of the Building Research Institute, Ministry of Construction, of the Japanese Government. He has also been a visiting fellow at the M.I.T.-Harvard Joint Center for Urban Studies as well as senior lecturer at the Institute of Social Studies, The Hague.

FRANK SMALLWOOD is Chairman of the City Planning and Urban Studies Program at Dartmouth College and Director of the Dartmouth College Public Affairs Center. Among his major publications are *Metro*

Toronto: A Decade Later and *Greater London: The Politics of Metropolitan Reform*. Numerous articles and reviews of his have appeared in such publications as *National Civic Review* and *Urban Affairs Quarterly*.

ERNEST WEISSMANN is Senior Advisor on Regional Development for the Office of Technical Cooperation of the Department of Economic and Social Affairs of the United Nations and a former director of that organization's Centre for Housing, Building and Planning. He has lectured widely on housing and on urban and regional development, and he is the author of a large number of reports for the United Nations.

Introduction

"Men come together in cities in order to live. They remain together in order to live the good life."[1] That was two thousand years ago. Aristotle was underlining for his audience the importance of the quality of life, as opposed merely to continued biological existence. At that time, his remarks were a commentary on the desirability of the then existing urban social system. Today, they are an exhortation.

What went wrong?

One can only wonder whether we are progressing far, or in the right direction, in pursuit of the good life. One must accept as inevitable man's dissatisfaction with his lot; it is but a reflection of his implacable urge to improve his condition. Yet, what we seem to be experiencing today is a growing dissatisfaction without a compensating facility to react.

Why is this so?

Three "explosions"—knowledge, population and technology—have presented our world with problems and possibilities. The physical development of Aristotle's city into today's metropolitan area has been made possible by an expansion of knowledge and technology. This growth has

[1]Aristotle, *Politics*, Book 1, Chapter 2; a rough but literal translation rendered by Tom Robinson of the Department of Philosophy, University of Toronto.

xvi

produced some benefits and has the potential for many more. However, it has also sharpened conflicts, increased dangers, multiplied complexity and, most important in this context, has posed a real challenge to the institutional arrangements whereby men have attempted to order their lives. Physical change is outpacing institutional change.

There is no reason why institutional change should not accompany physical change, but if frustration is not to mount further, there must be a response now, and this response must come from every sector of society. The problem is global and the sharing of ideas in the pursuit of solutions should be worldwide.

It was in response to this need that the Bureau of Municipal Research, in Toronto, decided that an appropriate way for it to celebrate Canada's Centennial in 1967 would be to initiate an international study of this phenomenon. As a result, forty metropolitan areas around the world embarked upon the Centennial Study and Training Programme on Metropolitan Problems. The study was designed to produce insights, through comparative analysis, into methods of coping with the difficult, complicated and interlocking sets of problems associated with the growth of the metropolis in modern society. The whole Programme extended over a period of almost two years.

This book draws upon the materials produced at various stages in the Programme and, as such, represents the major findings of those participating. The nature of these findings is influenced, quite understandably, by the process by which they were produced. Furthermore, the findings were only one of many products of the process. Others ranged from the establishment of new institutions to the more intangible result of improved working relations among and within institutions in the participating metropolitan areas.

It is because the methodology employed to approach a problem has so significant an effect on the chances of success and the nature of that success that this book is presented also as a commentary on the particular approach to metropolitan problems used in the Centennial Programme. It is for this reason that a brief introduction to the design of the Programme may be helpful here.

The Centennial Programme was organized in four stages:

Stage I: the production of the eleven basic study papers;
Stage II: the preparation of the review papers by the local study groups;
Stage III: the Seminar-Conference in Toronto;
Stage IV: the preparation of the findings and the follow-up by local groups.

Those organizing the Programme considered that increased understanding, that is, new perceptions coupled with knowledge, is a prime requisite for creative solutions and that only an interdisciplinary approach would yield such understanding. Thus, throughout the Programme, such an approach was utilized: in the selection of authors for papers; in the composition of the local groups in each participating metropolitan area; in the preparation of the review papers; and, in the organization of the Seminar-Conference sessions.

In the first stage, the Programme was initiated by the Toronto Bureau of Municipal Research. Participating groups in forty metropolitan areas in various parts of the world were brought into the study and basic study papers were commissioned. In the second stage the forty study groups prepared basic information on their metropolitan areas, reviewed one of the study papers, and selected and briefed their team of, in most cases, six people to attend a Seminar-Conference in Toronto in August, 1967. These two stages had three objectives.

One objective was to provide a comparative analysis of the main problems of metropolitan areas. The authors of the eleven study papers were selected from various disciplines and different countries. The papers were not, however, regarded as definitive. Consequently, these papers, which appear in chapters two to seven inclusive and chapters nine to thirteen inclusive have undergone considerable modification since their original presentation. Apart from one general paper, which appears as Chapter 1, each of the basic study papers dealing with a particular problem or topic was subjected to review by four study groups chosen so that differing cultural backgrounds, political systems and stages of development would be reflected in the review papers.[2]

The second objective was to establish in each of the participating metropolitan

---

[2]The forty metropolitan area study groups that participated in the Centennial Programme are listed here and grouped according to the paper that each had as its focal point for its studies.

The Functional Metropolis and Systems of Government—Montreal, Naples, Paris and Tokyo; Financing Metropolitan Government—Calcutta, Istanbul, Johannesburg and Winnipeg; Government Administration and the Political Process—Ibadan, London, Manilla and Vancouver; Intergovernmental Relations—The Netherlands, Philadelphia, Seoul and Sydney; Metropolitan Area Transportation—Athens, Bangkok, Milan, and San Francisco Bay Area; Planning and Urban Design—Buenos Aires, Hanover, Karachi and Warsaw; Education in the Metropolis—Detroit, Mexico City, Moscow and Prague; Utilities—Budapest, Edmonton, Nagoya, and Rio de Janeiro; Housing in Metropolitan Areas—Belgrade, Glasgow, Lima and Toronto; Health and Welfare—Cairo, Leningrad, Madrid and New York.

areas a study group composed of representatives of various areas of specialty and different interests from the local community. Thus, there would be representatives from different levels of government—both elected and appointed, the universities, the business sector and citizen organizations. In most instances these study groups consisted of fifteen to twenty persons. Thus in this stage alone, over six hundred persons throughout the world were actively involved in a common programme. These study groups were expected to initiate study in their own areas and to brief their team to be sent to the Seminar-Conference by discussion of the background papers. In addition, each study group prepared a review paper: an in-depth study of one of the basic study papers as it related to the other papers in the context of their own metropolitan area.

In this way an holistic approach was adopted toward the study of problems confronting the metropolis: holistic in terms of content—with all major problem areas being considered at one time; holistic in terms of space—the metropolitan region being considered within its national setting; holistic in terms of time—today's solutions must not act as constraints on a future society; and, holistic in terms of involvement—the responsibilities for solving problems lie not with governments alone but also with the leadership of the business community, citizen groups and the universities.

The third objective was to create the nucleus for a continuing organization engaged in the study of metropolitan problems. This objective has since been realized. Discussions took place at the Seminar-Conference on the feasibility and nature of an organization that would lend permanence to the working relations that had developed among and within the study groups. As a result of decisions made at that time, INTERMET (the International Association for Metropolitan Research and Development) has been established, with its Secretariat located in Toronto.

The third stage of the Programme was the Seminar-Conference held in Toronto in August, 1967. A memorandum to the leaders of the discussion groups, which constituted the working units of the Seminar-Conference, described the purpose and method of working of the groups as follows:

> The Seminar-Conference in Toronto is the third stage in the development of this Programme and its main function is to extend and deepen those changes in perception without which no creative or imaginative solutions will be found. In preparation for the Seminar-Conference, participants will have read the basic study papers, assisted in preparing a review paper, and engaged in discussion of metropolitan problems within their own study group. In the process, the participants will have learned a good deal but even more importantly will have discovered many new questions or extensions of old ones. All will have become increasingly aware that there are few experts and no ready-made answers to the issues posed by the growth of metropolitan areas.

> New questions reflect new perceptions. The purpose of the Seminar-Conference is to raise and explore new questions, new understandings, and new ways of thinking and acting on the issues involved. This may best be done in direct, face-to-face discussion and for this reason, the Toronto sessions are being held. They will bring together men and women from different disciplines, different cultures, and different social philosophies. The free exploration of these differences in experience and perception is the main purpose of the Seminar-Conference.

In this way, it was hoped that the Seminar-Conference discussions would lead to a greater understanding of urban management, administration and planning, and of the complex nature of comprehensive planning and its application in the context of the metropolis. To facilitate this understanding, it was necessary to examine first the more familiar areas: the major services required in the metropolis and the structure required to provide these services. The Seminar-Conference was divided into three phases. The Phase I discussions dealt with transportation, education, housing, utilities, health and welfare, and planning, although planning could perhaps have been regarded as a structural element, and, as will be observed, in the organization of this book this change has been made. Also, each of these services was considered in terms of its demands on the structural elements of governmental organization, intergovernmental relations, administration, public participation and finance. In Phase II, the discussions focussed on each of these structural elements. This provided the opportunity to consider the structure necessary to meet the demands expressed in Phase I. From these discussions the participants proceeded, in Phase III, to a consideration of how these elements may be integrated in a comprehensive planning process.

The composition of the discussion groups reflected the need to bring together men and women with different training and backgrounds. In each phase of the Seminar-Conference there were twelve discussion groups. Thus, in Phrase I, two groups would focus on any one of the six topics dealing with the services required in metropolitan areas. In Phase II, three groups would focus on any one of the four topics relating to the structural elements of the machinery that provides the services discussed in Phase I. To ensure the maximum exposure of individuals to new ideas generated at the Seminar-Conference, group membership was restructured between phases. Both phases I and II commenced with six and four concurrent general sessions respectively at which the author of a paper and a panel representing the study groups that had reviewed the paper discussed its substance in a general way. The purpose of these sessions was to provide an opportunity for clarification and corrections, and as far as possible to provide everyone with a common starting point for the detailed discussion that was to follow.

Two additional inputs, in the form of plenary sessions, were scheduled for the Seminar-Conference. The conference opened with Adam Andrejewski's[3] plenary address on social goals, which was designed to indicate the quality of human society that may be achieved in a metropolitan area. The second plenary address, by Bertram Gross,[4] which preceded the discussion by all groups on comprehensive planning, was expected to provide the framework for those discussions.

At every stage the design of the Programme focussed on growth in understanding as a prelude to creative action, and this growth in understanding was seen as a function of an interaction process. In the preliminary stages, apart from the interaction within each study group, the papers prepared by the authors were reviewed in

depth by the study groups. At the Seminar-Conference, the sessions were initiated by a further confrontation of the authors and representatives of the review groups. During the Seminar-Conference, the reports of each phase of the discussion were reproduced and made available to participants as they moved into the next phase. Within the discussion groups, participants were face-to-face with men and women from other parts of the world. The focus at the Seminar-Conference was on direct discussion, and the intense degree of involvement achieved was supplemented by written materials and by the supportive physical and social setting.

The organization of materials in this book attempts to take the reader through approximately the same process of adjustment and extension of horizons as that experienced by the Programme participants. Part I is a background piece on the nature of the contemporary metropolis. In Part II, the reader is introduced to the service needs of the metropolis and the demands that those services impose on institutional machinery. In Part III the reader is introduced to the responses that are open by way of institutional adjustments. Both parts II and III conclude with an evaluation of worldwide opinion on the issues discussed by the authors of the study papers. In Part IV a summary description of the comprehensive planning process provides a common point of reference for the earlier discussions in the book. Thus by a reading of the introduction to each part of the book and Part IV, the reader is able to obtain quickly an overview of the substantive issues involved, any one of which can then be pursued in greater detail by reference to the relevant sections.

---

[3]Professor Adam Andrejewski, *The Problems and Social Goals of Metropolitan Area Development.*
[4]Professor Bertram Gross, *The Changing Framework of Urban Planning: A General Systems Approach.*

Part I

The contemporary
metropolis:
A comparative view

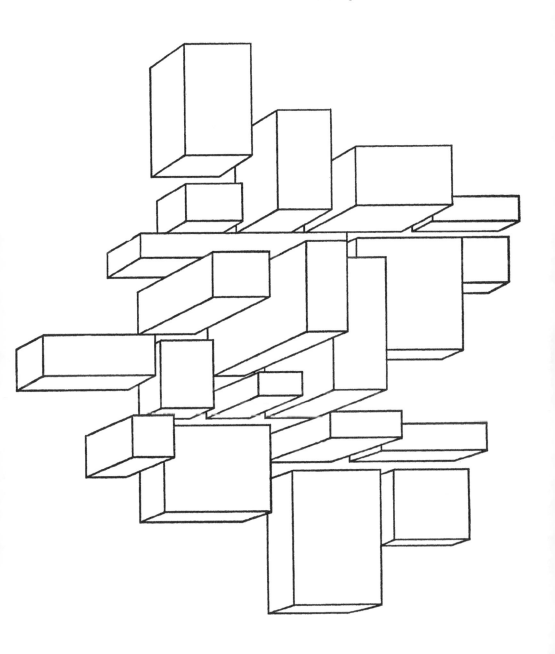

Introductory Note

What is the metropolis? What characteristics distinguish it from other cities? These questions require answers if we are to comprehend the distinctive nature of the problems confronting the contemporary metropolis and, more important, if we are to act to resolve these problems.

Not surprisingly, there is no one answer; in the following chapter the authors present the views of a number of distinguished authorities in addition to their own interpretations. In later chapters of this book further qualifications appear—a factor which is, in part, testimony to the need for further international study of the metropolis. Forstall and Jones, although writing from what is basically an American vantage point, are very fair to non-American views. For even though they give consideration to urban centres of 100,000 or more as metropolitan areas, and justify this, in part, by pointing to the fact that these are likely to become giant centres in the future, they do devote the greater part of their essay to metropolitan areas of population of 1,000,000 or more.[1] An extensive review of the common characteristics of 100 of these metropolitan areas is followed by

---

[1]Any substantive differences between North American and non-North American interpretations of terms are indicated in the footnotes.

an attempt to classify them, in as much as the metropolis is a focus of a related group of regional activities. Again, this classification is designed to draw out the common characteristics. The authors are somewhat sceptical of some of the factor-analysis techniques that have been applied in the classification of metropolitan areas; all too often they lack real analysis and present a misleading picture of inter-metropolitan disparity.

Having considered the major characteristics that relate to the size and functioning of the metropolis, the authors turn their attention to its internal political and administrative breakdown. While the creators of geographical units for scientific analysis can ignore these internal divisions, they are of vital significance for the political, social and economic well-being of the metropolitan community. As such, these internal units are part of the metropolitan enigma, and the problems of making modifications to governmental structures is well illustrated here.

Attention is drawn to the appendices. Appendix A provides some very useful comments on the problems of delimiting metropolitan areas, which is becoming of greater significance to central statistical offices due to the increasing rate of change of the configuration of urban areas. Again there is a very strong case to be made for more positive international cooperation if the comparative study of metropolitan phenomena, with its ensuing benefits, is to be facilitated. Appendix B provides some additional data, made available shortly before going to press, in the growth rates in major metropolitan areas in the period 1951 to 1968.

1

Selected Demographic, Economic, and Governmental Aspects of the Contemporary Metropolis

Richard L. Forstall
Victor Jones

Introduction

THE METROPOLITAN CITY

The metropolis, as a distinctive form of human settlement, consists of a large number of people living in and around one or more centres of high density. Within the metropolis, these centres are much more thickly settled and intensively used for urban activities than are any other sub-areas. Moving out from the centre one finds a graduated decline in the density of settlement and the intensity of use, until one leaves the metropolis altogether, either for open countryside or the outskirts of another urban centre or, perhaps, another metropolis.

How many people have to be settled in this fashion before the settlement can be called a metropolis? Is sheer magnitude a sufficient criterion, or should the term be reserved for the urban areas that contain well-developed and relatively autonomous social institutions? Or should the test of metropolitan status perhaps be cultural, economic or political dominance over an extensive hinterland?

A full examination of these questions is beyond the scope of this introductory chapter, in part because the Centennial Study and Training Programme on Metropolitan Problems has been concerned almost entirely with urban areas of more than

1,000,000 population. Nearly all urban centres of this size are metropolises, or possess metropolitan characteristics, by any definition. However, our discussion should begin by recognizing that there has been little consistency in the application of terms. In the United States of America, urban aggregations of 100,000 or even less are termed metropolitan areas if they possess a central city of at least 50,000 inhabitants. This practice has been laughed at by some commentators and criticized in a lower key by many others. For instance, William A. Robson has written that the

... criteria used in the United States are satisfied by a very large number of areas which cannot claim to be metropolitan centres in a proper sense of the term. The figure of 50,000 for the "core" city is so small that it robs the word "metropolitan" of any sociological and political significance; and the definition takes no account whatever of the functions which should be performed by a metropolitan area worthy of the name. It should be either a great political and governmental centre, or a commercial and industrial centre, or a cultural centre, or perhaps all of these, or at least two of them.

In our view, in a country as large and as highly developed as the U.S.A., only metropolitan areas with a central city of not less than 300,000 and a total population of at least 400,000 should be regarded as possessing metropolitan status. ... To place a vast mass of small fry in the same category as the great metropolitan communities of New York, Chicago, Los Angeles, Philadelphia, Detroit, Boston and so forth does not assist the serious study or understanding of the problem of the metropolitan city.[1]

Following an analysis of the economic relations between urban centres and their hinterlands, based on N.S.B. Gras's classic formulation, Otis Dudley Duncan and his associates conclude, more cautiously but no less certainly:

Perhaps it is not too wild an extrapolation to suggest that in the United States of 1950 an SMA [Standard Metropolitan Area] size of roughly 300,000 inhabitants marked a transition point where distinctively "metropolitan" characteristics first begin to appear.[2]

Hans Blumenfeld also believes that only very populous areas can be metropolises. In 1964 he would have restricted the term to "an area in which at least half a million people live within a distance not exceeding forty-five minutes travel time from its center by means available to the majority of the population." A year later, he considered raising the "critical mass that distinguishes a metropolis from the traditional city" to 1,000,000 inhabitants.[3]

Although our principal concern here is with metropolitan areas of over 1,000,000 inhabitants, it should be recognized that technological as well as organizational developments have not only spread out the few huge cities we have known until recently, but have also had similar effects on the structure, growth, and functions of the increasing multitude of smaller and medium-sized urban aggregations. Large metropolitan areas of the future will develop from the rapidly growing smaller urban areas. Decisions made now in both the

[1]W. A. Robson, ed., *Great Cities of the World* (London, George Allen and Unwin, Ltd., 1957), p. 33.

[2]Otis Dudley Duncan, et al., *Metropolis and Region* (Baltimore, The Johns Hopkins Press, 1959), p. 275.

[3]Hans Blumenfeld, *The Modern Metropolis*, Paul D. Spreiregen, ed. (Cambridge, The Massachussetts Institute of Technology Press, 1967), pp. 50-51, 61-62.

public and private sectors will, in large part, determine their future. Furthermore, some of them will be swallowed up by larger centres in the process of conurbation. Thus, these smaller metropolitan areas are important both to students of urban phenomena and to metropolitan planners and decision-makers.

A NOTE ON TERMS

At least part of the disagreement over the characteristics a city should possess if it is to be termed "metropolitan" is apparently due to semantic confusion. The term "metropolitan" has been used of cities in two distinct senses. First, the older, traditional sense relates to the metropolis as a major centre of urban activities, a gathering place of people, processes and ideas, and a centre of political or intellectual influence on surrounding areas.[4]

Most references to "metropolitan cities", "metropolises" and so on, carry the sense of the city as a major centre, a leader among other cities. But this significance has been complicated by the introduction of the term "metropolitan area", which is most commonly understood to mean "a city with its suburbs",[5] as distinct from

"city" or "city proper". The term "metropolis" or "metropolitan city", on the other hand, is most commonly used to distinguish between a given metropolitan area and a "lesser city" or "smaller city".

It is unfortunate that the term that has developed to denote "city-plus-suburbs" is one that uses the word "metropolitan". There is no reason why a "metropolitan city" (traditional sense) need have many suburbs outside its corporate limits and hence be possessed of a "metropolitan area" (new sense). Some quite small cities have extensive suburbs, either because of local administrative peculiarities or because of special functional characteristics. Examples include such small cities as Locarno (Switzerland) or Shamokin (Pennsylvania) and medium-sized cities such as Charleroi (Belgium), Bielefeld (Germany), Eindhoven (Netherlands), Halifax (Nova Scotia), Bakersfield (California), Flint (Michigan), or Wilmington (Delaware), none of which are more than moderately important as "metropolitan" centres in the traditional sense.

On the other hand, some undoubted "metropolises" have few, if any, suburbs. Shanghai is a notable example; it is certainly a "metropolis" by any standard relating to size, importance, or function. Its administrative limits are so extensive that they encompass all of the urban area and a good deal more besides. Hence, Shanghai has no "metropolitan area" in the sense of separately administered suburbs; its city and "metropolitan" government are presumably identical. Similarly, the extent of "metropolitanization"[6] in other cities varies

---

[4]This sense derives from the original meaning of metropolis, "mother city", a Greek *polis* or city-state from which colonies had been settled. The derivation would appear to have been through an adoption of the term by the Christian Churches for the city (and also the office) of a bishop with superior authority, a metropolitan over other bishops. In this ecclesiastical sense, the term is, of course, still in use. The original sense of "mother country" is rarely found today, although it has been retained in French usage to refer to Metropolitan France as opposed to overseas areas.

[5]Here and throughout this chapter, the term "suburbs" is used in its customary American connotation, that is of communities that form portions of the metropolitan area but that lie outside the administrative limits of the central city; such communities may be individually incorporated, but are not necessarily so.

[6]The term "metropolitanization" is used here to mean "the extending of the urban or metropolitan area beyond the boundaries of the central city". This extension could be measured in terms of either or both of population and physically built-up territory.

widely. The most significant metropolitan cities (old sense) are not necessarily those with the most extensive suburbs in the administrative sense. Naturally, since metropolitan cities are by definition large and complex, they are often cities in which the problem of the urban area spreading into a variety of administrative jurisdictions— and, consequently, the development of a "metropolitan area" (new sense)—is most apparent. The distinction between the two meanings becomes clearer by recognizing that the recent creation of an administrative Greater London, although it brought much of the metropolitan area under one administration and, therefore, tended to "de-metropolitanize" the area, in no sense reduced London's status as a metropolitan city.[7]

The two senses of "metropolitan" imply, when applied to "problems", that there are two distinct classes of metropolitan problems. One group is comprised of all the problems characteristic of major urban centres: rapid expansion, traffic difficulties, heterogeneous population, sanitary arrangements, and so on. Such problems of metropolises have no doubt existed in cities for centuries, though of course with changing character. A second group of metropolitan problems is an outgrowth of the incongruity between the city's socio-economic limits and administrative limits. With the urban area divided among several jurisdictions, the traditional metropolitan problems typically become compounded by the lack of coordination of efforts to solve them. A merging of administrative areas, or the creation of a new, all-encompassing administrative level for the entire urban area, may do away with this genre of "metropolitan problem". Such an administrative change will not, however, automatically solve the first type of metropolitan problem—that related to the city's status as a metropolitan centre or large city.

It would, no doubt, be vain to attempt to alter the meaning of a term in such widespread use as "metropolitan area". But some of the semantic confusion might be lessened if the abbreviated term "metro" could come into use for the "city-plus-suburbs" sense of "metropolitan". This short form is already in some use among business users of metropolitan-area data, and also in connection with metropolitan governments, such as that in Miami.

METROPOLITAN AREAS OF 100,000 OR MORE INHABITANTS

From the approximately 140 metropolitan areas of over a million inhabitants,[8] we have selected 94 for detailed analysis, and included as well the smaller areas—Vancouver, Winnipeg, Edmonton, Belgrade, Hanover, and Ibadan[9]—participating in this Study Programme. However, in order to place our discussion of these 100 metropolitan areas in the full context of actual

---

[7]It seems likely that the American misinterpretation of the terminology applied to the London area is, in a sense, responsible for the confusion of terms mentioned above. In the nineteenth century it became customary to refer in Britain to "the Metropolitan Area" (usually capitalized), meaning the entire area of London, including all the then numerous administrative jurisdictions existing in the urban area. As no other city in Britain was thought of as a "metropolis", "metropolitan area" could refer only to London. But officials and others in the United States, encountering this reference, apparently confused the sense of the term and understood that the inclusion of suburbs was implied by the word "metropolitan" when in point of fact it was the word "area" that was meant to imply this.

---

[8]There were approximately 140 such metropolitan areas in 1964; in 1968 this figure had increased to 153. (See Appendix B.)

[9]Locally defined metropolitan areas reported for Belgrade, Hanover and Ibadan approach or exceed 1,000,000 population.

TABLE 1-1
World Distribution of Metropolitan Areas with Populations of 100,000 or More.

Continent	Metropolitan Areas Listed in 1959*	Additional Areas by 1965**	Total as of 1965
Africa	48	52	100
North America	210	17	227
Middle America	28	11	39
South America	50	41	91
Asia	291	118	409
Europe	278	63	341
Oceania	11	4	15
U.S.S.R.	130	57	187
Total	1046	363	1409

Sources:
* International Urban Research, *The World's Metropolitan Area.*
**United Nations, *Demographic Year Book* (1965), Table 5.

and emerging metropolitan settlement of the world, we first present data on the number, distribution, and growth of all metropolitan areas of 100,000 or more inhabitants.

The first large-scale attempt to standardize the delimitation of metropolitan areas was made by International Urban Research at the University of California at Berkeley and reported in *The World's Metropolitan Areas* published in 1959.[10] On the basis of a modification of the Standard Metropolitan Area definition as used in the 1950 U.S. Census, 720 metropolitan areas of over 100,000 population were identified. In addition, 326 cities (that is, urban settlements that were not classified as metro-

politan areas) with populations of 100,000 or more in 21 countries were listed. Thus, in total there were at least 1,046 such urban concentrations in the world in the mid-1950's; and by 1965 this figure had grown to 1,400.[11] Table 1-1 shows the distribution, by continent, of large urban centres in 1959 and of additional places that had grown into this class by 1965.[12]

---

[11]This count is based upon a cursory comparison of such places as reported in the *United Nations Demographic Year Book* (1965), Table 5, with those reported in *The World's Metropolitan Areas.*

[12]For a useful summary of world population growth from 1800 to 1960 in urban communities of various sizes, see Gerald Breese, *Urbanization in Newly Developing Countries* (Englewood Cliffs, Prentice-Hall, Inc., 1966). Breese based these data on earlier compilations by Davis and Hertz and by Hoyt. Our population estimate of 2,500,000,000 for the world for 1950 (Table 1-2, below), is 100,000,000 larger than that cited by Breese because we have incorporated revised estimates for mainland China based on the 1953 census of that country.

---

[10]The principal collaborators were Suzanne R. Angelucci, Harley L. Browning, Kingsley Davis, Richard L. Forstall, Jack P. Gibbs, Gene B. Petersen and Thomas O. Wilkinson. See Appendix A of this chapter for a discussion of the theory and practice of defining and demarcating metropolitan areas.

Of the metropolitan areas of 100,000 or more, about one out of ten is of more than 1,000,000 inhabitants. This represents a very great increase over the numbers existing even in the recent past. It is widely believed that two or three cities in ancient times reached the one million mark, and the claims appear well founded, particularly in the case of Rome, although there are some critics who question the usually accepted figures.[13] Possibly one or two Oriental cities also reached this size. It is certain, however, that cities of 1,000,000 were extremely rare before the nineteenth century and that they never could have accounted for more than a very small percentage of the total population of their period. Even a population of 1,000,000 for Rome in the first century A.D. would still account for only one per cent of the roughly 100,000,000 population of the Roman Empire at that time.[14]

The first western metropolis to reach 1,000,000 was London, shortly after 1800. At that date, Tokyo, Peking, and Wuhan (Hankow) may also have had populations over 1,000,000. During the next seventy years, only three cities joined this group: Paris around 1835, New York around 1857, and Vienna around 1870. In 1870 these seven metropolises comprised a total population of about 13,000,000, or just one per cent of the total world population of that day and somewhat less than the

population of New York or Tokyo alone today.

The increase in the number of metropolitan areas of over 1,000,000 persons since 1870 is summarized in Table 1-2.[15] By 1900 there were twenty metropolitan areas of over 1,000,000, and by 1920, thirty, with a population of 70,000,000 or 4 per cent of the world's total. Expansion may have slowed somewhat during World War I, but it continued little abated through the 1920's and the widespread depression of the 1930's to reach a total of fifty-seven metropolitan areas of over 1,000,000 in 1939. The 1940's saw a distinct speeding up of the rate of increase in numbers; many non-European cities experienced more rapid economic and demographic expansion during Europe's involvement in World War II. By 1951 there were ninety-five cities over 1,000,000 and, by 1964 the number had further increased to 140. The total population in these 140 metropolitan areas in that year was 362 million, or over 11 per cent of the world total.

In less than a century, the world's population has increased by 150 per cent, while the number of large metropolitan areas has increased by 1,900 per cent, and their total population by about 2,700 per cent. As a result, they account for more than ten times as great a share of the world's total population now as in 1870.

At the present rate of increase in numbers alone, one could anticipate a total of more than 250 metropolitan areas over 1,000,000 by the year 2000, but any forecast of their total population would have to take into account how much larger the

---

[13]J. C. Russell, "Late Ancient and Medieval Population," *Transactions of the American Philosophical Society*, N.S. Vol. XLVIII, Pt. 3 (June, 1958), pp. 63-68. Russell concludes (p. 65) that Rome's population in the time of Augustus was about 350,000.
[14]Using Russell's figures (*ibid*, pp. 7 and 65) would reduce the ratio to less than one per cent.

---

[15]The data in this table and the accompanying discussion are based on a series of estimates by Richard L. Forstall, of the present and past metropolitan population of each major city. See also Appendix A for a further discussion.

TABLE 1-2
World Increase in the Number of Metropolitan Areas of 1,000,000 or More, 1870-1964

Year	Total Metropolises	Total Population (millions)	Total World Population (millions)	Percentage of World Population in Metropolises	Increase per year in Metropolises	
					Number	Population (millions)
1870	7	13	1,300	1.0		
					0.4	0.4
1900	20	24	1,600	1.5		
					0.5	2.3
1920	30	70	1,800	3.9		
					1.4	3.7
1939	57	140	2,250	6.2		
					3.2	5.4
1951	95	205	2,500	8.2		
					3.5	12.1
1964	140	362	3,200	11.3		

greatest of today's metropolitan areas were likely to become. It seems at least possible that New York and Tokyo are approaching a ceiling beyond which further population increase, even if prompted by economic expansion, might develop in nearby but separate metropolitan centres. A parallel development, which some already believe to be under way, is the aggregation of existing large metropolitan centres into even larger urban or metropolitan regions comprised of several semi-independent metropolises.[16] Such potential developments as these, together with the explosive metropolitan population growth of recent decades, would make any forecasts of maximum metropolitan populations in coming decades very uncertain. It is interesting to note, however, that the population of the world's largest metropolis (at any one time) has increased since 1870 by a factor of less than five. The largest city of that time, London, had about 3,800,000

inhabitants, whereas New York had a population of 16,325,000 in 1964.

THE 100 METROPOLITAN AREAS SURVEYED

Our analysis in the remainder of this chapter has concentrated on 100 metropolitan areas, which are listed in Appendix A. This limit was decided upon because of the problem of gathering reasonably comparable data from a number of cities internationally distributed. Though inclusion of all areas over 1,000,000 would have been desirable, the 100 metropolitan areas included do in fact comprise about 85 per cent of the total population of all metropolitan areas over 1,000,000, and there is no reason to believe that they do not represent them quite accurately.[17]

16See, for example, Jean Gottman, *Megalopolis* (New York, The Twentieth Century Fund, 1961).

17The group of 100 metropolises is also felt to be generally representative of those metropolitan areas that participated in the Centennial Study and Training Programme on Metropolitan Problems; for example, the median rate of the 1951 to 1964 population increase is almost exactly the same for the forty-one metropolises in the Study Programme (43.5%) and the fifty-nine additional metropolises (42.5%).

All the metropolitan areas participating in the Study Programme have been included; the total is forty-one, as we have treated Amsterdam and Rotterdam as separate metropolitan centres but have included them both. Six of these areas have metropolitan populations below 1,000,000, at least according to the data we have used. These are Vancouver, Winnipeg, Edmonton, Hanover, Belgrade, and Ibadan. To the forty-one we have added the fifty-nine metropolises ranking largest in a 1964 list of all metropolitan areas over 1,000,000. The result is the inclusion of all urban aggregations over 1,400,000, and of eight below that figure, the six noted above plus Prague and Rotterdam.

Of the 100 metropolitan areas so selected, 20 are in the United States and Canada, 9 in Latin America, 18 in northwestern Europe, 13 in southern or eastern Europe, 4 in the U.S.S.R., 5 in Africa or the Middle East, 6 in India or Pakistan, 10 in mainland China,[18] 13 in the remainder of east and southeast Asia, and 2 in Australia. Individual countries with four or more metropolitan areas include the United States (15); mainland China (10); Germany (8), including Berlin; the United Kingdom, Japan, and Canada (5 each); and the U.S.S.R. and India (4 each).

There are three metropolitan areas with over 10,000,000 population each. These are New York (16,325,000 in 1964), Tokyo-Yokohama (15,900,000), and London. Since 1964 Tokyo has passed New York and, within the boundaries as used here, is now the largest metropolitan area

in the world. As of 1964, there were 12 other cities over 5,000,000 and an additional 7 over 4,000,000, for a total of 22 of 4,000,000 or more.

100 METROPOLITAN AREAS: AGE AND GROWTH

We now turn to an examination of two specific characteristics of the 100 metropolitan areas: the age of the city as a major centre and the population increase in the period 1951-1964. Such analyses as these are admittedly somewhat rough, because for certain cities statistical data are available only in limited or partially estimated form. We believe, however, that this disadvantage is easily outweighed by the advantages of a broad survey of the characteristics of major world cities.

Other classifications could have been offered to illustrate general trends or characteristics found in a significant share of major cities. Four diverse examples are geographic or physical site, primacy (the degree to which the city has rivals in its country or region), ethnic makeup, and the political relations of the city administration with that of the country. Still further possibilities, such as classification in terms of climatic conditions, would be quite feasible but probably of less significance to an understanding of contemporary urban problems.

100 METROPOLITAN AREAS: AGE OF MAJOR URBAN CENTRES

The age of major urban centres is of interest both as an aid in recognizing the forces that produce great cities and for its practical relevance to the problems cities face today. The success of certain cities of long history in maintaining a continuing urban tradition is one of the most interesting and widely recognized

---

[18] The ten Chinese cities, none of which is in the seminar group, present a particular problem in obtaining data in almost every aspect, including total population in 1964 and growth since 1951. However, it has been felt desirable to include them even though much of the information for them represents rough estimates.

aspects of urban character. On the other hand, the problems and possible solutions facing administrators in a city whose street plan originated before 1700, or much of whose housing is more than a century old, are clearly in sharp contrast to those of a city laid out in the late nineteenth century and with only a few buildings dating from before World War I.

As a convenient generalization of metropolitan age, we have determined the period when each of the 100 metropolitan areas reached a population of 100,000, within the metropolitan limits of that time. (In some instances, a population of 100,000 within the *present* metropolitan limits was reached a good deal earlier.) The figure of 100,000 was long viewed as a criterion of large-city status. While it would be misleading to suggest that a population of 100,000 a century ago implied what one of 1,000,000 does today, there were relatively few cities over 100,000 in the world as late as 1850 and those that reached that size were certainly all places of great importance at the time. Moreover, some famous cities of unquestioned importance in the past never reached a size much greater than 100,000. A notable example is Venice, whose population at its height was under 200,000, though the city was then the leading commercial emporium for a large portion of the western world.[19]

Table 1-3 groups the 100 metropolitan areas by region and by period at which the size of 100,000 was reached. It should be noted that this table is by no means a summary of the total number of cities of 100,000 or more, either today or at earlier dates. There were many additional cities which, like Venice, reached that size early but did not go on to qualify for inclusion on the listing. Rather, the table reflects the

number of cities that attained 100,000 by a given date *and* were to grow much larger later.

For fourteen cities in the Middle East or Orient, it is very difficult to determine with any precision when they reached 100,000. These cities include nine of our ten Chinese cities (all except Harbin, which was not founded until 1897); Seoul in Korea; Cairo in the United Arab Republic; and Tokyo, Osaka and Kyoto in Japan. There is little doubt that all of these cities had a population of 100,000 by 1850 and some undoubtedly had reached that figure before 1500. More accurate statements, especially for the Chinese cities, cannot be made on the basis of the rough estimates available in Western sources. Such estimates, often derived from the descriptions of early visitors from the West, have often proved to be greatly exaggerated when actual censuses were taken. Perhaps data available to historians and demographers within the countries involved provides a basis for more accurately determining when these cities reached 100,000.

In the West, four of the present group had reached 100,000 by 1500. These were Istanbul, Paris, Naples and Milan. Among western European cities, Paris perhaps represents the oldest in continuing history as a large city; Istanbul's history is of course much longer still. Some cities, like Rome and Alexandria, which certainly surpassed 100,000 for considerable periods in ancient times, fell well below these figures in the Middle Ages; Alexandria did not regain 100,000 until after the opening of the Suez Canal and hence from one viewpoint is one of the newer metropolitan areas of the present day.

By 1700, another six cities had reached 100,000: London, Amsterdam, Vienna, Madrid, Rome, and Moscow. By 1800, the first American city, Mexico City, had made

[19]Russell, *op. cit.*, pp. 126-28.

TABLE 1-3

100 Metropolitan Areas by Region: Period in Which Population Reached 100,000

Region	Before 1500	1500-1700	1700-1800	(Before 1800-1850)	1800-1850	1850-1880	1880-1900	1900-1920	Since 1920	Total
United States/Canada	—	—	—	—	4	9	2	3	2	20
Latin America	—	—	1	—	2	3	3	—	—	9
Northwestern Europe	1	3	2	—	8	4	—	—	—	18
Southern Europe	3	2	1	—	—	—	1	—	—	7
Eastern Europe/U.S.S.R.	—	1	2	—	3	—	2	2	—	10
Subsaharan Africa	—	—	—	—	1	—	1	—	—	2
North Africa/Middle East	—	—	—	(1)	1	1	—	—	—	3
India/Pakistan	—	—	4	—	—	1	1	—	—	6
Mainland China	—	—	—	(9)	—	—	—	1	—	10
East/Southeast Asia (remainder)	—	—	—	(1)	1	3	2	1	—	8
Japan	—	—	—	(3)	—	1	—	1	—	5
Australia	—	—	—	—	—	2	—	—	—	2
Total	4	6	10	(14)	20	24	12	8	2	100

Note: Population refers to contemporary, not present-day, extent of urban area.

its appearance,[20] and Berlin, Hamburg, Barcelona, Warsaw, Leningrad, Calcutta, Bombay, Madras, and (more doubtfully) Delhi, had also reached 100,000. If all of the fourteen cities about which we are uncertain are assumed to have passed this figure before 1800, a total of 34, or just

[20]Some sources credit Mexico City with more than 100,000 population as an Aztec metropolis prior to the arrival of the Spaniards. Potosi, in Bolivia, may have had over 100,000 population in the early colonial period.

one third, of our 100 cities are accounted for by that year and were, therefore, already large cities at the opening of the industrial age.

By 1850, another 20 cities had reached 100,000, including for the first time four in the United States (New York, Philadelphia, Boston, and Baltimore). Thus by this date over half of the 100 cities had reached 100,000. Another 36 cities had reached 100,000 by 1900, leaving only 10 cities in our list that did not reach that size until the twentieth century. These include the four relatively small cities of Vancouver, Winnipeg, Belgrade, and Edmonton, and four large cities—Houston and Miami in the United States, Donetsk in the U.S.S.R., and Kitakyushu in Japan—which accordingly may be ranked as the newest among the world's major metropolitan areas. Donetsk and Kitakyushu are both major heavy-industry centres of recent development, closely associated with coal mining. Houston's rapid growth has been associated with the development of the large petroleum resources of Texas, whose exploitation began about 1900.

Miami and Edmonton are the only cities on the list that did not reach 100,000 until after 1920. Miami should perhaps be viewed as the harbinger of a new phenomenon among metropolitan centres, since it is a city whose economic development has been chiefly as a seasonal resort. The development of large resort cities (over 100,-000) dates back to the mid-nineteenth century, but Miami is certainly the first such urban centre to surpass 1,000,000.

Thus, most of the metropolitan areas of today are not new as cities, but only as giant cities. Indeed, in most of these centres the resources and circumstances encouraging the establishment of a large city have been exploited for at least a century or two. The average period is longer if we exclude cities in the Americas and other regions where large-scale settlement (and hence the establishment and development of new cities) has continued into recent times.

The pattern is thus one of the metropolitan areas of a given period emerging from the large cities of previous periods. It should be possible to distinguish the *city-forming factors* leading to the development of a city of significant size (which would vary somewhat from period to period) from the additional *metropolis-forming factors* that result in a few of the significant cities becoming relatively much larger than the others.

Whatever the metropolis-forming factors are, they sometimes produce a remarkable continuity of urban leadership. For example, it has been several centuries since London and Paris had any serious rival in their general vicinities. Likewise, the three largest cities of the eastern seaboard of the United States were the same in 1700 (when their combined population was less than 100,000) as today (with a combined population approaching 25,000,000), though it is true that their rank relative to one another has shifted several times during the interim.

While much of the increase in magnitude of these centres can be attributed to the great increase of the world's population in recent centuries, this does not explain why many cities have continued to be relatively the largest in their own areas of influence throughout the period. It is extraordinary, too, that the pervasive technological and other changes of the past few centuries have apparently not affected fundamental urban functions enough to alter significantly the original advantages of many earlier cities in favour of newer ones. Certain urban activities of today, notably factory industry, are "new" in the perspective of history and have resulted in the development of many new cities. But they have also augmented the size of many existing major cities. There are very few examples of factory-created cities "taking over" as metropolises of their regions from the former incumbents. The development of great industrial concentrations, with many new urban centres, in England, France, the U.S.S.R., the United States, and Japan, has apparently not seriously affected the leadership of the cities that were already the metropolises of these countries prior to large-scale industrialization.

In some other countries, the incursion of new economic forces has indeed altered the earlier pattern of major urban centres. The leadership of the mercantile and port cities of India, China, and southeast Asia did not develop at the expense of older centres (which often were inland cities) until large-scale contact by sea with other regions focussed city-forming factors on port locations. But the pattern of a high degree of continuity appears to be exhibited once again in the continuing leadership of these port cities, even when (as in mainland China) economic trends are in a decidedly different direction from what they were when Shanghai, Tientsin, and Canton rose to a high rank among the nation's cities.

As a basis for a more detailed analysis, it would be desirable to make a comparative examination of the later growth of the large cities of a given earlier period. For example, of the European cities of 20,000 or more in the fifteenth century, how many have an analogous relative importance today? How many other cities have emerged on the scene in the interim, and what relative levels of importance have they attained? What influencing factors can be discerned in each case? Such an inquiry, with similar ones for other regions for which reasonably complete population data could be obtained (for example, North America and Japan), would prove most helpful in casting additional light on the nature of metropolis-forming factors.

100 METROPOLITAN AREAS: POPULATION INCREASE, 1951-1964

To present an indication of the rate and volume of population increase for each of the 100 metropolitan areas, we have estimated the population as of January 1, 1951, within the same metropolitan boundaries as were delimited for 1964. The percentage increase for this thirteen-year period has then been computed. It should be noted that most large metropolitan areas have actually expanded somewhat in areal extent during the period, so that in one sense rates of growth have often been slightly higher than indicated.

The period has been chosen as reasonably representative of the period since World War II. During the 1945-1950 period, many cities were still in a phase of immediate recovery from the damage of war. For some, notably in Japan and Germany, this process was still continuing after 1950, and some of the relatively rapid growth shown by certain cities in these countries can be attributed to recovery of pre-war size. The choice of 1951 also makes possible relatively accurate estimates for most of the cities, since most countries took censuses in 1950 or 1951. Choice of a ten-year period, such as 1954-64, would often have required interpolating populations between censuses and therefore would have involved a possibility of considerable error for many individual cities.

Table 1-4 lists the metropolitan areas in descending order of 1951-1964 percentage increase. The total population of all 100 centres increased 42.5 per cent. This compares with a 28 per cent increase for the world's total population in the same period. The range covered is quite wide, with ten cities doubling or better during the period, while at the other extreme three grew by less than 5 per cent. Interestingly, no metropolis actually declined in population.

There is also a wide range in the percentage growth figures for regional aggregates of metropolises. (These are percentage increases for the total population of the cities in the region, not averages of the

TABLE 1-4

100 Metropolitan Areas: Percentage Increase in Population, 1951-1964

Metropolis	Percentage Increase 1951-1964	Metropolis	Percentage Increase 1951-1964
Bogotá	157.1	MAINLAND CHINA (10)	52.6
Lima[a]	142.1	Warsaw[a]	52.4
Miami-Fort Lauderdale	141.7	Madrid[a]	51.5
Sian	128.6	Montreal[a]	50.0
Caracas	125.8	San Francisco-Oakland-San Jose[a]	49.2
Seoul[a]	113.3	INDIA/PAKISTAN (6)	48.4
São Paulo	109.6	Vancouver[a]	47.4
Teheran	102.1	SOUTHERN EUROPE (7)	46.4
Karachi[a]	100.0	Buenos Aires[a]	45.3
Wuhan	100.0	Rome	44.3
Taipei	98.7	Athens[a]	44.2
Edmonton[a]	97.3	Melbourne	44.2
Djakarta	96.9	Belgrade[a]	43.5
Mexico City[a]	96.8	100 METROPOLISES (100)	42.5
Istanbul[a]	95.1	Gorky	42.5
Harbin	95.0	Ibadan[a]	42.2
Bangkok[a]	87.5	Donetsk-Makeyevka	41.7
Singapore	87.2	Milan[a]	41.6
EAST/SOUTHEAST ASIA (excluding China/Japan) (8)	83.7	Stuttgart	41.5
Manila[a]	81.3	Havana	40.9
LATIN AMERICA (9)	80.8	Barcelona	39.9
NEAR EAST/NORTH AFRICA (3)	79.0	Frankfurt am Main	38.1
Hong Kong	77.0	Cologne	37.8
Rio de Janeiro[a]	72.1	Canton	36.7
Cairo[a]	72.0	Madras	36.5
Santiago	71.4	Munich	36.4
Lahore	70.6	Moscow[a]	36.3
Alexandria	70.0	AUSTRALIA (2)	36.3
Los Angeles	69.8	Budapest[a]	35.6
Tokyo-Yokohama[a]	69.1	U.S.S.R. (4)	35.4
Houston	69.0	SUBSAHARAN AFRICA (2)	35.2
Delhi-New Delhi	65.7	Minneapolis-St. Paul	35.1
Nanking	65.0	Winnipeg[a]	34.2
Peking	61.5	Cleveland	33.7
Toronto[a]	58.1	Johannesburg[a]	33.3
JAPAN (5)	57.9	Mukden	32.5
Chungking	57.7	UNITED STATES (15)	32.0
Washington	57.4	Hanover[a]	31.6
Nagoya[a]	56.4	Kitakyushu-Shimonoseki	30.8
Tientsin	54.5	Saigon	30.8
Bombay	54.1	Sydney[a]	30.0
CANADA (5)	53.0	EASTERN EUROPE (6)	29.8
Osaka-Kobe	52.8	Katowice/Silesia	29.5
		Leningrad[a]	29.0

TABLE 1-4 (Continued)

Metropolis	Percentage Increase 1951-1964	Metropolis	Percentage Increase 1951-1964
Calcutta[a]	28.8	Rotterdam[a]	14.8
Shanghai	28.8	Bucharest	14.3
Detroit-Windsor[a]	28.4	Amsterdam[a]	13.4
Paris[a]	26.0	Boston	12.3
Baltimore	25.9	Brussels	11.3
Chicago	25.5	Birmingham (U.K.)	10.0
St. Louis	24.6	Prague[a]	9.9
Naples[a]	23.9	Pittsburgh	8.9
Essen/Ruhr	23.8	London[a]	7.0
Kyoto	23.1	Vienna	6.6
Philadelphia[a]	23.1	Liverpool	6.3
GERMANY (8)	21.7	UNITED KINGDOM (5)	6.3
New York[a]	20.9	Berlin	3.2
NORTHWESTERN EUROPE (excluding the U.K., Germany) (5)	18.6	Glasgow[a]	3.0
		Manchester	2.3
Hamburg	15.0		

Note: Regional aggregates of metropolises are designated in CAPITALS with the number of metropolises concerned noted in parentheses.

[a]Metropolitan areas represented in the Centennial Study and Training Programme on Metropolitan Problems.

individual city percentages. Thus they are weighted in the direction of the rates of increase of the larger cities.) Four groups of regions may be distinguished according to the following growth characteristics:

1. Regions with rapidly growing metropolitan areas: east and southeast Asia (excluding China and Japan), Latin America, Middle East/North Africa. Total metropolitan growth in each region was about 80 per cent. All three are regions of rapid increase in total population as well. There are, however, some other regions of rapid total increase (for example, India/Pakistan) whose urban centres do not register such high increases.
2. Regions with metropolitan growth at rates above the average (42.5 per

cent): Japan, Canada, China, India/Pakistan, and southern Europe. In each of these, the rate of increase was from 45 to 60 per cent. Three of these regions have also shown relatively rapid overall population increase during the period, but this has not been so characteristic of southern Europe or of Japan, where metropolitan areas are growing at a much faster rate than the total population.
3. Regions with metropolitan growth at moderate rates: Australia, U.S.S.R., Subsaharan Africa, United States, eastern Europe. Rates range from about 30 to 36 per cent. Four of these five regions are themselves areas of moderate to slow growth. The low metropolitan rate for Subsaharan

Africa is surprising; however, only two cities are involved, Ibadan and Johannesburg, and neither is very representative of the urban areas of this region as a whole, in which large metropolitan areas are practically absent.

4. Regions with slow metropolitan growth: northwestern Europe (18.6 per cent). This region includes Germany, which is perhaps in an intermediate position; the aggregate rate for all eight German cities is 21.7 per cent, but this rises to 28.1 per cent if Berlin, a special case in many respects, is excluded.

We now turn to the range of growth rates within individual regions. There was a broad range of individual growth rates in two otherwise dissimilar regions: the United States (from Miami, 141.7 per cent, to Pittsburgh, 8.9 per cent) and eastern Europe (from Warsaw, 52.4 per cent, to Prague, 9.9 per cent). In contrast, in the three regions with the most rapid increase and in southern Europe and Canada, the range was narrower, with essentially all metropolises in these regions growing rapidly. The range in Australia, the U.S.S.R. and Africa was also relatively narrow, with all metropolitan areas in each region growing at moderate rates. In Japan, China and India/Pakistan there was a moderate range, with a few cities growing at moderate rates (for example, Kitakyushu, 30.8 per cent, Shanghai, 28.8 per cent, and Calcutta, 28.8 per cent) while most showed rapid increases. Finally, in northwestern Europe, a few cities, notably in Germany, showed moderately rapid increases (Stuttgart, 42 per cent, Frankfurt and Köln each 38 per cent, and Munich, 36 per cent) while the others generally grew slowly (Essen/Ruhr, 24 per cent, Hamburg, 15 per cent, and London, 7 per

cent). Paris, with an increase of 26 per cent, was something of an exception, particularly since it differs from the German cities in that little of its 1951-1964 increase can be directly attributed to recovery from World War II.

The most rapidly growing individual metropolises are of special interest. Bogota, Lima, Caracas, Seoul and Teheran are all capitals of countries that are experiencing rapid population increases and that are still only moderately urbanized. (The high rate for Seoul reflects, in part, the abnormal situation in that metropolis in 1951.) Much of Karachi's growth is certainly attributable to similar circumstances, though it is no longer the official capital of Pakistan.

The growth of Sian, Sao Paulo, and Wuhan is probably attributable chiefly to the growth of manufacturing. As already noted, the figures for the Chinese cities can be considered only rough approximations, as no official data have appeared since the end of 1957. Sao Paulo's great size (5,450,000 in 1964) makes its continued very rapid growth (110 per cent) all the more remarkable, but Mexico City, a little larger, had almost as high a rate (97 per cent). Miami's increase may be explained by its already mentioned highly specialized character.

Nine metropolitan areas each increased by 2,000,000 or more in the period, and an additional twenty-two grew by more than 1,000,000 each. Far in the lead was Tokyo, whose gross increase amounted to 6,500,-000 in the thirteen years. The position of other metropolitan areas in the list of course reflects their size as well as their percentage increase; some large metropolitan areas showed substantial total increases, though their rates of increase were only moderate. Thus the second and third largest total increases were those of Los Angeles (3,185,000) and Osaka (3,075,000), each very large and with growth rates well above average. These and Sao Paulo and Mexico City all showed greater total increases than New York (2,825,000), which was the largest metropolis throughout the period. Similarly, Seoul actually increased more than Paris, Calcutta, or Chicago, though all of these increased by more than 100,000 per year. London's increase of 725,000, though certainly a significant number, was less than that of most metropolises in Latin America or the Far East.

Among the metropolitan areas with very low rates of increase, Liverpool, Glasgow, Pittsburgh and Manchester all had a relatively early start in industrialization and all have suffered relatively from the expansion of their industrial specialties in newer areas. Nor is overall population growth rapid in the regions in which they are located. Slow growth in its own region perhaps most readily explains Vienna's low rate of increase. Berlin's situation is obviously a very special one and perhaps the existence of any increase at all should be the occasion for surprise.

These summary comments do no more than suggest the wide variety of factors present in promoting or inhibiting the growth of major centres in the period since 1950. For example, it would be most illuminating to break down the increase for each metropolis into natural increase and net migration. This task would present so many problems of data collection and estimating that it has not been attempted here.

Certainly all the most rapidly growing metropolises are also experiencing high in-migration. But migration is also substantial to some metropolises that do not have a high rate of natural increase and whose overall growth is moderate. Actual migration rates to metropolises such as Milan and Paris may be as high as those to centres in some developing regions, where much of the population growth reflects natural increase alone.

100 Metropolitan Areas: A Functional Classification

Both the similarities and the differences among cities are evident to the most casual observers. The similarities encourage attempts to link together all cities, or most cities, in a theoretical framework that would "explain" their common features; but the differences present any single theory with so many exceptions that its universality is questionable. Perhaps an examination of actual cities for some common characteristics may aid us in discerning the broad outlines of a general theory.

We may first give brief attention to a definition of "city". Though a wide variety of definitions has been offered, Catherine Bauer has noted that the city exhibits "two opposite but reciprocal qualities: extreme complexity of parts on the one hand, but interdependence and a high degree of integration on the other."[21] Gerardus Wissink, in a concise discussion, concludes that the crucial urban characteristics are "diversity of functions within a relatively limited circumference and within an integrated whole."[22] This parallels closely the present-day concept of a geographic region, in which diverse human activities are carried on in a particular areal framework. A city is a special kind of geographic region, differing from other such regions chiefly in its highly concentrated character and usually in the relative complexity of its characteristics.[23]

[21]Catherine Bauer, "Do Americans Hate Cities?" *Journal of the American Institute of Planners,* Vol. XXIII (Winter, 1957), pp. 2-8.
[22]Gerardus Wissink, *American Cities in Perspective* (Assen, Van Gorcum, 1962), p. 14.
[23]Wissink also notes that "a city is primarily a geographical phenomenon" (*ibid.*)—primarily

A city, then, is a dense, complex and integrated human settlement.[24] Such a definition will include town- and city-type aggregations wherever they occur, irrespective of their specific functional orientation or their administrative arrangements; it will tend to exclude single-purpose agglomerations, for example of farmers, since these are generally not very complex; and, it will also tend to exclude districts or regions of dense settlement unless these can clearly be termed integrated, as, for example, urban areas with strong ties to a nearby centre.

As a form of settlement, cities are widely distributed and numerous. In most countries a significant, and in some a substantial, share of the total population resides in urban places, but the share of the total area, or even of the total occupied area, covered by urban development is always relatively small. This is true even in countries where the rapid outward expansion of metropolitan areas causes legitimate concern over urban consumption of the countryside. For example, in 1960 the urban population of the United States (corresponding, in essence, to the population of centres of 2,500 or more), constituting nearly 70 per cent of the total population and comprising over 125,000,000 persons, occupied only a little more than one per cent of the land area (40,238 square miles).[25] This census definition of "urban" omits considerable low-density suburban development on the fringes of large cities. Data are available for a somewhat broader definition, which includes the urban fringe of such centres, while limited to metropolitan areas of 100,000 or more. By this definition, in 1965 there were 189 such centres whose aggregate population was 119,420,700 or 61.8 per cent of the United States total, on an area of only 120,301 square miles—still less than 4 per cent of the national total.[26]

Even in densely settled countries, the urban areas constitute only a small fraction of the total area. In Japan, Densely Inhabited Districts, which are roughly analogous to the urban areas (as officially defined) of the United States, accounted in 1960 for 43.7 per cent of Japan's total population on just over one per cent of its land area (40,829,991 population on 3,865.2 square kilometres).[27] Clearly,

---

but, of course, not solely. For many activities, values, and institutions, the spatial aspects may be weak or non-existent. See the essays by Donald L. Foley and Melvin W. Webber in *Explorations into Urban Structure* (Philadelphia, University of Pennsylvania Press, 1964). Webber emphasizes the modern phenomenon of interest communities replacing or at least overlaying place-communities: ". . . it is now becoming apparent that it is the accessibility rather than the propinquity aspect of 'place' that is the necessary condition of community. As accessibility becomes further freed from propinquity, cohabitation of a territorial place —whether it be a neighbourhood, a suburb, a metropolis, a region, or a nation—is becoming less important to the maintenance of social communities" (p. 109). Yet the conditions under which today's metropolises evolved did in fact stress spatial relations, and most of the world's cities are still much more concerned with areas close to them than with areas far away. The geographic character of the metropolis is expressed in another sense in its internal organization, in which the different metropolitan functions and the space they require assume recognizable, often highly ordered, areal patterns.

[24] For a concise analysis of the urban place as an agglomeration, see Robert G. Spiegelman, *Analysis of Urban Agglomeration and Its Meaning for Rural People*, U.S. Department of Agriculture, Economic Research Service, Agricultural Economic Report No. 96 (Washington, D.C., 1966).

[25] *U.S. Census of Population: 1960*, Vol. I, U.S. Bureau of the Census, (Washington, D.C., Government Printing Office, 1961), p. xiii.

[26] Further data on these Rand McNally Metro Areas appear in Appendix A.

[27] *1960 Population Census, Densely Inhabited Districts*, Bureau of Statistics (Tokyo, 1961), p. 9.

urban places, though widespread in distribution, are a highly distinctive type of settlement sharply distinguishable in density from other types of settlement.[28]

It is also noteworthy that cities are among the oldest and most resilient of the works of man. Few human institutions or structures can match the vitality, longevity and nearly unbroken continuity of cities like Istanbul, Kyoto, Rome or Paris. Unlike a given structure, such as the Pyramids or the Pantheon, the continuity of these urban centres is expressed not in mere static durability but in a dynamic and steadily evolving function, yet always in the same location and, hence, in a fairly stable areal context. However, there have been other cities that were successful for a time and then dwindled, others that were established with high hopes that remained unrealized, and countless towns that never became cities at all. These facts suggest the presence of forces in the development and maintenance of cities which are remarkably persistent and pervasive when present and difficult to do without when absent.

Cities are alike because they are performing the same general functions under generally similar geographic circumstances. At the same time they differ because both function and geographic circumstances vary. Usually these are differences of detail, and the basic likeness of the city to other cities remains evident enough. Travellers to strange cities may find much to puzzle and amaze them in the local language, government and eating habits, but they will have little trouble recognizing the main thoroughfares, the market place, the poorer residential areas (if not always the richer ones), and usually the banks, administrative buildings and religious structures, not to mention such technologically international facilities as port facilities, railroad stations and airline terminals. These nearly universal likenesses among cities would appear to be much greater than those discernable in such alternative forms of human activity as farming practices, housing, or local government organization.

APPROACHES TO CLASSIFICATION

With these general points in mind, let us turn to a consideration of specific means by which cities can be classified by function.[29] A distinction may be drawn between classifications of cities according to specialization ("functional" classifications) and those according to importance ("hierarchic" classifications). In the United States most classifications have been chiefly of the "functional" type. The American literature using statistics for such classification dates from a 1937 study by Ogburn[30] and a 1943 paper by Harris.[31] Both used census data on the labour force to classify American cities in such categories as manufacturing or factory towns, retailing or trading towns, mining, diversified and so on. A number of later studies have introduced refinements in method but have continued

---

[28]This does not rule out the presence of relatively low-density areas in transitional zones between urban and non-urban settlements. The significant point is the much higher average density of the urban area.

[29]We have not attempted to discuss the types of functions carried on by different sections of a particular city or urban area, though these also can certainly be generalized in significant degree from city to city, and many of the factors discernable in their development are identical to or parallel to those we shall note in connection with the functions of the urban area as a whole.

[30]William F. Ogburn, *Social Characteristics of Cities* (Chicago, International City Managers' Association, 1937).

[31]Chauncy D. Harris, "A Functional Classification of Cities in the United States," *Geographical Review*, Vol. XXXIII (January, 1943), pp. 86-99.

to use labour-force data as the most convenient and widely available data indicative of the economic activities of cities.[32] Successive revisions of the Harris classification have appeared in the *Municipal Year Book* at intervals over the past two decades.[33] An early and extensive European study along similar lines presented international data in map form,[34] but international studies of this type have been hampered by difficulties in obtaining labour-force data, particularly for comparable industrial categories.

Recent classifications of this kind have been able to use more sophisticated statistical techniques such as factor analysis, made practicable by computers. But such analyses can simply multiply the volume of factual data available without necessarily contributing an interpretation that advances understanding of the extent to which generalizations about cities are valid.[35] Characteristically, the factor-analytic studies have eschewed such simple classifications as "manufacturing" or "commercial" for recognition of multiple functional specialities. In so doing they may have expanded the awareness of urban diversity more than the recognition of broad likenesses. Also, as there is only limited congruity between the census labour-force data and actual city economic function, the volume and detail of analysis sometimes result in a classification not of the city's economic activities, but of its labour-force or demographic characteristics. The extent to which the latter reflect the former must still be subject to careful interpretation.[36]

One of the present writers, while on the staff of International Urban Research at the University of California, Berkeley, gathered comparable labour-force data for about 550 world urban centres of 100,000 or more, mostly dating from censuses taken around 1950. However, cities in many countries could not be included because the labour-force data were not available or not comparable. As a new series of national censuses from the 1960 period appears, international comparability in this respect has increased, but it would still be very difficult to assemble comparable labour-force data for all cities over 1,000,000. Discussion that follows later has benefitted from the availability of labour-force data gathered in 1958 and 1959, but has not attempted to cite it in detail.

---

[32]Gunnar Alexandersson, *The Industrial Structure of American Cities* (Lincoln, University of Nebraska Press, 1956), is a particularly good example.

[33]The two most recent are: Victor Jones, Richard L. Forstall and Andrew Collver, "Economic and Social Characteristics of Urban Places," *Municipal Year Book, 1963* (Chicago, International City Managers' Association, 1963) and, Richard L. Forstall, "Economic Classification of Places Over 10,000, 1960/63," *Municipal Year Book, 1967* (Chicago, International City Managers' Association, 1967).

[34]William William-Olsson, *Economic Map of Europe* (Stockholm, Generalstabens Litografiska Anstalt, 1953). William-Olsson classifies more than three thousand places of 10,000 or more in Europe and its borderlands.

[35]As Ernest Weissmann has noted in a related context, "mountains of data on urban and regional development have been, and continue to be, gathered and published. Much of it misses the real issue at hand."

[36]C. A. Moser and Wolf Scott, *British Towns* (Edinburgh, Oliver and Boyd Limited, 1961), represents a careful example of the factor-analytic approach. Jeffrey K. Hadden and Edgar F. Borgatta, *American Cities: Their Social Characteristics* (Chicago, Rand McNally & Co., 1965), is a recent American example with an impressive array of data. Jack C. Fisher, *Yugoslavia—A Multi-National State* (San Francisco, Chandler Publishing Co., 1966), has used the same approach to classify both the cities and communes of Yugoslavia. Qazi Ahmad, *Indian Cities: Characteristics and Correlates,* Department of Geography, Research Paper No. 102 (Chicago, University of Chicago Press, 1965), is a similar study on an Asian nation.

The hierarchic approach to city classification relates to the relative importance of the city, as opposed to its functional specialization. The theoretical literature on this form of classification continues to expand,[37] but actual classifications of large groups of cities according to their position in the urban hierarchy remain relatively few.[38]

Though functional and hierarchic studies have been pursued to a large degree independently of one another, there is evidence that cities functionally classifiable in some single specialized categories are mostly smaller (that is lower in hierarchic position). Conversely, many large cities (high in hierarchic status) are "diversified" or "general-purpose" cities difficult to put into any single specialized category. An examination of the data available for our 100 metropolitan areas suggests that the functional approach using labour-force data, could not have extended much beyond recognizing most of the centres as diversified with greater or lesser amounts of manufacturing. Another approach would have been to attempt a hierarchic ranking of major world cities. To gather the symptomatic data for such a classification would encounter nearly insurmountable problems of data comparability and, indeed, of data significance from country to country and region to region; one may suspect that any such international study is still a number of years away.

In view of these considerations, the approach taken here has followed the general lines of the functional approach but has sought a further explanation of the large group of diversified or general-purpose cities and has interpreted it from a viewpoint more closely related to the hierarchic or central-place approach.

THE CITY AS A REGIONAL FOCUS

We have seen that the similarities among cities may be attributed in large degree to the recurrence in them of certain common urban functions. Furthermore, these activities, while widespread, focus upon a relatively limited number of locations at which they generate sufficient economic activity to produce highly concentrated and complex human settlements.

The most typical of these widely recurring focal activities is commerce, and it seems clear that most of today's large cities originated chiefly as market centres for a surrounding district, large or small, with the commerce carried on by land, by water, or a combination of the two. Many cities possess decided advantages of geographic situation for the role of a commercial focus, being located, for example, on good harbours adjacent to rich agricultural areas (San Francisco, Naples); at the mouths of major rivers leading into a continental interior (New York, Buenos Aires, Shanghai); at the inland ends of long estuaries (London, Hamburg, and many other northwestern European cities); at river or valley junctions (Lyons, St. Louis, Belgrade); in fairly flat regions, centrally on a river system (Paris, Moscow); centrally in lowlands, sometimes inland (Milan), sometimes coastal (Tokyo, Osaka, Nagoya); on straits (Detroit, Istanbul) or canals (Suez, Panama); at major "elbow" bends of rivers (Kuybyshev, Rosario, Cincinnati, Kan-

[37]Brian J. L. Berry, "Research Frontiers in Urban Geography," in Philip M. Hauser and Leo F. Schnore, eds., *The Study of Urbanization* (New York, John Wiley & Sons, Inc., 1965), is a concise summary and bibliography by one of the active writers in the field.
[38]See, for example, *Rand McNally City Rating Guide,* Richard L. Forstall, ed. (Chicago, Rand McNally & Co., 1964).

sas City); where rivers are crossed by major land routes (Winnipeg, Omaha, Novosibirsk, Omsk); or at heads of large lakes (Chicago, Toledo, Buffalo).

But such geographic factors alone are neither necessary nor sufficient for the development of a major focus. For example, there are major commercial centres such as Munich, Dallas or Denver which have no clear geographic advantages over various nearby cities. The history of the rise of individual cities demonstrates what a variety of other political and social factors may have been involved. In a number of European instances, such a non-geographic factor as the wish of a ruler appears to have been the chief determinant in the development of a large city; Madrid is a well-known example, but several cities in Germany have similar histories, including Karlsruhe, Munich and Stuttgart. However, it appears likely that such non-geographic or even arbitrary factors would not have been effective except in a generally permissive geographic context. For example, had the rulers of Württemberg chosen Esslingen or Heilbronn for their capital instead of Stuttgart, we may surmise that the city chosen, with situational advantages roughly equal to Stuttgart's, would have become the region's major city instead. However, had the rulers of Bavaria chosen Berchtesgaden instead of Munich, we may seriously question whether such a remote mountain town could ever have supplanted rival cities and achieved the economic domination of an extensive region.

We shall not pause further to inspect the geographic aspects of focal location, except to note that of course these may vary in significance as other circumstances change. One of the best-known examples, the situation of Venice, proved very valuable when much trade followed routes from Italy across the Alps and to the Near East, but had much less value after trade routes were diverted towards the northwest and the New World.

The identification of commercial activity as the original function of most cities is not intended to rule out the early presence of other functions. For example, most small cities today have at least some manufacturing and no doubt that has always been true. Though trading cities with almost no manufacturers are certainly found today and were found in the past, it is difficult to imagine a large city, ancient or modern, without considerable trade, at least some manufacturing or processing, some significant involvement with transportation (port, rail centre, caravan centre and so on), and some services (financial activities related to commerce; public services such as local administration, health and education; and personal services such as barbers and laundries). It would be most surprising not to find all these or their direct equivalents present in any city of significant size today, anywhere in the world, or in any city of the past.

Another group of urban functions is comprised of those found frequently but by no means so universally. Government over an extensive region (capital status) is an attribute of some cities but not of others, including some very large ones. Military or naval activities play a significant role in a few cities but are highly localized. Some cities of modern times (and a few ancient ones as well) have been popular as resorts, while others clearly have not. Mining has played a major role in a few cities while being absent from most. Religion certainly was the main function of some ancient cities, and higher education plays the major role in a few today. These specialized focal activities have been termed "sporadic" in contrast to the "ubiquitous" activities of commerce, transportation, manufacturing and local administration.[39]

The typical city, then, performs a variety of functions. It generally has commerce as an originating function, but also has manufacturing and transportation activities besides; in addition it may have various sporadic activities. Since cities constitute integrated wholes, these activities are carried on not in separate nodes or nuclei by separate groups, but in close geographic proximity and with elaborate overlapping institutional networks. To say that Paris is a commercial city, an industrial centre, an artistic and tourist centre of world renown, and the capital of France is meaningful only in the sense that it is *one* city that is being described, not four that happen to occupy the same site.

Since the commonest type of city is not solely commercial but combines an original interest in commerce with other closely related activities, we can generalize further about the nature and interrelations of these activities by recalling their focal character. Clearly the commercial activities cannot exist solely to provide the city itself with goods.[40] The commercial function predicates the existence of an area or region that is functionally served from the city as a focal point. Cities typically originate as

commercial foci for small areas; if and as they grow, it is concomitant with their becoming foci for larger areas. Normally, this results from the city expanding its trading area and growing as a result of the additional activities it thus draws to itself. Or the process may work in the opposite fashion, with the city growing as a result of other factors (the choice of a ruler; the development of specialized industry) and so becoming able to supply a wider range of commercial services to its surroundings and hence expand its area of influence at the expense of rival cities.

Usually associated with the commercial functions or services performed for a surrounding area are various other services, so that the typical large city is a regional focus for:

(1) the exchange of the various goods of the region among the region's inhabitants;

(2) the exchange of goods produced by the region with those of other regions;

(3) beginning the transport of goods to other regions (via land, water, or both);

(4) ancillary services relative to intra- and inter-regional trade (for example, wholesaling, banking);

(5) the processing of local goods for consumption within the city or region;

(6) the processing of regional goods for export to other regions; and,

(7) the manufacturing of specialized goods whose market in the region is limited and which use imported raw materials.

The performance of these closely related focal activities is both a natural outgrowth of the city's relation to the region and, at the same time, strengthens that relation. But there are additional less directly related

---

[39] Admittedly, these broad categories have their ubiquitous and sporadic aspects. For example, many branches of manufacturing are sporadic (iron and steel manufacturing, textile production, oil refining), whereas others are practically ubiquitous in large modern cities (baking, printing, brewing, garment making). See Alexandersson, *op. cit.*, p. 13.

[40] In theory this would be possible provided a continuous flow of money to pay for the goods entered the city from outside. Historians have hypothesized that Rome and some other ancient cities may in fact have largely lived in this way. Yet Rome at least was certainly always a major commercial focus for Italy, and for the entire imperial world.

focal activities that the city often performs, such as:

(8) the seat of political administration for the region or a major part of it;

(9) the seat of the religious authorities of the region or a major part of it;

(10) the seat of artistic, cultural, entertainment and higher educational facilities for all or part of the region;

(11) the headquarters of private business and social organizations (corporations, professional groups, trade unions, political parties) operating in the region; and,

(12) the centre for the provision of highly specialized professional services (for example, medical and legal services).

In an earlier period, provision of defence facilities for the region could have been added to this list. Though not primarily economic in character, these activities have important effects on the city's economy, both directly through the jobs they provide and indirectly by the atmosphere of political and cultural leadership created.

We may conclude this list of typical focal functions of major cities with some which are still less directly economic but no less significant to the urban milieu. The city is generally:

(13) the residence (for all or part of the year) of wealthy persons who derive their income from the region, through ownership of agricultural or other productive land or through trading, manufacturing and so on, carried on elsewhere in the region;

(14) the destination of surplus labour from towns and villages of the region, seeking work in a quantity or variety not available to them at home;[41]

(15) the destination of jobless persons seeking a livelihood or the wherewithal for living in any form, who believe (sometimes incorrectly) that they are more likely to find it in the city than in their rural villages; and,

(16) the destination of young, or individualistic, or disaffected persons (ranging from artists and intellectuals to alcoholics and criminals) seeking a livelihood or the where-variety, or anonymity than smaller communities in the region provide.

These diverse activities have the common characteristic of being performed by or in the city but for the region (which includes the city itself, of course). The list could no doubt be extended still further.

With the understanding that the specific boundaries of the region may vary for different functions,[42] the concept of region-serving may be extended to most or all of the functions of a large city. For example, Rome may be described as a local and regional commercial centre for much of central Italy and a manufacturing centre for perhaps a somewhat smaller area. As an administrative centre, however, its "region" (the whole of Italy) includes other commercial and manufacturing centres of at least equal importance over which Rome

---

[41]Juliusz Gorynski and Zygmunt Rybicki have emphasized the importance of migration to the metropolis in Chapter 9 in this book.

[42]Howard L. Green, "Hinterland Boundaries of New York City and Boston in Southern New England," *Economic Geography*, Vol. XXXI (October, 1955), pp. 283-300, is an interesting study of the position of the boundaries of influence in several different economic and social respects between two major American centres.

has little influence in purely economic terms. As a religious centre (and accordingly a centre for pilgrims and other visitors), Rome's "region" is much wider than Italy. Finally, as a cultural, artistic and tourist centre (based on a complex combination of factors, including its long history, some aspects of its climate, its religious association and certain of its handicraft industries), its influence and hence, in a sense, its "region", extend even more broadly than its religious role. Though such extended types of influence may seem more relevant to the examination of a city's "atmosphere" or "character" than to its economic function, certainly the economy of a city like Rome is significantly influenced by such roles and by the fact that they involve relations with very extensive regions.

Even in the case of Rome, however, we may note that the primary economic activities would appear to be related to a limited region, central Italy. Perhaps the majority of people in Rome are not greatly concerned economically with activities beyond Italy or even beyond central Italy. It is suggested that for most major cities, a relatively clear-cut region exists that the city serves with most or all of the regional functions already noted, with boundaries at least approximately the same in each case. To the extent that this is true and accounts for most of the city's economic activity, the city may be termed a "regional centre". Such a city will be a major commercial and transport focus, specializing in the products either produced by, or required within, the region. It will further be a manufacturing centre to a degree depending chiefly on the character of the region; a regional focus in an unindustrialized region may comprise most of the region's manufacturing yet still have relatively little (for example, Kinshasa, or Spokane, Washington), whereas the focus of an industrialized region will itself be an important industrial centre even though smaller cities within the region may be relatively more highly industrialized (for example, Boston, London, Stockholm).[43]

In addition to differing as to their degree of industrialization, regions also differ pronouncedly in the extent to which they are involved with interchange with the rest of the world. The chief cities of largely self-sufficient regions are generally small, not large enough, for example, to be included in our sample of 100 very large centres. Hyderabad might serve as a rare example of a large city still oriented primarily to serving as the focus of a region with relatively little extra-regional trade.

Regions whose dealings with the rest of the world rely largely on the export of a single product, usually an agricultural or mineral raw material, apparently also do not produce very large cities unless other factors are also present. Such a city as Rangoon remains of moderate size—it comprises perhaps five per cent of its nation's (region's) population and is largely oriented to activities related to the export of the nation's rice surplus. In contrast, the size of a city such as Sao Paulo can clearly not be attributed directly to its role as the handler of most of Brazil's coffee crop. Here the accumulation of capital and its application to industrial development has erected an extensive superstructure of manufacturing on the earlier and simpler role of a regional centre.

From the point of view of an assessment of the functions of metropolitan areas, the regional role is especially significant because of the basis it offers for comparisons —most cities do have major region-serving

---

[43]There appear to be few if any examples of relatively unindustrialized cities serving as major economic foci of regions with significant amounts of industry.

functions. We may thus recognize that the more specialized activities in a city are of different character without overlooking the fact that they, too, may be assessed from the point of view of regional service, differing in that the region served is much wider than the city's "average" region for other focal activities.[44]

THE 100 METROPOLITAN AREAS CLASSIFIED BY FUNCTION

The following pages outline a simple, functional classification of our 100 major cities.

Regional Cities—R. Cities whose economic functions have been and continue to be primarily regional. Each is the commercial focus and chief manufacturing and service centre of a region (greater or smaller) whose general limits are well recognized and within which it has little competition from rival centres. In addition to supplying many manufactured goods to the region, if in a highly industrialized region, the city generally produces some manufactures for export beyond the region as well. Of our 100 metropolitan areas, 74 are classed as regional cities. Sub-categories are:

Ra: A regional city which is also the administrative centre of a region or of most of it.

Rp: A city which is also a seaport of major significance to the region.

Rm: A city which is primarily a regional centre, but also exports manufactured products beyond the region,

on a scale sufficiently significant to make manufacturing of great importance in the city.

Manufacturing / Regional Cities—M / R. Each of these cities was originally a regional centre and continues to serve as a focus for its region, but its character today is chiefly that of a manufacturing centre, with markets far beyond the region served for other functions. The region served for general regional functions continues to be about the same as that served when the city was primarily a regional centre. The manufacturing may thus be viewed as having been superimposed on the original regional structure. Thus the *M/R* cities differ from the *Rm* cities chiefly in the degree of importance of manufacturing. Thirteen of the 100 metropolitan areas are in the *M/R* category.

P/R, X/R. These designations serve to classify occasional cities whose original region-serving character has assumed a specialized cast from the addition of activities other than manufacturing. The only such instances in our list of 100 cities are Singapore (*P/R*) and Johannesburg (*X/R*). Both these cities developed as specialized cities and regional centres essentially concurrently.

Manufacturing Cities with Limited Regional Functions—M(r). These cities were never regional centres of much significance, but came into existence primarily as manufacturing centres. (Many of these cities no doubt did originate as foci for a small area within which they were "regional centres".) They have carved out for themselves a region which they now serve, since they have become cities of sufficient size to offer much of the range of services of a more typically regional city. The region served by such a city is usually relatively limited in extent and over half the population may be within the metropolitan area of the city

---

[44]See the general discussion in the classic paper by Chauncy D. Harris and Edward L. Ullman, "The Nature of Cities," *Annals of the American Academy of Political and Social Science,* Vol. CCXLII (November, 1945), pp. 7-17. In Harris and Ullman's phrasing, "specialized-function cities perform one service such as mining, manufacturing, or recreation for large areas, including the general tributary areas of hosts of other cities."

itself. However, in contrast to the M/R cities, the region is larger than it was when the city's function was primarily that of a local service centre.

$P(r)$, $G(r)$, $X(r)$. These cities have developed primarily around functions other than manufacturing or regional services (such as sports, government, resort), but have similarly carved out regions for themselves, usually relatively small.

The specific classification for each of the 100 metropolitan areas appears in Appendix A. Since the classification is being applied only to 100 of the largest cities, we should anticipate that other types would be found if a longer list of cities were classified. In particular, every one of the 100 cities has been recognized as having at least a subsequent development of regional importance (r) if not a primary one—R—or an early one now made secondary—$/R$. Among cities of less than 1,000,000, some undoubtedly have so little regional significance that they would be classed as wholly manufacturing, port or mining cities or the like.[45] The "regions" served by such cities are barely larger than their own urban areas. It is not surprising that such circumstances are not found in any of the very large cities in our group of 100. Even an urban area so dedicated to industry as is the Ruhr or the Katowice area perforce serves as a fairly significant regional centre for its surroundings, simply because its great size, though produced by industry, results in its being able to provide a significant range of commercial and other services.

---

[45]Examples of purely manufacturing cities of considerable size (over 100,000) include Fall River and Lawrence in Massachusetts, Bolton and Huddersfield in England, Roubaix in France, Krefeld in Germany, and Kishiwada in Japan. Significantly, all these cities are close to much larger cities, and all are the result of early industrialization in the textiles industry.

As in any classification, certain decisions and allocations are essentially arbitrary but it is hoped that in this instance they are at least reasonably consistent. Cities have not been designated as ports unless the port is within the metropolitan area (for example, Caracas is counted as a port, but Seoul is not). Some cities that are incidentally ports have not been so designated because the role was not felt to be of sufficiently great importance: in this category are all inland ports, including the Great Lakes-St. Lawrence system cities except for Montreal.

Cities have been recognized as administrative capitals not only when national capitals, but also if they are capitals of a national subdivision that comprises most or all of their region of influence. In some instances (for example, Sao Paulo, Sydney, Melbourne) this seems clearly justified; in others (for example, Hanover, Bombay, Leningrad) the significance of the administrative function may be much smaller. Data on the numbers of persons employed in government would probably serve to determine a more accurate classification. Two cities, Karachi and Rio de Janeiro, have been denoted (a) because they were capitals until very recently and continue to serve some national administrative functions. Certainly, they owe much of their development to their past role as capitals. Several other cities could have been similarly designated on the basis of a role as capital in the fairly recent past: Nanking, Istanbul, Leningrad. Berlin and Vienna represent examples of cities much affected by a sharp reduction in the extent of the areas of which they formerly were capitals. Finally, Kyoto represents a special instance of a city developed originally as a capital and still preserving certain concomitants of that role long after the formal administrative role has ended.

A number of cities carry more than one subordinate designation, for example as both a port and a capital; in such cases no attempt has been made to list the subordinate activities in order of importance to the city.

Distribution of the 100 Metropolitan Areas by Function. Of the total of 100 cities, 74 are classed as chiefly regional centres *(R)*, most with various additional designations. In 15 others, the designation is *M/R* (13), *P/R* (1), or *X/R* (1). This leaves 11 of the 100 in which the regional role is seen as so secondary that it is relegated to the designation (*r*).

Of the total of 26 cities not designated as chiefly regional in character, 19 are accounted for by the United States (8), northwestern Europe (8), and Japan (3). In each of these three regions, non-*R* cities account for roughly half the cities in our list. All other regions together account for only 7 non-*R* cities. Two are in southeast Asia[46] (Hong Kong and Singapore), and one each in Subsaharan Africa (Johannesburg), the Middle East (Alexandria), the Soviet Union (Donetsk), eastern Europe (Katowice), and mainland China (Tientsin). The other five regions (Australia, Canada, Latin America, southern Europe, and India/Pakistan) have no non-*R* cities; all of their major cities are primarily regional in function. The very large non-*R* cities are thus characteristic of two types of regions, highly industrialized ones in which the non-*R* cities are usually manufacturing centres but may be of such specialized types as Washington; and relatively unindustrialized regions in which port activities have generated major cities not primarily viewable as regional in function. Even so, the large majority of metropolises in the developing regions on our list are regional in character.

Among the seventy-four *R* cities, only two are designated simply *R* without a subordinate designation as an administrative centre, port, or large-scale manufacturing centre. One of these is Edmonton, and it is probably no coincidence that it is also the smallest city on our list. As for Nanking, also designated simply *R*, it is possible that manufacturing is sufficiently developed to justify the designation *Rm*, or that the large size of the city should be attributed in some measure to its former role as China's capital.

Only twenty of the seventy-four cities are not designated as administrative centres. Eight of these are in North America, six in Europe and three in mainland China. Instances of very large cities that are not administrative seats have become frequent in the last century and a half, especially in countries settled originally from the British Isles, and reflecting something of a theory that the ordered progress of administration and government would work best if removed from the hurly-burly of economic life. Where this tradition is carried to its full extent, the national metropolis and other major cities may not even be capitals of their secondary political subdivisions, as in many instances in the United States. Apart from possible effects on the cultural and intellectual atmosphere of the city, clearly there is decided significance for the administrative position of cities like Los Angeles, Chicago, or New York, responsible to, yet not capital of, a political unit which they come close to

---

[46]Several writers have recognized the special character of the major cities of southeast Asia, a region in which most if not all major metropolitan development has been seen as a nonindigenous import. See Norton S. Ginsburg, "Urban Geography and 'Non-Western' Areas," in Philip M. Hauser and Leo F. Schnore, eds., *The Study of Urbanization* (New York, John Wiley & Sons, Inc., 1965), p. 332.

dominating but whose administration is centred elsewhere.

Thirty-three of the seventy-four *R* cities lack the subordinate designation *m*. These are fairly well scattered, but all the *R* cities in northwestern Europe and Japan and all but one (San Francisco) in the United States have sufficient manufacturing to carry the *m* designation.

Thirty-two of the *R* cities are designated as ports, while forty-two lack the designation. The two groups are relatively inter-mixed, but we may note the presence of eight non-port metropolises in north-western Europe, five in eastern Europe (comprising all those in our list in that region) and seven in mainland China.

World Cities. As was mentioned earlier, from an hierarchic point of view all of the cities in the list are viewed as placed very high and the classification as given makes no attempt to distinguish degrees of relative importance. However, five cities have an importance that in so many respects is of international scope that they might well be classified as "World cities". These cities are New York, London, Paris, Moscow, and Tokyo. In each, a number of functional aspects may be seen to be international. A number of additional cities, of course, have significant functional aspects of worldwide range.

Some specific comments may be offered about certain cities that present either un-usual configurations of activities or doubts about allocation, or both.

Port cities: P/R, Pm(r), P(r). In these six cities port functions are viewed as being more important than regional ones. In four, manufacturing for a wide market is noted and the designation is *Pm(r)*. The classi-fication seems indisputable for Hong Kong, whose regional functions have not been extensive on the adjoining Chinese main-land and which has had a short but varied career as a port with wide international connections. Its political independence of the mainland has also been of major sig-nificance to its role. Liverpool and Rotter-dam are great European ports with a long mercantile tradition and closely related large-scale industries (refining, chemicals). Both are close to cities that are older and larger, and neither has an extensive land-ward region over which they are dominant. Hamburg is generally parallel in function, though the land region served by the city is more extensive and no major rival city is close by. Yet Hamburg's regional role in northern Germany appears to be much less important than its size would suggest. Per-haps significantly, it has never served as capital of any extensive area but is, in fact, a rare instance of the retention of city-state administrative status into modern times.

Singapore, designated *P/R*, is somewhat analogous to Hong Kong but has not his-torically been so politically separate from the adjoining mainland area. Though its port functions would still appear to be dominant, it has important regional func-tions for Malaysia. Again it may be sig-nificant that Singapore has been of two minds about whether to join with Malaysia politically or not, whereas Hong Kong's desire to remain separate from China has been quite clear-cut.

Alexandria, designated *P(r)*, was cer-tainly a major regional centre in ancient times, when it was capital of Egypt as well as one of the great Mediterranean cities. Its modern size has certainly eclipsed its ancient magnitude but dates essentially from the opening of the Suez Canal. For some decades the city's economic emphasis was as an export port for Egyptian cotton and other products and as a regional ship-ping and financial centre, perhaps as much international as Egyptian. It does not seem to have served a very significant role for its densely populated landward region, possibly because of the nearness of the larger city of Cairo and Alexandria's own eccentric position at one edge of the Nile Delta. It would appear to have been a city that faced out as much as in, like Hong Kong and Singapore. Recent developments may have altered its role to such an extent that a different classification would be more accurate.[47]

Manufacturing cities of doubtful allocation. Kitakyushu, Japan, has been designated *Mp(r)*; the city comprises five recently merged port cities and forms one urban area with the sixth city of Shimonoseki. Several of these cities have major industry as well, but it does not appear that any have a wide regional significance.

The main designation of Kyoto as *M* is certain, and it has been designated *M/R* on the premise that something of a regional role continues. Kyoto's special character is certainly closely related to its former capital status and its continuing importance as the centre of Japanese traditional culture and production. It may be that from a regional point of view it can no longer be considered independently from Osaka, its giant neighbour. In fact, as of 1968, Kyoto has been recognized as part of a single Osaka-Kobe-Kyoto Metropolitan Area. (See Appendix B at the end of this chapter.)

Los Angeles, designated *M/Rp*, is another unusual phenomenon among major cities. Assessments of its functional role have not been easy to make, probably reflecting both a rather complex series of factors and the presence of some factors not previously seen, at least not on such a scale. Los Angeles is a very large city compared to the surrounding region and though certain regional connections are evident much of its great size must be attributed to other factors. Manufacturing was not especially prominent until after 1939, when the city had already reached substantial size (over 2,000,000). In some respects, the heavy concentration of regional population in one metropolis is reminiscent of Australia or southern South America; yet California has not really had a comparable history of export of raw materials through a single major port to the rest of the world. In recent years Los Angeles has become the dominant aircraft-manufacturing centre for the entire United States. The development first of motion pictures and then of aircraft was associated with the city's unusually temperate and sunny climate, within a large and wealthy nation that had few areas with such a climate. Much of its growth has been along specialized lines with a market or "region" of national extent, in contrast to development attributable to the serving of an immediate hinterland. There was also some development, especially before 1940, as a resort or retirement centre, another phenomenon not often found in very large cities before the twentieth century. In addition there was a period of active petroleum development, mainly in the 1920's and 1930's. These varied activities doubtless enhanced the city's attraction for migrants from other regions and the continually increasing labour supply in turn proved helpful in stimulating manufacturing expansion for a large national market which otherwise might have been difficult to serve from such

---

[47]As E. M. Forster wrote in 1922, comparing ancient and modern Alexandria, "then, as now, she belonged not so much to Egypt as to the Mediterranean." *Alexandria* (Garden City, Anchor Books, 1961; originally published in 1922), p. 13. This is an outstanding example of a non-scholarly work on a particular city that offers a dimension of understanding of its functions that a purely statistical analysis could scarcely achieve. It is indicative of the particular complexity of major cities that they have proved both so fascinating to and so effectively illuminated by the attention of major writers.

an eccentric location. These comments may serve to indicate the difficulty of attributing the growth of a metropolis of 7,000,000 to any single or simple cause.

Washington and Miami. Two other United States cities deserve comment for their unusual functional character. Washington remains as it has always been, primarily a governmental city, its growth reflecting quite closely trends in relative centralization or decentralization of the federal administration (for example, its most rapid growth has come during successive American wars). Now a very large city, it has developed a modest regional influence in an area originally dominated by Baltimore. Its national role in government has been increasingly supplemented by the addition of headquarters of associations of national scope, especially those which wish for various reasons to be close to government and its officials. However, the city shows few, if any, signs of assuming a broader role in the nation's economic life and still has very little manufacturing.

Miami represents the first instance of a resort city reaching a million population. Wholly a product of the present century, it reflects the presence within a large, rich nation of one small area of tropical climate. As with Washington, Miami's size has resulted in its assuming a significant and increasing regional role, though this is impeded by its situation near the southern end of peninsular Florida. Active attempts have been made to strengthen the city's regional economic functions as well as its port activities and manufacturing is of increasing importance though still very modest compared to other cities of its size.

Johannesburg and Ibadan. Finally, the two cities of Subsaharan Africa are of special interest. Major urban development in this region has been scanty in the past and the only three concentrations of cities today are the relatively new ones in South Africa and in the Copper Belt of Zambia and Katanga, and an older one in Western Nigeria. The two African cities in our list are representative of these areas. Johannesburg, designated X/R, represents a major regional centre that nevertheless owes its large size chiefly to the gold-mining industry. The other two major South African cities, Cape Town and Durban, neither included in our list, are both more familiar in type and are probably classifiable as Rpm and $Rapm$ respectively. Most of the Copper Belt cities are primarily mining centres.

Ibadan in Nigeria is the largest of the indigenous Nigerian cities, which have long been of great interest to western students of cities because of their relatively large size and close spacing. Ibadan has been classified as Ra, but the extent of its regional role may be less than this suggests. Since many of the population are agriculturists it is possible that a special classification for Ibadan would be more appropriate.

CITY AND REGION: CONCLUSION

This analysis of the 100 metropolitan areas by function has stressed their regional relations, chiefly because these constitute the most widespread common denominator in the functional characteristics of these varied urban centres. However, to the extent that a classification implies a series of explanations, this approach is probably most relevant to the development of metropolitan areas up to the present time. Most of the 100 centres are seen as the products of aggregative or agglomerating forces within their regions, operating in past decades or centuries to concentrate sufficient activity and wealth to provide permanent support for a large population.

Much of the literature on the city and the region reflects the dichotomy of past centuries between the city as a dense and distinct settlement and a surrounding region of much lower density and less intensive activity, the two closely linked by mutual interchange but sharply distinguished in appearance, life style and atmosphere. Today both the metropolis and the region are moving away from these sharp distinctions in three significant respects. First, the city's direct physical extent spreads more and more widely, as it grows in magnitude and as its citizens require more space-consuming housing and other facilities. This physical expansion, often out-stripping the evolution of adequate formal administrative machinery, brings the administrative problems and concerns of the city closer to, and less readily separable from, those of the region.

In a second and broader sense, the city's influence extends more strongly beyond its own physical borders into its region, as transport and communications become speedier and more efficient. The ties of the capital or metropolis to provincial centres and rural districts are much closer than they were a generation or two ago and there are fewer and fewer areas so remote as not to be under the dominance of some metropolitan centre.

Finally, life in the surrounding rural areas and lesser towns grows less sharply distinguishable from that in the city itself. Small-town dwellers and farmers can read city newspapers, listen to city radio, and watch city television almost as easily as the city dwellers, and what they read, hear and see erases more and more of the differences that so long have distinguished them from the citizens of the metropolis.

There is a further development that has begun to make itself visible in some of the developed nations, notably the United States. Many of the classic metropolitan functions of major urban centres can now be exercised quite readily over extensive areas, much broader than the "average region" that we have discussed and that was so prominent in their past development. It becomes increasingly difficult to attribute the functional patterns of individual large American cities to relations with a circumscribed surrounding region. For example, in their impressive study of America's metropolises and their regional relations, Duncan and his associates noted that "on the basis of our findings to this point there is little support for a characterization of the metropolis as a centre for non-resource-oriented industries whose markets are in its hinterland. Instead it would be more consistent to hypothesize that other metropolises are the major markets for metropolitan industries classified as second stage resource users. . . ."[48] In this sense, the "region" of any large American metropolis is comprised of most of the other metropolises of the nation.

It may be that American metropolitan areas are evolving from a "multi-regional system", in which each was involved primarily with its own immediate region, in the direction of a "single-regional system" in which individual metropolitan areas specialize in given functions for the whole nation. It may be, in other words, that metropolitan functions continue to be provided for the whole region by the same metropolitan areas as a group, but with the individual metropolitan centres specializing in providing particular functions to the entire region instead of nearly all functions to a particular region.[49] Such a development reflects, of course, the great impact of

---

[48]Otis Dudley Duncan, et al., *Metropolis and Region* (Baltimore, The Johns Hopkins Press, 1960), p. 226.

[49]The interpretation of a metropolis as specialized in certain functions for a market of nationwide extent appears to apply well to Los

advances in transportation and communication on the space and distance relations of cities. Yet, these developments have not yet altered the importance of the regional relations of most of the world's expanding metropolitan areas.

Additional important questions that our discussion has not answered may be raised. Why do some of the world's many nations and regions have large cities while others do not? Why do the proportions of total population living in cities or metropolitan areas vary so sharply from region to region? Why is it characteristic of many developing countries to have a high concentration of urban population in only one or two cities?[50] It seems clear that answers to these questions, too, must be sought in a further examination of the nature of the city's region-serving functions. Those who for purposes of urban, economic, or administrative planning conceive of the city and especially of the major city or metropolis as separable from its region are surely exhibiting a fundamental misconception of the city's nature and role.

The Metropolis and the Framework of Administrative Areas

So far our attention has been devoted to the characteristics of metropolitan areas as socio-economic and geographic units. However, there are significant respects in which the metropolitan area must be seen

___

Angeles, whose large size is very difficult to interpret as directly related to the performance of services for its immediate region. Los Angeles is the only very large American city to have developed almost wholly in the present century, in other words under contemporary conditions of transportation and communications efficiency.

[50]See UNESCO, *World Survey of Urban and Rural Population Growth*, Doc. E/CN. 9/187 (March 8, 1965), especially par. 48-53.

as being made up of many parts. This is particularly true in the political and administrative contexts, since the metropolitan area is rarely a single administrative unit. First we shall consider the nature of the different types of subdivisions to be found in the metropolis; secondly the role of the central city; and, finally, various aspects of the problem of defining the metropolis and its subdivisions as operating political and administrative entities.

INTRA-METROPOLITAN ZONES

Since a large metropolitan area consists of many parts differentiated from each other by economic function, land use, age, rate of growth, the socio-economic characteristics of its inhabitants and many other features, the demarcation of sub-regions or other subdivisions is of especial importance for the study of urban structure and function. The most commonly used subdivision is the political or administrative area. This is to be expected, since such areas are the most commonly used as building blocks for the construction of a metropolitan community.

The areal extent of political and administrative units varies enormously. It makes no sense, therefore, to compare the "central city" of Los Angeles (with an area of 459 square miles and a population of 2,660,-000) with the old City of London (less than 5,000 population living in one square mile), or with San Francisco (750,000 people living in 45 square miles). For the same reason, inter-metropolitan comparisons of suburban areas are unsatisfactory.

Suburbs are usually defined as either that part of the metropolitan area lying outside the central city or, frequently in the United States, as incorporated municipalities other than the central city. Sometimes the term "fringe area" is used as the equivalent of "suburban area". Frequent distinctions are made in American studies between the rural and urban fringes of metropolitan areas. One such distinction is to regard the urban fringe as being that part of the urbanized (built up) area of the Standard Metropolitan Statistical Area (SMSA) that lies outside the central city and the rural fringe as the remaining part of the SMSA.[51] The analysis of the 1950-60 growth rates for these three zones in SMSAs in the United States showed that both central city and rural fringe zones in most SMSAs grew less rapidly than did the urban fringe. Among those SMSAs with over 1,000,000 inhabitants, the central cities in only the Atlanta, Dallas and Houston SMSAs contained a larger percentage of the metropolitan population in 1960 than in 1950. In each of these, the area of the central city was considerably enlarged during the decade. Only in the Buffalo and San Francisco-Oakland SMSAs was there an increase in the percentage of metropolitan population rise in the outer fringe.

The Paris agglomeration has also been divided into zones designed to facilitate scientific investigation of metropolitan phenomena, as outlined in Table 1-5. This pioneer work of the *Institut national de la statistique et des études économiques* (I.N.S.E.E.) deserves detailed examination as an approach that could be followed by students of other metropolitan areas. It was also based upon the aggregation of administrative and political units (communes), even though "marginal communes are often not totally homogeneous and contain both urban and rural sections."[52] An elaborate system of criteria (proportion of population living from agriculture, number of inhabitants, density of population, population growth since 1936, commutation of workers out of the commune, commutation to the Department of the Seine and the existence of a railroad station or city bus line), each weighted, resulted in the ranking of several hundred communes according to scores ranging from 0 to 100. The classification of communes based upon these scores was used to produce a Paris region composed of six zones somewhat comparable to the central city, urban fringe and rural fringe in American usage. The authors express the hope that their method of delimiting zones within metropolitan areas will lead to a standardized classification of statistical subdivisions which can be used in comparative research. But this will be difficult to accomplish as long as the "building blocks" are administrative and political units as diverse in size and purpose as such units are in different countries of the world.[53]

[51]Victor Jones, "Metropolitan and Urbanized Areas," *Municipal Year Book, 1962* (Chicago, International City Managers' Association, 1962), pp. 31-43. A similar division of the Kinming metropolitan community has been made by Ernest Li, "A Study of Urbanism and Population Structure in a Metropolitan Community in China," in E. W. Burgess and Donald Bogue, eds., *Contributions to Urban Sociology* (Chicago, University of Chicago Press, 1964), pp. 419-28.

[52]*Délimitation de l'agglomération parisienne*, I.N.S.E.E. (Paris, December, 1959). Portions of the report have been translated into English by the Conference on Metropolitan Area Problems (now located at the University of the State of New York at Albany) under title "Delimitation of the Paris Urban Region". The quotation is from page 8 of the translation. The 1962 French census used the I.N.S.E.E. report as the basis for delimiting the Paris agglomeration (*complexe résidentiel*). See *Recensement de 1962: villes et agglomérations urbaines* (Paris, 1964), p. 5 and Table III.

[53]In the United States, where the building block is the country, which is far larger and more heterogeneous than the French *commune*, the difficulty is compounded. The New York Regional Plan Association divides the New York-New Jersey-Connecticut Metropolitan Region into a core, an inner ring, and an outer ring, each consisting of whole counties. See *The Region's Growth* (New York, Regional Plan Association, May, 1967), pp. 79-82.

TABLE 1-5
Subdivisions of the Paris Metropolitan
Area

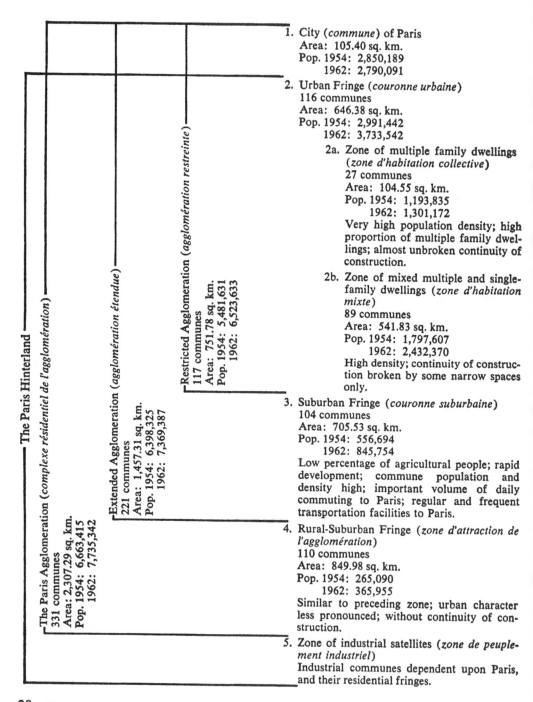

1. City (*commune*) of Paris
   Area: 105.40 sq. km.
   Pop. 1954: 2,850,189
       1962: 2,790,091

2. Urban Fringe (*couronne urbaine*)
   116 communes
   Area: 646.38 sq. km.
   Pop. 1954: 2,991,442
       1962: 3,733,542

   2a. Zone of multiple family dwellings
   (*zone d'habitation collective*)
   27 communes
   Area: 104.55 sq. km.
   Pop. 1954: 1,193,835
       1962: 1,301,172
   Very high population density; high
   proportion of multiple family dwellings; almost unbroken continuity of
   construction.

   2b. Zone of mixed multiple and single-
   family dwellings (*zone d'habitation
   mixte*)
   89 communes
   Area: 541.83 sq. km.
   Pop. 1954: 1,797,607
       1962: 2,432,370
   High density; continuity of construction broken by some narrow spaces
   only.

3. Suburban Fringe (*couronne suburbaine*)
   104 communes
   Area: 705.53 sq. km.
   Pop. 1954: 556,694
       1962: 845,754
   Low percentage of agricultural people; rapid
   development; commune population and
   density high; important volume of daily
   commuting to Paris; regular and frequent
   transportation facilities to Paris.

4. Rural-Suburban Fringe (*zone d'attraction de
   l'agglomération*)
   110 communes
   Area: 849.98 sq. km.
   Pop. 1954: 265,090
       1962: 365,955
   Similar to preceding zone; urban character
   less pronounced; without continuity of construction.

5. Zone of industrial satellites (*zone de peuple-
   ment industriel*)
   Industrial communes dependent upon Paris,
   and their residential fringes.

The following labels appear along the bracket structure:

— The Paris Hinterland

The Paris Agglomeration (*complexe résidentiel de l'agglomération*)
331 communes
Area: 2,307.29 sq. km.
Pop. 1954: 6,663,415
    1962: 7,735,342

Extended Agglomeration (*agglomération étendue*)
221 communes
Area: 1,457.31 sq. km.
Pop. 1954: 6,398,325
    1962: 7,369,387

Restricted Agglomeration (*agglomération restreinte*)
117 communes
Area: 751.78 sq. km.
Pop. 1954: 5,481,631
    1962: 6,523,633

It is natural for any observer of the data (or of the metropolis itself) to compare the characteristics of the central core with the rings surrounding the core. The man in the street as well as the social scientist is constantly comparing suburbia with the city, often in implicitly or explicitly pejorative terms. Often the much more careful and objective analyses by scholars are still gross and, as applied to social action, misleading. This is because the principal measure of similarities and differences is an average computed for the whole of a metropolitan zone, with the result that central city and suburbia are interpreted by many readers as homogeneous sectors of the metropolitan area.[54] As Robert C. Wood says,

If the figures for the indexes we have been quoting here are spread randomly throughout suburbia, then each suburb would tend to look like its neighbour and each would still contain a diverse mixture of races, occupations, degrees of education and family size comparable to the characteristics of the mammoth city population. Under these circumstances, the small town qualities would be so widely scattered that the existence of separate political boundaries would have no significance at all.[55]

Even the refinement in the definition of the areal unit of comparison which would result from breaking down large zones (that is a suburban fringe) into the political and administrative jurisdictions comprising the zone would be insufficient. The central city as a legal unit is too large to be compared with a much smaller suburban municipality. Furthermore, in the United States much of the fringe population lives in unincorporated territory.

In order to permit analysis of intra-metropolitan zones in a framework free of the limitations of political and administrative units, the United States Bureau of the Census has divided 180 of the SMSAs into fairly small and relatively permanent statistical areas for reporting certain population and housing data. It should be possible to devise a classification of these census tracts into typologically useful sub-areas of the metropolis.[56] Data by census tracts are used extensively by local planning agencies and by public and voluntary social welfare

[54]The most thorough study on this issue is Otis D. Duncan and Albert J. Reiss, Jr., *Social Characteristics of Urban and Rural Communities, 1950* (New York, John Wiley & Sons, Inc., 1956). The authors warn the reader about overlooking the variety of suburbs and the heterogeneity of suburbia (pp. 5-8).

[55]Robert C. Wood, *Suburbia: Its People and its Politics* (Boston, Houghton Mifflin & Company, 1958), p. 115. For other studies of suburbs, see Chauncy D. Harris, "Suburbs," *American Journal of Sociology* (July, 1943), pp. 1-13, and the classifications of suburbs and other cities of over 10,000 population, following the Harris approach, in the *Municipal Year Book* (Chicago; International City Managers' Association) by Kneedler (1950), Jones (1953), Jones and Collver (1959), Jones, Forstall and

Collver (1963), and Forstall (1967). Leo F. Schnore has contributed more than any other student to the comparative ecological sudy of suburbs in the United States. See his collected essays in *The Urban Scene* (New York, The Free Press, 1965). Duncan and Reiss (*op. cit.,* pp. 388-409) present a classification by economic function and income level (1950) of suburbs over 10,000 population, of central cities, and of cities outside metropolitan areas. Standard metropolitan areas are also classified. Hadden and Borgatta present a social profile of American cities in form of decile rankings of eleven socio-economic and demographic characteristics as a substitute for economic or functional classifications. *American Cities* (Chicago, Rand McNally, 1965), pp. 67-100.

[56]French scholars have frequently classified the subdivisions (*arrondissements*) of Paris, as well as its suburban *communes*, but usually with respect to one characteristic at a time. Among other examples are P-H Chombart de Lauwe and others, *Paris et l'agglomération parisienne*, Vol. II, *Méthodes de recherches pour l'étude d'une grande cité* (Paris, Presses Universitaires de France, 1952); Pierre George,

agencies. A few social scientists have studied the distribution by census tracts of land uses, social pathologies and demographic and socio-economic characteristics of the inhabitants of a few cities. The most intensive work has been done in Chicago.[57]

The term "social area analysis" has come to be associated with the work of Eshref Shevky, Wendell Bell and Scott Greer. They have constructed and applied indices of social differentation such as social rank, family life style and segregation among the various parts of a city or metropolitan area.[58] Some critics have viewed this procedure as little more than a method of manipulating census tract data. However, here again the paucity of applications of the classification, as of other classification schemes, leaves us uncertain of its theoretical utility. The results of the use of the indices in an analysis by Scott Greer of political attitudes and electoral behaviour in the St. Louis metropolitan area lead us to hope for a replication in similar circumstances in other metropolitan areas.[59] The availability of computers and of more sophisticated analytical techniques now make it possible to manipulate census tract data for all metropolitan areas.[60] Significant comparison and perhaps classification of cities and metropolitan areas could then be based on profiles of the spatial distribution of different types of sub-areas.

THE RELATIVE IMPORTANCE OF THE POLITICAL CENTRAL CITY

The relative importance of the central city in the metropolitan area as a whole is clearly of particular significance from the political and administrative point of view. The changing role of the central city and of suburbs also have economic, fiscal and social consequences of grave import not only to the inhabitants of a metropolitan area but to the nation.

The typical roles of suburbs and central cities as they seem to be stabilizing in North America are by no means universal structural patterns. In southern Europe, Latin America, and many other parts of the world it is typical for the elite to live in the core of the city and the disadvantaged newcomers live on the periphery, reversing the common North American pattern. However, transportation technology and economic development appear to be changing the physical structure of many metropolitan areas toward the North American and north European types, even in cultures placing a low value on dispersed settlements of single-family dwellings.

In the United States all statistical series for several decades have indicated that the periphery is growing more rapidly than the centre and that the increase of inhabitants

et al., *Etudes sur la banlieue de Paris: essais méthodologiques* (Paris, Colin, 1950); Pierre George and Pierre Randet, *La région parisienne* (Paris, Presses Universitaires de France, 1959). A study of metropolitan Hamburg classifying both sections of the city and suburban *communes* is Olaf Boustedt, "Siedlung und Wirtschaft im Raum Hamburg und Umland," *Hamburg in Sahlen*, 1967, Sonderheft 1.

[57]See Beverly Duncan, "Variables in Urban Morphology," in Burgess and Bogue, *op. cit.*, pp. 17-30.

[58]Shevky and Bell, *Social Area Analysis* (Stanford, Stanford University Press, 1954).

[59]John C. Bollens, ed., *Exploring the Metropolitan Community* (Berkeley, University of California Press, 1961), pp. 13-18, 221-314. See also Norberte Lacoste, *Les caractéristiques sociales de la population du Grand Montréal* (Montréal, Université de Montréal, 1958) and Oliver P. Williams, et al., *Suburban Differences and Metropolitan Policies* (Philadelphia, University of Pennsylvania Press, 1965).

[60]See Robert C. Tryon, *Identification of Social Areas by Cluster Analysis: A General Method with an Application to the San Francisco Bay Area* (Berkeley, University of California Press, 1955).

on the fringe has been followed by the location of new activities (and the relocation of many old activities) in the suburbs.[61] This has led many people to assert that the metropolis of the future will be without a centre—certainly without a central city. We do not have to accept this prophecy in order to recognize that there is a change in spatial distribution of people and activities, that the economic role of the central city is rapidly changing, and that these changes have political consequences. One consequence is that the central city will have to find ways of accommodating itself to the growing (and in some instances, superior) power and influence of people and institutions based outside the city. This adjustment, or lack of it, is already evident in the United States and Canada. Depending upon the political system, such adjustment need not be restricted to metropolitan and local governments. In highly urbanized nations, all higher levels of government will involve themselves or be involved by those seeking their support in metropolitan affairs.

At first glance, a metropolitan area in which the central city includes only a small proportion of the total population might be assumed to be one in which administrative diffusion is at a peak. In fact, this may not be so if special administrative arrangements have been worked out. For example, significant aspects of government may be handled by regional, provincial, or even national, administrations with authority

over the whole metropolitan area, leaving the central city itself with only limited powers in its limited area. A classification of metropolitan areas in terms of relative administrative and political roles of the central city and other areas would be very useful, although time-consuming to produce.

Here, the discussion of the relationship between metropolitan structure and the role of the central city will be based upon population data. In a few metropolitan areas, the extent of the administrative central city is not easily determined from available material. In some cases, the population and related data usually given for the "city" are more probably for a census-defined area that is not an administrative unit. In others, such as Hong Kong, there may not be any local administrative unit that really corresponds to a central city. Though most metropolitan areas have a single clearly dominant central city, an important minority have subsidiary centres in addition to the main city, or are binuclear or even polynuclear in structure. Since such variations evidently have implications for the role of the administrative central city, it will be convenient to examine the two aspects together.

From the point of view of simple metropolitan structure, our list of 100 metropolitan areas may be divided into five classes. (For the allocation of individual metropolitan areas, the reader is referred to Appendix A, where the proportion of population in the central city is also specified.)

(1) Uninuclear metropolitan areas: one urban centre and its suburbs, with no subsidiary centres of significant size relative to the main centre. This group includes 70 of our list of 100 cities.

(2) Metropolitan areas with one major

---

[61]See Raymond Vernon, *The Changing Economic Function of the Central City* (New York, Committee for Economic Development, 1959), and *Metropolis 1985* (Cambridge, Harvard University Press, 1960). The latter book is a summary and synthesis of the ten-volume economic study of the New York Region undertaken by the Graduate School of Public Administration at Harvard for the Regional Plan Association under Vernon's direction.

centre, but with one or more lesser centres of significant size. There are fifteen of these. Such subsidiary centres, while not large enough to be viewed as "twin cities" are usually places of independent origin and considerable importance that have merged with the metropolis as a result of its areal expansion. Typically they continue to operate as semi-independent "satellite" cities close to the metropolitan centre.

(3) Binuclear metropolitan areas, with two centres of comparable significance. These are "twin cities" like Tokyo-Yokohama or Minneapolis-St. Paul. Typically each of the central cities is a large municipality with many of the problems of a large metropolitan centre.

(4) Multinuclear metropolitan areas, with one centre dominant. Our list includes four instances, Birmingham (U.K.), Johannesburg, Manchester, and the San Francisco Bay Area. Significantly, the development of three of these has been closely associated with mining, which characteristically produces a diffused pattern of urban development and gives rise to multiple centres. The fourth metropolis, San Francisco, combines a large areal extent with peculiar physical circumstances that have promoted extensive development in areas away from the dominant central city, San Francisco itself.[62]

(5) Multinuclear metropolitan areas, with no dominant centre. There are three of these, two major European coal and heavy industry areas, Essen/Ruhr and Katowice/Silesia, and the somewhat similar Kitakyushu-Shimonoseki in Japan, where, however, the multinuclear character is due more to the exigencies of the site (a narrow coastal plain) than to diffused mining activity. In these centres, no one urban centre is clearly dominant and central cities must be noted simply in order of population.

Returning to the uninuclear group of seventy metropolitan areas, we may note that they range from those metropolitan areas for which there is only one jurisdiction—the central city—which thus includes 100 per cent of the metropolitan area population (and in the case of certain Chinese cities, may embrace an even larger area), to the three instances in which the central city includes less than 10 per cent of the total population.

The Chinese cities require a word of special explanation. The administrative areas of large Chinese cities are all quite extensive. Within these administrative areas the built-up urban areas are densely developed and of limited areal extent, but surrounded by densely populated agricultural areas. This suggests that few if any Chinese cities have either continuously built-up or other suburban development outside their administrative limits. For six of the ten Chinese cities, therefore, the metropolitan and city populations are identical in our list of metropolitan centres. For four of the largest (Shanghai, Peking, Tienstin and Chunking), available information indicates that the area included within the municipal boundaries is so extensive that its population considerably overstates that of the built-up urban area. If this seems surprising, it should be remembered that in the environs of a city like Shanghai rural densi-

[62]See James E. Vance, Jr., *Geography and Urban Evolution in the San Francisco Area* (Berkeley, University of California, Institute of Governmental Studies, 1964).

ties of 1,500 persons per square mile are found, so that a moderate expansion of the city boundaries could readily include an extensive agricultural population. Thus, as already noted, these four Chinese cities actually include 100 per cent of the metropolitan area population, and more besides, within their corporate limits. For all summary purposes in this chapter, however, we have treated them as having identical metropolitan and urban populations, like the other Chinese cities.[63]

Including the Chinese cities, twenty-three uninuclear metropolises have over 90 per cent of their population within the corporate limits of the central city. In certain instances (Ibadan, Teheran, Alexandria, Lahore) lack of detailed information may be partially responsible for the conclusion that there is little or no suburban development outside the corporate limits. The other cities (Madrid, Bombay, Rome, Prague, Bucharest, Edmonton) all have relatively extensive corporate limits, either of long standing (as in the case of Rome) or thanks to relatively recent expansion (notably Madrid and Bombay). Another twenty-one metropolitan areas in the uninuclear class have over 70 per cent of their

population within the corporate limits of the central city; twelve more have between 50 per cent and 70 per cent; eleven have between 25 per cent and 50 per cent; and three have less than 25 per cent. These latter three are Brussels, Sydney, and Melbourne, in each of which a very small central municipality is surrounded by numerous, populous, independent municipalities.

By and large, the rapidly developing cities of Latin America, southern Europe, and Asia have relatively extensive corporate limits; those of North America and Europe tend to have more restricted ones. Especially in North America the very great areal extent of most metropolitan centres further reduces the share of the total population within the central municipality. Only two of the fifteen largest United States metropolitan areas now have over half of their population in the central city and in one of these, Baltimore, the percentage is only 55. The remaining city, Houston (with an area of 339 square miles in 1964), has managed through extensive annexations to include 74 per cent of its metropolitan population. On the other hand, Los Angeles has long been characterized by a relatively high proportion of population outside the central city, although its corporate area embraces 459 square miles, because much of the corporate expansion of the central city was into largely unpopulated areas to the north and west rather than into the numerous, populous, suburban municipalities in other directions.[64]

Not surprisingly, the 30 metropolitan areas not classed as uninuclear have con-

---

[63]The limited information available on Chinese cities means that any meaningful distinction between the types of metropolitan areas is impossible. More extensive information is required not only on the land areas and population totals involved, but also on the systems of municipal administration. For example, it is not clear whether the vast area of 17,000 square kilometres reported for Peking is actually within the city limits or only under the administration of the city for some purposes. This latter situation is to be found in certain cities in the U.S.S.R. (for example, Leningrad) and also elsewhere (as in Tokyo). In these instances, our definition of "central city" is the more limited one of the area directly under the city government and thus excludes the more extensive area administered by the city government for some purposes.

---

[64]See Winston W. Crouch and Beatrice Dineman, *Southern California Metropolis* (Berkeley, University of California Press, 1963), pp. 149-179; and Richard Bigger and James D. Kitchen, *How the Cities Grew* (Los Angeles, University of California, Bureau of Governmental Research, 1952).

siderably lower proportions of population in the central city. The median for all 100 cities is 68 per cent, for the uninuclear cities 73 per cent. For the metropolitan areas with one or more important subsidiary centres, the median is 47 per cent. These metropolitan areas with the more important subsidiary centres noted in parentheses, include Delhi (New Delhi), Gorki (Dzerzhinsk), Caracas (La Guaira/Maiquetia) and Glasgow (Paisley, Motherwell), each with over half the population of the area in the central city; and, New York, (Newark, Paterson), Chicago (Gary), Frankfurt (Offenbach), Philadelphia (Wilmington, Trenton), Stuttgart (Esslingen), Calcutta (Bhatpara and other cities), Detroit (Windsor), Los Angeles (Long Beach) and Boston (Lowell, Lawrence, Brockton) each with less than half the population of the area in the central city. In Boston the proportion of metropolitan population in the central city is now below 20 per cent. Djakarta, also classed in this group, is a special case in which the separate centre, Tandjong Priok, is part of the administrative area of the central city and the latter apparently includes 100 per cent of the metropolitan area.

The eight binuclear centres have a median of about 35 per cent of the population in the chief central city. They include only two in which the proportion exceeds 50 per cent, Tokyo (second centre Yokohama) and Amsterdam (second centre Haarlem). The others are Donetsk (Makeyevka), Osaka (Kobe), Athens (Piraeus), Minneapolis (St. Paul), Miami (Fort Lauderdale) and Hong Kong (Kowloon). As already noted, the administrative arrangements in Hong Kong are different to equate with the customary definition of a central city, and for many purposes the entire urban area, including Kowloon, is adminis-

tered as a unit by the Crown Colony government.

In the four multinuclear centres with dominant central cities, a median of 25 per cent of the population is in the central city. The proportion is highest in Birmingham (U.K.) and lowest in San Francisco.

Finally, in the three multinuclear centres which do not have a dominant central city, the proportion in the largest centre is below 15 per cent in Essen and Katowice. In Kitakyushu it is now over 75 per cent, as a result of the recent administrative merger of five of the six central cities (Moji, Kokura, Tobata, Yawata, Wakamatsu) into one municipality, leaving only Shimonoseki as a separate city. In this instance, as in that of Djakarta, the administrative situation might justify considering the multiple centres as no longer effectively existing. However, as already noted in connection with the Chinese cities, administrative boundaries and urban or metropolitan limits often do not coincide.

THE PROBLEM OF MAJOR POLITICAL BOUNDARIES

Another factor of some significance to metropolitan political and administrative structure is the presence of higher-level political boundaries. Our list of 100 metropolises includes three that are divided by international frontiers, as in Detroit (where the situation appears to produce few problems), Berlin (where the opposite is clearly the case) and Rome (where the Vatican City is too small to constitute more than a technical instance). Hong Kong and Singapore both represent instances where small political units are comprised almost wholly of the urban area and relations with the neighbouring countries are obviously of significance to the administration of the metropolitan areas in question.[65] Finally, note should be made of Katowice which, though now wholly in Poland, underwent much of its development as an internationally divided area, first between Germany and Russia, then (and with a different boundary) between Germany and Poland —a circumstance which had significant effects on its development and characteristics.[66]

Another fourteen metropolises straddle the boundaries of states or similar political units with sufficient autonomy to present administrative problems. Seven of these are in the United States (New York, Chicago, Philadelphia, Boston, Washington, St. Louis and Minneapolis-St. Paul). The others, Mexico City, Rio de Janeiro, Caracas and Buenos Aires in Latin America, Hamburg[67] and Vienna in Europe, and Delhi in India, are all in small political units (in several cases termed federal districts), but all have spread into areas administered by adjoining states.

There are many other instances of metropolitan areas crossing boundaries of internal subdivisions like departments, but since such units typically have much more limited autonomy, the effect on metropolitan administration is usually much less pronounced. Moreover, boundaries of such units can often be changed, whereas those of autonomous units are much more difficult to alter.

POLITICAL AND ADMINISTRATIVE REGIONS

In delineating metropolitan and other regions, geographers, sociologists, economists and academic planners often ignore the existence of boundaries of governments which have existed for decades and sometimes for centuries. These bounded governmental jurisdictions have not only become symbols of man's "territorial imperative", but have become factors in the organization and maintenance of many enduring and vested structures of power, both public and private. Any modification or threatened alterations of governmental boundaries will be viewed as a means of rewarding some groups in the community and of injuring others. It is this differential effect, imagined or real, of changes in existing governmental areas and in the creation of new areas that distinguishes the political and administrative type of region from those discussed earlier in this chapter.[68]

---

[65]The French identify seventeen *agglomérations internationales,* in ten of which the French portion includes a majority of the population. *Recensement de 1962: villes et agglomérations urbaines, op. cit.,* Table II.

[66]See Norman J. G. Pounds, *The Upper Silesian Industrial Region,* Indiana University Russian and East European Series, No. 11 (Bloomington, Indiana University Press, 1959).

[67]Hamburg, a "city-state" since medieval times, has had identical city and state boundaries since 1937. Dr. Hans Peter Ipsen, "Hamburg," in *Handbuch der Kommunalen Wissenschaft und Praxis,* Vol. I, (Berlin, Springer, 1956), pp. 502-20.

---

[68]An excellent discussion of the politics of planning areas will be found in Cyril Roseman and William L. C. Wheaton, "Regional Planning and Policy Formation in the Philadelphia Metropolitan Area: A Political Analysis." Unpublished manuscript prepared for the Institute of Governmental Studies, University of California, Berkeley. Other studies in the Institute's second Franklin K. Lane Series will cover Bogota, Ibadan, Jerusalem, Kanpur, London, Los Angeles, Minneapolis-St. Paul, Montreal, New York, Rome and Milan, Ruhr, Sao Paulo, San Francisco Bay Area, Santiago, Stockholm, Tokyo, Toronto, Warsaw, Washington, D.C., and Zagreb (and other Yugoslav cities). The Institute of Public Administration, New York, has conducted an extensive study of urban governments. Published volumes arising from this study and available from Frederick A. Praeger, Inc., include: Annmarie Hauck Walsh, *The Urban Challenge to Government: An International Comparison of Thirteen Cities* (1969); Allan Austin and Sherman Lewis, *Urban Government for Lima, Peru* (1968);

Frank Smallwood has analyzed in detail the contest in the early 1960's among groups favouring and opposing the creation of the Greater London Council to replace the counties of London and Middlesex and to include portions of the counties of Essex, Kent and Surrey. Also involved was the Government's proposal to divide the Greater London area into thirty-two London Boroughs to consist of combinations of former metropolitan boroughs and other local authorities outside the former London County Council (LCC) area. A third aspect of the major reorganization was the delineation of jurisdictional areas for education. The final outcome was the creation of an Inner London Education Authority covering the area of the twelve "inner London" boroughs and the City of London. In the outer part of Greater London each of the twenty London Boroughs functions as a full local education authority. As Frank Smallwood says:

. . . from the time of its initial inception, the entire reform program contained a variety of highly volatile elements of political controversy.

The key question that remains is why this should have been the case. Why should a seemingly pallid program of structural reform spark off a protest drive to secure half a million signatures? produce the longest single sitting in the

history of the House of Lords? involve the consideration of more than 1,300 proposed Parliamentary amendments? arouse the enmity of teachers, doctors, and a host of other professionals? and force the Labour Party [at that time, the Opposition] to make an informal (if not, in actual fact, absolute) commitment to rescind the entire program if it won the next General Election?[69]

Whatever the French Government may have hoped to accomplish through the creation of the District of Paris in 1961 and especially by the current division of the departments of Seine and Seine-et-Oise into six new departments, the reorganization of the region will have the effect of diluting the electoral strength of the "red belt" of suburbs. The original proposal of 1959 for a strong administrative district was blocked by the opposition of communal officials. It was replaced in 1961 by what is formally an advisory regional planning agency. However, the Delegate General who acts as the chief executive officer is responsible to the Prime Minister.[70]

David T. Cattell, *Leningrad: A Case Study of Soviet Urban Government* (1968); Hans Calmfors, Francine Rabinovitz and Daniel Alesch, *Urban Government for Greater Stockholm* (1968); Eugen Pusic and Annmarie Hauck Walsh, *Urban Government for Zagreb, Yugoslavia* (1968); Babatunde A. Williams and Annmarie Hauck Walsh, *Urban Government for Metropolitan Lagos* (1968); and Annmarie Hauck Walsh, *Urban Government for the Paris Region* (1968). Forthcoming volumes will deal with Casablanca, Morocco; Karachi, Pakistan; Lodz, Poland; and Valencia, Venezuela.

[69]Frank Smallwood, *Greater London: The Politics of Metropolitan Reform* (Indianapolis, The Bobbs-Merrill Co., Inc., 1965), pp. 32-33.
[70]Annmarie Hauck Walsh, *Urban Government for the Paris Region*, The International Urban Studies of the Institute of Public Administration (New York, Frederick A. Praeger, 1968). For the politics of the reorganization of local government in metropolitan areas of the United States, see Victor Jones, *Metropolitan Government* (Chicago, University of Chicago Press, 1942), Chs. 9-11; Henry J. Schmandt, Paul G. Steinbicker and George D. Wendel, *Metropolitan Reform in St. Louis* (New York, Holt, Rinehart & Winston, Inc., 1961); Edward Sofen, *The Miami Metropolitan Experiment* (Bloomington, Indiana University Press, 1963); David A. Booth, *Metropolitics: The Nashville Consolidation* (East Lansing, Michigan State University, Institute for Community Development and Services, 1963); and, Brett W. Hawkins, *Nashville Metro: The Politics of City-County Consolidation* (Nashville, Vanderbilt University Press, 1966). Also enlightening is

These kinds of issues would never arise over an attempt by a social scientist to determine the area of the "real" city or of an urban region. The exceptions, of course, would occur if his definition were introduced into politics by his, or someone else's, effort to establish or extend governmental power or administrative activity over the metropolitan area he had defined. This is happening now in the United States as a result of the requirement in section 204 of the Demonstration Cities and Metropolitan Development Act of 1966 that all applications for federal loans or grants in ten major functional categories be submitted for review and comment to a metropolitan regional planning agency. The Secretary of the Department of Housing and Urban Development, subject to review by the Bureau of the Budget, designates these regional planning areas. In some metropolitan areas, especially smaller ones, the whole regional planning area will undoubtedly be the SMSA as previously designated by the U.S. Bureau of the Budget; other planning regions will consist of the SMSA and adjacent counties where urban growth is expected to occur in the next twenty years; still others will consist of a combination of SMSAs. Thus we see what is essentially a unit for statistical purposes becoming the "building block" for the establishment of a region for planning and perhaps for other governmental purposes. Already regional planning commissions or councils of local

governments are being created in most SMSAs (as modified) to meet the planning review requirements of the Act.

There has been widespread criticism of the SMSA on the grounds that, being composed of entire counties (except in New England), it contains extensive rural or uninhabited territory. Except for attempts to compare densities or gross ground area, there are probably few disadvantages in including whole counties in SMSAs even for demographic, social and economic analyses. Certainly, if the SMSA is to become an administrative, planning, or other governmental unit, even for voluntary intergovernmental collaboration, it is fortunate that counties with an established governmental and political base are maintained as whole units in the metropolitan region.

Political and administrative regions are seldom criticized because they are "overbounded". However, the frequently stated objective of planners and social scientists is a "true-bounded" metropolis—one with an area neither larger nor smaller (at any given moment!) than the area of the real city. The great anathema is the underbounded political or administrative area. Emrys Jones, for example, questions the realism of the boundaries of the newly created Greater London Council:

To sum up, there is an inner administrative region, a much wider planning region (but with no executive power), and a still wider geographer's region; and it is the latter which links up with other regions in England and which implies that no area now lies outside the influence of a major city. The critical changes brought about by the growth of London are now taking place in this wider region, and it is a great pity that the administration is so far from catching up with these events. It will probably

the synthesis and interpretation of the U.S. Advisory Commission on Intergovernmental Relations, *Factors Affecting Voter Reactions to Governmental Reorganization in Metropolitan Areas*, Report M-15 (Washington, D.C., A.C.I.R., 1962). The most recent and thorough review of sociological, economic, and political materials in the United States is John C. Bollens and Henry J. Schmandt, *The Metropolis: Its People, Politics and Economic Life* (New York, Harper & Row, Publishers, 1965).

be overcome only by some system of regional planning administration.[71]

Peter Hall, on the other hand, rejoices in the creation of the Greater London Council, for he sees that "from 1965, in London a situation obtains which is rare in any world city: physical reality, statistical reality and administrative reality are all approximately the same."[72]

There were powerful critics of the area originally recommended for Greater London by the Royal Commission in 1960 and only the reduction of this area by the Government permitted the enactment of the legislation in 1963. What Frank Smallwood said about the geographical politics of Greater London might be said of the determination of most boundaries of either *ad hoc* or general-purpose metropolitan agencies:

> It is all well and good to question the Royal Commission, and the Government, for not being bold enough in terms of their geographical thinking, but politics deals with the art of the possible. In the light of the battle that ensued over the current Greater London plan, despite its geographical modesty, it is difficult to conceive what would have happened if the area covered by this plan had been doubled or tripled in size.
>
> The geographical erosion that went into the making of the Greater London reform—the 25 per cent reduction in the size of the original review area—represented the great political concession to the second-tier authorities, especially those that were located on the outer fringes of the metropolitan area.[73]

The criteria which the Royal Commission used in its modest modification of the review area, as defined in the charge to the Commission, also illustrate the substantive differences between study areas and political-administrative areas. In the first place their overriding purpose was to delimit an area for local government and not for the "best" administration of any one service. Therefore, the Commission felt bound to consider how the reorganization of local government within a predetermined area would affect the "active participation in the work by capable, public spirited people elected by, responsible to, and in touch with those who elect them" as well as the efficient performance of professional administrators.[74]

Secondly, local government must be organized to meet the obligations imposed upon it by the central government. Imposition is probably a misleading term since, in many countries, irrespective of the form of government, central government decisions about what local governments may or shall do are the product of accommodation. Nevertheless, the final decision is likely to be with the central government if the central decision-makers are willing to make the decision. In any event, the Royal Commission recognized "that unless local authorities are so constituted as to be able to undertake all the functions appropriate to local government there will always be the risk that more and more functions will be taken away from local government and

[71]Emrys Jones, *Towns and Cities* (New York, Oxford Book Co., Inc., 1966), p. 102.

[72]Peter Hall, *The World Cities* (New York, McGraw-Hill Book Company, 1966), p. 31. Even so, Hall recognizes that "another, deeper, sort of reality" is developing beyond the Green Belt—and here the central government is moving in with a new plan for the South East. *Ibid.*, pp. 55-56.

[73]Smallwood, *op. cit.*, pp. 277-78.

[74]Royal Commission on Local Government in Greater London, *Report, Cmnd.* 1164 (London, Her Majesty's Stationery Office, 1960), p. 59 ff.

given to *ad hoc* bodies or to central government."[75]

The Commission was most concerned with the calibre of the local councillor and recognized that this would be affected by the population size and area of the new local government units.[76] In other words, the Commission did not regard

> social and economic factors as the sole criteria or even as the decisive criteria in making recommendations about local government boundaries within the Review Area. These factors often fail to give an unambiguous answer, and there are other criteria, such as tradition, administrative continuity and adequate scale for the maintenance of services, to which full weight must be given.[77]

However, the Royal Commission, and later the Government and Parliament were engaged in the relatively rare creation of a general metropolitan government for millions of people living over hundreds of square miles. Before completing this review of political and administrative regions it is desirable to look at the *ad hoc* approach to metropolitan administration and policy-making.

Special districts and authorities, so beloved in Anglo-Saxon countries, are only one organizational result of the *ad hoc* approach. A bureau in the central or intermediate government, or in a local government, can be autonomous either *de jure* or *de facto*. Charles Adrian sees a system of regional offices of the departments of, say, a state administration as being quite acceptable for the administration of a metropolitan area. Such an arrangement would permit the size of the administrative area for each function to vary according to its needs and, at the same time, ensure coordination through the department and the decision-making body.[78]

Professor Adrian's views on different *optimal* sizes of administrative areas for different functions raises one of the most difficult problems in all large complex organizations. If the objectives of metropolitan government are to establish certain representative relations between the governors and the governed and to coordinate the policies and administration of various interdependent activities, as well as to provide services and regulate certain activities effectively and efficiently, then optimal areas for accomplishing each objective may vary with respect to each function. If any one objective is *maximized* the other objectives will be unattainable.

Even though coordination is not an end in itself, a politically responsible and more rational planning, decision-making, and administrative process is likelier in a multi-purpose agency (whether of a local or regional government or an administrative bureau) than in a single-purpose autonomous agency. The very necessity of balancing conflicting claims for scarce resources among multiple purposes is the father of coordination and of political responsibility. This was recognized and cogently argued by James Madison in the Tenth *Federalist*.[79]

---

[75]*Ibid.*, pp. 60-61.
[76]*Ibid.*, pp. 62-64.
[77]*Ibid.*, p. 25.

[78]Charles R. Adrian, "Regional Services and Regional Government," *The State's Biggest Business—Local and Regional Problems*, Policy Papers for the Connecticut Commission to Study the Necessity and Feasibility of Metropolitan Government (Hartford, January, 1967), p. 122.
[79]See also Paul Ylvisaker, "Some Criteria for a 'Proper' Areal Division of Governmental Powers," and Robert C. Wood, "A Division of Powers in Metropolitan Areas," in Arthur Maass, ed., *Area and Power* (New York, The Free Press, 1959), pp. 27-49, 53-69.

Another problem in the governance of huge metropolitan areas is the delimitation of meaningful sub-areas. We do not mean census tracts, neighbourhoods, or "natural" communities, but political and administrative subdivisions of the larger metropolis. Most reorganizations of local government on a metropolitan basis have either retained existing units as "second-tier" governments (as is the case, for example, in Paris,[80] Miami, Toronto before 1967, and Tokyo), or have constructed new second-tier units as part of the creation of the metropolitan government (as, for example, in London, New York, and Berlin between the two World Wars).

In large part, this widespread trend results from the power of existing units of local government to protect themselves from annihilation or absorption into a unitary metropolitan government. But there are also grounds for believing that certain values can best be secured through units of government smaller than the metropolis, provided that the latter itself is organized to plan and act for the region. The search is for a compromise that offers local access, regional efficiency and intergovernmental viability. If we consider New York City a local government, then why should we fear any unitary metropolitan government, none of which (in the United States) would be as large? The search, uninspired though many believe it to be, continues in New York City for a means of sharing government with parts of the city without destroying its present unity. The five boroughs into which the city was divided at the time of the great consoli-

---

[80]We refer to the *communes* and not to the *arrondissements*.

dation of 1898 have not served this purpose. In fact, the power of the borough offices has been reduced periodically throughout the past sixty-five years. The same tendency over time to shift authority and functions from political subdivisions to the area-wide metropolitan government can be observed in the first decade of Metropolitan Toronto and in sixty years of the London County Council.

London is the prime example of a two-tier metropolitan government. Ten years after the London County Council was organized (1889), Parliament divided London into twenty-eight metropolitan boroughs largely along the lines of the old parish vestries, although over a hundred administrative units were abolished at the same time. The metropolitan boroughs were political bodies, not merely administrative subdivisions of the County of London. The organization of London government to enhance differences of opinion and objectives has been deplored. For instance, William A. Robson has stated that the London Government Act of 1899

. . . magnified the importance and independence of the minor authorities in every possible way. It gave each metropolitan borough council a mayor and aldermen with robes of office, gilt chains, a mace, and all the other insignia likely to enhance the feeling of their separate civic consciousness. Far from attempting to bring them into an organic relationship with the London County Council, the object was rather to engender jealousy and friction and to foster the parochial spirit.

The plan succeeded only too well, and conflict and dissension between the London County Council and the metropolitan borough councils has often detracted from the good government of London during the past fifty years.[81]

Professor Robson has raised perhaps the most difficult question to answer. Certainly conflict and differences among groups in a metropolis can become an end in themselves and actually be encouraged by governmental organization. However, if administrative convenience is the touchstone, then governmental organization can certainly be devised to smother dissent, smooth over differences and discourage meaningful discussion of alternative courses of action.

The Royal Commission on Local Government in Greater London refused to consider the abolition of the metropolitan boroughs. It felt strongly that "the health of local government requires their rehabilitation" at the same time that a council for Greater London should be established to perform "functions which can be or can better be performed over a wider area". It recommended therefore that the "primary unit of local government in the Greater London Area should be the borough". However, this raises two sticky questions: (1) which functions are local and which are regional in nature? and, (2) what is a function?

The allocation of responsibilities for governmental activities between the Greater London Council and the boroughs certainly does not exhibit the definitional neatness suggested by the Commission's assertion that "the functions to be performed by each type of authority should be as far as possible self-contained without overlapping or duplication and without the necessity for delegation from one to another."[82] Luther Gulick has reminded us that

> . . . a function is in fact made up of "aspects", much as matter is made up of atoms. The function of "public health" or of "public education" or of "recreation" is not a single indivisible thing. Each is a bundle of "aspects". From this it follows that we can discover within any function those aspects which are by nature or practicality national or state (or metropolitan) or local. This is done partly on the basis of tradition and experimentation, partly on the basis of the technology involved, partly on the basis of the interrelations required for good administration, partly to get a better fit between costs and available taxes, and partly on the basis of the constituency required to settle the intertwined policy questions effectively and fairly.[83]

The reorganization of the Municipality of Metropolitan Toronto introduced on January 1, 1967, reduced the number of constituent municipalities from thirteen to six. The trend, evident almost from the birth of Metro Toronto in 1954, toward the transfer of additional functions to the Metropolitan Corporation was continued by the reorganization. Nevertheless, both the Royal Commission on Metropolitan Toronto and the Government of Ontario rejected the proposal that the thirteen municipalities be amalgamated into a single metropolitan government. Commissioner Goldenberg relied principally on the report of the Royal Commission on Local Gov-

[81]William A. Robson, "London," in *Great Cities of the World* (New York, The Macmillan Company, 1957), p. 278. See also his *Government and Misgovernment of London* (London, George Allen and Unwin, Ltd., 1939), pp. 93-99, 359-70.

[82]Royal Commission on Local Government in Greater London, *op. cit.*, pp. 192-98.
[83]Luther Gulick, *The Metropolitan Problem and American Ideas* (New York, Alfred A. Knopf, Inc., 1962), p. 44.

ernment in Greater London to justify his decision that local government "as an instance of democracy at work" must be organized to permit wide participation and control by electors and, at the same time, the efficient discharge of its responsibilities. As Goldenberg says:

> . . . The requirements of both democracy and administrative efficiency will be better satisfied if the administration of such local services, as distinct from area-wide services, is as far as possible in the hands of local officials responsible to locally elected representatives in municipalities properly constituted to meet the needs of the Metropolitan Area.[84]

In order to strengthen local government at the municipal level, to remove some of the inequalities of financing public services, to raise the level of public services in some existing municipalities, and to provide a more equitable basis of representation on the Metropolitan Council, Commissioner Goldenberg recommended that the thirteen municipalities be consolidated into four cities. The government of Ontario modified his proposal by creating six reconstituted municipalities and six coterminous school boards.[85]

The U.S. Advisory Commission on Intergovernmental Relations[86] has applied seven criteria in assessing the optimum scale of operation of fifteen functions in the United States:

(1) The governmental jurisdiction responsible for providing any service should be large enough to enable the benefits from that service to be consumed primarily within the jurisdiction.

(2) The unit of government should be large enough to permit realization of the economies of scale.

(3) The unit of government carrying on a function should have a geographic area adequate for effective performance.

(4) The unit of government performing a function should have the legal and administrative ability to perform the services assigned to it.

(5) Every unit of government should be responsible for a sufficient number of functions so that it provides a forum for resolution of conflicting interests, with significant responsibility for balancing governmental needs and resources.

(6) The performance of functions by a unit of government should remain controllable and accessible to its residents.

(7) Functions should be assigned to that level of government which maximizes the conditions and opportunities for active citizen participation and still permits adequate performance.

---

[84]Royal Commission on Metropolitan Toronto, *Report* (Toronto, Province of Ontario, June, 1965), p. 177.

[85]Frank Smallwood comments: "The six-city modification enabled the government to create six coterminous education boards without being accused of running roughshod over the principle of local control over educational activities. Thus it is significant to note that the problem of educational administration appears to have caused the Ontario government to modify its commission's recommendations just as it caused the British government to modify the recommendations of its Royal Commission on Greater London a few years ago." *National Civic Review*, New York (November, 1966), pp. 590-91.

[86]*Performance of Urban Functions: Local and Areawide*, M-21 revised, U.S. Advisory Commission on Intergovernmental Relations (Washington, U.S. Government Printing Office, September, 1963).

Some of the problems that arise when attempting to apply these criteria to the allocation of functional responsibilities to different levels of government are brought to light by Stefan Dupré in the Appendix to Chapter 11 in this volume. Implicit in Dupré's comments is that the judgement of the Commission is value-based and should be taken for just that: a *judgement* of interested men of affairs based upon a conscientious attempt to arrive at a satisfying balance of inconsistencies and contradictions among the seven criteria, all leavened by the political realities of American federalism. It is, perhaps, most valuable to us as a demonstration of how political judgement can be based on rational analysis of a situation or problem.

The document, however, does examine and evaluate most data available in the United States on the optimum areal jurisdiction for the performance of governmental functions. It also demonstrates the utility and the influence of "facts", the analyses of facts, and theoreticians on public policies. This is why we have considered it appropriate to discuss, in this introductory chapter, the delimitation of metropolitan areas by geographers, sociologists and other social scientists, and their delimitation by politicians, administrators and interest groups.

Appendix A

The Delimitation of Metropolitan Areas

The delimitation of urban or metropolitan areas is a complicated task, in which the difficulties of defining boundaries for *any* geographic or socio-economic unit are compounded by the special complexities of urban areas. As metropolitan expansion continues, more and more countries are presenting data in their census publications for urban or metropolitan areas as well as cities. The various definitions and approaches to delimitation used by different countries are not necessarily comparable, however, no matter how justified each may be in terms of local applicability. In addition, there are a number of countries for which no official data on metropolitan areas are available. Consequently, there remains a problem of gathering worldwide data on metropolitan areas that are reasonably comparable in concept.

A pioneer attempt to outline the problems of comparable metropolitan delimitation and to delimit actual metropolitan areas on an international basis was *The World's Metropolitan Areas*, published by International Urban Research (IUR) at the University of California, Berkeley, in 1959.[87] In its introductory chapters, *The World's Metropolitan Areas* emphasizes the international non-comparability of cities administratively defined and the desirability of using comparable units for careful urban study. The general concept of the Standard Metropolitan Areas, or SMAs (now SMSAs), officially in use in the United States, is examined and accepted.[88] It is concluded that metropolitan areas for cities in other countries should be delimited by rules modelled as closely as possible on those used for the SMAs. A degree of comparability is judged to have been achieved by the adoption of this single approach, though it is recognized that the use of a variety of base data introduces some lack of comparability. The use of statistical criteria is stressed throughout, as is the usefulness of the resulting delimitations for the gathering of further statistical

---

[87]*The World's Metropolitan Areas, op. cit.*
[88]For the rules used for definition of the Standard Metropolitan Statistical Areas in the 1960 census, see *U.S. Census of Population: 1960*, Vol. 1, *op. cit.*, pp. XXIII-XXIV.

data.[89] However, *The World's Metropolitan Areas* gives relatively little attention to delimitations previously carried out in countries other than the United States.

A more recent study of the delimitation problem is that of G.J.R. Linge, prepared for the Australian Commonwealth Bureau of Census and Statistics, in connection with plans for recasting metropolitan areas used for the Australian census and other statistics.[90] Linge's approach is the reverse of that of IUR. His eventual concern is the delimitation of metropolitan areas in a single country, but he begins by examining with some care the approaches to the problem followed by a number of other countries, and his study offers a useful overview of the various approaches in use. Like IUR, he concludes that there is little comparability among such "official" definitions, but his prime purpose is not the development of an internationally comparable definition.

Both IUR and Linge[91] note an important distinction between the narrowly defined "urban area", corresponding to the continuously built-up area or a closely related concept, and the "metropolitan area" or socially. In the view of IUR, the built-up area *and* the surrounding areas that are less intensively developed but which are related to (the central area) economically and socially. In the view of IUR, the built-up area or "urbanized area"[92] is "theoretically the best for describing and understanding urban communities".[93] Its chief disad-

---

[89]In an effort to ensure greater comparability, stress is laid on the use of the county or its nearest equivalent as the unit of delimitation: "To apply the criteria of the SMA internationally, county equivalents in each nation must be established. This can best be done by thinking of the county as the administrative district immediately larger than the corporate city and generally surrounding it" (p. 22). This overlooks two significant points. First, the county was chosen as the unit for which extensive statistical data was available. Clear evidence of this is the fact that in New England, where the small units known as "towns" are more important administratively and socially than the counties and hence are often used for statistical tabulations, the official SMSAs use the towns instead of counties as "building blocks". (*The World's Metropolitan Areas* in fact substituted county-unit areas for the official SMAs in New England.) An adherence to the SMA principle by IUR should logically have resulted in selecting as the building block in each country the smallest unit for which the given range of data were available. Second, as will be noted in greater detail below, it is illogical to overstress the significance of a particular unit for which statistics happen to be available unless some relevance can be shown to exist between the nature of the unit and the area being defined. The choice of counties as building blocks for the SMAs was a recognized compromise between accuracy and statistical availability; to imitate it uncritically in a world-wide study unfortunately did not automatically increase comparability. Thus IUR's delimitations for "metropolitan areas" in Germany utilized *Kreise* (circles or districts) as units because of their supposed areal and administrative comparability to counties, although these units are much too large to be satisfactory building blocks for most metropolitan centres in Germany. Ample data are available to define German metropolitan areas in terms of the much smaller *Gemeinde (communes)*, and this

lengthy task has been accomplished by Olaf Boustedt in *Die Stadtregeionen in der Bundesrepublik Deutschland* (*Forschungsund Sitzungsberichte de Akademie für Raumforschung und Landesplanung*, Vol. XIV, Bremen, 1960); see also *Die Entwicklung der Bevolkerund in den Stadtregionen* (*ibid.*, Vol. XXII, Hanover, 1963). Alberto Aquarone, *Grande Città e aree metropolitane in Italia* (Bologna, Zanichelli, 1961), though avoiding specific definitions of his own, comments on the IUR delimitations for Italy (*e.g.*, pp. 13-14).

[90]G. J. R. Linge, *The Delimitation of Urban Boundaries for Statistical Purposes with Special Reference to Australia*, Publication G/2, Research Institute of Pacific Studies, Australian National University (Canberra, 1965).

[91]*Ibid.*, p. 15.

[92]The "urbanized area" is the term used by the U.S. Census for this narrowly defined type of metropolitan area. *U.S. Census of Population: 1960*, Vol. 1, *op. cit.*, pp. XVIII-XIX.

[93]*The World's Metropolitan Areas, op. cit.*, p. 13.

---

vantage is the difficulty or impossibility of defining it in terms of existing administrative and statistical areas. IUR tends to distinguish between the two approaches primarily on this basis, rather than on any theoretical distinction between the urban area narrowly defined as a physically continuous zone, and the metropolitan area more broadly defined in terms that are not limited to purely physical or density criteria. Linge, in contrast, indicates a recognition that both concepts are theoretically justifiable ways of viewing the urban area.

A few countries recognize two sets of areas, one corresponding to the more broadly defined "SMSA" approach and another, more narrowly defined, corresponding to the "urbanized area" approach.[94] As we have noted earlier, the delimitations for Paris reflect a recognition of the appropriateness of both a narrow definition and a broader one for the same city.

In an effort to ensure the greatest possible accuracy, delimitations of the "urbanized area" type may use census enumeration districts as building blocks, as do both Australia and the United States.

The advantages of accuracy are, of course, offset by the disadvantage of having practically no other statistical data available in terms of the resulting area. In countries like France and West Germany, where the basic administrative unit, the commune, is relatively small in area, a delimitation of an urbanized area can be made in terms of whole communes with relatively little loss in accuracy (especially in terms of people involved); both the French (I.N.S.E.E.) and German (Boustedt) studies previously cited have followed this approach. In fact, the communes of the Paris area are probably not more extensive on average than the typical United States enumeration district or the Australian collector's district on the outskirts of the urban area.

Such differences in the "grain" of local administrative and census units are often very great. Naturally the size of the units for which data are inspected will determine the accuracy of the final boundary established. To be sure, when the boundary follows the boundaries of individual administrative units, it is possible to divide those units for which there is evidence that one part is clearly not a part of the metropolitan zone being delimited.[95]

Units of the commune, parish, or even the township type are relatively small areas averaging from one or two square kilometres to 100 square kilometres each. Delimitations that begin with much larger

---

[94]Linge's own conclusions are reflected in the practice of the 1966 Census of Australia, which presents data for both "urban centres" (the narrow definition, to be redefined at the time of each census) and "statistical divisions" or "statistical districts" (the broader definition, generally made up of whole administrative areas and viewed as extensive enough to encompass urban growth for some years to come). Commonwealth Bureau of Census and Statistics, *Census . . . 1966, Population: Principal Urban Centres of Australia*, Field Count Statement No. 4 (Canberra, 1966), pp. 2-4. It is noteworthy that Linge avoids use of the term "metropolitan", and that in Australian usage the "metropolitan areas" are the urban areas of the six State capitals only, which preserves the original sense of "metropolitan" far better than the common United States usage (see *supra* Ch. 1, Sec. B).

---

[95]For example, the *commune* of St. Germain-en-Laye, west of Paris, has a total area of 49.18 square kilometres and a 1962 population of 34,621. But over 80 per cent of this area is made up of the extensive Forêt de St. Germain-en-Laye. This fact is perfectly evident from a detailed map and a delimiter might quite logically omit this part of the commune, perhaps estimating the area and the population (negligible or nil) so omitted. See I.N.S.E.E., *Délimitation de l'agglomération parisienne, op. cit.*, supplementary map no. 1.

units, such as the countries employed in the SMSA definition, encounter a greatly exaggerated version of this problem. Such large units obviously comprise many separate communities and treating them as wholes for delimiting or building-block purposes clearly constitutes a large-scale generalization of the reality being observed. This is the most significant deficiency of the definition of SMSAs, which any careful examination will recognize to be a very gross delimitation.

Over a period of more than ten years, one of the present writers (Richard L. Forstall) has delimited metropolitan areas for United States cities on a township and community basis,[96] primarily for use in various publications of Rand McNally & Company.[97] As of 1965, this delimitation recognized a total of 189 "Metro Areas" of 100,000 or more in the United States, with a total 1965 population of 119,420,-700 and a total land area of 120,301 square miles or 311,578 square kilometres. This may be compared with a population of 128,261,300 and a total land area of 357,489 square miles (925,892 square kilometres) comprised in Standard Metropolitan Statistical Areas. The SMSAs are somewhat more numerous than the Metro Areas, but the general comparison is valid; the difference between the two population figures is only 7 per cent, but the SMSAs contain almost three times as much land as the Metro Areas. The overall density of the SMSAs is 359 per square mile (139 per square kilometre), that of the Metro Areas 909 per square mile (383 per square kilometre).

For many purposes, the SMSAs are quite adequate as a metropolitan delimitation. They are particularly useful for broad comparisons between major metropolitan areas, since a wealth of data is available for the grouping of countries comprising each area and since the inaccuracy of the county-based system is generally less in the cases of larger metropolitan areas. The SMSAs are also generally satisfactory for the examination of some dynamic aspects of metropolitan activity such as population growth. However, as the comparison with the Metro Area data show, they are quite inadequate for any serious comparison of metropolitan densities. They are probably somewhat inaccurate for an examination of central city versus suburban growth, since the suburban "ring" is obviously rather crudely defined when whole counties are used. Perhaps the salient point is that the delimitation, no matter how useful for some purposes, is not useful for others, and this must be taken into account in utilizing resulting data. Exactly the same point can be made for any type of delimitation. In general, the finer the detail desired, the less convenient the area for statistical compilation; the broader the statistical data available, the more gross the delimitation.

---

[96]These delimitations are generally in keeping with the Metropolitan Districts recognized in United States censuses prior to the introduction of the SMAs in 1949, though the criteria used are more detailed and most individual metropolitan areas have greatly expanded since the last definitions in terms of metropolitan districts were published. See Henry S. Shryock, Jr., "The Natural History of Standard Metropolitan Areas," *American Journal of Sociology*, Vol. 63 (September, 1957), pp. 163-70; Warren S. Thompson, *Population: The Growth of Metropolitan Districts in the United States: 1900-1940* (Washington, D.C., U.S. Government Printing Office, 1948); and Bureau of the Census, *Census Population: 1950*, Vol. 1, (Washington, D.C., U.S. Government Printing Office, 1952), pp. XXXV-XXXVI and Table 30.

[97]Chiefly the *Rand McNally Commercial Atlas and Marketing Guide*, 88th edition, 1957, through 98th edition, 1967 (Chicago, Rand McNally & Co.).

Assuming that larger areas, like counties, are to be used in a delimitation because of their statistical convenience in compiling other statistics, a more scientific approach would be first to delimit the metropolitan area in as accurate terms as possible and then to convert the delimitation to county terms according to specific rules. This second stage may, for example, lead to such refinements as an acceptance of county boundaries for all counties with (for example) less than 10 per cent, or more than 90 per cent, of their population in the area delimited in greater detail in the first stage. Counties more evenly divided, with between 10 per cent and 90 per cent of their population in the detailed area, could perhaps be divided into two quasi-counties and an effort made to compile other statistics for the two sections separately, so as to permit aggregating for the metropolitan area. This would represent a reasonable compromise between the inconvenience of a delimitation in terms of very small units, and the rigidity of adherence to inappropriately large units. Also, the more detailed delimitation would be useful in itself for many purposes.

The Rand McNally "Metro Areas" of the United States include communities if they meet the specific criteria listed below. The individual delimitations have usually been made in terms of townships and similar minor civil divisions, but these minor administrative units have been split if it appears that there is a significant portion that should not be included. The same criteria have been observed in delimiting the major metropolitan areas outside the United States for which data are presented in this chapter. For these metropolitan areas the delimitations are in terms of specific minor civil divisions (*communes, municipios*, etc.) in the countries for which fairly detailed maps and census material

are available. Where such materials are lacking, the delimitations are necessarily more approximate. However, in broad terms the extent of suburban areas outside the municipal limits tends to be greatest in the developed countries for which census and map data are also most readily available and complete. Hence, there are only a few metropolitan areas for which the use of an approximate delimitation involves a possibility of significant error in the total population of the metropolis.

In delimiting the metropolitan areas referred to in this chapter, communities were included according to the following criteria:

(1) Communities or areas connected to the central city by a continuous extension of the urban area. (By "continuous extension" reference is not necessarily meant to include areas of narrow ribbon development.)

(2) Communities, whether or not connected by this unbroken urban extension, with a majority of their population supported by commuters to the central city or to areas already included in the metropolitan area. In practice, this means that at least 15 to 20 per cent of the labour force must be commuters. Communities are not included, however, if the agricultural labour force is greater than the number of commuters.

(3) Certain special kinds of communities that may not meet either of the above criteria, but for which evidence indicates that they should be

considered parts of the metropolitan area. Examples of these special types of communities include:

(a) Outlying industrial communities which may have few commuters but are almost linked by urban development and have a clear economic tie with the metropolis. Often there is significant outward commuting from the metropolis to the outlying industrial community. Examples include waterfront communities like Tilbury on the Thames or Zaandam and adjacent communities near Amsterdam.

(b) Small satellite mining communities if the urban area is almost continuous and if it is clear that the central city serves as a social and economic focus. Examples occur in the Scranton and Wilkes-Barre areas in Pennsylvania and around Lens and other centres in Northern France.

(c) Outlying resort communities, on a somewhat similar basis to mining communities. The common feature in these otherwise disparate types of community is the emphasis on an urban or quasi-urban function that is not strongly focal, but instead is related to a dispersed resource—mineral deposits for mining areas, seashore for resort communities. Metropolitan areas with important mining or resort development are often notably dispersed in character.

(4) Adjacent cities of large size (over 50,000) are included in a single metropolitan area if there is little or no unurbanized area remaining between them, if there is evidence of active urban growth in the interstitial area, and if other links between the two appear to be strong. Inter-commuting, for example, should amount to a total of 5 per cent of the combined labour forces. However, it is very difficult to develop satisfactory, objective criteria for the consistent treatment of such adjacent or closely spaced urban areas as those of Lancashire or Yorkshire in England, eastern Massachusetts, Connecticut, or northeastern Ohio in the United States, or the Ruhr area in Germany.

Were more or less uniform statistical data for small areas (particularly on commuting and the agricultural labour force) available for all countries, the specific delimitations for metropolitan areas would be more consistent than they actually are. Nor can a single student pretend that in every case the accuracy of his delimitation will be equal to what could be achieved by a scholar in the city itself, aided both by local knowledge and more often by more detailed or more up-to-date statistical information. The delimitation of metropolitan areas represents a topic for which there is surely *much* scope for organized international cooperation.

Meanwhile it is hoped that the chief goal, international consistency and comparability, has been generally achieved in the data presented in this chapter. These data are summarized in Table 1-6.

TABLE 1-6
The 100 Metropolitan Areas: Summary Data

Metropolis	Metropolitan Area Population 1964	Metropolitan Area Population 1951	Percent Increase 1951-64	Central City, 1964 Population	Central City, % of Total	Structure Class	Passed 100,000 by	Functional Class
Montreal**	2,250	1,500	50.0	1,210	53.8	A	1880	Rpm
Toronto**	1,960	1,240	58.1	665	33.9	A	1880	Ram
Vancouver**	840	570	47.4	400	47.6	A	1920	Rp
Winnipeg**	490	365	34.2	260	53.1	A	1920	Ra
Edmonton**	365	185	97.3	345	94.5	A	1942	R
CANADA (5)	5,905	3,860	53.0	2,880	48.8			
New York**	16,325	13,500	20.9	8,085	49.5	B	1850	Rpm*
Los Angeles	7,750	4,565	69.8	2,660	34.3	B	1900	M/Rp
Chicago	7,090	5,650	25.5	3,525	49.7	B	1880	Rm
Philadelphia-Trenton-Wilmington**	4,800	3,900	23.1	2,025	42.2	B	1850	M/Rp
Detroit-Windsor**	4,320	3,365	28.4	1,610	37.3	B	1880	M/R
San Francisco-Oakland-San Jose**	3,730	2,500	49.2	750	20.1	D	1880	Rp
Boston	3,480	3,100	12.3	635	18.2	B	1850	Rapm
Washington	2,400	1,525	57.4	785	32.7	A	1880	G(r)
Cleveland	2,240	1,675	33.7	835	37.3	A	1880	M/R
St. Louis	2,155	1,730	24.6	720	33.4	A	1880	Rm
Pittsburgh	1,955	1,795	8.9	575	29.4	A	1880	M/R
Baltimore	1,700	1,350	25.9	930	54.7	A	1850	M/Rp
Minneapolis-St. Paul	1,540	1,140	35.1	470	30.5	C	1900	Ram
Miami-Fort Lauderdale	1,450	600	141.7	315	21.7	C	1930	S(r)
Houston	1,420	840	69.0	1,045	73.6	A	1920	Rpm
UNITED STATES (15)	62,355	47,235	32.0	24,965	40.0			
Buenos Aires**	7,700	5,300	45.3	2,950	38.3	A	1880	Rapm
Mexico City**	6,100	3,100	96.8	3,050	50.0	A	1800	Ra
Sao Paulo	5,450	2,600	109.6	4,425	81.2	A	1900	Ram
Rio de Janeiro**	5,250	3,050	72.1	3,600	68.6	A	1850	Rp (a)
Santiago	2,400	1,400	71.4	640	26.7	B	1880	Ra
Lima**	2,300	950	142.1	1,975*	85.9	B	1880	Rap
Bogotá	1,800	700	157.1	1,650	91.7	A	1900	Ra
Caracas	1,750	775	125.8	1,225*	70.0	B	1900	Rap
Havana	1,550	1,100	40.9	875	56.5	A	1850	Rap
LATIN AMERICA (9)	34,300	18,975	80.8	20,390	59.4			

Metropolis	Metropolitan Area Population 1964	1951	Percent Increase 1951-64	Central City, 1964 Population	% of Total	Structure Class	Passed 100,000 by	Functional Class
London**	11,025	10,300	7.0	7,975	72.3	A	1600	Rapm*
Manchester	2,850	2,785	2.3	652	22.9	D	1850	M/R
Birmingham	2,640	2,400	10.0	1,115	42.2	D	1850	M(r)
Glasgow**	1,885	1,830	3.0	1,030	54.6	B	1850	M/Rp
Liverpool	1,685	1,585	6.3	738	43.8	A	1850	Pm(r)
UNITED KINGDOM (5)	20,085	18,900	6.3	11,510	57.3			
Essen/Ruhr	5,200	4,200	23.8	729	14.0	E	1880	M(r)
Berlin (East and West)	4,025	3,900	3.2	2,186	54.3	A	1800	Ram
Hamburg	2,300	2,000	15.0	1,855	80.7	A	1800	Pm(r)
Cologne	1,550	1,125	37.8	836	53.9	A	1850	M/R
Munich	1,500	1,100	36.4	1,166	77.7	A	1850	Ram
Frankfurt am Main	1,450	1,050	38.1	692	47.7	B	1880	Rm
Stuttgart	1,415	1,000	41.5	635	44.9	B	1880	Ram
Hanover**	750	570	31.6	567	75.6	A	1880	Ram
GERMANY (8)	18,190	14,945	21.7	8,666	47.6			
Paris**	8,000	6,350	26.0	2,800	35.0	A	1500	Ram*
Vienna	2,025	1,900	6.6	1,635	80.7	A	1700	Ram
Brussels	1,975	1,775	11.3	171	8.7	A	1850	Ram
Amsterdam-Haarlem**	1,730	1,525	13.4	866	50.1	C	1700	Rpm
Rotterdam**	1,010	880	14.8	732	72.5	A	1850	Pm(r)
N.W. EUROPE (excl. U.K., Germany) (5)	14,740	12,430	18.6	6,204	42.1			
Milan**	2,775	1,960	41.6	1,658	59.7	A	1500	Rm
Madrid**	2,575	1,700	51.5	2,450	95.1	A	1700	Ra
Rome	2,525	1,750	44.3	2,379	94.2	A	1600	Ra
Barcelona	2,175	1,555	39.9	1,650	75.9	A	1800	Rpm
Istanbul**	2,000	1,025	95.1	1,675	83.8	A	1500	Rp
Athens**	1,975	1,370	44.2	650	32.9	C	1900	Rap
Naples**	1,765	1,425	23.9	1,205	68.3	A	1500	Rpm
SOUTHERN EUROPE (7)	15,790	10,785	46.4	11,667	73.9			
Budapest**	2,265	1,670	35.6	1,920	84.8	A	1850	Ram
Katowice/Silesia	1,975	1,525	29.5	283	14.3	E	1900	M(r)
Warsaw**	1,600	1,050	52.4	1,222	76.4	A	1800	Ra
Bucharest	1,400	1,225	14.3	1,265	90.4	A	1850	Ra
Prague**	1,110	1,010	9.9	1,011	91.1	A	1850	Ram
Belgrade**	660	460	43.5	620*	93.9	A	1920	Ra

Area								
Moscow**	8,450	36.3	6,200	80.5	3,218	A	1800	Rapm
Leningrad**	4,000	29.0	3,100	46.7	794	C	1920	M/R
Donetsk-Makeyevka	1,700	41.7	1,200	74.8	1,066	B	1900	Ram
Gorky	1,425	42.5	1,000	73.2				
U.S.S.R. (4)	15,575	35.4	11,500		11,403			
Johannesburg**	2,200	33.3	1,650	26.1	575	D	1900	X/R
Ibadan**	640	42.2	450	100.0	640	A	1850	Ra
SUBSAHARAN AFRICA (2)	2,840	35.2	2,100		1,215			
Cairo**	4,600	72.0	2,675	82.6	3,800	A	p-1850	Ra
Teheran	2,425	102.1	1,200	95.9	2,325	A	1850	Ra
Alexandria	1,700	70.0	1,000	97.1	1,650	A	1880	P(r)
NEAR EAST/N. AFRICA (3)	8,725	79.0	4,875	89.1	7,775			
Calcutta**	6,700	28.8	5,200	44.8	3,000	B	1800	Rapm
Bombay	4,700	54.1	3,050	95.7	4,500	A	1800	Rapm
Delhi-New Delhi	2,900	65.7	1,750	88.8	2,575	B	1800	Ra
Madras	2,150	36.5	1,575	84.9	1,825	A	1800	Rap
Karachi**	2,100	100.0	1,050	73.8	1,550	A	1900	Rp(a)
Lahore	1,450	70.6	850	93.1	1,350	A	1880	Ra
INDIA/PAKISTAN (6)	20,000	48.4	13,475	74.0	14,800			
Hong Kong	3,275	77.0	1,850	22.1	725*	C	1880	Pm(r)
Seoul**	3,200	113.3	1,500	97.7	3,125	A	p-1850	Ra
Djakarta	3,150	96.9	1,600	100.0	3,150	B	1900	Rap
Manila**	2,900	81.3	1,600	41.0	1,190	A	1880	Rap
Bangkok**	2,250	87.5	1,200	73.3	1,650	A	1850	Rap
Singapore	1,825	87.2	975	60.3	1,100	A	1880	P/R
Saigon	1,700	30.8	1,300	79.4	1,350	A	1900	Rap
Taipei	1,550	98.7	780	66.3	1,028	A	1920	Ram
E./S.E. ASIA (excl. China, Japan) (8)	19,850	83.7	10,805	67.1	13,318			
Shanghai	7,600	28.8	5,900	100.0	7,600*	A	p-1850	Rpm
Peking	4,200	61.5	2,600	100.0	4,200*	A	p-1850	Ram
Tientsin	3,400	54.5	2,200	100.0	3,400*	A	p-1850	M/Rp
Mukden	2,650	32.5	2,000	100.0	2,650	A	p-1850	Ram
Wuhan	2,600	100.0	1,300	100.0	2,600	A	p-1850	Ram
Chungking	2,050	57.7	1,300	100.0	2,050*	A	p-1850	Ra
Canton	2,050	36.7	1,500	100.0	2,050	A	1920	Rap
Harbin	1,950	95.0	1,000	100.0	1,950	A		Rm
Nanking	1,650	65.0	1,000	100.0	1,650	A	p-1850	R
Sian	1,600	128.6	700	100.0	1,600	A	p-1850	Ram
CHINA (MAINLAND) (10)	29,750	52.6	19,500	100.0	29,750			

TABLE 1-6 (Continued)

Metropolis	Metropolitan Area Population 1951	1964	Percent Increase 1951-64	Central City 1964 Population	% of Total	Structure Class	Passed 100,000 by	Functional Class
Tokyo-Yokohama**	9,400	15,900	69.1	8,700	54.7	C	p-1850	Rapm*
Osaka-Kobe	5,825	8,900	52.8	3,120	35.1	C	p-1850	Rpm
Nagoya**	1,375	2,150	56.4	1,850	86.0	A	1880	M/Rp
Kyoto	1,300	1,600	23.1	1,350	84.4	A	p-1850	M/R
Kitakyushu-Shimonoseki	1,040	1,360	30.8	1,030	75.7	E	1920	Mp(r)
JAPAN (5)	18,940	29,910	57.9	16,050	53.7			
Sydney**	1,800	2,340	30.0	166	7.1	A	1880	Rapm
Melbourne	1,425	2,055	44.2	75	3.6	A	1880	Rapm
AUSTRALIA (2)	3,225	4,395	36.3	241	5.5			
TOTAL, 100 METROPOLISES	218,490	311,420	42.5	187,155	60.2			

*World City.

**Participant in the Centennial Study and Training Programme on Metropolitan Problems. Metropolitan Area populations for 1964 and 1951 are for metropolitan areas delimited by Richard L. Forstall. (See Appendix A, supra.) Both figures refer to the 1964 extent of the metropolitan area.

Note: Central city populations refer to the municipal area of the first-named city (in metropolises with more than one central city). In several instances (marked *) the area given may be a census area rather than a municipality, for lack of precise information. The populations for a few cities (including Edmonton and Nagoya) include certain suburbs annexed after 1964 but before the next census. The population of London refers to the present area of Greater London. The Structure classes are as follows (see also p. 41-42, supra):

A Uninuclear metropolises with no subsidiary centres of significant size

B Metropolises with one major centre but with lesser centres of significant size

C Binuclear metropolises

D Multinuclear metropolises, with one centre dominant

E Multinuclear metropolises, with no dominant centre

The data on the year by which the metropolitan area had passed 100,000 refer to the most recent passing of 100,000 and to the termini of the following periods:

Before 1500
1500 - 1600
1600 - 1700
1700 - 1800
1800 - 1850
1850 - 1880
1880 - 1900
1900 - 1920
1920 - 1930

A p- preceding 1850 indicates exact date unknown, but prior to 1850.

In totalling data for the metropolises by region, the entire Detroit-Windsor Metropolitan Area has been included in the United States total for convenience.

Appendix B

Immediately prior to going to press, revised and updated data have been prepared on the world's metropolitan areas with a population of one million or more, as of January 1968. As the chapter makes extensive use of the data presented in tables 1-1 to 1-6, it is not possible, in the time available, to incorporate the new material in the text itself. These new data are presented here in the belief that they will prove of interest to the reader.

Table 1-7 lists all 153 metropolitan areas of one million or more population in order of rate of population increase. This table, therefore, may be compared to Table 1-4. The metropolitan areas marked * in Table 1-4 are similarly marked in Table 1-7. The only change in this regard is the omission of the six metropolitan areas of less than one million population that are included in Table 1-4.

Another point to note about Table 1-7, is that the metropolitan areas marked ** are those that are not listed in Table 1-4. On the average, these areas have experienced a significantly higher growth rate from 1951 to 1968 than the areas listed in Table 1-4. Ankara, which does not appear on the list in Table 1-4, had the highest rate of growth of any city over one million from 1951 to 1968.

TABLE 1-7
Metropolitan Areas of 1,000,000 or More

Percentage Increase in Population, 1951-1968	Metropolis	Percentage Increase in Population, 1951-1968	Metropolis
250.0	Ankara**	132.0	Wuhan
214.3	Bogotá	128.6	Karachi*
200.0	Anshan**	127.5	NEAR EAST/NORTH AFRICA
198.9	Guadalajara**		
180.5	Belo Horizonte**	126.8	Istanbul*
177.8	Taipei	125.0	Bangkok*
177.4	Caracas	121.1	Harbin
176.3	Monterrey**	120.6	Cairo
171.4	Baghdad**	115.4	San Diego-Tijuana**
170.0	Taiyuan**	113.9	LATIN AMERICA
169.2	Sian	113.3	Taskent**
168.4	Lima*	110.0	Dacca-Narayanganj**
166.7	Medellín**	106.3	Recife**
160.0	Teheran	104.0	EAST/SOUTHEAST ASIA
153.8	Sao Paulo		(excl. China, Japan)
152.5	Miami-Fort Lauderdale	103.1	Singapore
145.6	Seoul*	100.0	Manila*
145.5	Fushun**	100.0	Kunming**
142.3	Pôrto-Alegre**	100.0	Rio de Janeiro*
134.4	Djakarta	99.2	Dallas**
132.3	Mexico City*		

TABLE 1-7 (Continued)

Percentage Increase in Population, 1951-1968	Metropolis	Percentage Increase in Population, 1951-1968	Metropolis
97.4	Delhi-New Delhi	60.5	Havana
97.1	Lahore	60.5	Montevideo**
96.7	Pusan**	60.0	Changchun**
95.8	Algiers**	60.0	Kanpur**
94.6	Santiago	59.2	Barcelona
93.0	Alexandria	59.1	Tientsin
92.0	Kuybyshev*	58.8	Tsingtao**
89.2	Hong Kong	58.4	Fukuoka**
88.2	Chelyabinsk	57.9	ALL METROPOLITAN AREAS
87.9	Los Angeles	57.5	Melbourne
87.5	Houston	57.1	Hiroshima-Kure**
87.1	Chengtu**	55.7	Kharkov**
86.4	Tokyo-Yokohama*	53.5	Gorky
86.0	Dnepropetrovsk**	53.4	U.S.S.R.
85.6	Sverdlovsk**	53.3	Athens*
83.8	Kiev**	51.7	Seattle**
82.7	Peking	51.6	Milan*
82.0	Toronto*	50.3	AUSTRALIA
79.2	Baku**	50.0	Canton
77.8	Casablanca**	50.0	Saigon
76.9	Atlanta**	50.0	Mukden
76.7	Washington	49.2	Madras
76.0	Tsinan**	49.1	Johannesburg*
75.3	Madrid*		SUBSAHARAN AFRICA
75.1	CANADA	49.1	Kansas City**
75.0	Nanking	48.4	Calcutta*
74.0	JAPAN	48.0	Lyon**
72.5	Bombay	47.7	Munich
72.4	Denver**	47.1	Cologne
72.1	Nagoya*	47.0	Stuttgart
72.1	Turin**	46.5	New Orleans**
71.5	CHINA	44.8	Frankfurt am Main
70.5	Donetsk-Makeyevka	44.4	Sydney*
69.3	Montreal*	44.3	Bangalore**
69.0	Bandung**	44.1	Surabaya**
67.7	Chungking	43.4	Marseilles**
65.7	INDIA/PAKISTAN	42.4	Düsseldorf**
65.1	Osaka-Kobe-Kyoto	41.7	Minneapolis-St. Paul
64.0	San Francisco-Oakland-San Jose*	41.7	UNITED STATES
63.4	Novosibirsk**	41.2	Dairen**
62.4	Ahmadabad**	40.7	Budapest*
62.3	Buenos Aires*	39.4	Paris*
61.9	Warsaw*	37.6	Moscow*
60.8	SOUTHERN EUROPE	37.4	Detroit-Windsor*
60.6	Rome	36.4	Lisbon**
		36.3	Milwaukee**

TABLE 1-7 (Continued)

Percentage Increase in Population, 1951-1968	Metropolis	Percentage Increase in Population, 1951-1968	Metropolis
35.4	EASTERN EUROPE	23.4	Hyderabad**
34.5	Cincinnati**	22.6	Essen-Dortmund-Duisburg (The Ruhr)
34.3	Cleveland	19.7	Rotterdam*
33.2	Kitakyushu-Shimonoseki	16.8	Hamburg
32.8	Katowice-Bytom-Gliwice	16.6	Brussels
32.2	Baltimore	16.5	Amsterdam*
32.2	Shanghai	15.3	Boston
31.6	Chicago	13.9	Prague*
31.6	Naples*	13.1	Copenhagen**
30.3	St. Louis	12.4	Antwerp**
29.4	Stockholm**	11.0	Birmingham (U.K.)
28.8	Philadelphia-Trenton-Wilmington*	7.9	Liverpool
28.7	Mannheim-Ludwigshafen-Heidelberg**	7.0	London*
		7.0	Pittsburgh
26.8	NORTHWESTERN EUROPE (excl. U.K., Germany)	6.4	UNITED KINGDOM
		6.3	Vienna
26.5	Bucharest	5.2	Leeds-Bradford**
26.1	Leningrad*	4.9	Newcastle-Sunderland**
25.2	New York*	3.8	Manchester
		1.9	Berlin
24.5	GERMANY	1.6	Glasgow*
23.6	Buffalo-Niagara Falls**		

Note: Regional aggregates of metropolises are designated in CAPITALS.

*Metropolitan Areas represented in Centennial Study and Training Programme on Metropolitan Problems.

**Added since 1964 list appearing in Table 1-4.

Table 1-8 provides a ranking by population size as of January 1, 1968. This listing is comparable in most ways with Table 1-6 with the following exceptions.

In a number of cases the boundaries of individual metropolitan areas have been revised. This results both from continuing expansion of the areas, and from the availability of new or improved information.

The most notable expansion is the addition of Kyoto to the Osaka-Kobe area, reflecting the rapid expansion of all of these urban centres and the large volume of daily commuting which now takes place from Kyoto to Osaka. The 1951 data in this table always refer to the same territorial area as the 1968 data. Thus the metropolitan areas whose boundaries have been altered since 1964 will have 1951 population figures differing from those given in Table 1-6.

The populations have been updated to 1968. In a few cases a census or other official figure has become available since Table 1-6 was compiled which has resulted in a substantial correction in the previous estimate (examples include Teheran and Seoul). Thus, even for a metropolitan area whose territorial extent has not changed, and which therefore is listed with an identical 1951 population, comparison from 1964 to 1968 may not be valid.

TABLE 1-8
153 Metropolitan Areas: Summary Data

Rank 1968	METROPOLITAN AREA	Estimated Population 1/1/1968	Estimated Population 1/1/1951	Percent Increase 1951-68	CENTRAL CITY Population 1/1/1968	CENTRAL CITY Rank 1968
1	Tokyo-Yokohama, Japan	20,500,000	11,000,000	86.4	8,950,000†	2
2	New York, United States	16,900,000	13,500,000	25.2	8,000,000	3
3	Osaka-Kobe-Kyoto, Japan	12,300,000	7,450,000	65.1	3,100,000	17
4	London, United Kingdom	11,025,000	10,300,000	7.0	7,865,000	4
5	Moscow, U.S.S.R.	9,150,000	6,650,000	37.6	6,475,000†	6
6	Paris, France	8,850,000	6,350,000	39.4	2,819,000	25
7	Buenos Aires, Argentina	8,600,000	5,300,000	62.3	2,900,000	23
8	Los Angeles, United States	8,455,000	4,500,000	87.9	2,810,000	26
9	Calcutta, India	7,900,000	5,325,000	48.4	3,000,000	20
10	Shanghai, China	7,800,000	5,900,000	32.2	10,700,000*	1
11	Chicago, United States	7,435,000	5,650,000	31.6	3,460,000	14
12	Mexico City, Mexico	7,200,000	3,100,000	132.3	3,200,000	16
13	Sao Paulo, Brazil	6,600,000	2,600,000	153.8	5,200,000	7
14	Rio de Janeiro, Brazil	6,100,000	3,050,000	100.0	4,000,000	10
15	Cairo, United Arab Republic	5,900,000	2,675,000	120.6	4,500,000	9
16	Bombay, India	5,650,000	3,275,000	72.5	5,000,000	8
17	Essen-Dortmund-Duisburg, West Germany (The Ruhr)	5,150,000	4,200,000	22.6	705,000	
18	Philadelphia-Trenton-Wilmington, United States	5,025,000	3,900,000	28.8	2,015,000	34
19	Peking, China	4,750,000	2,600,000	82.7	7,800,000*	5
20	Detroit-Windsor, United States-Canada	4,625,000	3,365,000	37.4	1,575,000	50
21	Leningrad, U.S.S.R.	4,350,000	3,450,000	26.1	3,340,000†	15
22	Seoul, Korea	4,175,000	1,700,000	145.6	3,950,000	11
23	San Francisco-Oakland-San Jose, United States	4,150,000	2,530,000	64.0	710,000	
24	Berlin, Germany (West and East)	3,975,000	3,900,000	1.9	2,163,000	30
25	Djakarta, Indonesia	3,750,000	1,600,000	134.4	3,750,000	13
26	Delhi-New Delhi, India	3,750,000	1,900,000	97.4	3,000,000	18
27	Boston, United States	3,575,000	3,100,000	15.3	595,000	
28	Victoria, Hong Kong	3,500,000	1,850,000	89.2	700,000**	
29	Tientsin, China	3,500,000	2,200,000	59.1	3,900,000*	12
30	Milan, Italy	3,365,000	2,220,000	51.6	1,684,000	45
31	Teheran, Iran	3,250,000	1,250,000	160.0	2,875,000	24
32	Manila, Philippines	3,200,000	1,600,000	100.0	1,400,000	55
33	Mukden, China	3,000,000	2,000,000	50.0	3,000,000	19
34	Madrid, Spain	2,980,000	1,700,000	75.3	2,900,000	21
35	Nagoya, Japan	2,925,000	1,700,000	72.1	2,100,000	31
36	Wuhan, China	2,900,000	1,250,000	132.0	2,900,000	22
37	Manchester, United Kingdom	2,890,000	2,785,000	3.8	611,000	
38	Rome, Italy	2,810,000	1,750,000	60.6	2,631,000	27
39	Santiago, Chile	2,725,000	1,400,000	94.6	635,000	
40	Bangkok, Thailand	2,700,000	1,200,000	125.0	1,975,000	37

TABLE 1-8 (Continued)

Rank 1968	METROPOLITAN AREA	Estimated Population 1/1/1968	1/1/1951	Percent Increase 1951-68	CENTRAL CITY Population 1/1/1968	Rank 1968
41	Washington, United States	2,695,000	1,525,000	76.7	815,000	
42	Birmingham, United Kingdom	2,665,000	2,400,000	11.0	1,102,000	82
43	Chungking, China	2,600,000	1,550,000	67.7	2,600,000	28
44	Sydney, Australia	2,600,000	1,800,000	44.4	156,000	
45	Lima, Peru	2,550,000	950,000	168.4	335,000	
46	Johannesburg, South Africa	2,550,000	1,710,000	49.1	600,000	
47	Montreal, Canada	2,540,000	1,500,000	69.3	1,225,000	69
48	Barcelona, Spain	2,475,000	1,550,000	59.2	1,665,000	46
49	Karachi, Pakistan	2,400,000	1,050,000	128.6	1,700,000	44
50	Madras, India	2,350,000	1,575,000	49.2	1,975,000	38
51	Budapest, Hungary	2,350,000	1,670,000	40.7	1,985,000	35
52	Hamburg, West Germany	2,335,000	2,000,000	16.8	1,833,000	41
53	Istanbul, Turkey	2,325,000	1,025,000	126.8	1,850,000	40
54	Melbourne, Australia	2,300,000	1,460,000	57.5	76,000	
55	Toronto, Canada	2,275,000	1,250,000	82.0	690,000	
56	St. Louis, United States	2,255,000	1,730,000	30.3	675,000	
57	Saigon, South Vietnam	2,250,000	1,500,000	50.0	1,640,000	47
58	Canton, China	2,250,000	1,500,000	50.0	2,250,000	29
59	Cleveland, United States	2,250,000	1,675,000	34.3	780,000	
60	Bogotá, Colombia	2,200,000	700,000	214.3	2,075,000	33
61	Caracas, Venezuela	2,150,000	775,000	177.4	1,450,000**	53
62	Harbin, China	2,100,000	950,000	121.1	2,100,000	32
63	Athens, Greece	2,100,000	1,370,000	53.3	660,000	
64	Alexandria, United Arab Republic	2,075,000	1,075,000	93.0	1,875,000	39
65	Brussels, Belgium	2,070,000	1,775,000	16.6	169,000	
66	Katowice-Bytom-Gliwice, Poland	2,025,000	1,525,000	32.8	290,000	
67	Vienna, Austria	2,020,000	1,900,000	6.3	1,640,000	48
68	Taipei, Taiwan	2,000,000	720,000	177.8	1,275,000	65
69	Singapore, Singapore	1,980,000	975,000	103.1	1,150,000	74
70	Pittsburgh, United States	1,920,000	1,795,000	7.0	540,000	
71	Baghdad, Iraq	1,900,000	700,000	171.4	1,200,000	70
72	Doneck-Makejevka, U.S.R.R.	1,875,000	1,100,000	70.5	855,000	
73	Naples, Italy	1,875,000	1,425,000	31.6	1,263,000	66
74	Glasgow, United Kingdom	1,860,000	1,830,000	1.6	950,000	
75	Amsterdam-Haarlem, Netherlands	1,805,000	1,550,000	16.5	858,000	
76	Baltimore, United States	1,785,000	1,350,000	32.2	915,000	
77	Havana, Cuba	1,765,000	1,100,000	60.5	990,000	
78	Sian, China	1,750,000	650,000	169.2	1,750,000	42
79	Nanking, China	1,750,000	1,000,000	75.0	1,750,000	43
80	Liverpool, United Kingdom	1,710,000	1,585,000	7.9	701,000	
81	Warsaw, Poland	1,700,000	1,050,000	61.9	1,285,000	64
82	Lahore, Pakistan	1,675,000	850,000	97.1	1,575,000	49
83	Cologne, West Germany	1,655,000	1,125,000	47.1	855,000	

TABLE 1-8 (Continued)

Rank 1968	METROPOLITAN AREA	Estimated Population 1/1/1968	Estimated Population 1/1/1951	Percent Increase 1951-68	CENTRAL CITY Population 1/1/1968	CENTRAL CITY Rank 1968
84	Munich, West Germany	1,625,000	1,100,000	47.7	1,244,000	68
85	Minneapolis-St. Paul, United States	1,615,000	1,140,000	41.7	445,000	
86	Ahmadabad, India	1,600,000	985,000	62.4	1,350,000	60
87	Kiev, U.S.S.R.	1,590,000	865,000	83.8	1,457,000	52
88	Houston, United States	1,575,000	840,000	87.5	1,150,000	73
89	Bucharest, Romania	1,550,000	1,225,000	26.5	1,400,000†	56
90	Gorky, U.S.S.R.	1,535,000	1,000,000	53.5	1,140,000	79
91	Leeds-Bradford, United Kingdom	1,530,000	1,455,000	5.2	507,000	
92	Montevideo, Uruguay	1,525,000	950,000	60.5	1,325,000	62
93	Frankfurt am Main, West Germany	1,520,000	1,050,000	44.8	662,000	
94	Miami-Fort Lauderdale, United States	1,515,000	600,000	152.5	320,000	
95	Lisbon, Portugal	1,500,000	1,100,000	36.4	820,000	
96	Turin, Italy	1,480,000	860,000	72.1	1,132,000	80
97	Recife, Brazil	1,475,000	715,000	106.3	1,050,000	89
98	Pusan, Korea	1,475,000	750,000	96.7	1,475,000	51
99	Stuttgart, West Germany	1,470,000	1,000,000	47.0	614,000	
100	Kitakyushu-Shimonoseki, Japan	1,465,000	1,100,000	33.2	1,055,000	88
101	Bangalore, India	1,450,000	1,005,000	44.3	1,025,000	93
102	Chengtu, China	1,450,000	775,000	87.1	1,450,000	54
103	Hyderabad, India	1,450,000	1,175,000	23.4	1,025,000	94
104	Tashkent, U.S.S.R.	1,440,000	675,000	113.3	1,295,000	63
105	Kharkov, U.S.S.R.	1,425,000	915,000	55.7	1,148,000	78
106	Newcastle-Sunderland, United Kingdom	1,400,000	1,335,000	4.9	250,000	
107	Copenhagen, Denmark	1,385,000	1,225,000	13.1	655,000	
108	Buffalo-Niagara Falls, United States-Canada	1,372,000	1,110,000	23.6	460,000	
109	Milwaukee, United States	1,370,000	1,005,000	36.3	745,000	
110	Fushun, China	1,350,000	550,000	145.5	1,350,000	58
111	Taiyuan, China	1,350,000	500,000	170.0	1,350,000	59
112	Tsingtao, China	1,350,000	850,000	58.8	1,350,000	61
113	Baku, U.S.S.R.	1,335,000	745,000	79.2	785,000†	
114	Dallas, United States	1,325,000	665,000	99.2	795,000	
115	San Diego-Tijuana, United States-Mexico	1,325,000	615,000	115.4	670,000	
116	Cincinnati, United States	1,325,000	985,000	34.5	495,000	
117	Guadalajara, Mexico	1,300,000	435,000	198.9	1,125,000	81
118	Atlanta, United States	1,300,000	735,000	76.9	510,000	
119	Stockholm, Sweden	1,275,000	985,000	29.4	768,000	
120	Bandung, Indonesia	1,225,000	725,000	69.0	1,150,000	75
121	Surabaja, Indonesia	1,225,000	850,000	44.1	1,150,000	76
122	Kansas City, United States	1,215,000	815,000	49.1	550,000	

TABLE 1-8 (Continued)

Rank 1968	METROPOLITAN AREA	Estimated Population 1/1/1968	Estimated Population 1/1/1951	Percent Increase 1951-68	CENTRAL CITY Population 1/1/1968	CENTRAL CITY Rank 1968
123	Mannheim-Ludwigshafen-Heidelberg, West Germany	1,210,000	940,000	28.7	324,000	
124	Medellin, Colombia	1,200,000	450,000	166.7	910,000	
125	Kuybyshev, U.S.S.R.	1,200,000	625,000	92.0	1,016,000	95
126	Kanpur, India	1,200,000	750,000	60.0	1,060,000	87
127	Dnepropetrovsk, U.S.S.R.	1,200,000	645,000	86.0	837,000	
128	Changchun, China	1,200,000	750,000	60.0	1,200,000	71
129	Casablanca, Morocco	1,200,000	675,000	77.8	1,170,000	72
130	Dairen, China	1,200,000	850,000	41.2	1,150,000†	77
131	Algiers, Algeria	1,175,000	600,000	95.8	1,000,000	97
132	Porto Alegre, Brazil	1,175,000	485,000	142.3	850,000	
133	Sverdlovsk, U.S.S.R.	1,160,000	625,000	85.6	981,000	
134	Novosibirsk, U.S.S.R.	1,160,000	710,000	63.4	1,080,000	86
135	Belo Horizonte, Brazil	1,150,000	410,000	180.5	1,000,000	96
136	Prague, Czechoslovakia	1,150,000	1,010,000	13.9	1,040,000	91
137	Lyons, France	1,125,000	760,000	48.0	560,000	
138	Hiroshima-Kure, Japan	1,100,000	700,000	57.1	530,000	
139	Seattle, United States	1,100,000	725,000	51.7	550,000	
140	Tsinan, China	1,100,000	625,000	76.0	1,100,000	83
141	Kunming, China	1,100,000	550,000	100.0	1,100,000	84
142	Rotterdam, Netherlands	1,095,000	915,000	19.7	711,000	
143	Fukuoka, Japan	1,085,000	685,000	58.4	805,000	
144	Dusseldorf, West Germany	1,075,000	755,000	42.4	689,000	
145	Monterrey, Mexico	1,050,000	380,000	176.3	900,000	
146	Ankara, Turkey	1,050,000	300,000	250.0	1,035,000	92
147	Dacca-Narayanganj, Pakistan	1,050,000	500,000	110.0	425,000	
148	Anshan, China	1,050,000	350,000	200.0	1,050,000	90
149	New Orleans, United States	1,040,000	710,000	46.5	655,000	
150	Antwerp, Belgium	1,040,000	925,000	12.4	240,000	
151	Chelyabinsk, U.S.S.R.	1,035,000	550,000	88.2	851,000	
152	Marseilles, France	1,025,000	715,000	43.4	900,000	
153	Denver, United States	1,000,000	580,000	72.4	480,000	
	TOTAL, 153 metropolitan areas of 1,000,000 or more, 1968	419,532,000	265,735,000	57.9	248,112,000	

Note: The central city total is exclusive of 6,350 in outlying rural portions of three Chinese cities, Shanghai, Peking, and Tientsin, technically within the city limits but not considered to be within the metropolitan area. The figures given in the table above for these three cities include these rural areas in the central city figures but not in the metropolitan area figures.

*Municipal boundaries include extensive rural areas, which have been excluded in estimating the metropolitan population.

**Population for a census district that does not correspond to an actual administrative unit.

†Population within city limits, excluding areas beyond limits but administered by the city government.

Part II

Services in the metropolis

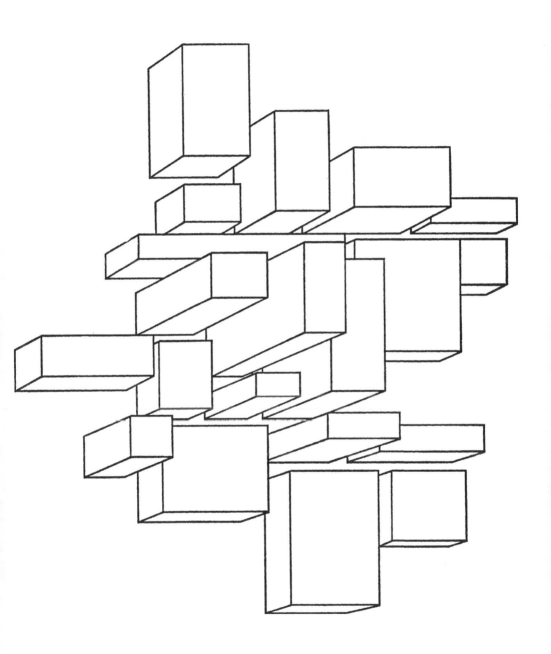

Introductory Note

The major objectives of the following examination of services in the metropolis are the identification of the nature of the demands being made upon these services; the ways in which these separate demands can best be satisfied, given existing resource constraints; and the demands that are subsequently imposed upon the politico-administrative machinery and other institutions responsible for providing services. These various and often conflicting demands are then picked up in Part III, which examines the various functional and structural aspects of the institutions most suitable for the provision of services.

As the Introduction to this book has already made clear, the separate treatment of the services provided in the metropolis is an academic exercise. The interdependency of services is such that each has to be planned and administered in relation to the others. It is doubtful whether man as yet fully accepts this basic tenet, judging by the present organization of his life. There is still the desire to identify simple cause-effect relationships in seeking solutions to organizational and environmental problems; the traffic jam for example, tends to be seen as a product of too many cars in one place at one time. Unfortunately, life is not this simple.

In any given metropolitan area these interdependencies are complicated further in degree by the great size of the metropolis, both in terms of its physical extent and of its areas of influence (which are of a more abstract nature), and by the continuing and generally rapid change in all sectors of activity. Thus, because these interdependencies are more difficult to conceptualize and to express in any rational organization, the provision of services is similarly affected. Furthermore, the given state of these characteristics in any metropolitan area is extremely difficult to alter except over a very long period of time.

In contrast, there are two additional variables that affect the complexity of the interdependencies that are subject to much more control by man: the stage of technological development reached, and the politico-administrative structure existing in the country and, especially, in the metropolis. Such technological developments as the application of computers to the decision-making process facilitate not only the physical integration of services but also assist in the more organized provision of those services. Truly holistic models of urban systems have not yet been developed, although partial models, pertaining to any one service area, such as transportation, have seen more progress. There is, however, a great danger that the advances made in one service area may be used as a basis for the replanning of a city or the restructuring of its governmental machinery without due consideration being given to the implications for the other services. Hence the necessity of stressing the interrelated nature of these services.

The six service areas discussed here are transportation, utilities, housing, education, health and welfare. In each of the following chapters the same major issues recur. They are essentially products of the combined consideration of the variables discussed above: size, technological development, politico-administrative structure, and change. Most important, consideration is coloured by the imposition of local values which not only results in different answers to the issues posed but also in different problems created. Part of the answer to many problems involves a change in value structure. Service standards should reflect a quantitative and qualitative equality in the enjoyment of services, but this needs to be balanced against the costs of providing this equality and the question of who is to pay for the provision of equal services. The limitations of resources may impose priorities on the satisfaction of demand, and the needs of the country may have to be put before those of the metropolis. This is especially so, for example, in the case of the education of persons for certain jobs vital for the leadership and economic development of the country. Yet, at the same time, these needs may best be satisfied in the metropolis, and the question arises as to where the resources, especially financial, should come from to meet the provision of the services.

Another basic question is that pertaining to the assignment of functional responsibility to governments or other institutions for the performance of tasks. It may be far more efficient for a central government to be responsible for the financing and administration of housing in a metropolis, but the desire to facilitate citizen access to ensure that citizen housing needs are being met may deem that a metropolitan or local government be responsible for some aspects of the planning and jurisdiction of the service. The increasing intervention of government in the provision of services and the involvement of a greater number of authorities in the provision of any one service calls for integration of governmental machinery, and compromises have to be struck if the planning and administration of individual services are to be performed both efficiently and fairly. These compromises affect not only structural arrangements but also the goals sought and the means by which they are realized.

However, although these questions arise in the consideration of each of the services examined here, they require separate treatment in each instance and often make different demands in their resolution. Understandably each author has his own focus of interest. But one of the advantages of the process followed in the Centennial Programme is that these biases tend to be counter-balanced by the reactions of the study groups. For example Kain, in his essay on transportation, puts heavy emphasis on the use of modern technology and pricing techniques in developing new metropolitan transportation systems. This quite naturally sparks debate on the socio-economic objectives of transportation and the need for greater consideration of the social impact of transportation systems in the metropolitan region and, indeed, the priorities that should be attached to transportation. On more specific issues, such as road versus rail, there are good cases made by proponents of both sides. Kain emphasizes the key role of roads for any mode of transportation. What is agreed upon is the need for a greater integration of all of the elements of a transportation system, and Kain does provide some valuable pointers in this regard.

The ever-present problem of how to satisfy the dictates of efficiency and citizen access is perhaps best illustrated in the debate on utilities introduced by Hanson in Chapter 3 and continued in Chapter 8. Hanson's impression of the existing situation is that the concern for efficiency in the provision of utility services is paramount in most parts of the world. And although utilities are regarded as fringe activities of government, Hanson concludes that the case for government ownership is based not so much on ideology as upon the desire to achieve savings through the technological integration of services and economies of scale. To be weighed against these considerations, as both Hanson and the study group realize, is the apparent inefficient management of public authorities. Furthermore, the publicly-run utility often provides us with the most obvious example of the extensive *ad hoc* authority, which is frequently subject to less public questioning and influence than the utility run by a private company, sensitive to its public image.

Another field in which the degree of government intervention varies greatly is housing; but in this instance ideology, or public values, can be regarded as a major factor influencing the intensity of governmental activity. However, both Sazanami, in Chapter 4, and the study groups, leave one with the impression that the assumption of more responsibility on the part of government is highly desirable. Obviously enough, this may require a change in public attitudes—a point well illustrated by the need to direct housing policy goals to satisfying social need as opposed to effective economic demand. Yet, the satisfaction of social needs or, more appropriately, the fulfillment of socio-economic possibilities depends on several factors. It requires substantial governmental legislation and has implications for the restructuring of governmental authorities in the interest of regional planning. The initiative of private enterprise is also necessary to provide cost reductions through improved technology. Most important, however, is the desire of the individual to improve his condition. Indeed, the choice should remain with the individual to select his priorities when it comes to satisfying social needs. The possibilities of this being realized, even in a resource-scarce situation, are exposed by the study groups.

In Chapter 5 Philp introduces us to the what, who, how, where, when and why, as regards the provision of an education system. All but the first of these questions, which calls for an elaborate examination of curriculum, Philp attempts to answer in a discussion that focusses upon such issues as administration, financing, physical plant and the education of minority groups. In his examination of administration Philp pays great attention to the nature and location of governmental control, as it is this more than any other factor that reflects the national attitude toward education. As this condition is largely financial, the question arises as to what should be the nature of the relationship between the spending authority and the tax-raising authority. Judging by the opinions of the study groups, there appears to be a general acceptance that more central control is inevitable, although the metropolitan-wide authority is favoured for handling most aspects of education administration.

It is also of interest to note, especially in comparison with the other services mentioned thus far, that education is not seen as a service in which the public necessarily wishes to economize. Although qualified by the recognition that standards must be related to societal needs, the greatest concern is for quality. Equality is regarded, in this instance, as the opportunity to optimize the potential of the individual, rather than the provision of the same conditions for all, as is the case with, say, housing or utilities. This differential treatment of individuals is easier to justify in some instances, as with handicapped minority groups, than it is with others, such as the education of a leadership. Yet, this type of differential treatment should be seen as for the good of society as well as for the individuals

concerned. There is no denying, however, that some tough decisions frequently have to be made, and the best example is perhaps that of investments in education, as cited above.

With health and welfare the situation is different again. Everyone should be offered access to the same standard of services. Yet demand will vary greatly, and the question then arises as to who pays—society or the consumer? With environmental health services the answer is easier to provide: in many cases the responsibility for maintaining high standards may rest with the international community. For welfare services, the question is more debatable. In Chapter 6 Bakács introduces us to the problems that have arisen as a result of the neglect of our natural environment. We have fouled our nest and have been unable to adjust. Indeed, many have suffered mentally or physiologically in the process. As we cannot throw out the nest with its occupants and start again, we have to think in other terms. The study groups provide illustrations of the values of new technologies, administrative change and planned development as preventative measures.

Rose, too, in the final essay on welfare services stresses the need to move to the preventative approach from that of maintenance. Apart from accessibility, Rose is similarly concerned with standards of welfare services and the integration of the welfare services. The changing distribution patterns of the needy within the metropolis pose new problems in providing the services both efficiently and fairly. Inasmuch as the experiences of the study groups bear out what Rose has to say, they are also a silent commentary on the problems remaining to be solved. Unfortunately, the same has to be said of most other services in many of the world's great metropolitan areas.

2 Transportation in Metropolitan Areas

John F. Kain

Generalizing about the urban transportation problems of the hundreds of widely different cities and metropolitan areas in the United States is a gargantuan task. Total population, spatial distribution and density of population and employment, family income, timing of physical development, characteristics of existing transportation facilities, size and strength of the central business district, automobile ownership, transit use, topography and climate are but a few of the factors that contribute to urban transportation problems and at the same time influence the nature of policies appropriate to a particular area. As these factors differ greatly from city to city, one might fear that transportation policies appropriate for New York City have little relevance for Boston, Massachusetts, let alone Laredo, Texas. In 1960, less than one per cent of the employed residents of Laredo used transit and the total population was only 60,000. In New York City, 61 per cent of employed residents used transit and the total population was 7,800,000.[1]

The variation in size of central city populations and metropolitan area populations

---

[1] *County and City Data Book, 1962*, U.S. Bureau of the Census (Washington, D.C., U.S. Government Printing Office, 1962), Table 6.

for the urban centres participating in the Centennial Study and Training Programme on Metropolitan Problems is brought out quite clearly by referring to Table 1-6 prepared by Richard Forstall and Victor Jones for the introductory chapter to this volume.

Far more disparate and just as significant are the differences in incomes between the participating metropolitan areas. While statistics on family or per capita incomes are not available for most of these metropolitan areas, national per capita income figures, which can be used as a rough index of metropolitan per capita incomes, are available.[2] To the difficulties inherent in generalizing about solutions or policies for such diversity must be added the difficulty of obtaining data for most of the participating metropolitan areas. Whereas there are convenient and reasonably consistent sources of data for urban areas in the United States, which describe many of the dimensions of metropolitan social and

physical structure that appear to be relevant to transportation analysis, planning and policy, there is no handy reference such as the *County and City Data Book* for the world's metropolitan areas. Although descriptive statistics of this kind leave out much that is crucial to the analysis and planning of transportation for a particular place, they do provide considerable understanding of a community's structure and a rough delineation, at least, of the nature of its transportation needs. Similar data, however, are unavailable, difficult to obtain, or nearly impossible to interpret for the widely diverse areas represented in the Centennial Programme. Thus, only the most general kinds of analysis and prescription can even be attempted here and these must be regarded as highly speculative.

This chapter attempts to extrapolate research findings obtained from American experience to metropolitan areas in other countries. This approach may have merit, if the nature of metropolitan transportation problems is to any great extent related to the level of economic development or the level of per capita income. If this is the case, United States experience could provide valuable insights into probable transportation problems in the metropolitan areas of the less developed countries of the world. With the benefit of hindsight these areas may be able to profit from the mistakes of American cities. This view is stated explicitly by Wilfred Owen:

> A closer examination of the transportation difficulties that seem so intractable in Chicago and New York may help to mitigate the problems of Bangkok and Beirut, and of new cities in the developing world yet to feel the effects of the urban-industrial revolution.[3]

---

[2]*Business International, Weekly Report to Management on Business Abroad* (New York, October 20 and October 27, 1961), pp. 4-5. National per capita income figures for most of the other countries represented in the Centennial Study are available from this source. It should be noted, however, that per capita incomes of the participating metropolitan areas are almost certainly greater than those of their respective countries. Although inherent weaknesses in international comparisons of per capita incomes of this kind are well known, they do provide a rough index of the level of development in several metropolitan areas. One of the greatest difficulties is encountered in obtaining a satisfactory index of the purchasing power of the different countries. The figures given in *Business International* use official exchange rates, which are considered a poor measure of relative purchasing power. For a discussion of the problems in making international comparisons of per capita income *see* Milton Gilbert and Irving B. Kravis, *An International Comparison of National Product and the Purchasing Power of Currencies* (Paris, O.E.C.D., 1954), and Milton Gilbert and Associates, *Comparative National Products and Price Levels* (Paris, O.E.C.D., 1958).

---

[3]Wilfred Owen, *The Metropolitan Transportation Problem,* rev. ed. (Washington, D.C., The Brookings Institution, 1966), p. 230.

Alternatively, current conditions in the United States may be considered a forecast of the probable course of metropolitan development in less developed countries, barring any radical changes in technology, or significantly different (and effective) government policies on metropolitan development and transportation. Studies conducted in the United Kingdom, for example, have found the approach to have considerable merit. The concept of United States experience as a forecast of future conditions pervades the much acclaimed British report *Traffic in Towns* (also known as the Buchanan Report) prepared by a study group of the Ministry of Transport. The summary report of the Steering Group states:

> These matters are not purely speculative. We can draw on the mass of evidence available from the United States and Canada, which are, broadly speaking, a generation further into the Motor Age than Britain . . . and, though there are obvious differences between North America and Britain, we think they are differences of degree only and not such as to invalidate a comparison. . . . It would be a great mistake to say, "Things won't be like that here". In more respects than not, they are likely to be very much the same.[4]

This same approach has been followed by P.H. Bendtsen in his monograph, *Town and Traffic in the Motor Age*:

> On the strength of the development already experienced in American cities, the attempt has been made to forecast the

development in our less motorized part of the world [Denmark], and to consider possible remedies. . . . In the present book, the attempt has been made to arrive at an understanding of these problems on the strength of research data already available. As the requisite surveys have hitherto mainly been carried out in American cities, the book is largely based on American research. But data available from European cities have also been included where possible.[5]

Intermodal Competition and the Urban Travel Market

ROAD AND RAIL: A FALSE DICHOTOMY

The choice for urban transport policy is invariably phrased as being between investment in roads for private automobile use or a rail rapid transit system. This view implies that there are but two urban transport alternatives—a gross oversimplification of the technological alternatives.

In contrast to this pervasive view of the technological alternatives, only a small part of all urban travel is by grade-separated rail systems. Moreover, as the following discussion of urban travel markets illustrates, grade-separated rail systems account for only a small percentage even of public transit usage. Road transit provides the overwhelming majority of public transport services. Yet the potential for technological improvement of the performance of road transit has been virtually ignored by transit planners.

Comparative cost analyses strongly indicate that under conditions in the United States road transit has a commanding cost

---

[4]*Traffic in Towns: A Study of the Long Term Problems of Traffic in Urban Areas*, Reports of the Steering Group and Working Group appointed by the Minister of Transport (London, HMSO, 1963). The Steering Group, paragraph 10.

[5]P. H. Bendtsen, *Town and Traffic in the Motor Age*, Trans. by E. Rockwell (Copenhagen, Danish Technical Press, 1961), p. 6.

advantage over rail.[6] Only at extremely high densities, or when the investment is already in place, do the costs of grade-separated rail rapid transit systems become competitive with new road transit systems. Although similar comparative cost analyses have not been performed for other countries, it is probable that they would favour highway-based systems even more strongly in less developed countries where wage rates are lower. This is because bus transit systems tend to be labour-intensive, whereas rail systems tend to be more capital-intensive.

The confusion about road and rail arises from an oversimplified view of urban transport systems, which stems from a tendency to equate "transit" with "rail" and, in the same way, "private automobile" with "highway". Yet, as noted above, only a small percentage of transit users travel by rail systems; the majority rely on road transit vehicles. Similarly, roads carry a varied assortment of vehicles including private automobiles, a wide variety of trucks, scooters, buses, taxi cabs, jitneys and other more exotic passenger and freight vehicles.

Discussions of urban transportation also confuse existing operating conditions of road transit with conditions that are technologically feasible. Most of the shortcomings of existing road transit systems are due to the way in which urban streets and highways are used and managed rather than to any inherent technological disadvantages. Very large improvements in existing road transit services are usually possible with only small monetary outlays.

Most analyses of the comparative performance and cost of rail rapid transit and road transit suffer from a common weakness. Typically the existing highway-based system is compared with the most technologically advanced, most sophisticated and heavily subsidized rail rapid transit system that the transit planners are able to devise. This sophisticated and heavily subsidized rail transit system is never compared with technologically advanced, sophisticated and equally subsidized road transit systems. In varying degrees, the practice is to compare a Model T Ford and a new Cadillac, without reference to price.

Perhaps more importantly, the widespread construction of high-performance, limited-access urban highways in an increasing number of countries provides excellent opportunities to create high-performance, high-speed highway transit systems. To appreciate this, it must be understood that there is not one, but rather an almost infinite number of road-based transit systems of differing speeds, coverage, levels of service and so on. Adaptability and divisibility are among the most important advantages of road transit systems. These attributes may also be their undoing, since imaginative and sophisticated planning and coordination are needed to achieve their full potential. By comparison, rail transit solutions to urban transportation problems are simpler and make fewer demands on the imagination.

Too little attention is being given to road-based transit systems in solving, or at least improving, the urban transportation problems of the world's metropolitan areas. This is almost certainly due to the tendency to think of transit and roads as competing systems. Frequently, private auto systems and public transit do compete for investment dollars and for existing road capacity. They may also complement one

[6]See John R. Meyer, John F. Kain and Martin Wohl, *The Urban Transportation Problem* (Cambridge, Harvard University Press, 1965). This is one of the few systematic, comprehensive studies of the comparative costs of alternative passenger technologies available.

another, however, by sharing the heavy cost of structural investment. As is discussed more fully below, the key question is the nature of the rules for deciding how roadway space is to be allocated among private and public users. There may indeed be large, densely populated, low-income metropolitan areas where careful cost analyses of all feasible systems strongly favour the construction of high-capacity rail lines for urban passenger transportation. It is clear, however, that careful comparative analysis of all alternative systems is not the rule. There is almost no consideration of new high-capacity road transit systems or improvements to existing road transit systems.

URBAN TRAVEL MARKETS

Substantial intermodal competition exists in urban travel markets. In the United States and in most developed countries, the principal intracity passenger modes competing for the patronage of urban residents are private automobile, transit bus, streetcar or train, rail rapid transit, commuter railroad, taxi cab and walking. As intermodal competition is greatest and has its most serious consequences during peak hours and for journeys to and from work, this, and most other discussions of urban travel markets, emphasizes commuting. This does not imply that other kinds of urban travel are less important or that there are no special problems related to the provision of adequate urban transportation services for other purposes. It is peak hour commuter trips, however, that invariably lead to new investment demands and for which the problems of intermodal competition are most serious. Moreover, off-peak and non-work urban transportation services generally would benefit from measures leading to improvements in peak hour services.

In the urban areas as a whole, the private automobile is overwhelmingly the most important mode of transportation for journeys to and from work, and its dominance for other kinds of trips is even greater. Table 2-1 shows that for all United States urban areas in 1960, 67 per cent of daily work trips were made by private automobile. Somewhat surprisingly, walking was the choice of 10 per cent, with an additional 3 per cent working at home. Bus transit accounts for just over 11 per cent of work trips and commuter railroads for just under 2 per cent. Transit remains far mor competitive for work trips than for social or recreational, shopping, personal, business and school trips. Travel time and costs appear to be the most important considerations to the commuter.[7] By contrast, privacy, schedule flexibility and cargo-carrying capacity apparently are far more important for shopping, social or recreational, personal business and other kinds of non-work travel. Private automobiles have a commanding advantage for these purposes. Non-work trips generally are made during off-peak periods when transit services are less frequent, and are made to diverse and widely scattered locations poorly served or not served at all by transit. Moreover, as several family members may accompany the driver, private automobile costs are frequently less than transit costs.

As with any averages, modal-use statistics for the entire United States hide a great deal of important variations. In addition to trip purpose, the extent of automobile dominance, or conversely the importance of high-density modes (transit and walking) depends upon: the size of the urban area, the level and distribution of income,

---

[7] Meyer, et al., op. cit., pp. 83-107.

TABLE 2-1
Number and Percentage of U.S. Workers Using Each Commuter Mode

Means of Transportation	Entire U.S. (Thousands)	Percentage	Urban U.S. (Thousands)	Percentage
Private auto or carpool	41,368	66.8	30,296	67.4
Railroad, subway or elevated	2,484	4.0	2,437	5.4
Bus or streetcar	5,323	8.6	5,143	11.4
Walking	6,416	10.4	4,718	10.5
Other	1,620	2.7	1,029	2.3
None (Work at home)	4,663	7.5	1,357	3.0
Not reported	2,782	—[a]	2,186	—[a]
All workers	64,656	100.0	47,166	100.0
Percentage of all vehicle trips by transit		15.8		20.0
Percentage highway transit of all transit		68.2		67.8

[a]The percentage not reported has been allocated proportionally to the remaining travel modes.

Sources: U.S. Census of Population: 1960 (Washington, D.C., U.S. Bureau of the Census, 1962); and *United States Summary*, Final Report PC (1)-16 (Washington, D.C., U.S. Government Printing Office, 1962).

the density of urban development, the geographic distribution of employment, the extent of transit investment, the age of the city, and the extent of subsidies made to each mode.

Although private automobile travel is less important in most countries other than the United States, its significance is increasing rapidly as incomes and car ownership rise. The greater dependence on public transit for commuter travel previously noted for the United States is also characteristic of countries with lower levels of automobile ownership. For example, the London Traffic Survey of 1963 reported that while 35 per cent of all work trips were made by car or cycle, cars or cycles were used for 54 per cent of other home-based trips.[8]

Table 2-2 presents the transit-riding ratio (daily transit passengers divided by population) for twenty-four large metropolitan areas. Transit riding per capita appears highest in the largest cities of developed countries. It tends to be somewhat lower in many American cities reflecting the greater use of private cars. It is also relatively low in several of the large cities in less developed countries. These low transit-riding ratios in the large cities of developing countries are probably due to low incomes. Residents of these lower-income countries more often walk, and frequently work at home, therefore making fewer and shorter trips. As incomes rise the population becomes more motorized, uses public transit for a larger proportion of total trips, and makes more trips. With further increases in incomes, automobile ownership increases rapidly and automobile trips are increasingly substituted for transit trips.

---

[8]Freeman, Fox, and Partners Engineering Services Corporation, and Wilbur Smith and Associates, *London Traffic Survey: Future Traffic and Travel Characteristics in Greater London*, Vol. II (July, 1966), p. 72.

TABLE 2-2
Mass Transit Usage in Twenty-four Major Metropolitan Areas

Metropolitan Area	Population (thousands)	Mass transit passengers, daily (thousands)	Highway mass transit passengers %	Riding ratio
Berlin	2,198	2,532	72	115.2
Bombay	4,500	1,915	100	42.6
Calcutta	2,549	2,750	100	107.9
Chicago	3,550	1,700	85	47.9
Hamburg	1,800	1,224	63	68.0
Hong Kong	3,265	2,663	78	81.8
London	8,347	9,500	69*	113.8
Los Angeles	2,600	700	100	26.9
Madrid	2,374	1,500	80	63.2
Manila	1,139	1,040	100	91.3
Melbourne	1,912	799	100	41.8
New Delhi	2,659	625	100	23.5
New York	7,782	7,558	38	97.1
Osaka	3,012	2,928	70	97.2
Paris	5,600	7,047	NA**	125.8
Rome	2,250	2,713	NA	120.6
San Francisco	750	670	100	89.3
Seoul	2,910	1,698	100	58.4
Sydney	2,183	1,530	67	83.3
Taipei	1,082	194	82	17.9
Teheran	1,800	1,500	87	83.3
Tokyo	10,000	3,736	75	37.4
Toronto	1,566	903	88	57.7
Houston	1,352	156	100	11.5

Note: It should be noted that both the population and transit use statistics in this table are subject to considerable error; thus no great credibility should be assigned to any single figure.

*Not available from Owsley; estimated from data published in Freeman, Fox, and Partners and Wilbur Smith & Associates, *London Traffic Survey* (July, 1966), p. 83.

**NA—not available

Source: Clinton Owsley, *World Wide Transit Study* (Houston, Public Service Department, 1963).

LAND USE AND INTERMODAL COMPETITION

As the preceding discussions suggest, the success of competing intracity passenger modes in attracting urban passengers (particularly commuters) depends to a great extent on population and workplace density and related aspects of urban form and development. As a result, the relative costs of the several modes depend in large part on workplace and residence densities, and their juxtaposition. It follows that the best, or most efficient, urban transportation system cannot be determined without specifying the existing and probable forms of urban development. Use of each intra-city passenger mode appears to be more dependent on urban structure than on the level of transit subsidies, transport investment decisions, and other pure transportation policy variables. Transportation policies in turn affect urban structure and urban development. There are long historical lags, however, and metropolitan development is affected by many forces and policies besides transportation. Within a given urban structure or environment, transportation policies may have far less effect on the use of alternative travel modes than is commonly believed. This strong dependence of urban travel behaviour and modal efficiency on urban structure and land use is widely recognized and is the basis for nearly all urban transportation analyses, forecasting and planning.

In North America and western Europe urban development appears to be becoming less dense as industry and other employers disperse from their previous locations in dense central cities. A movement of households to the suburbs has accompanied this employment dispersal. Suburbanization of population is one of the most talked about aspects of United States metropolitan development since 1945. Changes in inter- and intra-city transport and communications systems appear to be the principal forces underlying this trend toward lower density and less concentrated metropolitan development. The automobile's role in this process has been widely observed and generally exaggerated. In fact many who find this lower-density form of

metropolitan development undesirable consider the automobile to be the villain and charge it with creating ugliness and destroying our cities. More dispassionate analyses point to changes in the technologies of intercity freight and passenger transport, changes in production methods, rising incomes, and a growing taste for lower-density forms of housing as the more fundamental causes of lower-density metropolitan development.

It seems possible that the rapidly growing metropolitan areas outside North America and western Europe may never develop the highly centralized and concentrated employment centres found in more developed countries. The industrialization of these less developed countries is taking place during a period when there are alternatives to rail and water transport. As a result, the concentrated employment patterns of the nineteenth century industrialized city may never occur in the rapidly growing cities of less developed countries. If this different pattern of metropolitan development occurs, its consequences for transportation planning and the choice of urban transportation systems are very important. As cost analyses have illustrated, rail transit systems are at their best in serving a few high-density workplace centres.[9] If employment is highly dispersed, more ubiquitous urban transport systems appear more promising.

The Case for Coordination

THE RELIANCE OF ALL URBAN TRANSIT FORMS ON ROADS

Joint use of common facilities by several competing transportation modes is a fundamental aspect of urban intermodal competition. Of the urban commuter modes,

[9]See Meyer, et al., *op. cit.*, especially Part II and Appendices.

only commuter railroads and rail rapid transit operate on "private" or exclusive rights of way. Nearly seven-tenths of daily transit work trips made by United States residents were made entirely on local streets (see Table 2-1). While all transit modes combined are less able to compete for off-peak and non-work trips than for peak and work trips, the proportion of off-peak transit travel by highway modes is even greater than the peak hour proportion. This reflects the much greater specialization of grade-separated, high-capacity rail facilities, which typically provide service to only a limited number of centrally located high-density workplace areas. Thus, in terms of the proportion of transit passengers carried, road transit modes are an even more important segment of public transit systems in the United States than the 68 per cent figure for work trips suggests.

The statistics in Table 2-2 on the percentage of transit passengers using road transit systems illustrate that the world's largest metropolitan areas are also heavily dependent on road transit modes. Clinton Owsley presents data on the total number of passengers carried by mass transit and the number carried by rapid transit. This distinction appears to correspond roughly to the distinction between highway-based and grade-separated transit modes. If a bias exists, it appears to understate the importance of highway modes. Only in New York (of the twenty-two metropolitan areas for which data are available) does the percentage of all transit travel by highway modes fall below 60 per cent. In eight of these areas, all transit travel is by highway modes. In addition, most railroad commuters drive or take a bus at one or both ends of their journey. For example, although 56 per cent of all workers employed in Chicago's Loop in 1956 arrived

at work by commuter railroad or rail rapid transit, only 26 per cent originated their trip on one of these grade-separated facilities.[10] This use of highway modes for residential collection and/or central area distribution is characteristic of all rail commuter systems. Finally, in all but a handful of large American metropolitan areas, all urban public transit is provided by road transit systems.

This heavy dependence of transit on urban roads is frequently overlooked and has been all but ignored in discussions of urban transportation policy and planning. Yet, it is overwhelmingly the most important reason for the coordination of all urban transportation planning and policy. Much of the so-called urban transportation problem (in the United States at least) results from attempts to patch up public transit systems without regard to the environments in which they operate and the facilities upon which they rely.

Presently, in the United States and in most other countries, private automobiles and transit buses largely compete in their use of urban roads. Better coordination of transportation policies and, more particularly, disciplined use of urban highways could greatly increase complimentarity between automobiles and buses and allow a much improved and wider range of urban transportation services. Coordination would also provide economies in the construction of these facilities.

Transit's largest, actual and potential markets are found in dense built-up areas where highway space is least plentiful and most expensive to provide (and where its provision is most disruptive to existing land

[10]John F. Kain, "The Commuting and Residential Decisions of Central Business District Workers," *Transportation Economics* (New York, National Bureau of Economic Research, 1965), p. 269.

uses), and during those times of day when the demands on transport capacity are greatest. Since transit is more economical per rider than private automobiles in its use of highway capacity and since the demand for transit is least for non-work and off-peak hour travel, an urban transportation system that would allocate proportionally more highway space to public transit vehicles in high density, congested locations and during peak travel periods would economize on expensive urban highway facilities.[11] To some extent, this is the existing situation. Transit has a much larger part of the urban travel market in larger, older and higher-density urban areas having more centralized employment locations and in this way serves a larger proportion of work trips and peak hour travel. This result, which reflects consumer adaptation to congestion and other transportation costs, occurs in spite of an uncoordinated and inconsistent urban transportation policy.

STRIKING A BALANCE

Metropolitan areas where automobile ownership has not yet reached high levels have the opportunity of devising arrangements for disciplining the anticipated higher levels of automobile use and of providing for a better balance between the use of urban street space by public transit vehicles and by private vehicles of various kinds. This is especially the case if joint use of urban streets by bus transit and private vehicles is contemplated at the time capacity-increasing traffic engineering and street improvement schemes (which are certain to accompany higher levels of automobile ownership) are planned and carried out. Opportunities of this kind are being lost almost every day in metropolitan areas all over the world. The numbers of vehicles using the streets of the world's metropolitan areas are increasing rapidly and these increases are by no means limited to the most developed countries. For example, between 1953 and 1961 motor vehicle traffic on several main streets in Taipei, Taiwan, increased between 46 and 810 per cent per year.[12] Similarly, between

---

[11]A recent report on developments in The Philippines points out that there are substantial differences in the efficiency of different kinds of public transit units. In commenting on steps being taken to improve traffic flow in The Philippines it notes: "Another step being taken by the Philippine Government to ease traffic congestion is the gradual elimination of jeepneys and small time operators as a means of public transit, by substituting the standard-size buses operated on a big scale. It has been found through studies undertaken by the Philippine Bureau of Public Highways that a passenger in a jeepney occupies about twice as much space as a passenger in a bus. Obviously the bus is a better mode of transportation in moving people en masse than smaller units insofar as road occupancy is concerned. Most of the people of The Philippines are more dependent upon the use of mass transit vehicles in their daily journeys than on privately owned automobiles. Traffic problems would have been inconceivably worse if the whole travelling public or a major portion thereof used private cars in its everyday trips. The same studies of the Bureau of Public Highways also show that a passenger in a private car occupies about four times as much space as a passenger in a bus." United Nations, Economic Commission for Asia and the Far East (ECAFE), *Transport and Communications Bulletin for Asia and the Far East*, No. 37, (New York, 1964), p. 12.

[12]The number of motor vehicles per day (VPD) using Roosevelt Road increased from 2,123 to 66,649. VPD on Nanking E. Road increased from 1,231 to 11,215, the number on Chungshan S. Road from 3,881 to 15,897, and similar increases were experienced on other main streets during the period. "Railways, An Elevated Rapid Transit Railway System in Taipei Area," in United Nations, ECAFE, *Transport and Communications Bulletin for Asia and the Far East*, No. 38, (New York, 1965), p. 16.

1950 and 1961 the number of motor vehicles registered in Taipei increased from 3,358 to 14,358 or at an annual rate of 15 per cent, and the number of other vehicles increased from 97,000 to 228,000 or by 8 per cent a year. During the same period, the number of city buses increased from 117 to 409 and the number of yearly passengers from 24 million to 184 million. These increased demands on Taipei's urban transportation capacity have led to proposals for the construction of an elevated rapid transit railway system. Authors of an article describing the proposed system comment as follows on these developments:

> ... Traffic volume in Taipei has increased rapidly and transportation has gradually become inadequate.... It is an undeniable fact that the streets in urban areas have become increasingly crowded, especially during rush hours, and traffic accidents have increased every year.

> The present crowded traffic conditions in downtown Taipei, if not relieved, may become even more serious in the future. According to some estimates, the population of Taipei in 1968 will be 1.34 million; motor vehicles will be up to 40,000. Traffic in Taipei on streets such as Chungshan North Road, Yenping North Road, and Chungcheng Road would then be completely snarled.[13]

The cost of the proposed elevated transit railway system, which includes improvements of existing Taiwan Railway Administration lines serving urban and suburban areas and the construction of a twenty-two kilometre, elevated, rapid transit railway system, is estimated at approximately U.S. $25 million, using the official exchange rate.

_____

[13]*Ibid.*, pp. 19-20.

Exploiting the Potential of Transportation Systems

Having examined the nature of the urban travel market, the characteristics of the different modes of transportation available to the market, and the need for a coordinated approach to the provision of transportation services, we must now examine the various means by which coordinated, efficient transportation systems can be realized. This is best done through a three-part examination of the planning, management and pricing of transportation systems.

PLANNING TRANSPORTATION SYSTEMS

Private and public road transportation services, while not the only competitive urban transportation modes, are the most important ones. Yet, interdependence is especially important because of the interaction of the performance characteristics and efficiency of the two systems. When road space or capacity is plentiful, as in off-peak periods or at low densities, there are no serious conflicts. During peak periods, however, and particularly in high-density areas, increased use of each system adversely affects the performance of the other system. The private automobile system, since it is usually the greater user of road space, will generally more seriously affect the performance of the public system. The high costs and low performance of most existing surface transit systems are due in large part to this negative interaction during peak hours. Moreover, this interaction is generally more adverse to system performance at those locations where, and during periods when, urban road transit ought to be most competitive. In high-density centres during peak hours speeds decline precipitously, system reliability deteriorates and operating costs rise steeply. What might be an economically served transit market becomes an inefficient, marginal, declining and unprofitable operation.

The alternative of improved coordination of road-based transit and private automobile systems is a largely uncharted path. There are, however, several obvious directions that might be pursued. Improvement upon the current use of already extensive metropolitan expressway systems is the most obvious. The existing practice of undisciplined free access probably reduces effective vehicular capacity or flow in many instances. In many cases, the provision of entry and exit controls and the metering of flow would increase the vehicular flow during peak hours and reduce accidents. Recent experiments with expressway surveillance systems in Detroit, Chicago and Houston have yielded much valuable information on the behaviour, performance and methods of managing limited-access urban highways. Control of limited-access highways to provide better flow characteristics should be relatively simple, since these highways have relatively few entry and exit points and in many urban areas constitute a reasonably complete and independent system. If separate ramps were provided, transit vehicles could be given priority access; and if access controls were also available, peak hour vehicular speeds could be set at any level desired. Buses could enter uncongested expressways in all parts of the metropolitan area and operate at high speed (50 to 60 miles per hour) on uncongested highways into the downtown area or to regional commercial or employment centres. While access controls might actually increase hourly flow during peak hours, very high operating speeds could be achieved only by some reduction in hourly vehicle capacity. These modifications, although involving engineering, organizational and other technical issues, are reasonably simple and have enormous potential for improving the quality, operating speeds and efficiency of highway-based transit systems.

Although great possibilities exist for improving the quality of public transit systems in United States cities through the use of existing or currently planned urban expressways, even greater opportunities exist in those countries only now beginning planning and construction of urban expressway networks and in those where construction is not yet contemplated. Urban expressways in the United States were not designed for the joint and simultaneous use of automobile and transit systems. It seems certain that the designs, locations, alignment and ramp locations of many urban expressways in the United States would have been very different if this joint use had been anticipated and planned. Some cities already may have too much urban road capacity serving dense centres. However, even if high-performance, lower cost road transit systems had been available, high-income American motorists might have decided to pay the higher costs of private automobile commuting to dense central areas. Yet, it is impossible to be certain, since this alternative was never provided. It would be unfortunate if, as incomes and automobile ownership rise in other countries, urban motorists are not given a richer set of urban transportation alternatives through the better management of urban road systems. Even if the current level of road construction in United States cities is justified, these considerations had they been taken into account, would have altered greatly the timing of urban expressway construction and increased benefits provided by these systems.

As a caveat, however, it is worth pointing out that many traffic engineering "improvements" may affect transit bus operations adversely. One-way street systems may increase the vehicular speed and flow on existing streets, but they also may make transit buses more inconvenient by forcing greater indirectness of routing and by making bus riders walk further to less convenient bus stops. A news story in the *New York Times* of May 18, 1966, included the following comments on the effect on transit use of converting Fifth and Madison Avenues to one-way operation.

> Cars are moving faster, but buses are going slower on Fifth and Madison Avenues since those streets were made one way on Jan. 14. . . . The heavy loss in short-haul passengers has meant the abandonment of 103 of the 971 runs on the Fifth and Madison routes, most of them shortly after the new traffic regulations were introduced. . . . Bus drivers say they often need 10 or 20 minutes longer to cover routes along Fifth and Madison Avenues than they did before the one-way regulations were applied. . . . The Transit Authority and the bus passengers are agreed that thousands of persons who used to take buses for eight or ten blocks now prefer to walk that distance because of the inconveniences caused by the conversion of Fifth and Madison to one-way traffic.

In the United States, at least, it appears traffic engineers determine local circulation patterns primarily on the basis of vehicular operating speeds and flow. This results in an inefficient use of expensive and scarce downtown street space.

THE MANAGEMENT OF TRANSPORTATION SYSTEMS

The lack of control over the use of highways frequently gives rise to the overuse of certain facilities, such as urban expressways, during peak hours. In many instances, volumes become so heavy that instability results, the facilities break down, and traffic comes to a standstill with actual reductions in vehicular capacity. Frequent accidents aggravate these situations, creating still greater congestion and further reductions in vehicular capacity. This wasteful use of expensive and high-performance highway facilities is unnecessary. Yet, it is merely symptomatic of a more general economic waste resulting from the undisciplined use of urban roads.

Much of this poor management and organization arises from a division of responsibility by mode, and a narrow perception of the tasks of providing urban transportation capacity. In the United States, for example, responsibility for urban transportation is highly fragmented. Vast numbers of local governments, state highway departments, private transit companies, public and semi-private authorities, regional planning bodies and other private and public agencies have substantial roles in the provision of transportation in United States metropolitan areas. While some progress has been made in coordinating and planning urban transportation services, existing institutional structures remain hopelessly inadequate. No agency has the responsibility for urban transportation as a whole. The result is great economic waste, substantial deterioration of the quality of urban life, and massive duplication of investment. This situation is by no means peculiar to the United States. Ernest Davies devotes most of his pamphlet, *Transport in Greater London*, to a discussion of these inadequacies in London's transport and land-use planning.

> The present position can be summed up as follows: (1) Responsibility for the planning of roads and traffic is spread over a large number of separate and uncoordinated authorities. (2) The division of responsibility for town planning

and for traffic has resulted in inadequate consideration being given to transport needs. This has caused a serious deterioration of the transport system. (3) No overall roads and traffic plan has been prepared in relation to the development plans for the Greater London area, nor would it have been possible to do so to the best advantage owing to the lack of data on which to base such a plan. (4) The steady deterioration in public transport is largely due to traffic congestion which has increased costs and led to an increase in the use of private transport.[14]

Further it is alleged in the United States, at least, that great public hostility exists toward any efforts aimed at better management of road systems, and particularly toward the use of tolls and other methods for disciplining road users and rationalizing the use of urban roads, although there is no evidence that this has been tested.

Thus, the failure to make more rigorous attempts to improve public transit services through better road management appears to be due to fragmentation of responsibility for urban transportation policy, and a tendency to think of urban roads and mass transit as separate programmes with different aims and objectives and serving different functions. Road planners have exhibited almost no concern for transit operation in the design, planning and operation of urban roads. Similarly, transit planners have shown remarkably little interest in transit systems that would more fully exploit the inherent complementarity of private and public road transportation systems, and little imagination about modifying road design and operation to improve the quality of urban transit services. Rather than attempting to coordinate and rational-

ize transit and automobile use on urban roads, transit planners have concentrated their efforts on the construction of expensive, capital-intensive, and often redundant rail rapid transit systems.

Improvements can be obtained, however, from controls on city streets and highways, particularly in high-density central areas. Limited actions of this kind have already been tried in some downtown areas with the provision of special bus lanes and other measures. Various licensing arrangements, tolls for entering the downtown area and other measures have been proposed, but seldom used or even tried. Bridges, tunnels, toll gates, on-street and off-street parking, and the like, provide still other opportunities to manage existing systems and to bring user charges and the costs of providing expensive peak hour road capacity more closely into line. These and more sophisticated control devices, such as computer-operated signal and control systems, provide opportunities for improving the quality of both private and public urban transportation services.

PRICING TRANSPORTATION SYSTEMS

The charges for using each transportation mode only partially reflect the differences in urban environments, cost of road space and the demand for urban road space. All too frequently they bear little or no relation to the full, long-run costs of providing these urban transportation services. As might be expected, these low charges result in overuse of these facilities during peak hours and in high levels of congestion. This is certainly true in the metropolitan areas of the United States and probably applies to any of the world's major metropolitan areas. Moreover, the discrepancy between the payments by peak hour automobile users for urban roads and the full costs of providing them are almost certainly greatest

---

[14]Ernest Davies, *Transport in Greater London* (London, London School of Economics, 1962), p. 9.

at those urban locations where urban transit operations are now most significant, where the potential market for commuter transportation is greatest, and where road transit is most adversely affected by private automobiles. Many current automobile users would not use their automobiles for commuting to central areas during peak hours if they were required to pay the full costs of urban road facilities. If significant numbers were so discouraged, congestion levels would decline, vehicular speeds would increase and the reliability and efficiency of road transit would improve markedly. Of course, urban motorists might be willing to pay these higher costs for peak hour use of urban roads in high-density urban areas. If so, this would be a strong argument in favour of increased construction of urban roads. Current practice provides no basis for resolving this most crucial issue of public policy.

The foregoing observations suggest that a major goal of public policy for metropolitan transportation should be to ensure that the cost of urban travel by alternative modes reflects as nearly as possible the full costs of providing these services. Moreover, if this is not done, there will be a serious impact on competing modes using the same rights of way. Short-run urban transportation policy and planning should focus on those circumstances where the most serious imbalances between charge to users and costs of facilities exist, and especially where these affect the operating efficiency of competing modes. Thus, in high-density urban areas policy should emphasize an improved, more efficient use of underpriced road capacity.

Road pricing has long seemed attractive as a means of achieving the desired discipline of urban road facilities. It appears to have the virtue of both providing a means of rationing existing scarce street space

and of providing a means for determining the appropriate level of urban road investment. The suggested test is the same as that prevailing generally in market economies: the supply of road capacity should be expanded so long as users are willing to pay the full incremental costs of these increases in capacity. Generally, monopoly profits would be earned in the short run until the capacity could be expanded, and the argument is that these be used to expand capacity until the monopoly profits disappear.

Although there has been considerable interest in the use of road pricing schemes of various kinds in the United States, an even greater interest currently exists in the United Kingdom where there has been much discussion of road pricing in recent years. Beesley and Roth provide a detailed discussion of several means of restraint or disciplining highways. Their suggested methods include restraint by administrative edict, restraint by price, and restraint by parking regulation. Beesley and Roth appear to favour strongly restraint by price.[15]

Road pricing schemes differ widely in complexity, sophistication and sensitivity. More complex and sophisticated systems would better provide for fluctuations in demand and benefit to specific users at particular times of day, but administrative costs would be relatively high. The simpler the pricing scheme becomes, the more it begins to resemble various physical controls in its operation, system effects and incidence. The choice of rationing schemes appears to depend principally on technical considerations. More sophisticated road pricing schemes would have higher initial and higher operating costs. In many

---

[15]M. E. Beesley and G. J. Roth, "Restraint of Traffic in Congested Areas," *Town Planning Review*, Vol. 33, No. 3 (October, 1962), p. 9.

instances the costs of highly sophisticated systems may make elaborate road pricing schemes inefficient. When this is true, the alternatives are either the use of physical controls or the expansion of transport capacity until the desired levels of congestion are achieved. The latter practice is the one that has been followed in the United States for the most part. Despite many assertions to the contrary, levels of congestion in most American cities have declined, travel speeds have increased and the duration of the morning and evening peaks has been reduced. In cities such as St. Louis, Denver and Phoenix, congestion has reached low levels already (except for some very localized conditions). It appears very little is to be gained from either elaborate control systems or sophisticated road-pricing systems.

Thus the feasibility of sophisticated road-pricing systems ultimately depends on technological and cost considerations. Many people have argued that a sophisticated road pricing system could be developed at a cost low enough to justify its widespread use. Beesley and Roth identify two kinds of devices—self-contained recording units *on* vehicles (these can be compared to taxi meters); and remote control units actuated *by* the vehicles but situated at a central exchange (these can be compared to telephone meters). They further comment on these systems as follows:

Although investigations into the costs and benefits of different systems are being carried out, not enough is known for any definite recommendations to be made. The requirements of the ideal system are fairly easy to specify: it must allow for variation in charges over a wide range, so as to relate charges to road use in an equitable manner. It must be cheap to operate, easy to understand and yet difficult to evade by fraud. Avoidance of centralized billing would be an advantage, as would be the possibility of paying in advance in small amounts.[16]

CONCLUSION

The foregoing presents some of the possibilities for improving our transportation systems in metropolitan areas. These opportunities generally will be realized only when planning for public transit, highway and other kinds of urban transportation are coordinated under a single agency with a broadened mandate for coordinated urban transportation planning. There is not much hope for improving urban transportation as long as the present situation exists —where road planning is thought of as something different from transit planning and policy, and both are considered distinct from urban and metropolitan planning.

---

[16]*Ibid.*, p. 11.

3 Utilities in Metropolitan Areas

A. H. Hanson

The broad scope of municipal involvement
in the provision of public utilities and the
wide variations found in the degree of involvement bring to the fore the problems
of discussing this issue in meaningful terms.
To reduce this problem to manageable proportions, only three services—water, transportation and electricity–are discussed here
and only in terms of ownership, economic
and financial inputs and geographic extension. To ensure that this will not lead to too
confined a conception of municipal involvement in the field of public enterprise
it is appropriate to commence with a more
general review of the scope of municipal
enterprises.

The Scope of Municipal Enterprise

It might be suggested that the more economically advanced the country the more
important is municipal enterprise in its
economic system and the greater is the
variety of the municipal enterprises that it
displays. There is, however, no hard and
fast correlation between economic development and municipal trading. Much depends on national tradition. If, among the
advanced countries, the widest range of
municipal enterprises is to be found in Germany and Austria, the probable reason is
that both the strength of municipal patriotism and the nature of the legal system have

facilitated their creation, expansion and diversification.[1]

At the other end of the scale are Laos, Vietnam and Northern Nigeria. Laos, perhaps the least advanced of the three, has no municipal enterprises at all,[2] Vietnam has only two, both in Saigon.[3] Northern Nigeria made what was apparently a premature experiment by giving the native authorities responsibility for water undertakings. These bodies lacked both technical efficiency and financial strength, with the result that the regional government was compelled to nationalize the water services in all but two of the localities.

However, it is not possible to draw a positive correlation between municipal enterprise and economic development. Thailand and the United States, for example, destroy the neatness of the picture. Although Thailand is a developing country and its local government institutions are not strong, it has a considerable variety of municipal enterprises. The United States has many more but, although economically the most advanced country in the world, it still leaves a rather high proportion of its municipal utilities in the hands of concessionaires and is much less distinguished than, say, Germany and Austria or even Britain and France for the entrepreneurial initiative of its city councils. The reason for the United States' "backwardness" in this field is obvious and well known. The country is distinguished for the buoyancy, self-confidence and achievement of its predominantly private enterprise system, and there is a strong feeling among most of its people that public undertakings, whether municipal, state or federal, should be created only when private enterprise has failed to provide, or failed to provide satisfactorily, a service that is generally considered essential. Elsewhere, however, such comparatively simple explanations of exceptional cases are not so readily available, and much further study is needed to discover precisely why a given country has a given level of municipal enterprise.

LEGAL CATEGORIES

Whether municipal enterprises are frequent or rare, they must be run in a business-like way. The difficulty is to reconcile the requirements of business freedom with those of public control. This dilemma has been fully documented with regard to nationalized industries, but comparatively little study has been given to it at the municipal level, where it is only slightly less acute. The whole subject is often discussed from a

---

[1] In Austria, according to a recent enquiry, the 119 municipalities with more than 5,000 inhabitants run a total of some 870 municipal enterprises. These municipalities account for half the population of the country. The enterprises include gas companies; water companies; electric power companies; tram, trolley and omnibus services; haulage services; funeral undertakings; bathing establishments; market gardens, farms, forestry enterprises; slaughterhouses; building and building requirements enterprises; woodworking shops; chemical companies; food and drink manufacturing companies; catering establishments; publishing houses; weighbridges; cable railways and lifts; ports, shipping and ferries; pharmaceutical enterprises; technical research institutes; wine-producing companies; brickyards; travel agencies; credit, finance and insurance undertakings; laundries; mineral baths and hydropathic establishments; theatres and cinemas; and airport companies. This list is not exhaustive.

[2] Vientiane, the capital, is supplied with electricity by La Société d'Electricité du Laos, a state enterprise, and with water by La Société Centrale des Eaux du Laos, a "mixed" company. Although the municipality of Vientiane has shares in both enterprises, it is not, at least according to the relevant decrees, represented on their Conseils d'Administration.

[3] In South Vietnam, transport, electricity and water services are mostly publicly owned, but run by the national, not the municipal, authorities.

strictly legal point of view, the supposition being that once the correct legal form for the enterprise has been discovered, the correct *modus operandi* will almost automatically follow. This is a complete fallacy, there being no observable positive correlation between the "correctness" of the enterprise's legal form and the quality of its performance. Nevertheless, the legal category within which a municipal enterprise is placed is important, for the enterprise will thereby receive certain opportunities, of which it may or may not take advantage, and become subject to certain limitations, from which it may not find escape easy.

In broad terms, there are two ways of constituting a municipal enterprise. The first is to set it up as a department of municipal government, subject, like other departments, to administrative control by the mayor, the city manager or the relevant committee, and to ultimate policy control by the council. Such an enterprise may be either self-contained or so mixed up with the rest of municipal administration as to lack specific identity. As a rule, the bigger enterprises in the bigger municipalities are more or less self-contained, whereas the smaller ones in the smaller municipalities are heavily dependent on services provided them by other municipal departments such as finance, personnel, legal and stores. When an enterprise is set up as a department, normal department structure may be modified to facilitate the commercial functions of the enterprise. For example, a separate, commercial-type budget or a distinct personnel system may be set up. Delegation of authority to the general manager may also be considerably wider than that usually made to other top-level officials. The extent of such modifications depends partly upon the intelligence with which the municipality approaches the task of setting up a municipal enterprise, but also upon what the law relating to municipal enterprise permits it to do.

The second method is to establish a specialized agency or *ad hoc* authority, with a distinct legal personality (separate from that of the municipal body) and a considerable measure of operational autonomy. Such an agency may be given the form of a joint-stock company, in which the municipality (or municipalities) concerned hold all or most of the shares; or it may be constituted by virtue of special legislation or (in countries which have "divided" legal systems) under certain rules of public law. Whatever the precise legal device employed, the purpose of the exercise is clear: to endow the enterprise with autonomous powers calculated to enable it to perform its commercial duties with the greatest possible efficiency while at the same time preserving ultimate public control over the policies that it pursues.

The alternatives have obvious parallels in national enterprise. The first corresponds with a departmentalized undertaking, staffed by civil servants and wholly under the control of the relevant minister; the second with a public corporation or state company. In national enterprise, however, the search for suitably autonomous forms of enterprise has generally proceeded with more vigour than it has in municipal enterprise. With certain outstanding exceptions, municipalities have been reluctant to abandon the close conciliar control which characterizes the enterprise forming an integral part of the municipal administration. This reluctance may be partly due to conservatism or inertia and partly to the fact that, in general, municipal enterprises are smaller, simpler and easier to run than national enterprises. But it also reflects a belief in the virtues of local democracy and a tendency to emphasize the "public service" as distinct from the commercial element in the management of local utilities. Whichever of these factors predominates in particular cases, local enterprises are usually integrated as much as possible with the municipal administration even when the law permits and perhaps encourages more autonomous forms of organization. Of the countries reporting for this survey, the following have kept a majority of their municipal enterprises in the former of the two classes defined above: Austria, Brazil, Ceylon, India, Italy, Japan, New Zealand, the Scandinavian countries, South Africa, Spain, Thailand and the United Kingdom.

The use of other forms of enterprise, however, is quite common and in certain countries becoming increasingly so. The joint-stock company or the partnership, for instance, proves a particularly useful device when a municipality wishes to join with other municipalities in the establishment of an intermunicipal enterprise or to enlist the participation of private or cooperative capital. It can also be used (sometimes at the cost of placing a rather strained interpretation on the company law) for enterprises wholly owned by a single municipality. In the latter case its main justification is the greater managerial flexibility that it may afford, especially when the laws regulating the integrated forms of municipal enterprise are excessively restrictive. In countries where corruption and nepotism have become a feature of municipal administration, the adoption of the company form may help keep local politicians at arm's length from the undertaking.

This last justification, however, should not be regarded as applicable wherever the company form has achieved a certain popularity; for this type of organization is widely used in certain countries, such as Norway and the Netherlands, where municipal administration is exceptionally "clean". In the Netherlands the joint-stock company is used when priority has been given to managerial flexibilty and the public-law form of enterprise when self-administration and democratic control are regarded as of paramount importance.

A further advantage of the company form is that it enables the municipalities to extricate themselves from certain legal rigidities in which they would otherwise be involved, for instance, when they wish to raise a loan. In Sweden, the joint-stock company is sometimes used as a means of avoiding the rule that all the transactions of Swedish government offices (with certain exceptions) must be open to inspection by the public; but this country's municipalities have shown a stronger preference for another type of *ad hoc* agency, known as the foundation. This form of enterprise enables municipalities to enter into association with private industries, trade unions, cooperatives and building societies and at the same time to avoid not only the rules of publicity but those of the Companies Act.

The company is not the only alternative to the integrated enterprise. Indeed, throughout the world municipal enterprise has become subject to a considerable variety of special legal forms, each one aimed at separating the undertaking, to some degree, from the rest of municipal administration and at giving it a measure of operational independence. The most familiar of these is the French *régie*, which has been widely imitated, particularly in those countries which were formerly ruled by France or which have been strongly influenced by French institutions. According to French law, public services of an industrial or commercial character may be provided either by the local authority itself or by way of concession from the local authority to a private enterprise. If the authority chooses the former, it may establish a *régie municipale,* a form of undertaking subject to well-defined rules, thus summarized by Brian Chapman:

> It must have the prior approval of a central authority, usually the *Conseil d'Etat.* Three-quarters of its board of administration are appointed by the Mayor and the remainder by the Prefect, and the director of the *régie* is appointed by the Mayor with the Prefect's approval. The Mayor represents the *régie* in legal actions, and the *conseil municipal* must approve major questions of policy and pass its accounts.[4]

The *régie* may or may not have separate legal personality, according to whether it is of the "type 1926" or the "type 1955". It is administered by the local authority's own staff, under the control of the mayor; it has its own budget, separate from the municipal budget, and is required to balance revenues and expenditures.

Municipal enterprises of the *régie* kind, reproducing some but not all of the characteristics of the French prototype, are found in many countries. In Italy they are known as *enterprises municipalisées*; in Spain, as public establishments; in Brazil, as autarchies. Frequently, the degree of legal autonomy they enjoy is not accompanied by a corresponding degree of operational autonomy, either because the municipal authorities are jealous of any signs of independence, or because the tutelary authorities exercise powers of approval and direction too persistently and too fiercely.

Another specialized and semi-autonomous form of municipal enterprise is the public corporation. This is more frequently used at the national and regional levels than at the local level. There is some difficulty, moreover, in defining the term public corporation in a way that will be internationally accepted, and one is often confronted with corporation-type bodies which are difficult to categorise. Perhaps the clearest example of the public corporation in local government is the French *établissement public municipal*, more usually known as the *office municipal*. This is run by a board that includes representatives of the participating local authority or authorities, and has not only a separate budget but a separate legal personality. It is used mainly for quasi-commercial services in the fields of housing and health. Corporations at the municipal level are also reported from Japan (under the Local Housing Supply Corporation Law) and from Iran (Teheran United Bus Corporation and Teheran Regional Power Station); but in default of further information it is difficult to say precisely how corporate these organizations are. In Yugoslavia, the unique self-administered enterprises founded by the communes to provide their citizens with public utility services might conceivably be termed public corporations without stretching the meaning of the term too far. Elsewhere, as in India and Finland, corporation-type bodies are sometimes used for enterprises in which adjoining local authorities collaborate. An instance from India is the recently formed Calcutta Metropolitan Water and Sanitation Authority.

As might be expected, intermunicipal enterprises, which have become increasingly common as changes in technique have made existing municipal boundaries look increasingly arbitrary, display the greatest variety of legal forms. Sometimes, as we have already noted, the public corporation is used, but more frequently the joint-stock

---

[4]Brian Chapman, *Introduction to French Local Government* (London, George Allen and Unwin, Ltd., 1953), p. 47.

company (with the adjoining local authorities as shareholders) has proved convenient. Such intermunicipal companies are found in countries such as Austria, Belgium, Denmark, Japan, the Netherlands and Sweden.

In some countries, such as France, the practice has developed of forming joint-stock companies (particularly for purposes of regional development) in which private parties and the central government, as well as municipalities, participate. One of the earliest examples of these was the *Cie. Nationale du Rhône*, created, by the Law of May 21st, 1921, with the object of controlling the waters of the Rhône from the Swiss frontier for purposes of providing electrical energy, facilitating navigation and irrigating the Comptat Venaissin and Languedoc. Among its various shareholders were—and are—the *communes* and *départements* of the Rhône valley, those of the Paris region and the Municipality of Paris itself. More recent bodies of this type are organized under the terms of the Law of May 24th, 1951. Their establishment requires a

> decree in *Conseil d'Etat* on the report of the minister responsible for housing and reconstruction, of the Minister of Finance, Economic Affairs and the Plan and of the Minister of the Interior. . . , after the *Comité nationale d'Urbanisme* has given its opinion.[5]

Mixed companies for smaller-scale projects (such as a coach station or a low-cost housing enterprise) may attract municipal

participation under the terms of the Laws of March 20th, 1951.

In the Scandinavian countries, however, the law relating to cooperative societies is frequently preferred to the joint-stock company law as a basis for collaboration.

Many forms of intermunicipal enterprise fall right outside the categories into which public enterprise is conventionally divided. In Britain, for instance, there is the joint board, which can be used for the provision of any municipal service, non-commercial as well as commercial, and which in certain defined cases can be imposed upon local government authorities by the ministry concerned. Frequently charged with the duty of managing a water or transport undertaking, the joint board is a "strong" form of intermunicipal cooperation. As the leading authority on English local government law says, joint boards

> are entities altogether separate and distinct from the authorities which appoint their members. They are almost invariably bodies corporate with a perpetual existence which does not permit of their dissolution merely by the withdrawal of one of the appointing authorities from further participation in the scheme they control. Their powers and duties are in consequence determined once and for all at their creation, and cannot be revoked or altered merely at the wishes of one or more of the appointing authorities. . . . Joint boards . . . are their own financial masters and obtain the moneys they need by issuing peremptory precepts on the authorities which appoint their members.[6]

Comparable to the English joint board is

---

[5]Margerita Rendel, "How the *Conseil d'Etat* Supervises Local Authorities," *Public Law* (London, Autumn, 1966), pp. 226-30. This excellent article is a particularly useful source of information about municipal enterprise in modern France.

[6]*Hart's Introduction to the Law of Local Government and Administration*, 6th ed. (London, 1957), p. 138.

the French *syndicat intercommunal*. A *syndicat*

> is formed when two or more Communes agree to associate together to provide a particular service. . . . Once formed, the *syndicat* acquires the status of a corporate personality with its own property and revenue, and it can sue and be sued in its own name. The general direction of its affairs is in the hands of an administrative committee, composed of two delegates elected by each *conseil municipal*, regardless of size, making up the *syndicat*. There is no restriction as to who should be the communal representatives, but one is generally the Mayor.[7]

A *syndicat* can be formed to organize any service lying within the jurisdiction of the participating local authorities. Most frequently it is a means by which small, rural and individually weak communes collaborate to supply themselves with water or transport.

Forms of intermunicipal enterprise comparable with the English joint board and the French *syndicat intercommunal* are found in many countries. In Italy, for instance, there are the *entreprises municipalisées consortiales*, governed by the Municipal Law of 1925; in Germany there are the associations, governed by the Administrative Association Law; in the Netherlands, the public law joint boards; in Sweden, the federations.

Belgium displays a unique form of intermunicipal cooperation for the management of public enterprises. In that country there is a considerable number of *personnes morales de droit public*, some national, others regional, which enjoy the participation of, but do not fall exclusively under the control of, the local authorities. These, created by special law or royal decree, are of the public corporation type, but they are not local public corporations nor even intermunicipal corporations in the ordinary sense—yet no account of municipal enterprise in Belgium which omitted them could pretend to be complete. They possess some of the characteristics of public corporations, of joint-stock companies and of cooperative societies.

Finally, one must mention the existence of enterprises which, although local, are not part of the general structure of local government and enjoy independent existence under boards of directors elected to office by those who consume the services provided. In the United States, local water, power and transport undertakings are sometimes organized in this way.

From the above examples, it will be readily seem that any trans-national legal classification of municipal enterprises would constitute a very difficult undertaking, so great is the variety of forms and so ambiguous the status of some of the enterprises concerned. However, exactness of legal classification is not of major importance, since we are primarily concerned with the manner in which municipal enterprises operate. This, it is true, is affected by their legal form, but certainly not determined by it. Studies of national enterprise show that there is no correlation between organizational types and operational efficiency. International experience, as suggested previously,

> goes to show that whatever legal immunities are granted to a public enterprise, ministers and parliamentarians who are determined to violate them will usually find ways and means of doing so. On the other hand, even a throughly bad law can be less than disastrous to the

---

[7]Chapman, *op. cit.*, pp. 50-51.

performance of the public enterprise if its provisions are liberally interpreted and sensibly applied.[8]

Far more depends on the quality of the personnel running the enterprise than on the wisdom of the law constituting it. This is fully and frankly admitted in the Brazilian report, and accounts for the failure of the new and improved legal form—the autarchy—to improve the performance of the municipal enterprises that were brought within it:

> It was thought at first that the administrative autonomy granted to the autarchies would be a remedy against the spoils system and other defects that afflict the directly-run utilities, such as their total subordination to administrative red tape and strict adherence to regulations.

> It was soon found out, however, that the forces of the spoils system in a developing society were strong enough to penetrate the supposed shield of autarchy, for it began to show the same distortions as direct administration, often in a much more acute form. The unwillingness of high government officials (mayors and councilmen) to relinquish full control of operations to the autarchies (a resistence which is part and parcel of the spoils system syndrome) also annulled some of the basic principles of the autarchy concept, which is liberation from administrative red tape and strict political control.

One should not, of course, go to the opposite extreme and say that the law is irrelevant to the performance of municipal enterprises. On the contrary, it is very relevant, but not decisive.

CHANGES IN LEGAL STRUCTURE

The structure of municipal enterprise has remained remarkably stable in many countries for a long period. No significant changes have been experienced in such countries as Austria, Belgium, Ceylon, India, New Zealand, South Africa, Thailand and the United Kingdom over the last twenty years. Elsewhere, the characteristic tendency is toward more flexible forms of organization. Norway, for instance, has made Bergen's municipal housing enterprise into a joint-stock company and is considering the same form of organization for passenger transport in both Bergen and Oslo. Sweden reports increasing use of both the joint-stock company and the foundation. In Denmark, growth in the popularity of the joint-stock company is associated with the amalgamation of enterprises, particularly in electricity.

In Italy, the use of the joint-stock company for metros, airports, seaports and electricity has been provoked by the unsatisfactory condition of the Municipal Law. According to the Italian report, the need to simplify and rationalize the law's provisions regarding municipal enterprise has been frequently urged, but no action has been taken; hence the municipalities have taken every opportunity to escape its restrictions. In Brazil, the failure of the autarchy to facilitate much-needed improvements in the management of municipal enterprises has caused reformers to seek a solution through the mixed company, municipally owned but with symbolic participation by private capital in order to meet the requirements of the commercial law.

---

[8]*Government Organiaztion and Economic Development*, Organization for Economic and Cultural Development (Paris, 1966), p. 121. To adapt this quotation to municipal enterprises substitute "mayors and councillors" for "ministers and parliamentarians".

In France, too, the search for flexibility has led away from the *régie* and the *syndicat*, and toward the joint-stock company, usually in its mixed form which permits the company to raise its own loans and have access to private as well as public capital.

Japan has sought greater flexibility through reform of the municipal law itself. The Municipally Owned Public Utilities Law of 1952 facilitated the adoption of more business-like methods by local enterprises. In Iran, likewise, there is a new Municipal Law that authorizes the municipalities to establish independent organizations for water, power, transport and the like if they wish to do so.

On the other hand, there are a few instances of movement in the opposite direction. In Germany, for example, for unspecified managerial, organizational, financial and fiscal reasons there is a trend away from the more autonomous type of municipal enterprise, while in the United States there appears to be less enthusiasm than formerly for independent commissions and committees. In one case—Finland—reversion to the more integrated type of enterprise has occurred for a special reason, quite unconnected with considerations of managerial efficiency or democratic control. Finnish municipalities have shown a tendency to abandon the joint-stock company, not because it has failed to give satisfaction, but because it is subject to both state and municipal taxation, whereas the traditional form of municipal enterprise is free from both. One may reasonably suggest that if municipalities feel compelled to adjust the forms of their enterprises to the peculiarities of the system of taxation, it is the latter that is in need of reform.

Changes in this direction, however, must be regarded as atypical. Most structural reforms, in most countries, have been in the direction of greater autonomy. These are not solely the product of the search for greater managerial flexibility. They have also been made necessary, particularly in some of the more developed countries, by the need to create local enterprises with areas of jurisdiction wider than those delimited by municipal boundaries. The inadequacy of the size of the average local government unit has in some cases provoked the nationalization of services previously under municipal ownership; but in others it has stimulated the expansion of intermunicipal entities, of the joint-stock company, public corporation or special type, and the development of new institutional forms of collaboration between national and municipal authorities. Sometimes, as in France, such collaboration is largely fictitious. In theory, the nationalized *Electricité de France* is, in respect of its distribution services, the concessionary of the local authorities; in practice, distribution is managed, albeit with the advice of the local authorities concerned, by a *direction* in Paris which operates through regions, departments, subdivisions and districts. It was originally intended to establish regional boards on which local government authorities would have representatives; but this form of decentralization, although actually provided for by the electricity and gas nationalization law of 1946, never came into operation, much to the discontent of the local authorities.[9]

LEGAL INSTRUMENTS

As municipalities are generally subordinate governmental agencies which enjoy only such authority as has been granted to them

---

[9]See, for instance, *Local Government in the XXth Century*, International Union of Local Authorities (The Hague, Martinus Nijhoff, 1963), pp. 137-39.

by the law, they are necessarily restricted both as to the kinds of enterprises they may create and as to the manner in which they may create them. In these matters, a broad distinction exists between two groups of countries.

The first group is characterized by the conferment on local government authorities of a general power to create municipal enterprises, limited only by the conception of the public interest. How significant such a general power may be depends first on the specific sanctions (particularly those relating to loan raising and property purchase) which are applied to, and second on the width or narrowness of the interpretation given to public interest by superior administrative bodies and by the courts of law.

The second group is characterized by the lack of any general power of creation. In such countries a municipal enterprise may exist only by virtue of specific legislative authority. The municipality will then know, by consulting the relevant statutes, precisely which enterprises it is permitted and which it is forbidden to promote. The same statutes will prescribe a procedure that must be followed in the process of creation. In England, for instance, the local authority is in some cases obliged to organize a preliminary consultation of the electorate (by way of a town meeting and, if demanded, a ballot) and is always required to convince the relevant ministerial body, and ultimately Parliament itself, of the necessity and viability of the proposed enterprise.

Critics of the restrictions imposed on English local authorities have sometimes expressed admiration of the relative freedom enjoyed by their counterparts in France, by virtue of generality of the powers existing in that country. In practice, however, the differences between the two systems are much less than they at first appear. In each, although it is the local authority that normally takes the initiative, it is a national authority that decides whether and on what conditions that initiative shall be allowed to come to fruition. In England, the decision is in the hands of the ministry and the legislature, both of which employ quasi-judicial procedures. They are certainly bound by legislation in a way that does not apply in France; but on the other side of the Channel the manner of interpreting public interest has an equally—and until recently had a more—restrictive effect. Such interpretation is ultimately in the hands of the *Conseil d'Etat*, which, at least in pre-war days, gave public interest so narrow a meaning that municipal undertakings were in fact rarer in France than in England.[10] Since then, the modification and liberalization of the *Conseil's* former doctrines have permitted a considerable efflorescence of communal enterprise. In the old days, the *Conseil* would permit a local authority to undertake a service only if the authority could prove that private enterprise had failed to provide it, or that provision by private enterprise had been inadequate or inefficient, or that the service was run as a monopoly or that it ought to be run as a monopoly. During the period of the Fourth Republic, however, the concept of public welfare applied by the *Conseil* became so wide that communal authorities regarded themselves as having a general warrant to intervene whenever they thought that the needs of their inhabitants were not being met. In addition, specific legislation authorized them to establish services such as abattoirs, *pompes funèbres*, low-cost housing and restaurants.

Of the countries that accorded general

---

[10]See Felix, *L'Activité économique de la Commune* (Paris, 1932).

powers of enterprise-creation to their municipalities, it was pre-Nazi Germany that succeeded in conferring the widest measure of practical freedom. The only formal limitations were an obligation to give financial priority to compulsory services and the confining of the enterprise to the area under the municipality's jurisdiction. Local authorities, under this regime, became internationally famous for the variety of trading services that they undertook—milk-supply, restaurants, theatres, concert halls, pawnshops, breweries, bakeries, banks, hotels and so on. Restrictions on this proliferation were first imposed by the Local Government Order of 1935, which stated:

> The municipality may establish or substantially extend economic undertakings only (1) if the undertaking is justified by public ends; (2) if the undertaking stands in proper relationship to the capacity of the municipality to operate and to foreseeable needs; (3) if the object is one that cannot be better and more economically satisfied by some other agency.[11]

These restrictions still apply. For the Order, expressing a bias toward private enterprise, is in close accord with the economic policies of the present federal government, and is not specifically "Nazi".

By way of generalization, one may say that in most countries, whether they belong to the first or to the second of our original groupings, the following is something like the standard procedure, prescribed by law, for the creation of a municipal enterprise: (a) the passing of a resolution by the municipal council (with or without consultation of local public opinion; (b) the

[11]H. Finer, *Municipal Trading, A Study in Public Administration* (London, George Allen & Unwin, Ltd., 1941), p. 69.

approval of this resolution by higher administrative authority and in some cases by the legislature; (c) the granting of permission by similar authority for the raising of the necessary loan funds (on prescribed conditions) and for any necessary invasion of private rights (for example, compulsory purchase and rights of way).

The law, as we have seen, not only prescribes the kind of municipal enterprise that may be created, together with its manner of creation, but also governs the *modus operandi* of those municipal enterprises already in existence. In this connection, one must distinguish between laws of three different levels of generality: (1) a law that regulates one particular enterprise, such as the Calcutta Tramways Act in India or the Metropolitan Water Board Act in England; (2) a law that regulates a particular class of enterprise, either generally, such as South Africa's Water Act, or in certain particulars, such as the same country's Motor Carrier Transportation Act; and (3) a law that applies to all classes of municipal enterprise, whether the regulation of such enterprise is its specific purpose (as with the Japanese Municipally Owned Public Utilities Law) or is incidental to a wider purpose, such as prescribing the structure and defining the power of local authorities (as with the English Local Government Act of 1933).

Among the countries surveyed there seems to be no doubt that general legislation of some kind is necessary, in order to secure a measure of regularity, uniformity and probity, although in some countries such laws are thought to go too far into matters of detail, over-meticulously prescribing forms of organization and procedure. However, much depends on the intelligence with which the relevant provisions are conceived. The Municipal Law of the Netherlands, for instance, contains a number of rather detailed provisions about such matters as budgeting, reporting and auditing; yet the reporter of that country registered no complaint against it. Nevertheless, one may reasonably suggest that municipal laws should err, if at all, on the side of liberality and flexibility, for the manner of organizing a municipal enterprise, as indeed of any other municipal service, will have to change as expansion takes place and new techniques are discovered.

Consequently, legal regulations can quickly lose their originally rational justification and become old-fashioned and restrictive. The required amendments can be made of course, but only if the legislature has the time and the energy to give such matters its attention. If it fails to do so, municipalities (as in Italy) will tend to seek escape by giving their enterprises legal forms to which the municipal law does not apply. While this is obviously far better than passively waiting for long-delayed legislative improvements, it does tend to reduce the law to a state of incoherence and to give the national authorities an excuse to substitute administrative controls for legislative ones that are becoming ineffective.

The same considerations apply, *a fortiori*, to laws that regulate a specific class of enterprise. There are several reasons for the abundance of such laws, although some reasons are little more than excuses. That the state should attempt to determine in such detail the manner in which a local council provides services mainly for the benefit of the people who have elected it, and to whom it is therefore responsible, certainly betokens a certain lack of trust in the discretion of local authorities and in their fitness to govern even the limited areas that have been placed under their jurisdiction.

There are, however, valid reasons for the state's legislative expression of its anxiety about the quality of local administration. Everyone is bound to recognize, nowadays, that the running of a municipal enterprise is of interest to those who live outside the municipality as well as to those who live within it. Visitors as well as citizens can be poisoned by impure water, and long-distance motorists are not less frustrated by bad traffic planning than local ones. Moreover, there is an increasingly strong public demand for a certain uniformity in the quality of basic services throughout the different areas into which a country has been rather arbitrarily divided for local government purposes—a demand which intensifies as people become more mobile geographically. Legal regulation there must be, and the problem is to find a suitable point of balance between national uniformity and local diversity—one that does not make nonsense of the whole conception of local self-government.

Within the tendency toward stricter central control, which seems almost universal, the factors which affect the liberality or restrictiveness of the laws regulating municipal enterprise are the traditions of the country concerned, the nature of its legal system and the extent to which its legislators, supported by public opinion, are committed to the establishment and the strengthening of the more decentralized forms of public administration.

The greatest freedom from restrictive legal provisions appears to be enjoyed by the municipalities of the Scandinavian countries (where, significantly enough, they are endowed with local sources of finance more adequate than those enjoyed by municipalities in most other countries). In France and Italy, on the other hand, the relevant laws are drafted tightly enough to give rise to considerable frustration among councillors and officials and bitter complaint from people who place a high value on municipal freedom. These two countries are specifically mentioned because (unlike many other countries, particularly in the developing areas of the world, where centralizing tendencies are also strong) they possess local authorities competent and politically conscious enough to resent what seems to them the lack of adequate freedom. They are also countries whose legal systems embody the principle of administrative *tutelle*. It is by virtue of this overriding principle that so many of the discretionary powers conferred on local authorities are subjected to the veto of higher authorities.

Yet it must be admitted that *tutelle* is not the necessary condition for such subordination, which is hardly less complete in, for instance, the United Kingdom, where no such principle exists. England, indeed, achieved very much the same result through the conferment of powers of approval and direction on ministries by specific legislation, together with the considerable dependence of local authorities on central government grants. To understand the manner in which responsibility for municipal enterprises is shared between central and local government, we must therefore leave the formal provisions of the law and consider the actual behaviour of the national administrative authorities in the exercise of the powers they derive from law, convention and custom.

ADMINISTRATIVE SUPERVISION

In considering how the national administrative authorities supervise plans and operations of municipally owned public utilities, one has to take into account the formal range of powers possessed by such authorities; the extent to which and the way in which such powers are actually used; and the extra-legal influence that these authorities exert by virtue of their superior knowledge, their closer proximity to the major centres of political decision and the financial sanctions they may have at their disposal.

In almost every case of intervention by the national administrative authorities, all three classes of factors are inseparably intertwined. Take, for instance, a proposal by an English local authority for the creation of a new municipal enterprise. The procedure laid down by the law may involve the promulgation of a Provisional Order or a Special Order by the relevant ministry. This means that the municipality has to convince the civil servants who advise the minister of the necessity and good sense of what it has resolved to do. There will almost certainly be a formal local inquiry, conducted by the minister's inspector, but more important than this will be the long and detailed technical and financial discussions between local officials and national officials. In such discussions, the national officials will necessarily endeavour to bring the local proposals into conformity with current government policy, but may also consider the possible political repercussions, both national and local, of a too rigid insistence on the government's point of view.

The local authority, on its side, will be aware of the dangers of a quarrel with the ministry, the complications of which extend far beyond the ministry's possible refusal to promulgate the Order under discussion. If the Order is refused, or if the procedure by way of Order is not available for the particular enterprise concerned, the municipality may have the alternative of proceeding by way of private bill, a complicated, time-consuming and rather expensive method. In matter of strict law, it does not then have to convince the ministry of the virtues of its proposal, but only Parliament itself. Such a bill, however, has to pass through the usual stages in both Houses, and consequently can be killed if the government, secure in its party majority, considers it contrary to the public interest. Hence the promoting authority will be no less anxious to obtain ministerial approval and the preliminary negotiations between national and local officials will be no less complicated and no less vital.

In France, the legal sanctions employed, though different, are no less severe. If a *commune* wishes to organize a service *en régie*, it must obtain the approval—which will not be lightly granted—of the relevant tutelary authority. In some cases recourse must be had to the Minister of the Interior, Minister of Finance and *Conseil d'Etat*. Tutelary permission is also required for the creation of a *syndicat*, and here again the *Conseil d'Etat* may have to be consulted. The legal hurdles that have to be jumped before a mixed company can be brought into existence have already been referred to briefly. A large mixed company with regional responsibilities usually requires the assent and cooperation of several different ministries. If there is disagreement among them, and if the *Conseil d'Etat* is unable to resolve such disagreement, then no action can be taken.

It is true that in England and France formal ministerial powers are considerably greater than they are in some other countries. Consequently, the sanctions that hover over the financial-technical discussions are severe. But even in countries where local authorities appear to have much greater freedom to create enterprises, the *de facto* situation is often not dissimilar; for, as we have noted, the raising of capital from external sources (which is usually necessary) almost invariably requires the approval of higher authority, and

this in itself is a sufficiently powerful sanction to enable the relevant ministry to modify the local authority's plans as it considers technically or financially necessary and politically feasible. Whatever the state of the law, the national authorities will usually ensure that a project is, from their point of view, soundly conceived. The local authority, after all, is proposing to use scarce national resources for a public purpose, and in doing so it may be risking the government's own financial credit. Such considerations cannot be neglected by any government, and they necessarily imply a certain measure of administrative supervision.

Whether the government is equally justified, when vetting proposals, in adopting the role of the consumer's protector, as it so often does, is much more open to question. Does the local consumer require protection against schemes promulgated by the councillors he himself has elected? Is not local government itself a form of consumers' democracy? These questions are real. On the one hand, it can be argued that if councillors are stupid, ignorant or corrupt the electorate has its remedy at the next election. On the other hand, it cannot be denied that the citizen-consumer is not in a position to judge until after the event, by which time irremediable damage may have been done to his interests. As with so much else in local government, the issue here is not a simple one of local discretion versus central control. "How much of each?" is the question that must always be asked. One can only hope, if one values local government, that it will be answered in a way that does minimum violence to democracy at the grass roots.

So far we have looked at this matter solely from the standpoint of the creation of an enterprise. The interest of the national administrative authorities in it, however, is a continuing one. Long-term plans, budgeting, capital financing and perhaps pricing will usually be matters of state concern. Such concern is likely to become particularly intense if the government is providing an enterprise with a subsidy either directly or through a general contribution to municipal revenues. Even in the United States, where local authorities enjoy far more autonomy than they do in most other countries, aid from the federal government is inevitably accompanied by supervision of such things as plans and specifications by a federal agency. Central government interest in an enterprise will also be enhanced if the capital market is tight and the enterprise is therefore in sharp competition with other public undertakings for available funds.

How far national authorities actually go in their continuing intervention is not always clear. What one needs for the understanding of central-local relations in respect of all kinds of services is rather less legal analysis and generalized complaint and considerably more case study. This is perhaps particularly important for countries with comparatively new systems of local government, in which central-local relations are still fairly fluid.

It is true that most of the countries reporting for this survey said that their national authorities do not interfere with the day-to-day operations of municipal enterprises. "Day-to-day", however, is a highly ambiguous expression, and it is by no means certain that all national reporters meant the same thing by it. Whereas for some it may mean that a transport committee is free to decide whether to replace trolleys by buses, for others it may mean that the committee can do little more than arrange running schedules or appoint drivers.

The extent to which a municipal enterprise comes under the supervision of the national administrative authorities will obviously depend partly on its purpose and character. The heaviest supervision is normally reserved for those enterprises that clearly come within the field of utilities and provide services of real infrastructural importance. Enterprises of a less essential kind, created to make a profit for the municipality, to attract tourists, to provide private enterprise with a little public competition or to improve the neighbourhood's cultural amenities will normally attract little supervision unless they make significant calls on capital resources or involve the exercise of compulsory purchase powers. Such enterprises, when permitted by law, often provide the municipality with a field in which it can take real untrammelled initiative. When well run, they may also constitute a legitimate source of municipal pride.

Supervision of municipal enterprises by the national authorities is sometimes rendered over-complex and even incoherent by multiple responsibility. The municipality has to go to the subject ministry (for example, Industry, Power or Transport) for one kind of approval, to the Ministry of Finance for another and to the Ministry of the Interior (or whatever other ministry has a general responsibility for local government affairs) for yet another. One possible way of avoiding the muddles and delays inherent in this kind of arrangement is to channel all communications through the Ministry of the Interior, which then acts as the sponsor of local authorities with the other departments concerned. Alternatively, the subject ministry may act as sponsor for all matters concerning the particular class of municipal enterprise for which it is responsible. Reforms of this kind are not difficult to introduce or to operate, provided there is some spirit of give and take among central ministries, and usually they soon prove their worth in speeding up and rationalizing the administrative process.

PRIVILEGES AND OBLIGATIONS

As we have seen, municipal enterprises are subject to forms of control from which private enterprises—even those with concessionary status—are exempt. They are also burdened with extra obligations. They must, for instance, show no favour to any consumer or class of consumers, unless specifically permitted to do so by law. If public utilities, they must regard service rather than profit as their major objective. If a profit is made, law, custom or the force of public opinion dictate that it shall be passed on to the consumer in the form of lower charges or better services, or to the local taxpayer by way of contribution to the general municipal funds.

In return for the shouldering of these obligations, however, they enjoy certain privileges, some of which are also enjoyed by concessionaires. Monopoly, of course, is itself a privilege, although nowadays its value to the enterprise is much less than in the days of "gas and water socialism". In water supply, admittedly, it is usually complete; but power enterprises, particularly in the more developed countries, now encounter powerful competitive pressures owing to the fact that gas, electricity, coal and oil are in some respects mutually substitutive. Even less valuable, in these days of the ubiquitous private automobile, is a local monopoly of public passenger transport.

Another frequently enjoyed privilege is exemption from taxation. Where such exemption, complete or partial, is granted, an irrational distinction is often made be-

tween different forms of municipal enterprise, with the result that one form (for example, the joint-stock company) pays taxes which another (for instance, the integrated enterprise) avoids. (As we have seen, in certain countries this anomaly affects the frequency with which municipalities choose to adopt a particular form.) Among those countries granting a form of tax exemption, Belgium gives fiscal exemption to all *régies* and *intercommunales* (although her right to do so has apparently become affected by Article 90 of the Treaty of Rome); Brazil exempts directly administered enterprises, autarchies and public corporations from all taxes except the social security tax; Denmark exempts all electricity, gas and water undertakings from the corporation tax; Finland and Norway exempt all "pure" municipal enterprises from the property, income and assets taxes; Iran frees municipally-owned utilities from all tax payments; so does the Sudan; New Zealand exempts them from the profits tax; while South Africa exempts them from the income tax and the company tax. Among the countries making no tax concessions to municipal enterprises are Germany, India, Spain and Italy.

To exempt or not to exempt is a matter for the fiscal policy of the country concerned. It is difficult to lay down any general principles on the subject. In so far as taxation is imposed on profits or net earnings, it is simply a way of allocating among public authorities a surplus which is publicly generated. Indeed, the very fact that a tax payment by a municipal enterprise is in the nature of a transfer payment within the public sector of the economy offers justification for according different fiscal treatment to public and to private utilities. Although the later may be restricted in the amount of profit they may make, the actual sum, whether large or small, must necessarily be treated like any other income accruing to private individuals as a result of investment in trade or industry. But there is no justification at all, as we have already suggested, for making the fiscal treatment of a public utility depend on its form, as is the practice in several of the reporting countries.

Another privilege that municipal enterprises sometimes receive is cheap capital. Sometimes they may obtain specially favourable terms from municipal financing institutions such as the Turkish Iller Bank. Sometimes a government guarantee is used to provide them with financial easement. There has been controversy about the economic propriety of this kind of privilege. Some say that the enterprise (or the municipality on its behalf) should go to the market just like any other concern. To this there are several replies. First, municipal enterprises, not being normal profit-making concerns, cannot engage in self-financing to the extent that most private enterprises can; secondly, they are subject to rules and regulations from which private enterprises are exempt; and thirdly, they are not free, as private enterprises are, to raise money from whatever source they can find, on whatever terms they can negotiate, and at any time they like. On the contrary, their capital-raising activities are closely controlled by the national authorities. One may therefore confidently state that this particular privilege is not, in principle, an unreasonable one.

As has been explained already, the remainder of this chapter will concentrate exclusively on the three public utilities of water, transport and electricity, with the object of discovering to what extent and in what manner the general principles we have elucidated may be applied to them.

Ownership

PUBLIC AND PRIVATE

Although, as we have seen, controversy about the legitimate extent of municipal enterprise is still active, there are few countries in the world where anyone would question the legitimacy of the ownership, by local authorities, of the three utilities: water, electricity and local public passenger transport. It is widely recognized that some form of public ownership of these utilities is desirable; whether it is municipal ownership or another kind will depend partly upon such practical considerations as whether local government areas are suitable for the purpose and whether a given municipality has the financial strength and administrative competence to provide an economical and efficient service.

The general case for the municipal ownership of utilities has already been stated. In advanced countries it has an almost irrefutable validity, provided that municipalities are large enough or can be made large enough by local government reform. There would be no point in listing the reasons why they are likely to do better than the concessionaire; experience has proved that they actually do better, on almost every count.

In the less advanced countries (or in the less advanced municipalities in the advanced countries), the situation is however rather different. Competence and public spirit in local government circles may be so low that the municipality's incapacity to run a public utility economically and efficiently is obvious. In such circumstances there is a clear case for not giving it the responsibility, at least not until the prestige of the local government service has been raised sufficiently to attract able and honest people, both as councillors and as officials.

One of the alternative courses is to nationalize the service. If it is already in the hands of local authorities, they may have to be deprived of it. This has actually happened in Northern Nigeria, where, as we have seen, most of the so-called native authorities have been required to hand over their water undertakings to the Ministry of Works. Nationalization, however, is based on the assumption that whereas the local government authority is incompetent to run a public utility, the national government is competent to do so. Such is not invariably the case.

The other way out of the dilemma is to make over the service to a private firm on concessionary terms. Again in Northern Nigeria, the municipality of Kano did this after it had demonstrated its own incapacity to run a bus undertaking. "Privatization", however, assumes the existence of an entrepreneur willing and able to undertake the service on behalf of the municipality. Also the municipality itself must have the competence to exercise the necessary supervision. These two conditions are not always present.

At the very worst, none of the three possible methods of operation may succeed in producing an even moderately satisfactory service. This is illustrated by the depressing history of Khartoum's public passenger transport, which has been operated at various times, with equal lack of success, by private firms, the municipality and the provincial government.

Assuming, however, that a viable choice exists between a publicly owned and a privately owned utility, ideological factors will to some extent influence the way in which it is made. Socialist-minded governments will have an initial bias in favour of the public sector; those favouring free enterprise will have one toward the private sector. Such biases are rarely absolute, except in the Communist countries, where a decision of principle has been taken to eliminate the private sector, either suddenly or gradually; and even Communist countries may yield to immediate considerations of convenience and practicality. National pride may also produce a countervailing bias. In Brazil, for instance, free enterprise tends to be favoured, but there is strong hostility toward the foreign private capital (particularly from the United States and Canada) that has been invested in many of the country's public utilities. If, therefore, the government or the municipalities could spare the resources to "buy out the foreigner" (thereby reconstituting these utilities as national enterprises, municipal enterprises, or mixed enterprises with indigenous private participation) it would no doubt be doing a popular thing.

Such biases concerning the three public utilities under consideration are still faintly evident in a few advanced countries outside the Communist group, and indeed are fairly strong in one of them. Austria, to judge by its report, is predisposed toward public enterprise, national or municipal, whereas Finland tends to regard such enterprise as a "gap-filler", to be used only when the private sector or the cooperative sector is unwilling or unable to provide the service concerned. It is partly for this reason, no doubt, that transport services are public only in the three major cities of Helsinki, Turku and Tampere, that numerous small privately owned electricity - generating plants still exist, and that in rural areas water is still supplied by private companies or by cooperatives. Sometimes, even in advanced countries, ideological predispositions are stronger among members of the government than among members of the public, with the result that the changes in ownership of municipal utilities favoured by the government are not put into effect.

In the German Federal Republic, the federal government, believing that economic "miracles" can be accomplished only by private initiative, has been selling many of its own public enterprises. It has not, however, been very successful in persuading municipalities to imitate its example.

The only country where opinion rather strongly—and sometimes almost fanatically—favours the private ownership of at least two of the three utilities is the United States. Even there, however, an exception is made in favour of water-supply undertakings, only 14 per cent of which (in terms of population served) are in private hands.

In general, one must point to technical development rather than to ideology as the main factor now responsible for changes in the pattern of ownership and control of the three public utilities. This is particularly evident in electricity supply, which, when it reaches a certain level of technical sophistication, demands regulation, coordination and development on at least a national scale. This demand can be met either by outright nationalization (as in Britain and France) or by control so detailed and ubiquitous that the "rights" of the owners of generating stations and distribution networks (whether private or municipal) do not amount to very much. The latter situation is well illustrated in Italy, so far as the municipal sector is concerned, particularly since the creation of the Italian State Electricity Corporation (ENEL) in 1962. The private companies were then nationalized, but the municipal undertakings were given the alternative of being absorbed by the new national agency or accepting the status of ENEL's concessionaries. Most chose the latter, since when their autonomy has been so substantially reduced that they are no longer free even to devise their own tariffs.

In transport, economic factors (themselves a product of technical development) have made the choice between public and private ownership increasingly unrealistic. Urban passenger transport, having recently recovered from the replacement of the tram by the bus and the trolley, is now generally faced, at least in the more developed countries, by the impossibility of profitable operation, as a result of the interconnected factors of increasing urban congestion and competition from the private automobile. Hence few services could be less attractive to the private investor and entrepreneur, and municipal operation becomes essential for the continuation of the service itself. The intimate connection between the provision of public transport and highway maintenance, traffic planning and the general planning of the urban environment provides strong additional support for the municipal ownership of this utility.

In some countries, the choice between public and private ownership has been bypassed by the creation of mixed enterprises. This is a prominent feature of the Belgian municipal scene, where it has been used for both electricity and transport. In the Scandinavian countries, as we have already noted, municipalities often form such enterprises in association with a cooperative partner. In Denmark, for instance, some of the comparatively small number of large generating stations in which electricity production is now concentrated are jointly owned by municipalities and cooperatives.

In respect of changes of ownership during the course of the last twenty years, there is a variety of experience. In Austria, Brazil, Finland, India, Iran, Italy and the Netherlands there has been a trend toward the public ownership of the three utilities; in Germany, the trend has been in the opposite direction, although much less strongly than the federal government would like.

Thailand has experienced changes in both directions: while its electricity has been nationalized, its buses have been "privatized". In the United States, there has been an increase in the public ownership of power companies but a certain "privatization" of municipal transport, through the device of the management contract. Among the countries reporting no perceptible movement either way are Denmark, Japan, New Zealand, Norway, South Africa, Spain and Sweden.

Possible motives for these changes have already been suggested. Generally there is a combination of the practical and the ideological, with the former having the greater weight.

PRIVATE OWNERSHIP

When a public utility is operated by a private company, the term concession is normally employed to denote the legal relation between the company and the public authority which has been granted, by law, the right to negotiate the agreement under which the former operates. However, in England and in countries with the English system of law, the term, although sometimes used, is of no legal significance. Concessionary companies in these countries have normally obtained their franchises direct from the legislature by way of a private act of parliament or of a type of administrative order that requires specific parliamentary approval to acquire validity. Having made this reservation, we shall continue to employ the terms concession and concessionary company as applying to all situations where private parties have obtained permission, from whatever source, to operate public utilities.

In the present context, concession means the granting to a private party (usually a limited company) the exclusive right to supply, on specified conditions and for a specified number of years, transport, water or electricity over a given geographical area. More particularly, we are concerned with a grant of this kind that is made either by a municipality or by some other authority in respect of an area which a municipality, either on its own or in collaboration with other municipalities, might otherwise supply by the establishment of a public enterprise.

To reduce to some kind of classificatory order the immense variety of arrangements falling within the scope of this general definition, we shall first take the two extreme or limiting cases. On the one hand, the granting and administration of the concession may be placed entirely at the discretion of the municipality, which is restricted only by the terms of the law of contact. This would appear to be the normal situation in only one of the reporting countries—Sweden. On the other hand, the role of the municipality (to the extent that it has any role at all) may be confined to that of administrative agent of the central government. This would appear to be the normal situation in Thailand, where concessions of transport undertakings (which are the only concessions of any interest) are negotiated under the terms of the Transportation Act of 1954 by the Ministry of Communications, which also regulates prices, standards of supply, routes and almost all other matters of any importance. In the first case the role of the municipality is decisive; in the second, almost nugatory.

In between these two extremes, there are many possible situations. Perhaps the simplest, or at least the most intelligible, is that in which the municipality is free both to negotiate the contract and to supervise its performance within the terms of a general concessions law or a law regulating a specific class of concession, or both. Examples of this are found in the Netherlands and Italy. Such freedom, however, is often illusory, since in almost every instance the law gives specific powers of negotiation and administration both to the municipality concerned and to some authority forming part of the central government or emanating from it. In such cases, the role of the municipality is usually a subordinate one.

Examples of the sharing of powers are plentiful. The Sudan provides one of a singularly straightforward kind. There the local authority tells the concessionary bus company what routes its vehicles must follow and at what places they must stop, while a national Licensing Authority gives it permission to operate and draws up regulations governing the safety and the quality of the service. In Belgium, the normal tutellary control over the concession-granting communes is supplemented by the requirement that charges must have specific government approval. In Finland, bus services are not only licensed by the central authorities but controlled by them in respect of safety, quality and frequency of service. In India, the companies are subject to state of Union supervision of charges and profits as well as of the safety, quality and frequency of service. Often the regulatory powers of the national authority apply equally and impartially to municipally owned as well as privately owned utilities as is the case, for example, with all road passenger transport undertakings in South Africa.

From the above necessarily rather sketchy information, it will be clear that to generalize about the status and control of concessionary undertakings even in the comparatively limited field of our three chosen utilities is extremely difficult. Useful transnational comparisons would have to be preceded by a series of studies of individual national systems. These themselves often display considerable complexity. In England, for instance, a concessionary company often owes its basic franchise to a private act of parliament, is certainly subject to regulation by a government department charged with the administration of the acts that apply to the utility concerned, and often comes within the jurisdiction of *ad hoc* licensing authorities, such as the Traffic Commissioners.

The basic objectives of national regulation are easy to list however. They are, briefly, (a) to ensure that the terms on which concessions are made do not vary, from one locality to another, more than may be required by particular circumstances; (b) to ensure that the services provided by the concessionaires conform with nationally acceptable minimum standards and display whatever measure of uniformity may be currently regarded desirable; and, (c) to protect the consumer against exploitation. It is again significant that municipalities are rarely regarded as capable of securing the last of these objectives by their own unaided efforts.

Although the concession is not a popular method of providing a utility except in the United States and a few other countries, there can be little doubt that further study of the ways in which it is used would be profitable. In most of the advanced countries, admittedly, its role is likely to diminish; in some of them it is already of very small importance indeed. In the developing countries, however, it will continue to be employed, as either a temporary or a permanent expedient—to what extent depends partly on the attitude of the government concerned toward private enterprise and partly on the relative abilities of the private and public sectors to run utilities economically and efficiently. There are some governments that prefer to make use of private enterprise, when it is available;

and there are others that are virtually compelled to make use of it as a result of their own and their municipalities' incompetence in the field of economic administration and management. As long as these views and these situations prevail, the concessionary company will have a role to play.

Economic and Financial Aspects

MONOPOLY

That municipal water, transport and electricity enterprises should display economic and financial features different from those characteristic of ordinary private businesses is due to the fact that these services are essential to the continued existence of the community as a community and to the fact that they usually possess some monopoly characteristics and are therefore unamenable to the automatic economic controls provided by free competition. In this context it matters little whether an enterprise is a *de jure* monopoly so long as it possesses *de facto* monopoly features. Of the three enterprises concerned, the only one which is occasionally subject to competition is transport, where examples can be found of public and private enterprises operating side by side. Even in this case, however, the licencing system usually involves the suppression of any effective competition. As all three services must be provided, whoever is given the responsibility for providing them, and as it is normally sensible and convenient to organize each one as a monopoly, there is clearly a case for stressing certain unique features of their economic operation and financial status.

PROFITABILITY

Declared public policy is to provide services at cost except where subsidization

is a matter of public policy, and respondents to the questionnaire suggested that these enterprises were not expected to make a profit in order to subsidize other municipal services. In general, this is good policy, for there is no reason why the local taxpayer should be subsidized by the users of the municipal enterprises (nor indeed the users of municipal enterprises by the local taxpayer). What constitutes a profit in a municipal enterprise, however, is by no means clear. One has to consider, for instance, how far the enterprise is expected to be self-financing, whether the municipality looks for a "reasonable return" on the money it has invested in the undertaking and, if so, whether it treats any such return as a profit or as a factor cost.

Disregarding these complexities for the moment, one may say that the possibility of making a profit in two of the utilities is remote. In the supplying of water the "social service" element is very strong and consequently anything more than an accidental profitability would be regarded as inappropriate. A water enterprise normally aims at breaking even, but there is certainly no objection to its showing a loss if social objectives are thereby served. As for public passenger transport in cities, a municipality will usually consider itself fortunate if it succeeds in breaking even, for the obstacles to the making of a profit in this type of enterprise are, as we have seen, considerable. Moreover, the subsidization of transport services on social grounds is often more than legitimate; it is absolutely necessary. Only in electricity, therefore, can profitability be regarded as a reasonably attainable aim. Whether it is legitimate is another matter—but it is sometimes pursued.

COSTS AND CHARGES

The normal principle in all three enterprises is that the consumer should pay the cost of the service which is provided him. However, some degree of cost subsidization for the benefit of particular classes of consumers is almost universal. It may arise accidentally, as a result of the nature of the charging system, and may indeed be necessary up to a point in order to avoid a multiplicity of charges which would be confusing to the consumer and uneconomic to operate; but it may also be deliberately engaged in as a matter of social policy. Political factors, of course, also play a large part in the determination of charging policy. Once cost subsidization has been started it is very difficult to stop, even when any rational social justification it once possessed has disappeared.

In transport the most usual form of deliberate cost subsidization has as its aim the removal of real or alleged hardship. Invalids, children, commuters and workers have the right to special concessionary fares; and sometimes similar concessions are extended to other classes of consumers such as police, soldiers and students. Electricity enterprises sometimes charge below cost for the benefit of large families, and the subsidization of the rural by the urban electricity consumers is a common phenomenon.

Another frequent practice is the subsidization of all consumers using a particular service. Thus, transport services to developing suburbs are often provided below cost, partly for promotional and partly for social reasons, while a minimum regularity of service at constant charges is maintained on all lines, despite the fact that lack of passengers at particular times of the day means that the services provided at those times can never pay their way. Sometimes a distinction is also made between domestic consumers and institutional consumers, usually for the benefit of the former. This is

the case with the supply of water in countries such as Japan and Denmark, while in Yugoslavia it appears currently to apply to all municipal services, although attempts are now being made to rectify the wide discrepancies that have arisen.

In all the above instances the element of hardship is the decisive one, but the subsidization of services on the basis of cost and benefit calculations that take into consideration economic gains and losses extending far beyond those of the individual enterprise is also practised. Sometimes it is hoped that the enterprise will be able to recoup any resulting losses at a later stage, as when stimulatory rates are charged for electricity supplies. In other situations, however, there may be no expectation that the service concerned will eventually break even or make a profit. Passenger transport fares, for instance, may be set at a level calculated to secure the maximum use of facilities (irrespective of whether they cover the cost of the service) so that the economic losses accruing from road transport congestion may be reduced to the minimum.

Any such special arrangements may be entirely justifiable in particular places and at particular times. Indeed one of the advantages of municipal enterprise is that it can be more easily used for the attainment of wider social purposes, without regard for immediate economic gains and losses, than can private enterprise. The use of municipal enterprise for purposes of indiscriminate individual and collective subsidization, however, must be carefully avoided. Departure from the normal principles of commercial operation should be made only for specific and defined reasons. Even if commercial considerations do clash with other terms of reference, they should not be sacrificed indiscriminately. One must also deplore as inimical to sound principles of operation the almost automatic readiness of some municipalities to cover losses of municipal enterprises whether these arise from the social obligations of the enterprise or from other causes, such as avoidable inefficiency. The correct practice is to calculate carefully the costs of the uneconomic services imposed on the enterprise for social reasons, and then grant them specific subsidies of an appropriate amount. This appears to be the established practice among the municipalities of countries such as Finland and Germany.

LOSSES

While it may be possible, in theory, for a municipal enterprise to go bankrupt, and particularly so in the case of those enterprises which have been given the joint-stock company form, in practice it never occurs. Losses may be carried forward or capitalized or covered by loans or met by drawing on reserves. When all such expedients have been exhausted (and perhaps before some of them have been even tried) the municipality will come to the rescue of the enterprise with a grant. This is inevitable as the service must be maintained, come what may. Although different ways of covering losses may be legitimate in different circumstances, there are some which are bad in principle. One of these, reported from Ceylon, is the transfer to the losing enterprise of the surpluses of profitable ones.

PROFIT UTILIZATION

When profits are made they can arise deliberately or accidentally. It may not be easy, however, to know whether a profit in any real sense of the word has actually been made. A genuine profit is a surplus that arises after interest has been paid on loan capital and provision has been made for depreciation and reserves. The measurement of the profitability of a municipal enterprise therefore requires the application of commercial principles of accounting. As such principles are not invariably applied, particularly in the more integrated enterprises, the sum frequently described as profit (that is, the sum paid into the municipal treasury at the end of the annual exercise) is of highly ambiguous significance. Unfortunately, the national reports do not always make clear the sense in which the word "profit" is being used.

The manner in which profits should be used depends partly on the source from which they arise. Where there is monopoly they can arise from the exploitation of the consumer, in which case they should be returned to the consumer in the form of lower charges. If, on the other hand, they are the result of efficiency, it is generally good policy to use a proportion of them to reward the personnel responsible for the efficiency, by way of higher wages, or bonuses, or fringe benefits. Identifying the source of the profit, however, generally requires sophisticated accounting techniques which only the larger and more up-to-date municipalities are likely to have at their disposal.

What is bad in principle is to use the profits of a municipal utility to relieve the burden on the local taxpayer. That part of the profit not used to improve conditions among the personnel of the enterprise should be returned to the consumer either directly, by the application of a break-even policy or indirectly, by reinvesting the profit in the enterprise itself. However, the extent to which a municipal utility should be self-financing is still a matter of considerable controversy.

In practice, the disposition of profits frequently depends upon the status of the enterprise concerned. The more autonomous the enterprise and the clearer the separation of its accounts from the general municipal accounts, the freer the directing body is likely to be in the disposal of any profit that may accrue. The temptation to use profits to subsidize municipal funds is obviously strongest when the enterprise is of the more integrated type, although, as we have noted, the real extent of such subsidization is frequently unknown. The removal of such temptation is undoubtedly one of the reasons for giving an enterprise autonomous status, although autonomy by no means invariably assists in the attainment of this objective.

The formal manner in which profits are treated is not really of great importance. For instance, while profits may be paid into the general municipal fund they will not necessarily be used to subsidize general revenues. The important thing is what the municipal treasury, acting on instructions from the council, actually does with them. Where the municipality is free to decide on the disposal of profits accruing from its enterprises, prevalent practices differ widely from enterprise to enterprise, from municipality to municipality, and from country to country. Reports from many countries (Brazil, Finland, Iran, Italy, Sweden and Thailand, for instance) indicate that such profits are often treated simply as local revenues. In other countries, such as Ceylon, Denmark, New Zealand and Norway, profits are often ploughed back into the enterprise that makes them.

In several countries the councils are denied this freedom, the disposal of their enterprise profits being regulated by law. This is true, for instance, of the *sociétés de transport urbain* in Belgium. In Austria the by-laws of the municipalities themselves usually state that the net proceeds of municipal enterprises shall be used for investment purposes.

However, except in the case of electricity, municipalities are rarely worried by the problem of profit disposal. Their more usual concern is the avoidance of loss.

PRICING POLICIES

At a given elasticity of demand, the profit or loss made by a municipal utility depends on the pricing that it adopts. A surplus or a deficit in the accounts does not, however, automatically indicate whether a correct pricing policy has in fact been adopted, since the charges made by municipal utilities may have wide economic and social implications. For this reason intervention by national authorities in the pricing policies of municipal utilities is widely practised Even when, as in the Netherlands, the enterprise or the municipality has formal freedom to fix prices, the national authorities may, in practice, make the final decision. On the other hand, in some countries (such as Denmark, India, New Zealand, Sweden and the United States) there is complete freedom of price determination by the management or the council in respect of all three utilities. A second group of countries reported that national price determination applies to some of the utilities but not to all. In Austria, Brazil, Ceylon, Germany, Iran and Italy it applies only to electricity. In Finland it applies to transport (except tramways) and in Japan to both transport and electricity. Another group of countries reported that the prices charged by all municipal utilities are subject to national regulation. Thailand appears to be the only country where price fixing for municipal enterprises is an actual ministerial responsibility. In Belgium, Norway, South Africa and Spain, central approvals must be sought. How rigorous this requirement is, and how seriously the central authorities treat their price-approving duties, does not generally emerge from the reports. South

Africa, however, suggests that in practice the municipalities retain considerable freedom of manoeuvre. In Yugoslavia, the federal executive council does not possess any special powers over enterprises founded by the communes, but these enterprises, like others providing basic services or commodities, are prevented from raising their prices above a maximum level.

Thus, with the exception of a very few countries, the discretion of the enterprise or council in fixing the charges for utility services is severely limited. It is difficult to say what exactly is the effect of central price fixing. In general it seems that pressures on behalf of uncommercially low prices are likely to be stronger at the local level than at the national level, since the citizen (who is generally more conscious of his role as consumer than as local taxpayer) can bring more effective influence to bear on the municipality than on the central government. (Stefan Dupré elaborates upon this thesis in Chapter 11.) For this reason the national government may be able to view the pricing policies of municipal enterprises with greater objectivity than can the municipalities themselves and it will certainly be able to consider the issues in a much wider context. To this one may add that the national government will certainly be reluctant to fix prices at a level that will result in constant loss thus requiring the national exchequer to subsidize the enterprise. The question of how much freedom the municipalities should possess in this field is, therefore, a complicated one that does not admit of any simple straightforward answer.

EXTERNAL SOURCES OF FINANCE

The adoption of break-even policies, together with the shaky economic condition of many municipal enterprises and the prevalence of what are sometimes called political prices, mean that the possibilities of self-financing are strictly limited. New capital funds for expansion, re-equipment and improvement have therefore to be raised predominantly from sources external to the enterprise itself. Such funds may be obtained from various sources, of which the most important are the municipality, special financing institutions, the capital market and the national government or its regional and provincial agencies.

Municipal grants and loans are reported from Austria, Belgium, Brazil, Denmark, Ceylon, Finland, India, Italy, Korea, South Africa, Spain, Sweden, Thailand and Yugoslavia. For the more integrated enterprises these are normal sources of capital finance. Little information has been provided, however, about the circumstances in which grants are preferred to loans and vice versa, or about the returns, if any, that the municipality normally expects of its investments. The issue of bonds to the public and other forms of recourse to the capital market are also well-established methods of fund raising, particularly in the more advanced countries. Possibilities in this direction are obviously limited by the terms the municipality or the enterprise is capable of offering. Obviously the terms will be directly related to the commercial viability of the enterprise concerned and consequently are likely to be used more freely for electricity, than for transport or water. This type of financing is generally limited to the larger municipalities and enterprises. Normally recourse to it requires central government approval and sometimes the assistance of a government guarantee is provided. Recourse to the capital market may be an expensive method when funds are in short supply. In some circumstances it may even be forbidden by the central government in order to prevent "overheating" of the capital market. This was the situation in the Netherlands in 1966.

When access to the capital market is limited, either by government decree or by considerations of cost, a particularly important role may be played by the special municipal financing institutions, of which many of the reporting countries provide examples. Belgium has a *Crédit Communal*; the Netherlands, a Bank of Municipalities; Japan, a Finance Corporation for Public Enterprises; Spain, a Local Credit Bank; Thailand, a Municipal Loan Fund; and the United Kingdom, a Public Works Loans Board. Such institutions are sometimes organized on a cooperative basis by the municipalities themselves, with state assistance. The value of their special facilities obviously depends on the volume of funds they have at their disposal and the terms they are able to offer the municipalities that borrow from them.

Loans and grants to municipal enterprises from the central government, without the intervention of any special financing institution, were reported from a great variety of countries including Belgium, Ceylon, Germany, India, Japan, Korea, Spain and Thailand.

In many countries the capital finances of municipal enterprises are in bad condition. The reason for this is often that the enterprises themselves are in bad condition, but poor organization for the supply of capital funds is also a factor. One would expect to find the latter defect in the less advanced countries, where funds are short and the capital market is underdeveloped, but the more advanced countries seem to have their troubles, too. One of the most bitter complaints about poorly organized financing of municipal enterprises came from Italy, a country that has made most striking economic progress in recent years. Here the unhealthy situation of most of the communal budgets, together with the obsolescence of the law relating to municipal enterprise, appears to be the main source of the trouble. Possible remedies may be found in the amendment of the law and in the establishment of stronger and better-endowed municipal financing institutions. Indeed, these remedies seem to be relevent to the problems experienced by many of the reporting countries. In particular, the specialized institution seems to have a very important role to play if municipal utilities are to be financed in an orderly way, on reasonable terms, and without disorganizing the capital market.

It should be added that availability of capital finance may depend, rather arbitrarily, on the legal form that the enterprise has been given. In France, for instance, a *régie* that lacks legal personality may often find itself short of funds, particularly as *communes* are not allowed to support their *régies* from the proceeds of taxation. The *syndicat* is sometimes as badly off, since it is dependent on members who may not see eye to eye on the subject of its financial requirements. It is for reasons such as these that French local authorities have preferred to set up joint-stock organizations whenever possible and appropriate. This form certainly offers a way out of the capital-raising impasse; yet it seems anomolous that an enterprise's ease of access to capital finance should be determined by its legal status rather than by its social importance and commercial performance.

Geographical Extension

THE PROBLEM

Deciding upon the optimum size of a public utility, in relation to the existing areas of local authorities, is one of the most acute problems with which we are confronted. Its general nature has already been made clear. Broadly, technical progress has rendered many of the local government units which were formerly responsible (and which often remain responsible) for the provision of utility services unsuitable for this purpose, usually because they are too small. An additional complication arises from the fact that the optimum area of provision usually differs from one service to another, with the result that, even though a particular utility may continue to be economically and efficiently provided by a given municipality, the provision of a whole range of services, in a coordinated manner, becomes increasingly impracticable.

Possible solutions to these problems have also been indicated. The most radical is a thoroughgoing reform of the whole structure of local government, with the object of bringing into existence a series of units with population catchment areas suitable for most, if not all, of the functions which local government authorities have been created to perform. The application of this solution, however, meets with considerable difficulties. Each local government authority has a vested interest in its own self-preservation, and the totality of these vested interests can have sufficient political significance to make a government hesitate about initiating major reforms. Oddly enough, the frequently deplored indifference of the public (as distinct from the councillors and officials) toward local government affairs can have a similar effect. The restructuring of local government will probably be met with the same indifference, gaining no mass popularity for the government initiating it. Why, therefore, undergo the danger of alienating a series of vested interests when this will have no compensating political advantages?

These inhibitions are sometimes given greater respectability and plausibility by talk about the need to preserve local democracy, the assumption being that only the small local authority can be really democratic. In most countries the result has been a piecemeal approach to the task of reform. The number of small local authorities has been reduced and, simultaneously, where a two-tier structure exists, there has been a tendency to remove control of the more important services from the lower tier to the upper one. But these processes have rarely gone far and fast enough to meet the requirements of service expansion and technical progress. Normally, only a highly authoritative government has the power to push through the necessary changes. Paradoxically, it is precisely this kind of government whose will to do so is particularly weak, not because (like less authoritative governments) it is afraid of vested interests, but because it lacks enthusiasm for local democracy. Even in Japan, where the post-war occupation authorities were anxious to give the country grass roots democratic institutions such as it had never before possessed, the opportunity was not fully seized.

Nationalization of services, either partial or complete, offers an easy alternative to the difficult and dangerous task of local government reform. Britain and France, where municipalities have had to yield up some of their most important trading undertakings, such as gas and electricity, to national corporations, provide classic examples of this policy. Elsewhere, deprivation has usually been rather less complete. Municipalities have often retained even their electricity generating plants, on condition that they operate them in accordance with national regulations, and usually continue to be involved in distribution, either as distributing authorities or through representation on regional distribution agencies.

As we have seen, there are also other forms of cooperation between municipalities and the central government, such as mutual participation in *ad hoc* agencies of the joint-stock company or cooperative type. Here the Belgian example is one that may well repay study. But all these solutions involve a marked diminution of municipal autonomy. Such a diminution is quite clearly inevitable, whatever solution is adopted. In default of general local government reform, the solution that appears to diminish municipal autonomy least is intermunicipal cooperation. However, when most of the important services are in the hands of a series of intermunicipal, *ad hoc* agencies, on which the citizens are only indirectly represented through delegates appointed by their local councils, effective inter-service coordination, unless initiated by the national authorities, becomes impossible, and local democracy all but disappears.

Even radical local government reform, which would create a series of viable regional agencies, each with its own elected council, does not offer the perfect solution to these problems. Given the conditions of modern life, there is no solution that can offer the degree of municipal autonomy that would satisfy the more enthusiastic advocates of local self-government. Such reform, however, offers by far the most satisfactory approach to a solution in those countries where political and social conditions make it possible. Elsewhere, if local government is to continue to play any significant part in the management of those utilities that are straining against its boundaries, there is no alternative to the somewhat untidy arrangements we have described.

LEVELS OF OPERATION

The search for areas of operation wider than that of the municipality has been at its most intense in the supplying of electricity. Even in comparatively backward countries, or countries where distances between the main centres of population are great and communications poor, national control and national and/or regional administration of this utility have sometimes made marked progress. In Brazil, for instance, side by side with a multiplicity of small and purely local thermal and hydro-electric plants, there are regional enterprises with both generating and distributing responsibilities. Iran has now established a number of regional power stations, the first regionally organized utilities that country has so far seen.

The newer forms of cooperation between municipalities and the national and regional authorities controlling the electricity grid may be illustrated from Scandinavia.[12] In Sweden the municipalities now generate only about 7 per cent of the supply, 43 per cent of it is produced by the State Power Administration (operating through seven regional agencies) and the remainder by private power companies. Coordination of supply is in the hands of a central operating management.

Denmark is unusual in that the government plays no part in the actual generation and supply of electricity, the major part of which is undertaken by eleven private power companies in ten electricity regions. These transmit bulk supplies to local authorities and cooperative societies, which are responsible for distribution to the consumers in the urban and rural areas respectively. The situation thus appears to be one of private enterprise in generation and

---

[12]Most of the information contained in this paragraph is derived from the very useful brochure entitled *Power Supply in Scandinavia*, issued on the occasion of the 13th Congress of UNIPEDE (Stockholm, 1964).

wholesale transmission and of public or co-operative enterprise in retail distribution. Participation of the distributors in some of the power companies (organized as partnerships or cooperatives), however, makes the total picture a little more complicated, as does the fact that the municipalities of Randers and Copenhagen organize both production and retail distribution within their respective areas. Overall control is exercised by an electricity council responsible to the Ministry of Public Works, and there is an Association of Danish Electricity Works to conduct negotiations with the public authorities on behalf of the power companies. Recently, this body has established a research department to help solve the technical and economic problems which have a bearing on the improvement and development of electricity supplies in Denmark. Again, a rather untidy looking system not only gives reasonable satisfaction to the consumers but also involves the municipalities in the administration of the utility, mainly on the distribution side but to some extent on the producton side as well.

The technologically dictated trend toward the wider area of management and administration does not, therefore, automatically eliminate the municipalities from the business of supplying and distributing electricity; but it does considerably reduce their role and limit their independence. The same is true, although by no means to the same extent, for water and transport. As the consumption of water increases and the problems involved in long-distance transmission are solved, even the watershed, which appears to be the natural area of supply, may cease to have much administrative relevance. As whole districts and provinces become built up into vast conurbations, the division of passenger transport responsibilities (now no longer local in any

meaningful sense) between the series of contiguous municipalities into which the area is divided becomes inconvenient and uneconomical. Again, the municipalities do not need to be driven from the field, but they have to exercise whatever responsibilities may remain to them in relation to a concept of the public interest that they themselves cannot be allowed to determine alone.

The factors influencing the actual size and shape of the administrative unit are varied and obviously differ considerably from one service to another. Hitherto we have stressed the technical one, which certainly tends to predominate in the supply and distribution of electrical energy, but there are many others each of which has to be given due weight. The size of a country clearly counts for a great deal. In a small country it may prove feasible to provide and administer services on a nation-wide basis. On the other hand, in very large countries even the utilities in which the central government is interested are necessarily subject to a large measure of regional decentralization. How much decentralization is necessary (as distinct from desirable) depends largely on the condition of the communications system. Topography is also influential, if not decisive, to the extent that it facilitates or impedes communications and creates natural areas of supply. Even more important are the distribution and density of population. Several reporting countries have not yet felt the pressure to widen their areas of utility administration partly because, having a comparatively sparse and dispersed population, they do not need the regional, still less the national, unit for the time being. Yugoslavia, despite the remarkable economic development it has undergone during the course of recent years, appears still to be in this position.

Finally, and perhaps most important, there are always political factors. Often municipalities are determined to keep what they have by way of control. In Brazil, the principle of local autonomy, which entitles each municipality to maintain its own services, inhibits regional organization or even straightforward intermunicipal cooperation. The Netherlands may be taken as illustrating a more general "political inertia" principle, insofar as the tendency is to retain control of a public utility in the hands of the body that initiated it, whether a municipality, intermunicipal group, or province.

Such political obstacles to a widening of the area of operation are not confined to national or provincial capital cities. Indeed, a large town may be eager to participate in a wider association, expecting to be able to dominate it. A small municipality, on the other hand, may be almost fanatical in its pursuit of a "what we have we hold" policy, and a number of small towns, jointly determined to defend their privileges against encroachment by the centre, may well succeed in deflecting the government away from courses of action which appear to be dictated by economic and administrative rationality. One may freely admit, however, that these political factors are usually of greater strength at a level higher than the municipal. It is when a country is divided into federal units which are highly self-conscious politically that the area problem can become really acute. India provides many examples of the difficulties arising from this source. Interstate quarrels about the use and control of river water have been frequent and economically damaging. This subject does not directly concern us here, but it is relevant to note that similar antagonisms have sometimes held up the rationalization of electricity distribution. The interstate zonal councils, formed to stimulate and organize cooperation between adjoining states in this matter and in others, have not proved a great success.

JOINT OPERATION

As we have already suggested, one of the most immediately practicable ways of solving the area problem is that of intermunicipal cooperation. Municipalities do not always need to be persuaded of its advantages to them—although many countries now have laws giving the government discretionary powers to compel them to cooperate, if need be, in the provision of certain services. Where such cooperation is freely undertaken, it is good for the reputation of local government and it has the additional advantage of basic conformity with the principles of local democracy—even though democracy may become somewhat diluted by the removal of the centre of effective decision-making to an indirectly elected body. Some observers decry joint boards as a very inferior alternative to the necessary radical reform of local government areas, and so indeed it is to a considerable extent. In the British White Paper of 1944 on the future of local government the potentialities inherent in the joint board were used as an excuse for evading the whole issue of reconstruction. Nevertheless, even if local government areas were as rational as they could be made, they would still not be suitable for all the functions that local authorities might be required to perform, and hence the various forms of intermunicipal cooperation would probably still have a role to play.

The simplest form of intermunicipal cooperation is the contract, whereby a given municipality undertakes to provide another or others with certain services in return for payment. Normally, this is practicable only when a large and strong municipality with well-developed utilities of its own, is contiguous with comparatively small and weak authorities which find the provision of such services for themselves impossible or uneconomical. It may also be the only practicable form of cooperation when the contiguous local authorities are divided by mutual antagonisms and jealousies. Despite its limitations, the contract is certainly a widespread form of intermunicipal cooperation, being reported from countries as dissimilar as Belgium, Brazil, Ceylon, Germany, Japan and South Africa.

Its disadvantage is that it is an unequal form of cooperation, in which the recipient authorities have very little effective control over the service provided. Hence its supersession by the reform of areas or by more sophisticated cooperative arrangements is usually welcomed. In Calcutta, for instance, the replacement in 1966 of contractual arrangments for water supply between the Calcutta municipality and adjoining authorities by a Calcutta Metropolitan Water and Sanitation Authority, with jurisdiction ranging over the whole conurbation, was an important step forward. This body also provides a good example of the *ad hoc* regional agency largely controlled by the local authorities participating in it—a stronger, more permanent and more egalitarian method of cooperation than a mere contractual arrangement.

The Calcutta Metropolitan Water and Sanitation Authority, which is unique of its kind in India, has been constituted by a separate act of the legislature; but many countries have general legislation (relating either to local government as a whole or to the provision by local authorities of a particular service) permitting similar forms of cooperation to be undertaken at the discretion of the authorities themselves, subject to whatever kind of administrative supervision happens to be in force. The

type of organization we have already looked at in connection with the British joint board and the French *syndicat intercommunal* has, in fact, a wide provenance and is frequently used for the administration of the three utilities with which we are here concerned. In Italy, for instance, there are consortia of communes; in Spain, joint associations; in Sweden, federations; in the United States, joint operating agencies.

Sometimes, however, looser forms of intermunicipal associations not requiring specific legislative sanction are adopted. In Japan, although there is legal provision for the establishment of joint boards, none have actually been created. Cooperation usually takes the form of agreement among the municipalities concerned to participate in Associations of Local Public Bodies, with representative councils and autonomous managements. In Austria, one of the many forms of cooperation available is the so-called administrative community, which possesses no separate legal identity, its constitution being drawn up in accordance with the agreement reached among the participating municipalities. In Yugoslavia, communal assemblies are free, by way of contractual arrangements among themselves, to establish joint public utilities. On the other hand, in some countries, such as Belgium, legislation not only prescribes the precise legal form that the intermunicipal agency shall take (joint-stock company, cooperative, public corporation and so on) but supplements municipal representation on its governing council by central government representation.

These distinctions can be important. It is necessary, also, to distinguish organizations such as those listed above, which are governed (wholly or partly) by representatives of municipal councils, from other *ad hoc* organizations that cover several municipal areas but are governed by the directly elected representatives of all persons in the relevant district. Such organizations, as we have seen, are frequently found in the United States. Further examples, from New Zealand, are the electric power boards and the Auckland Regional Authority for Water and Transport. It is not easy to justify such direct election of a regional or district *ad hoc* body. The municipalities are thereby deprived of control over a service that is very much their concern, and the electorate is called upon to deliver its verdict on the administration of one isolated service—a function which no electorate, however well educated, is really suited to perform.

However, the solutions to the geographical problem that we have surveyed here all have their merits and demerits, depending on the circumstances in each particular country. It is impossible, therefore, to offer any general prescription. The important thing is that each country, and indeed each municipality, should be fully aware of the breadth of international experience in this field, and be prepared to select from the various possibilities available those best adapted to its particular needs.

4

Housing in Metropolitan Areas

Hidehiko Sazanami

All cities today, in developing and developed countries alike, are in a state of flux. Despite the growing sophistication of available planning techniques and the organizational and institutional structures for planning, the pressures of urban growth continue unabated and the complexities of urban form brought about by social, economic and technological development increase. Most of the proposed and administered remedies for the decline of central areas and uncontrolled expansion of the suburbs are nothing but temporary, makeshift solutions.

The problems are serious, for rapid urbanization has disrupted the traditional human scale of life. Traditional human values and political, social and family institutions have been shaken to their foundations. The emergence of the so-called affluent society, while marking unprecedented improvement in living standards generally, has given rise to a plethora of crucial social dilemmas, not the least of which can be seen as in some sense spiritual. In less developed countries, rapid population growth has aggravated problems stemming from an already low standard of living, often threatening even the most basic sources of livelihood.

As economies become increasingly consumer-oriented, feelings of economic inferiority become distressing to more and more people. In the developed, free-enterprise countries, innovations in ways of life come less from the spontaneous demands of people than from manufacturers eager to market new commodities, which further accentuates unbalanced standards of living. In developing countries, top priority is given to economic development while social development, including housing and city planning, is left far behind. Yet, as Charles Abrams points out, it is obvious that in the long run the failure to provide social facilities, including housing, affects economic growth.[1]

The future city will inevitably be a metropolitan region in which city and countryside are closely linked in every way. Home, place of work and recreational facilities will be functionally linked by communication and transportation networks, and a new pattern of human settlement will emerge. The spatial structure of this pattern will be based on a new concept of social values. Thus the existing urban land and housing policies require radical re-examination. Past proposals for the ideal city have ranged from those of Dante and Thomas More which stressed the religious and ethical elements; the socialistic proposals of Fourier and Robert Owen; to those that emphasized physical form, whether it was Ebenezer Howard's "Garden City", Le Corbusier's "City for Three Million", Frank Lloyd Wright's "Broadacre City", or Doxiadis' "Dynapolis". Current thinking is directed toward the need to accommodate the dynamism of cities. The pace of change promises a life worth living to some and disruption of values to others.

This, however, is the long-term outlook. In the short run we are faced with the need to identify specific goals and execute programmes to achieve them. This not only calls for an improved linkage between physical and economic short range plans, but also between all short- and long-term plans, which in turn depends to a large extent on the development of tools for the scientific analysis of urban growth trends.

It is within this context of change and the growing metropolitanization of our environment that this chapter explores some aspects of the housing situation in the metropolis of today.

Housing Supply and Demand

The quantity of housing, and therefore the density of occupancy, are products of the social and economic level within a given country. The more industrialized and urbanized a country is, the lower its occupancy density. For example, the countries of Europe, North America and Australasia experience favourable densities, while those of Asia and Latin America show high levels of occupancy. In western and central Europe the average density is as low as 0.8 persons per room.[2] Density in Canada and the United States is 0.7,[3] as it is in Australia and New Zealand. However, in Panama an average of 2.4 people occupy one room,[4] while in Hong Kong the average is 12.76 persons, and in Djakarta two or three families (a total of up to 45 people) may live in a three-room house.[5]

Improvements in the quantity of the housing supply in developing nations de-

---

[1]Charles Abrams, *Man's Struggle for Shelter in an Urbanizing World* (Cambridge, The Massachussetts Institute of Technology Press, 1966), pp. 105-113.

[2]*U.N. Report of the Latin American Seminar on Housing Stastistics and Programmes* (Copenhagen, United Nations, 1962), p. 16.
[3]*U.N. Statistical Yearbook*, Vol. 16 (New York, United Nations, 1964), Table 185.
[4]*Ibid.*
[5]*Review of the Housing Situation in the ECAFE Countries* (Bangkok, United Nations, ECAFE, 1965), p. 5.

pend on the rate of new housing construction. The outlook is not encouraging. In countries with serious housing shortages the number of new housing units per thousand per year is smaller than it is in the more prosperous countries. In 1964, for example, European countries constructed an average of 8.5 new housing units per 1,000 population,[6] while the Asian rate was less than 5 per thousand (except Hong Kong with a rate of 11.0 per thousand and Japan with 7.2).[7] This is due, in part, to the scarcity of human and material resources and to the high cost of construction as related to average family income. It is further compounded by the high rate of population growth in developing countries and the inadequacy and instability of economic and political institutions.

Further disparities between the developing and developed countries are revealed if one examines housing construction investments in terms of gross fixed capital formation in housing as a percentage of GNP. Such an analysis shows that this capital formation is about 5 per cent in western Europe,[8] 4.1 per cent in the United States,[9] 4 per cent in Japan, but less than 2 per cent in India.[10]

Thus, while the housing supply may be regarded as adequate in the developed countries, the shortage in developing countries is so severe that the goal of one house to one household, let alone one person to one room, will be difficult to achieve for many years to come.

Worldwide variations are to be found also in any survey of the statistically quantifiable indices of housing quality. These indices are floor space per housing unit; rooms per housing unit; units supplied with piped water, electricity, flush toilets and bathrooms; and the age of the buildings. Unfortunately, statistics for all indices are not available on a worldwide basis.

In Europe, floor space per unit averages 50 square metres although this ranges from 86.4 square metres in Denmark to 38.5 square metres in the U.S.S.R.; the number of rooms per dwelling may be over five in western Europe and yet in parts of eastern Europe may be below three. By contrast, in Asia, over 50 per cent of the population live in one-room dwellings.[11] Piped water is available in almost all dwellings in the United Kingdom, and about 50 per cent of the dwellings in urban areas throughout the rest of Europe, but this figure again drops to about 20 per cent for many Asian countries and about 10 per cent for Latin America generally (although there are some exceptions such as Costa Rica with 94 per cent availability).[12] The supply of electricity, flush toilets and bathrooms follows a similar pattern. Post World War II housing accounts for about 33 to 44 per cent of the stock in the Scandinavian countries, about 33 per cent in the United States and a far smaller percentage in eastern Europe (for example, East Germany, 11 per cent).[13] Statistics on age of structure are not available for the Asian countries, but

[6]*Major Long-Term Problems of Government Housing and Related Policies*, Vol. 2 (New York, United Nations, 1966), Table B-9.

[7]*Review of the Housing Situation in the ECAFE Countries* (Bangkok, United Nations, ECAFE, 1965), p. 2.

[8]*Major Long-Term Problems of Government Housing and Related Policies*, Vol. 2 (New York, United Nations, 1966), Table B-3.

[9]*Housing Statistics*, United States Department of Housing and Urban Development (Washington, D.C., 1966).

[10]*Review of the Housing Situation in the ECAFE Countries* (Bangkok, United Nations, ECAFE, 1965), p. 5.

[11]*Ibid.*

[12]*Ibid.*

[13]*Major Long-Term Problems of Government Housing and Related Policies*, Vol. 2, (New York, United Nations, 1966), Table A-5.

the buildings tend to be of poorer materials generally and often temporary in nature. For example, in Indonesia, only 5 per cent of the dwellings are permanent, 65 per cent are semi-permanent and the remainder temporary.[14]

The demand for housing stems from a combination of three forces: population growth, rate of housing replacement, and the improvement of living standards. In all countries these forces are exerting pressure on the housing market.

The fundamental factor is the 1.9 per cent per annum increase in world population. Also, although a common characteristic of modern society is the reduced size of families, the concomitant acceleration of the family-formation rate due to early marriage is, in itself, creating an increased demand for housing.

With current trends toward not only urbanization but metropolitization, the demands become more and more pronounced. Whereas in 1960 only 10 per cent of the world population lived in cities of one million or more, for the year 2000 the comparable figure is estimated at 21 per cent.[15] This pressure upon urban areas results in demands for improved services and changes in land use, which in turn often mean the demolition of residential areas. This, plus the decay of buildings and environments, creates further demand for the replacement of residential buildings.

Finally, the improvement of living standards is a never-ending source of pressure for more and better housing. One housing unit per household and one room per person are the minimum standards that every country would like to set, although these are quite unrealistic goals for many. For the developing countries an annual construction rate of eight to ten units per 1,000 population is necessary if present needs alone are to be met. This is approximately the rate of construction to be found in European countries. This also requires that the developing countries invest 5 to 6 per cent of their GNP in housing construction each year. In fact the average target for construction is about seven units per 1,000 and only Japan with a target of 13.4 units per 1,000 population is on a par with western European countries and the U.S.S.R.[16]

Housing Standards

Three major variables affect the establishment of housing standards: the stage of economic development relative to other countries; regional location; and household income. Consideration of the stage of development is relevant if, for example, one were to consider the demand that the advent of the automobile has made upon additional space for a garage. At almost the same time, electricity, gas and piped water have enabled the kitchen to be reduced in size. The value of private external living space, in the form of a patio or a balcony, will vary according to the climatic zone, it being of greatest significance in warm climates. Again, according to the income available, a household is able to exercise greater or lesser choice over its housing. Since World War II, governments have had

---

14*Review of the Housing Situation in the ECAFE Countries* (Bangkok, United Nations, ECAFE, 1965), p. 4.

15Homer Hoyt, *World Urbanization: Expanding Population in a Shrinking World*, Urban Land Institute Technical Bulletin No. 43 (Washington, D.C., Urban Land Institute, 1962), Table 16, p. 49.

16*Review of Housing Situation in the ECAFE Countries* (Bangkok, United Nations ECAFE, 1965), p. 2; *Major Long-Term Problems of Government Housing and Related Policies*, Vol. 2 (New York, United Nations, ECAFE, 1966 and 1965), Table B-13.

to provide housing for an increasing percentage of the population on the basis of their income.

Qualifications must also be made of the term "housing standards". First, with regard to "standards", there is the minimum standard below which houses may be selected for demolition; there is the maximum standard to which governments relate their present housing codes, although such housing is unlikely to be supported by public assistance; and, there is the optimum standard which is generally the target for future housing policies. Secondly, the concept of the term "housing" also needs clarification. While the basic function of housing is to offer shelter and seclusion and the facilities that will make home life pleasant, any consideration of housing standards must also include the environmental factors. These include general land use, schools, recreation areas, shopping facilities, infrastructural services, and the nuisances of pollution, noise and land subsidence. This concern for the regional environment is beginning to be reflected in housing codes, such as that of the American Public Health Association.

It is with these qualifications to the term "housing standards" that we can now examine the economic, social and environmental aspects of housing standards.

THE ECONOMIC ASPECTS

An economically sound housing policy should reflect a close association between the level of the national economy and housing standards set. From what has already been said of existing conditions this close association might be thought to exist now. However, it is the lack of a fairly fine adjustment that can create problems, especially in the poorer countries. There are many instances, in Asia for example, of tenants of a two-room dwelling unit sub-

letting one room in order to supplement their low income; such a situation is far worse than if well-designed, but smaller dwelling units were provided. Thus revision of housing standards is called for to bring them into line with economic conditions. Malaysia provides a good example of a country with modest housing standards: per capita living space is set at 4.7 square metres, a bedroom for three persons is 11 square metres, a living-dining room becomes a master bedroom at night, and corridors are eliminated.[17]

This example of Malaysian housing standards is in contrast to past experience in the Latin American countries where housing standards for middle-income families were based on those of the United States and the western European countries. These standards, however, did not reflect economic reality, with the result that the middle-income families moved into housing designed for low-income families, and the low-income families were thus forced to live in substandard housing in marginal settlements. To rectify this situation, five Central American countries adopted a new housing standard in 1960 that set the size of a three-bedroom housing unit at 52 to 69 square metres and that of a two-bedroom unit at 46 square metres. Recently, these standards have been lowered again and yet, even now, 63 per cent of the urban dwellers cannot afford to pay their rent.[18]

European countries tend to be fairly rational in establishing their housing stan-

[17]*Report on Minimum Standards for Low Cost Housing*, Ministry of Local Government and Housing (Kuala Lumpur, Malaysia, 1966), p. 10.

[18]Anatole A. Solon, *Housing Conditions of Urban Low Income Families in Relation to Levels of Social and Economic Development in Latin America* (Pittsburgh, University of Pittsburgh Graduate School of Public and International Affairs, 1966), p. 18.

dards. The 1960 Köln Standard of the International Federation of Housing and Planning (IFHP) stipulates that 70 square metres is desirable for a household of five. Most countries, however, seem to be guided more by their own economy than by this standard. For comparable sized households the United Kingdom's standard is 80 to 90 square metres whereas that of the Netherlands is 50 square metres.[19] In all the countries, however, great efforts have been made to create detailed standards that relate the size of a family to the number of bedrooms, the amount of living space, the size of each room, plumbing facilities and so on. The United Kingdom is well known for its especially high housing standards and for its efforts to further improve them. In the past, however, even the United Kingdom has not been free from the problem plaguing developing countries today—the dilemma arising from the conflict between too optimistic a standard and the economic pressure working against its realization.[20]

While examining the economic aspects of housing standards it is appropriate to comment on the role of housing standards in the development of a housing supply policy. There is a very great danger that too much attention is given to household income as opposed to household size when determining housing standards. This may appear appropriate but in fact it may lead to the construction of only a few, albeit high standard, units. A somewhat different situation arose in Caracas, where the city built high-rise accommodation for low-income families but had difficulty in collecting the rents, which had been set to match the high standards as opposed to the incomes of the occupants. As a result, the expected funds for further construction dried up.[21] A policy of using limited funds to produce the maximum effects is urgently needed in countries such as Venezuela.

In developed countries more extensive subsidization is feasible, given their higher GNP and higher rates of housing investment. In 1957, 91 per cent of the housing units in France and 58 per cent in the United Kingdom were receiving public subsidies, whereas in Italy, where the GNP is lower and housing investments are smaller, the corresponding rate was 21 per cent.[22] Italy is also of interest in that although the amount of government-subsidized housing is small, it is of a high standard and public housing is in this way expected to stimulate the construction of similarly high-standard housing by private enterprise. Another means by which subsidies can be made available to a greater number of housing units is to accept lower standards. Although the Netherlands compared favourably with Italy in terms of its GNP in 1957, its housing standard was lower, and yet the ratio of houses receiving public assistance totalled 95 per cent.

THE SOCIAL ASPECTS

The nature of the demand for improved housing becomes more specialized only with the general improvement of housing, although these specialized demands vary from one part of the world to another.

Where housing conditions are extremely

[19]For the United Kingdom, see Herbert Ashworth, *Housing in Great Britain* (London, Thomas Skinner, 1957), p. 107. For The Netherlands, see Ministry of Housing and Building, *Housing in The Netherlands* (The Hague, 1964), p. 56, Table 1.

[20]A. W. Cleeve Bar, *Public Authority Housing* (London, B. T. Batsford, Ltd., 1958), pp. 53 and 54.

[21]*1963 Report on the World Social Situation* (New York, United Nations, 1963), p. 138.

[22]*Financing of Housing in Europe* (Geneva, United Nations, 1958), p. 14.

bad, the universal demand from all is for basic necessities such as land, water supply and sewers. The demand for a housing unit per se is nothing more than a demand for a single room where the family can eat and sleep together. The previous references to the high density of occupancy and the common practice of sub-letting a second room indicate that the provision of a second room is not the most immediate requirement in many countries.

An example of a very low degree of specialized demand is provided by the Malaysian housing standards, previously referred to, where, although there are several rooms to a unit, rooms are shared and may have several functions. For example, the dining- and living-room space as common space is small, with one of the bedrooms being expanded to fourteen or seventeen square metres for such functions.[23] The balcony also plays an important function in the Malaysian plan; on summer nights it serves as both sleeping and dining space. The cost of such outdoor space is half or one-third that of an indoor room.

Japan, on the other hand, provides us with an example of a more specialized demand which is growing and changing with the economy. Although the housing standard is low (for example, the size of an apartment unit for a five-member family built by the Japan Housing Corporation (JHC) is only about forty-five square metres), the ownership of durable consumer goods is high. The sophisticated attitude of JHC tenants, whose educational level tends to be higher than average, demands that there be separate rooms for dining and sleeping and that parents and children sleep in separate rooms. As a result of such demands, the master bedroom may be as small as 8.5 square metres and the dining kitchen as small as 7.5 to 8.5 square metres. In view of the fact that one fourth of the households in Tokyo and Osaka live in one-room units in wood-framed apartment houses, those built by the JHC may appear to be adequate so far as Japanese housing standards are concerned. However, the great dissatisfaction expressed by JHC tenants has been focused on their smallness. Since surveys hitherto conducted show that an overwhelming majority of tenants prefer a larger number of small rooms rather than a smaller number of large rooms, effort has been directed toward increasing the number of bedrooms. With the improvement of consumer life in recent years, however, the tenants' demand for larger rooms has also increased. Future policy, therefore, will be to allocate any increase in the total floor space of a unit for the creation of a living room as common space, for additional rooms and for the expansion of the size of each room.

The improvement of standards of living and the shortage of housing space in Japan are likely to accelerate the tendency of urban households to become nuclear families. The national average size of households was 5.03 in 1935, and 4.56 in 1960[24] and a survey conducted by the JHC in 1966 showed that the average size of tenant households of its apartment buildings was as small as 3.4.[25] As a result of the nuclearization of families, the problem of the aged is becoming serious; another survey indicated that most of the aged living with their

[23]*Report on Minimum Standards for Low Cost Housing*, Ministry of Local Government and Housing (Kuala Lumpur, Malaysia, 1966), pp. 10-12.

[24]M. Honjo, *Monograph on Japanese Housing* (Tokyo, Preparatory Committee on Housing and Planning for IFHP Congress in Tokyo, 1966), p. 3.

[25]*Periodic Report of J.H.C. Residents* (in Japanese), Japan Housing Corporation, Building Division (Tokyo, 1965), p. 38.

families in the JHC apartment blocks expressed satisfaction with this way of life as opposed to being by themselves.

In contrast to the Japanese demand for private space is the demand pattern shown by the Latin American countries where housing layout is closer to the European pattern, with the size of common space (including a kitchen, dining room and living room) totalling twenty-five square metres on the average. This compares favourably with the United States Public Housing Administration Standard of 20 to 30 square metres, Canada's 22 to 26 square metres and the Netherlands' 20 to 24 square metres.[26] In view of the extremely large number of dwellers in marginal settlements, it is difficult to say that such a generous standard for common space is realistic; hence, the Latin American attitude that its public housing is becoming "showroom" housing.

Moving further up the scale of the specialization of demand we find greater effort being expended to relate the demand for expanded common space to the varied lifestyles of each family by providing different types of housing plans. For example, the United Kingdom's Dudley Report of 1944 recommended that special consideration be given to the lack of variety in dwelling types, to cramped and ill-adapted living space and, inadequate, shoddy and badly placed out-buildings.[27] The goal of providing for private space within the house has already been achieved in the United Kingdom. The present problem is how to provide better living space to meet such specialized needs as those of the aged, single men and women and childless families.

THE ENVIRONMENTAL ASPECTS

As has been stressed already, satisfaction with housing should be measured not only in terms of the house but also in terms of the quality of the community environment. This consists of both physical and socio-economic factors. Here consideration is given to the more important aspects of the physical planning as they impinge upon the community environment.

The Scale of Residential Community Development

The optimum size of any one residential community is determined by its location relative to the major cities which it both serves and is served by. Small-scale housing communities are acceptable if located within or close to a built-up area and if incorporated in a metropolitan planning scheme. Suburban housing estates or new towns that are more or less isolated from the built-up area, however, will face a lack of community service if the scale of development is too small.

The experience of the United Kingdom in community development is of interest. Much attention has been given to the development of towns and to the expansion of existing suburban communities of the large conurbations. In the early stages of development, close contact with the countryside was emphasized, and the size of the new towns was determined with this objective in mind. The final report of the Reith Commission on New Towns in 1946 recommended that the built-up area of a new town be 30,000 to 50,000 in population and that the total population dependent on

[26]*Standard for Planning and Design of PHA-Aided Low-rent Housing*, U.S. Public Housing Administration (Washington, D.C., 1958); for Canada, see *Apartment Building Standards*, Central Mortgage and Housing Corporation (Ottawa, 1956), p. 8; for The Netherlands, see *Housing in The Netherlands*, Ministry of Housing and Building (The Hague, 1964), p. 56, Table 1.

[27]Barr, *loc. cit.*, p. 53.

the new town be 60,000 to 80,000.[28] It was argued that if a new town were smaller than this it would be difficult to provide necessary urban services and facilities and if it were larger the distance to the town centre would be too great and contact with nature would be lost.[29] In recent years, however, the United Kingdom has found it more economical to build much larger new towns of about 100,000 with much larger attractive civic and shopping centres. In the U.S.S.R. about 800 new towns have been constructed since 1917. Some of these are entirely new communities, others are expansions of existing settlements. While the range in size is from 30,000 for an expansion of a residential community to almost half a million for an established industrial city, the average size tends to be about 250,000. In Japan, several new towns with populations of 100,000 to 300,000 are being planned as satellite towns for the large metropolitan areas of Tokyo, Osaka and Nagoya.

In the United States, large-scale housing developments have been undertaken by private developers. Among such developments are Levittown (Long Island) with a population of 50,000, and Levittown (Pennsylvania) with a population of 100,000. Don Mills, a large residential development with a population of 250,000 in the suburbs of Toronto, is also a private corporation development.

Outside Stockholm eighteen neighbourhood units were built between 1952 and 1963, and five others were under construction as of 1965. Each unit has a population of about 10,000 and is dependent on Stockholm for community facilities to a far greater degree than originally planned.

Consequently, much larger residential new towns—Vallingby in the west with a population of 25,000 and Farsta in the south with a population of 35,000—have also been built. The latest concept is that new centres of about 25,000 persons would serve several neighbourhood units each of about 10,000 population.[30]

The Neighbourhood Concept

Early new town development throughout the world incorporated the concept of the neighbourhood community from 5,000 to 10,000 persons centred around a primary school, shops and public halls for promoting friendship among the residents. In Harlow, England, for example, the concept of community units is very evident. About 150 to 400 housing units comprise an informal community which maintains a common meeting place; several such communities comprise a neighbourhood; several neighbourhoods comprise a neighbourhood cluster; and several clusters form a town.[31] Community facilities are provided at each level. In France the *Commission de la vie dans les grands ensembles* (of the Ministry of Construction) established four categories of community as follows: residential group (200 to 500 units); neighbourhood (800 to 1,200 units); quarter (2,500 to 4,000 units); and city or town. For each of these certain standards of community facilities have been established.[32] In the U.S.S.R. the comparable pattern is of a number of small housing groups comprising a micro-raion of about 6,000 to 13,000

---

[28]Reith Commission, *Final Report* (London, HMSO, July, 1946), B. II, p. 20.
[29]*Ibid.*, pp. 17-19.

[30]Goran Sidenbladh, "Stockholm: A Planned City," *Scientific American* (September, 1965), p. 114.
[31]Frederick Gibberd, *Town Design* (New York, Frederick A. Praeger, Inc., 1959), p. 306.
[32]Commission de la vie dans les grands ensembles, "Grille d'équipement," *Urbanisme*, Vols. 75 and 76 (Paris, 1962), p. 27.

population with schools, shopping and other community services. The next larger unit is the residential-raion (from 25,000 to 60,000 people), which in turn is part of the city-raion (with a population from 120,000 to 200,000).

Essentially this neighbourhood community concept suggested a closed community, and it is of interest to observe that a new approach is now emerging that refutes this conventional idea. It does not attempt to enclose people's lives within a small area, but aims at providing greater choices among life styles. It is based on the idea that a modern city is an open one, where opportunities for friendships and contacts with a variety of people are abundant. Such a city requires a dynamic centre and is centripetal. This more open-ended approach is apparent in the planning of Toulouse-le-Mirail, France; in Kozoji, Japan; in Cumbernauld, a new town near Glasgow; and in Hook, which was planned by the London County Council, but which did not materialize.

Residential Density and Housing Type
In the suburbs of American cities the most common form of housing unit is the single-family house which generally occupies a plot of 650 square metres. The resultant low density of ten units per hectare is made possible only by heavy dependence on the automobile. In contrast to this is the "garden city" plan of the United Kingdom, with "12 to the acre" (30 units per hectare), which has been regarded, until recently, as ideal.[33] Now it is recognized that this density is too low to give residential areas an urban character. The density of 75 to 125 persons per hectare was adopted for the early new towns and the recent tendency is toward even higher densities. In the Hook

plan a three-level density was adopted: for the central area, 250 persons per hectare; for the inner ring, 175; and for the outer ring, 100.[34] In the same way, Cumbernauld adopted densities of 160 to 300 persons per hectare. In France, Italy, Germany and the Netherlands medium- and high-rise apartment houses are predominant, resulting in high residential densities.

In the micro-raion of the U.S.S.R., most housing is prefabricated and takes the form of four- or five-storey apartment buildings. The density is very high, being about 400 persons per hectare. In Japan, too, four- or five-storey apartment buildings are in a majority. The Japan Housing Corporation built two-storey terrace houses until 1960 but since then the shortage of residential land has forced the Corporation to build as many high-density apartment buildings as possible. The distance between apartment buildings is such that each apartment receives sunlight for at least four hours a day at the winter solstice. The distance thus determined is about twice the height of the apartment blocks. There may be playgrounds or other facilities between them. The resultant net density is about 100 units or 300 to 400 persons per hectare.[35]

In Japan most publicly assisted housing is medium-rise housing, while single-family houses with gardens are built by owner-occupiers and private developers. This pattern of construction does not reflect the Japanese people's desire for houses with gardens, however, as was shown by a survey conducted by the Asahi Newspapers in 1965. Those polled in the Tokyo-Yokohama area favoured single-family housing by an overwhelming mar-

[33]Barr, *op. cit.*, p. 37.

[34]*The Planning of a New Town*, London County Council (London, 1961), p. 37.
[35]K. Usui, *Planning for Apartment House and Housing Estate* (in Japanese) (Tokyo, Gihodo Ltd., 1966), p. 395, Tables 3.2.8. and 3.2.9.

gin. A mere 4 per cent favoured high-rise apartment buildings.[36] The fact that most dwellers in high-density, huge metropolitan areas desire single-family houses, or at least the characteristics that these houses possess, cannot help influencing the future policy for publicly assisted housing.

In recent residential planning, mixed development techniques have been widely employed. The merit of these techniques is to help create a desirable social environment by providing different types of housing designed to meet different needs based upon age, taste and social class. Physically, the diversity of architectural forms also helps to create an urban environment of greater interest.

Harlow is one of the early new towns that attempted to translate such techniques into practice. In Vallingby and Farsta, high-rise apartment buildings are located close to the town centres and surrounded by three-storey apartment buildings; in the outer ring there are single-family houses and terrace houses. In the plans for Toulouse-le-Mirail and Kozoji, high-rise apartment blocks integrated with pedestrian decks form the town centres, which are ringed by medium- and low-rise buildings. Mixed development techniques are also employed in Lambertseter (Norway), Tapiola (Finland) and Brondbyoster (Denmark). In the past, northern European countries have favoured high-rise apartment buildings in that they facilitate the use of such techniques as a neighbourhood-wide heating system, but recent technological advances have made it possible for low-rise buildings to receive equally good services. Hence mixed development appears to be more feasible.

---

[36]*Challenging the City* (in Japanese), Asahi Newspapers (Tokyo, 1965), p. 378.

The Economic Aspects of Housing Problems

HOUSING INVESTMENTS AND THE NATIONAL ECONOMY

Investment in housing is of several types with the most basic distinction being made between private enterprise and government as the source of investment.

Private investment in housing takes three forms; the first type being owner-occupied housing. This is the simplest form of housing investment and is motivated primarily by two factors: the individual's need for housing and his economic ability (including credit) to pay for it. A second type is investment in housing for rent. Unlike the first, the return from the investment is the only motivating factor. A third type is an investment by employers in housing for their employees, with a view to providing employees with better housing to secure labour and to increase labour productivity. As with the second type, it is based on a profit motive, but indirectly so.

Government investment in housing, unlike the private forms of investment, is motivated by national economic and social policies. The national economic motive assumes that better housing for the labour force will contribute to the growth and development of the national economy. The social motive is so obvious that it needs no explanation.

The relative proportion of these various types, as percentages of the total housing investment in the national economy, varies according to the stage of socio-economic development and the ideology of the socio-economic system. As such housing investments are based on different motives and determined by different principles of behaviour, it is not possible to consider them as one and the same. Yet they do tend to be complementary and their sum total is determined by people's desire or need for housing, both quantitatively and qualitatively, and by certain principles of resource allocation based on the economic rationale. The more important issue, therefore, is to determine what is an acceptable allocation of resources for investment in housing, from the viewpoint of the national economy, assuming the resources to be available. This, unfortunately, is no easy question to answer and in many countries proves to be a basic problem.

As a general rule, except in those countries where per capita income is less than $100 per annum, the more advanced the stage of economic development the greater is the rate of housing investment to GNP.[37] However, this is a generalization. To gain greater understanding of the existing situation in different economies, it is of help to envisage the ratio of housing investment to GNP as being broken down in the following manner:

$$\frac{\text{Housing Investment}}{\text{GNP}} = \frac{\text{Gross National Capital Formation}}{\text{GNP}} \times \frac{\text{Housing Investment}}{\text{G.N.C.F.}}$$

The proportion of housing investment to gross national capital formation depends not so much on the stage of economic development as on the rate of gross national capital formation. Yet it is doubtful whether such a relationship obtains in a highly developed "mass consumption" economy. As the propensity to consume increases and the rate of gross national capital formation decreases markedly, investments in housing remain very high. For example, in the United States the rate of gross national capital formation is extremely low but the proportion of housing investment to gross national capital formation is high and therefore the rate of housing investment is high. In sharp contrast and also running contrary to the hypothesis is Japan, where too much emphasis has been placed on investments in production, plant and equipment. Although Japan has achieved a high rate of economic growth (often regarded as miraculous), housing problems have been aggravated year by year.

The competition for resources for housing and production facilities does not appear to be so keen in countries that have already achieved a high standard of living, but in reality a large investment in housing is needed to promote building and other related industries. This is because the prosperity of the building industry, like that of other consumer-oriented industries, is essential to the prosperity of the economy as a whole. Also, in the developed countries, the public's desire to raise the housing standard to a level commensurate with other standards of consumer life is strong.

Thus, while there may be similar needs for investment in housing, the situation in the developing countries, where the present standard of living is low and a considerable sum of investment in basic production facilities is needed, is radically different from that in the developed countries. However, in some developing countries, where resources are scarce and the need for investment in industrialization great, the rate of housing investments is as high as in developed countries. In such countries it is likely that priority has been given, by tradition, to investments in housing rather than in production, plant and equipment in

---

[37]*Statistical Yearbook, 1960* (New York, United Nations, 1961), Tables 166 and 175.

spite of the low rate of capital formation. Also, generally speaking, it is easy to invest resources in housing although it is sound economic policy and effective housing policy to invest in housing only those resources unsuitable for investments in production and other facilities.

In the event that it is necessary (and possible) to curb housing investments, restrictive measures should aim at a balanced pattern of consumer life as well as a balance between consumption and investment in production facilities. It is, however, an extremely difficult task to strike a balance between investment and consumption. Too high a rate of capital formation, including that in housing, may give rise to too great a demand for consumer goods (which cannot be easily met by domestic production alone), to risks of inflation and to a crisis in the international balance of payments.

Housing investments consume capital, material, labour and other resources. They also indirectly affect the growth of investments in production facilities. Subsidies are needed to fill the gap between overhead expenses for maintaining modern housing and average incomes and to meet effectively rising housing demands. Such subsidies necessitate higher tax rates, and higher tax rates in their turn exert negative effects on personal and corporate savings which are the sources of investment money for production facilities. This problem is all the more apparent in developing countries where large subsidies are needed to bring new low-rent housing within the reach of the general public.

CONSTRUCTION COSTS AND CONSUMER
PURCHASING POWER

In order to bridge the large gap between housing cost and the individual's ability to pay for it, today's housing policy in most countries extends assistance in one form or another to individuals wishing to secure their own housing. A simple indicator of just how much public assistance is needed is the ratio of income level to housing cost; whereas the per capita GNP serves as an effective index of the national economic effort required to provide modern housing designed to meet the appropriate standard.

In most western European countries, housing construction cost is over 3.5 times the annual income of the average male worker in manufacturing.[38] The higher the workers' wages, the better the housing demanded; thus the increase in wages is offset by the improvement in the quality of housing. If the socially desirable ratio of housing rent to income is (as is commonly believed) set at 20 per cent as a maximum, the annual rent paid by the tenant for housing should not exceed 5 per cent of the housing construction cost. In some less developed countries of western Europe, the cost of modern housing is about eight times the annual income of the average male worker in manufacturing; in such instances a worker's annual expenditure on housing should be maintained at less than 2.5 per cent of housing construction cost to keep the cost within his reach. In the United Kingdom, on the other hand, construction cost of public housing is about three times the annual income of the average male worker in manufacturing, and the annual rent can be increased to about 7 per cent of the construction cost.

The figures cited indicate not only important relations between construction cost and income but also the need for both stabilized interest rates on borrowed housing funds and long depreciation periods. Assuming that the maximum rent-paying

[38]*Financing of Housing in Europe* (Geneva, United Nations, 1958), pp. 40-41, Table 37.

ability of average working-class households in most western European countries today is about 5 per cent of housing construction cost, the stabilized interest rates should be kept at 2 to 3 per cent per annum, allowing for slight fluctuations due to the difference in depreciation periods, overhead cost, and maintenance and repair expenses. However, such a low interest rate fund is nowhere available in these countries.[39] Thus, while housing construction funds are secured through private financial markets under normal conditions, it is impossible in most countries to keep the annual household expenditure to construction cost ratio lower than 8 per cent or even 12 per cent when the interest rate is high and the amortization period short.

The situation is far more serious in Asia and the Far East. The gravity of the problem is better appreciated if one compares per capita national income of these countries with those of the more developed countries and the combination of high construction costs further complicates the matter. In these areas, with the exceptions of Singapore and Malaysia, three to seven years' household income is needed to build a small house with a floor space area of 30 square metres.[40] It is obvious that both direct and indirect government subsidies are needed to provide housing for large numbers of urban households in these areas.

Thus far we have been examining these economic problems with average income levels in mind. A crucial question relating to the provision of public assistance is the percentage of households, of all income levels, that can afford to pay rent.

Unfortunately, most countries do not have adequate data on this matter.

Increase in income is a big factor in bridging the gap between the cost of housing and an individual's ability to pay for it. In developed countries, in circumstances of normal economic development, the cost of housing has apparently been decreasing relative to income and if this tendency continues, the chances are fairly good that workers will be able to purchase new housing. Prospects for workers in developing countries, however, are dismal. Moreover, the gap between housing cost and rent-paying ability may be even further widened because of the rapidly increasing cost of both housing construction and land, along with the improvement of average housing standards.

The increase in construction cost has been caused by shortages of construction materials and skilled labour, the slow pace in increasing productivity in the building industry, and the growing demands for housing. Needless to say, the rise in land prices has resulted from increasing demands for residential land, especially in countries now undergoing rapid urbanization. Accordingly, rationalization of building techniques and measures for controlling land prices are necessary requisites of any housing policy.

Improvements to the quality of housing are desired by every social stratum. As has already been illustrated, with an increase in income, demands will be made for more interior space, an increased number of rooms, better facilities, increased common space and lower densities—all of which will accelerate the increase in housing costs. While progress in building technology may result in a decrease in the real cost of housing construction in the future, the basic economic fact that the cost of housing is high relative to income levels will not

[39] *Ibid.*, p. 5, Table 4.
[40] *Current Housing Policies and Programmes in the Countries of the ECAFE Region* (Bangkok, United Nations, 1965), p. 4.

change. If there were to be any significant change, it would be due to income increases resulting from increased productivity in fields other than housing. Any reduction in construction costs based on a rationalization of building techniques, will be offset for quite some time by demands for improved quality.

URBAN LAND DEVELOPMENT COSTS

The never-ending increase in urban land development costs is a major problem confronting all free enterprise economies and certain aspects of the problem are also shared by socialist countries. The more important factors contributing to the increase in development costs are: the increase in land prices; the increase in costs of servicing land with improved facilities; and the increase in public regulation of land development.

The increase in land prices is basically due to the increase in population and, more particularly, to the urban population or, looking at it another way, the concentration of that population in urban areas. Although the already highly urbanized countries of Europe are not experiencing the large-scale internal migration typical of many other parts of the world, other factors, such as the previously mentioned increase in the number of households and the desire for improved standards of living, are contributing to the growing pressure on urban land. Certainly, the more extreme instances of the increase in urban land prices are found in the developing countries undergoing rapid urbanization. For example, between 1950 and 1965 some of the undeveloped parts of Calcutta increased in value by 900 per cent.[41] A similar increase in the value of residential land took place

in six major cities in Japan between 1955 and 1965.[42]

Speculation, more than any normal factor of the economy, accounts for this increase. In countries in the early stages of economic development, comparable opportunities for investment in production facilities are lacking. The concentration of population in cities creates a great demand for urban land, and the investment of funds in purchasing land for speculation and land ownership in urban areas is a great attraction for many people. The lack of effective land use planning in these countries results in residential and commercial land alike being affected by the high price of land. This poses a serious obstacle to all forms of urban development, both directly and indirectly, in that it also encourages and perpetuates the fragmentation of land ownership.

Speculation has also been an important factor in the increase in land prices for western European cities, where the increases have been quite remarkable. For example, between 1957 and 1963, the price of urban land in Denmark increased by 278 per cent. In the United States, between 1948 and 1962, the average increase in urban land prices was only 259 per cent.[43]

In nearly all cities of the more developed countries, the provision of such facilities as roads and sewers is considered the responsibility of public authorities. It is largely the municipalities that have to bear the increased burden of cost arising from the higher standards of living and the desire

---

[41]YOJANA, Indian National Planning Commission (26 January, 1966), p. 56.

[42]Heihachiro Adachi, "Present Housing Problems & Home Land Development in Japan". A paper circulated at the Tokyo Seminar of the International Federation of Housing and Planning, 1966.

[43]*Proceedings of the Seminar on the Supply, Development and Allocation of Land for Housing and Related Purposes* (New York, United Nations, 1965), p. 215.

for better facilities. Many countries, therefore, resort to such measures as special funds, low interest loans, national subsidies and so on, for securing funds for urban land development and extending assistance to municipalities. The question of how and by whom the cost of development should be shared is also being experimented with (for example, the development-charge system in West Germany).[44]

The regulation of the land market by the public sector can take one or more of three forms: first, the use of incentives to be applied to the private sector; second, the application of regulatory codes and so on to the public and private sector; and third, the aquisition of land by public agencies.

One means of encouraging the development of urban land is to give a financial stimulus to landlords. Among such measures adopted in western Europe are development charges and land increment taxes or vacant land taxes. These are aimed at prompting reluctant landlords to develop their land in order that development programmes may be realized. The land increment tax has been adopted in cities in West Germany, Italy and Denmark.[45] Tax relief for landlords who cooperate in development of land or in the advance acquisition of land by public agencies is another incentive offered, and often accompanies the use of eminent domain. Complicated by land appraisal and land price problems, however, these measures do not necessarily prove effective when there are shortages of urban land or at times when the land prices are increasing rapidly.

Among the means of public regulation of urban land development are building codes, subdivision regulations, zoning codes, and building permit systems. These measures have been widely adopted in western Europe and, to some extent in Latin America, but far less widely in Asia. Subdivision regulation is not practised in any of the Asian countries. In view of the lack of power to enforce regulatory codes, and because of the low standards of living and housing, these codes in Asian countries reflect low standards and are not very effective.

In Asia and Latin America, where there is a high percentage of the population employed in primary industry and the concentration of large populations in cities is expected to continue, public urban land policies will shift from the negative regulation of, and taxation on, urban land to the positive measures of providing urban land on a large scale.[46] Acquisition of urban land by public agencies has many advantages in that it can help to design the whole community environment and channel urban development in the appropriate directions. It also facilitates land development that is difficult for private developers to undertake. Since increases in land prices and land speculation are largely the result of urban land shortages, many public agencies have recently started large-scale programmes for providing urban land. In Bangkok, 24 per cent of the land in the city is owned by public agencies; in Phnom Penh, Cambodia, 50 per cent of the land is publicly owned; and Karachi and Delhi have launched long-range programmes for acquiring public land.[47] Generally speaking, however, the acquisition of public land in Asia and Latin America is faced with many difficulties because land

---

[44]*Ibid.*, p. 53.
[45]*Ibid.*, p. 53 (West Germany), p. 77 (Italy), and p. 50, 199 (Denmark).

[46]Abrams, *op. cit.*, p. 290.
[47]*Land Policies for Urban and Regional Development in the Countries of the ECAFE Region* (Nagoya, United Nations, 1966), p. 34.

ownership is not only fragmented, but also intimately related to livelihood. Many landlords, if bought out, would not be able to make a living. Also, public agencies often lack sufficient funds and competent administrative personnel. Unlike western European countries, Asian and Latin American countries find it extremely difficult to obtain, in advance, public land for future development. As to the method of obtaining land, voluntary sale by landlords is the most common whereas eminent domain is rarely involved. Eminent domain is generally used only for limited kinds of public works. In many Asian and Latin American countries, there is a great need for streamlining legal procedures concerning eminent domain, for rationalizing compensation and for establishing independent real estate appraisal agencies.

In Europe, cities are making great efforts to provide the right amounts of urban land at the right times. Private ownership of land is common and public agencies are giving top priority to the advance acquisition of land. In the eastern European countries, too, land ownership is not nationalized outside of the U.S.S.R., and efforts similar to those in the western European countries are being made to provide urban land. Slight differences in land ownership are observed between a group of countries which have inherited ancient Greek and Roman traditions concerning communal ownership of land, and a group of countries where private ownership of land is the rule. In both groups, however, the idea that public agencies should intervene in land-use planning for the maximum public good is widely supported. Even in countries oriented to a market economy, public agencies do not encounter strong resistance in their efforts to acquire urban land.

Housing Policy

The emergence of a housing policy, as such, is a fairly recent phenomenon in all parts of the world. It is essentially the clearer enunciation of certain social and economic objectives, relating to housing, which have emerged with the organization and growth of the welfare state. The objectives are common to all countries in that all housing policies aim to meet the quantitative demands for housing and the demands for the improved quality of housing and its environment in addition to ensuring a balanced investment in the housing sector as it relates to other sectors of the economy.

The enormous demands made upon governments faced with the need to rebuild their cities after World War II was probably the first occasion for the development of a comprehensive social and economic housing policy. Since then it has been difficult to separate the objectives of, for example, anti-unemployment housing construction programmes, employer-provided housing, the use of housing investment as a shock-absorber in the economic cycle, and the use of housing as social capital to facilitate labour mobility. In socialist states, the integration of social planning and economic planning gives formal recognition to this all-important relationship. Here we turn to a brief examination of two major adjuncts of housing policy: housing plans and the financial means by which the policy is realized.

In most eastern European countries, long-term housing plans and town planning programmes are coordinated with economic planning by virtue of their being formulated under the guidance of the central governments. In western European countries, comprehensive housing plans aspire only to check the increase in the prices of building materials and to expedite construction projects. Most Latin American conutries have short-term public investment plans, and some countries have national long-term housing programmes under which public agencies and private enterprises cooperate. Many Asian countries have only independent ten-, five- or three-year housing programmes aimed at solving housing shortages by the target years and which specify the amount of housing to be constructed, the ratio of public versus private construction and the method of financing this construction.[48] Whereas these plans anticipate the building of ten to thirteen housing units per 1,000 population per year, these figures far exceed the past construction records (often less than five units); great efforts are required to achieve their goals. In these countries, industrial development policy is also very important for the provision of building materials.

The tools for financing housing policies must be expected to overcome two important problems which have already been elaborated upon. The first of these is the fact that the housing market, left to be controlled by the rent-paying ability of households, would never attract sufficient funds away from more lucrative investments. The second is the fact that the maximum amount which households can pay for housing is one fourth to one fifth of their income in terms of rent and four to six years' income in terms of housing construction cost. Even in the United States, where people enjoy the highest standard of living, it is estimated that one sixth of the population cannot afford to own standard housing.

To overcome these two problems, various measures are being used. In those countries where there is evidence of lack of rent-paying ability, there exist arrangements for rent control, rent supplement, long-term, low-interest loans and publicly subsidized housing. Measures to provide housing funds include state insurance for home mortgages, tax relief, public finance and public housing. Individuals, firms, cooperatives and state or local authorities are involved in the actual construction of the housing.

During World War II almost every country adopted rent-control measures. It was found, however, that the lack of market flexibility in such measures had not stimulated investment in housing construction and maintenance. With the recovery of economies, following the war, adjustment or abolition of rent-control measures has been considered by some countries.[49]

Rent supplements are a device for encouraging private construction of rental housing for low-income families while maintaining the quality of private rental housing. Under this system, public authorities pay the difference between the economic rent demanded by landlords and the amount that tenants can afford to pay. This system has recently been instituted in the United States for low-income families. The federal government pays the difference between one fourth of the family income and the rent.[50] The premium system in France has a similar purpose; the government provides subsidies at the rate of N.F. 10 per one square metre of floor space for a period of twenty years.

The system of state insurance for home mortgages arose from the necessity to facilitate the flow of private funds into housing

---

[48]*Review of the Housing Situation in the ECAFE Countries* (Bangkok, United Nations, 1965), pp. 6-30.

[49]*Major Long-Term Problems of Government Housing and Related Policies*, Vol. 2 (New York, United Nations, 1966), pp. 19-20.

[50]*1965 Housing and Urban Development Act*, United States.

investment and to reduce risks for lending institutions. In the United States after World War II it played a vital role in maintaining a high level of housing construction. The system, however, will not work except in a highly matured economy where funds for housing construction are abundantly available and demands for investment in production facilities are relatively low.

Tax relief can be designed to channel more funds into housing investment. In West Germany, for example, taxable incomes may often be significantly reduced through provisions for accelerated write-offs for personal investment in housing, and up to 30 per cent of corporation profits may be deducted from taxable incomes if the corporations use such money for interest-free loans for social housing construction. These measures have succeeded in diverting more personal and corporate money to investment in housing.[51]

In rapidly industrializing countries, demands for investment in production facilities are high, and low-interest, long-term loans for housing investment become scarce. Under such circumstances, national governments are forced to invest their funds in housing. France has a housing fund for low-income families through which builders of rental housing can borrow up to 75 per cent of the housing cost at the rate of one per cent per annum with an amortization period of forty-five years, and with no payment in the first three years. Government low-interest, long-term loans for housing are common not only in such western European countries as Norway, Sweden and West Germany, but also in Asian countries such as Japan, the Republic of China, India, the Philippines and Hong Kong, and in Latin America where government funds are supplemented by large amounts of housing funds mainly in the form of aid from foreign countries.

After World War II, the ratio of public to private housing construction among the western European countries was highest in the United Kingdom (90.4 per cent in 1948).[52] In recent years this has dropped considerably (42 per cent in 1964)[53] but still remains the highest among western European countries. On the other hand, the percentage of such housing in Switzerland, the United States and Greece is very low.[54] In the socialist countries of eastern Europe, the proportion of public housing is extremely high: the highest being 62.3 per cent in the U.S.S.R. (1964) and the lowest being 23.2 per cent in Bulgaria (1964).[55]

Public funds comprise a varying percentage of total investment in housing from country to country. In 1955, this proportion was 90 per cent in Ireland, 65 per cent in the Netherlands, and 45 to 55 per cent in Belgium, Denmark, France, Norway, the United Kingdom and Spain.[56] In such countries as Greece, Portugal and Turkey, the proportion was as low as one to 5 per cent.[57] Again, in Asia there are great variations. For example, in Japan the ratio of publicly aided versus private construction is 2:3, and in Ceylon the ratio is 1:2, whereas in countries such as the Republic of Korea, Malaysia and the Philippines, housing construction is almost entirely in the hands of private individuals and firms.

[51]Paul F. Wendt, *Housing Policy* (Berkeley and Los Angeles, University of California Press, 1963), p. 134.

[52]*Ibid.*, p. 50, Table III-9.
[53]*Annual Bulletin of Housing and Building Statistics for Europe, 1964* (Geneva, United Nations, 1964), pp. 30-33, Table 7.
[54]*Ibid.*
[55]*Ibid.*
[56]*Major Long-Term Problems of Government Housing and Related Policies*, Vol. 2 (New York, United Nations, 1966), p. 17.
[57]*Ibid.*

Housing Problems in the Metropolis

In principle, most of what has been said thus far on the supply and demand for housing, on housing standards and on the economic and social aspects of housing problems, can be applied on a nation-wide scale to housing in small urban settlements and in the metropolis alike. There are, however, certain conditions prevailing in the metropolis that give rise to special problems in the provision of housing. Most of these conditions relate directly to its major characteristic of both absolute and relative size, others may relate more to its age. Here we examine some of the implications, for the provision of housing, of these characteristics and the conditions to which they give rise.

The significance of the size of the metropolis relative to other urban centres in any one country has been remarked upon by others in this study. However, it is pertinent to comment here on the attraction of the largest urban centres as the most likely source of employment, whether hoped for or real, and the problems that the generally poor, rural migrant faces when he arrives in the metropolis. The lack of housing available for the unskilled rural migrant often results in his having to find shelter in overcrowded housing in slums or shanty towns on the fringe of the metropolis where he is often without transportation to the few places of employment, generally in the centre of the metropolis. However, the rate of growth of the largest metropolises continues unabated. For example, between 1950 and 1960 the population increases in selected cities in Latin America was as follows: Buenos Aires (5,100,000 to 7,000,000); Rio de Janeiro (2,303,000 to 3,200,000); Caracas (694,000 to 1,250,-000).[58] In Asia increases are similarly

marked. In these countries the largest centres are always the strongest magnets for the migrant and yet it is this growth arising out of the attraction of the large centre that in turn causes the problems of the shortage of land, water and housing; the traffic congestion; the pollution of air, soil and water; and the conditions of ill-health; all problems that destroy the attraction that the city may have at one time possessed.

In the more developed countries, the rate of urban growth may not be so pronounced since many of them have passed through the initial stages of urbanization with the industrialization of their economies. However, unless otherwise controlled, there is a tendency for a continuing movement of the already urban population to the largest urban centres and within these urban centres a change in the pattern of settlement. In some countries the growth of one large centre can result in an extremely unbalanced urban settlement pattern. For example, Copenhagen in 1960 had a population of one million, which comprised 22 per cent of the Danish population, and in the year 2000 the anticipated 2.5 million will represent about 41 per cent of the national population.[59] The intra-metropolitan changes taking place as a result of internal migration of the residential population are of great interest for those responsible for providing housing. In London, for example, the core of the metropolis is losing its population while the increase is taking place in the outer fringes. The 1961 census figures for the London region, by zones, are as follows[60]:

[58]Marcia N. Koth, Julio G. Silva and Albert G. H. Dietz, *Housing in Latin America* (Cambridge, The Massachussetts Institute of Technology Press, 1965), p. 16.

[59]Eric Reade, "A New Giant Finger for Copenhagen?" *Journal of the Town Planning Institute* (London, December, 1961), p. 335.

[60]Peter Hall, *London 2000* (London, Faber & Faber Ltd., 1963), p. 18.

TABLE 4-1
1961 Census, London Region, by Zones

Zone	Area (sq. km.)	Population
London Region (total)	11,427	12,453,000
Outer Ring	9,557	4,282,000
Suburban Ring	1,567	4,977,000
Inner Ring	277	2,925,000
Central Area	26	270,000

While the population decreased by 176,000 in the area comprising the core, the inner ring and the suburban ring, from 1951 to 1961, the outer ring increased by about 964,000. About a quarter of this increase in the outer ring was due to a national increase and the remaining three-quarters to migration.[61] A similar pattern of the outward migration of the urban population can be observed in other major metropolitan areas.

This shifting of the population creates problems beyond the provision of housing. It creates new demands for improved transportation facilities since the general tendency is for much of the working population to continue to commute to the centre for employment. However, this needs qualifying to the extent that in some large metropolitan areas, especially in North America, there has been a decline in the status of the central area as a location for business and there has been a strong tendency for many businesses to move to more suburban locations. There are many reasons for this outward movement of business, some economic, some sociological, but in large part they are conditions that result from the characteristic of the greater age of the metropolis and especially its centre. Where the urban infrastructure has

not been maintained, as is all too often the case, the central area becomes run down and blighted. Businesses and residents able to do so move out of the area thus further reducing its viability. However, both governments and the private sector are aware of the need for urban renewal and much has been done to try to halt the decay of central areas merely because they are undergoing change.

The objectives of renewal will vary according to the type of conditions that exist. For example, the cities in underdeveloped countries are more concerned with providing decent shelter than anything else, whereas in North America there is more of a concern for improving the centre of the metropolitan area as a business centre of status and in Europe there is the additional concern for preserving the historic features of the core. In the underdeveloped countries all that is possible, with the limited funds available, is "spot" slum clearance. More often rehabilitation is a cheaper way of improving existing shelter, especially as it is only one of many needs, along with the supply of water, the provision of sewers and streets, and so on.

In many developed countries large-scale slum clearance has been carried out, although this often gives rise to difficult problems in the relocation of residents. There is a greater need for a metropolitan-wide

---

[61]*Ibid.*, p. 80, Table 6.

perspective to be adopted for urban renewal. The relocation of displaced individuals far from their place of work is too frequently a product of poorly planned renewal schemes. The large scale separation of land uses, which is another condition that prevails in the large urban areas in developed countries, means that people may not be able to live close to their place of work. Still, it is undesirable to move them further away than necessary. If possible, the construction of satellite communities should also involve the provision of employment nearby.

Reference has already been made to many examples of different types of new town and satellite developments. However, it may be appropriate to comment briefly on the way in which urban development appears to be leading to the "megalopolis" first described by Jean Gottmann.[62] Gottmann applied this term to the eastern seaboard of the United States, where he observed the merging of urban areas from Boston to Washington. Later megalopolitan developments seem to be appearing on the west coast of the United States and also from Chicago through to Detroit, Toronto and Montreal. In Japan, the Tokaido megalopolis is emerging along the Pacific Coast between Tokyo and Osaka and in Europe the Ruhr industrial complex is beginning to merge with the "Randstad" in Holland to produce a north-west European megalopolis.

[62]Jean Gottmann, *Megalopolis* (New York, The Twentieth Century Fund, 1961), p. 7.

In the developing countries populations concentrate in a few large metropolitan areas, but thus far there has been little sign of megalopolitan development. Indeed the development of Chandigarh, Brasilia and Islamabad as three new capitals may be a sign of the potential that the developing countries have, despite the scarcity of funds, to develop fine cities that will assist in the decentralization of growth of the older cities.

In the future, as in the past, urban form will be the product of the interplay of centrifugal and centripetal forces. Many different types of forms can be produced from the single-centred, high-density metropolis, through the multi-nuclear cities with garden suburbs with moderate densities, to a region-wide, uniformly low-density settlement. Further variation appears in the linear pattern and the similar corridor pattern. There is a great need to evaluate on a more scientific basis the advantages and disadvantages of these urban patterns and their functions. The often antagonistic desires of the public and private sectors regarding, for example, the location of industry must be reconciled in a comprehensive plan. Metropolitan problems are enormous and require the cooperation of all if they are to be tackled logically. Now is the time not only for physical planners but also for persons from all other interests involved in the development of our environment to cooperate in the establishment of a fundamental policy for metropolitan settlement in the future.

5

Education in Metropolitan Areas

Hugh Philp

Every society is faced with the need for the socialization of its children, that is, for making them acceptable members of the society. In preliterate systems this process is carried out by the family and by various informal institutions such as kinship groups, "the old men", the Impi. With increasing social complexity, sometimes including literacy, more and more formal institutions develop and many of the functions of socialization devolve upon them. In this sense an "education system" becomes a formal agency created within the society, sometimes by the government, sometimes by other major institutions such as religious organizations, to undertake certain aspects of socialization. Some functions are taken away from the family and other groups, and are formally passed to the school system. In most modern societies these functions are exercised by agencies of government at a national, provincial or local level. The functions themselves do not change, although the method of exercising them may take many different forms. The functions remain:

    (a) to provide enough people with the necessary skills, knowledge, attitudes, and values to fill the economic and social roles necessary to maintain and develop the society

149

(this is the "social function" of socialization and obviously includes manpower aspects);

(b) to meet the intellectual, emotional, social, and physical needs of the human organisms of which the society is composed (this is the function in relation to individuals as such, called loosely the psychological, or personality, function);

(c) to give to the potential members of the society the values and attitudes characteristic of and essential for the preservation and development of the mores of the society (this is the cultural function in the anthropological sense of "culture").

This formulation is derived from the theory developed by Parsons[1] and others. In all systems of education these sanctions must be fulfilled, whether administered by a metropolis or by a local board, in a small village or by a great nation, as a central activity of government. In the attempt to fulfil them, however, a set of basic problems arises. We are concerned here with the methods and techniques of solving these problems in relation to the development of large cities.

The Basic Problems

An education system must solve certain basic problems if a society is to develop and cater adequately to the needs of its children. Kipling's well-known little jingle provides some useful pegs:

> I have six honest serving men
> They taught me all I knew
> Their names are what, and why and
> when
> And how and where and who.

It is, in fact, impossible to consider other problems of education without careful consideration of what is to be taught to children, and although such a consideration lies outside the scope of the present study, it is important to realize that all the other questions—whom to teach, how to teach, where to teach, when to teach, and even why teach—can be answered only in terms of curriculum. This cannot be said too often nor too strongly, and any discussion of the nature of the provision of education must never lose sight of the fact that children come to school to learn something. Schools are built, equipment is purchased and teachers are trained in order to facilitate learning. The problems of urban schools, like those of rural schools, are essentially problems of curriculum and learning. Although in this chapter we will not be talking about curriculum as such, the basic philosophy underlying everything to be discussed turns upon what children have to learn in terms of curriculum. This said, we shall turn briefly to each of the other questions, describing in outline issues with which the chapter will be concerned in some detail.

More and more in the twentieth century the groups to whom education is to be directed are expanding both in age range and in nature. The concept of life-long education is in line with the familiar anthropological concept of continuous socialization, but adds to it a series of notions concerned with the changes imposed upon a society by rapidly developing technology. There is a growing need for education for change, particularly in urban areas where change is more immediate in its effects, not only in employment but also in social life and recreation. The idea that the community is responsible for a formal system of education is not quite a hundred years old, but the last ten years or so have seen almost every country accept the responsibility for

---

[1] T. Parsons, *Towards a General Theory of Action* (Cambridge, Harvard University Press, 1951).

the compulsory and, in most cases, free education of specified groups of children. The pressure for this has been partly humanitarian but mainly economic. With increasing industrialization, the developing countries since World War II, like the developed countries in the late nineteenth and early twentieth centuries, have extended both the compulsory range of education and the voluntary range, although this has seldom been done on much more than an *ad hoc* basis. Most countries have been obliged to make choices in terms of groups for whom provision can be made. It is easy to point to the more obvious choices—urban or rural, girls or boys, the gifted or the rich, the black or the white, but there are more subtle distinctions which are often typical of metropolitan areas: access to transport is one, employment opportunities another. Consideration of such factors frequently leads to educational opportunity for certain groups in the community and denial of this opportunity, or its offer in a limited form, to the other groups. *Who shall be educated?* is still a real and critical issue in most parts of the world.

The third set of problems is concerned with how to educate. How should the curriculum be brought to children? What facilities have to be provided? This question embraces the economics of education. Which should be used: teachers or television or programmed material linked with correspondence education? Is there a different answer for different groups at different stages of their educational careers? How should the system be run? How can it be financed?

Clearly, this question is linked with the "where" problems. Should the traditional school buildings or a complex of buildings be used? Where should they be located in relation to each other, to habitation areas, to transport, and to other related facilities? If the traditional pattern of schools is not used, where else should education take place?

The "when" group of problems separates into two distinct sets, one related to the division of the day and the year, and the other to the division of the school life of the child. The first sub-set is concerned with questions about length of school day, with staggered hours, with efficient use of expensive and sometimes scarce plant and equipment, with seven-day and year-long use as opposed to use for six hours a day, two hundred days a year. The second set poses questions about the length of the compulsory school life, about optimal ages of beginning and ending different phases of education and, critically, about ages of the transition from school to work.

Finally, there is a set of questions that begin with "why". These are essentially philosophical and, like those concerned with curriculum, they are crucial. They are concerned with the aims of education and also with the relation between other specific issues and the stated aims.

For those concerned with education in metropolitan areas these questions remain valid. The rest of this chapter is concerned with showing how the special conditions of the twentieth-century metropolis places restrictions on the kind of answers that can be found to these questions. It is also concerned with the kinds of solutions that are possible because of the special facilities, as well as the special disadvantages, of the metropolis.

The Education System and the Metropolis

In education, as in many other spheres, services must meet the needs not only of the metropolis itself but also of the wider community of which it forms part. The degree of responsibility varies greatly according to the specific place, social and economic as well as geographic, which each particular metropolis occupies vis-à-vis its hinterland. Metropolises like Sydney or Bangkok or Paris have a very different set of responsibilities and functions in education from those such as Toronto or Kobe or New York. The first group are not only state or national capitals, but also the centres of administratively centralized education systems. Their education authorities are responsible for planning to meet the needs of the metropolis and other large cities, as well as the smaller urban centres and rural areas. The Ministry of Education in Paris, for example, has to make careful and detailed provision, financial and administrative, for education in cities like Lyons and Marseilles, as well as for metropolitan Paris itself. Differences of this kind impose many restrictions, some of which greatly affect the kind of system which can be evolved to meet the economic and social needs of the community. It is clear that even within a more or less uniform culture like that of Australia, the social and economic needs of a metropolis like Sydney differ in kind as well as in degree from those of a small town like Milparinka. Yet, where there is centralized control, as in this case, there is an obligation to provide equal opportunity and facilities to all children. This is frequently interpreted as meaning the same opportunity and the same facilities. Thus in systems like the Australian or the French, the planning and provision problems differ somewhat from those of heavily decentralized systems like the Canadian or English. It will be necessary to return to this issue again, particularly when considering the appropriate structure for administration of education in cities. It has important implications for many other issues such as teacher education and teacher employment.

More generally, however, the education system of a metropolis is faced with the major tasks described earlier: meeting the personality needs of children; giving to prospective citizens the necessary skills and knowledge to fill the enormous range of social and economic roles essential to the life of the community; and assisting and furthering the attitudes and values of the society. For the moment we are most concerned with the social and economic functions of the system. If the education system is to provide an adequate flow of manpower for industry and commerce at all levels and at the same time meet the social requirements of its existence, then educational planning must be integrated with social and economic planning. In the twentieth century it is useless, and indeed it can be dangerous, to look on education as an entity in itself or as concerning only the psychological and social needs of the child as an individual. The phenomena of the educated unemployed are a steadily growing illustration of the danger of such a point of view. To provide "education" without considering how children endowed with this "education" are to use their knowledge and skills in the community is to plan for tragedy. In recent years events in many North American cities, in Tunisia, in the Philippines, in Ghana, in Egypt and all over Latin America are evidence enough that education systems not geared to employment opportunities can lead to disaster. As Conant argues with devastating clarity, the problems created are directly proportional to the size of the city:

> Out of work and out of school since they turned sixteen, these youth [urban children in slum areas] behave in ways that may have serious political consequences; similar behaviour of youth in smaller cities would be far less serious. It is a matter of geography in the last

analysis. Three factors are significant: first, the total size of the group to whom I am referring—the larger the group, the more dangerous; second, the density of the population—the number of frustrated youth per block; third, the isolation of the inhabitants from other kinds of people and other sorts of streets and houses. If one compares the slum areas in the largest cities with similar districts in small cities, the difference as regards those three factors is clearly evident.[2]

This is partly a function of the segregation of distinct economic and social groups in the metropolis, exacerbated in some countries by segregation of an "ethnic" or religious nature; partly a function of a failure to relate school programmes to social and economic demands; and partly a function of migration to the metropolis, both internal and external, of young people with an educational background ill adapted to the needs of urban living.

Social and economic opportunities within the metropolis impose restrictions on the groups to whom educational facilities are and can be offered. It is all very well to argue on theoretical, psychological and educational grounds that in the mid-twentieth century all children should stay at school for twelve or thirteen or even fourteen years, but if the economic structure of any given metropolis is such that most of the industrial and social tasks can be adequately filled by young people after six or seven years at school, then longer programmes are not only uneconomic but can lead directly and inevitably to social disintegration and chaos. In this context "equality of opportunity" surely means a great deal more than the right to go to schools that provide the same curriculum and facilities with equally well-trained

teachers in comparable surroundings. It means also the right to an education that will prepare them for the kind of employment and social life that their society offers. The education system must prepare young people to take advantage of opportunity in economic and social terms as well as in educational and psychological ones.

In a real sense the nature and efficiency of the formal education system determines the "success or failure" of the city as a viable human environment. This is most obvious in cities in basically agricultural developing societies. In the rural areas in these countries the patterns of family and community living still provide agencies of socialization that are in themselves almost as self-sufficient and viable today as they were two or three hundred years ago. Here the education system can be conceived more narrowly and is less closely related to the day-to-day needs of the community than it is in the cities. This is changing, and rightly so, but it remains true that planned, formal education systems are more crucial to the viability of a developing metropolis than to its rural hinterland. In the heavily industrialized countries the statement is true in a somewhat different sense. Planned programmes of education are clearly essential for rural areas and small towns, but there is less need for diversity and less demand for complexity than in the metropolis. If the education system is not carefully geared to the growing needs of the city then the city collapses. One could argue historically that part of the reason for the collapse of the Greek city-states was their failure to develop educational programmes to meet the growing diversity of their social and economic needs. The same problem faces the cities of the twentieth century, most acutely in the rapidly growing metropolises of the developing countries.

---

[2]James B. Conant, "Social Dynamite in our Large Cities," *Vital Speeches* (New York, 1961).

Acceptance of the view that educational planning is integral to social and economic planning has a number of interesting and important implications. Not least of them is the fact that the school complex is, in the twentieth century, the only institution common to all members of the community for a considerable sector of their lives. Just as in western Europe and North America the church or chapel used to be the common geographic factor in the lives of men, so today is the school.[3]

In the medieval cities the castle, the cathedral and the churches were the foci around which the urban complex developed. Even today in Buddhist communities the temples represent the planning centre, whether this planning is conscious or unconscious. In the eighteenth and nineteenth centuries, as industrial cities began to grow, groups of factories or mines became the foci. Today there is sound reason for arguing that the centres around which urban planning should take place are the educational institutions. In a city such as Accra or Bangkok rather more than 30 per cent of the inhabitants are children of school age. The proportions are somewhat smaller for London or Moscow, but they are still more than 20 per cent. School-age children, then, are the largest single group with a common set of geographic objectives and if we add the number of adults whose activities are directed toward the school for at least part of every day, it may well be postulated that urban development if it is to be planned should be planned around the complex of schools and other educational institutions.

The sheer intricacy of a modern school system lends weight to this view. For instance, Istanbul is a metropolis of some 1,500,000 inhabitants of whom about 250,000 are students in full-time educational establishments. There are more than thirty different types of such institutions maintained by the government, not including private schools, special schools for atypical children of various kinds, nor a large group of part-time trade schools and Koranic schools for adults. In a really large metropolitan area like London the number of different types is correspondingly greater. The task of providing physical access to the complex is immense. Admittedly, the problem is greater than it need be because of the lack of integrated planning of the education system itself, but even with careful rationalization the complexity remains staggering.

At the primary level it is fairly easy to solve the problem of access by using a rectangular block access pattern, as in Moscow, or a hexagonal access like that being planned for a number of new and developing cities. Solutions are not even particularly difficult in older, more heterogeneous cities where transport lines and land use are already firmly established. At the secondary level, however, access depends largely on the administrative school pattern and on the basic physical shape of the city, its existing transport facilities and so on. The problem of access, therefore, becomes more difficult. For example, with the pattern that has evolved in Turkey, where there is such a proliferation of school types at the secondary level, the problem of providing equality of access to all children of all types is well nigh impossible, and the tendency is at best to locate specific types of schools in relation to particular types of industry, thus exacerbating any existing tendency to segregation. Even in communities where secondary schools are "comprehensive" in type, with a variety of

[3]For a detailed discussion of this thesis, see, for example, Lewis Mumford, *The Culture of Cities* (London, Martin Secker & Warburg, Ltd., 1938).

programmes in the same building or set of buildings, problems of access are sometimes difficult to solve. They are worst in older metropolitan areas where the institutions have to fit an existing urban form rather than being part of a planned whole. The task, for example, of transforming the Sydney system from one of proliferation, diversification and selection to one of comprehensive neighbourhood secondary schools has taken rather more than ten years to complete. The difficulties of solving this problem have been heightened by the fact that Sydney is almost archtypical of a large area of dormitory suburbs surrounding a relatively small commercial centre with industry tending to spread in only two directions (in this case, west and south) from the city centre. Here the pattern of educational institutions at the secondary level has tended to be set by the habitation pattern, although there are some imaginative signs of secondary schools being established at key transport access points before major housing development has taken place. At the other extreme, in many large areas of Moscow a housing block system is directly and functionally related to a pattern of primary and secondary schools. However, it is pertinent to ask whether such a planned pattern may not, under specific social conditions, lead to and intensify segregation of children into tight neighbourhood groups.

At the tertiary level the problems are extremely difficult, particularly when there is a spread pattern of secondary schools. They are further complicated in cases where a small number of universities, teachers' colleges and other institutions serve a large area. In some instances tertiary institutions for an entire state or even country are located within only one or two metropolitan areas.

The Practical Issues

ADMINISTRATION

In considering problems of education administration, it is useful to look first at the basic types of political organizations that control education, and then at the administrative structures that serve them. It is possible to distinguish four major types, although mixtures of them are far more common.

(a) *A single government nation state with a central control responsible for all education within its borders.* In such a system finance, building construction, teacher training and supply, equipment, and all other resources are controlled by this central political unit. In practice, except for a few countries with somewhat specialized political and social systems, such as Albania or Kuwait, pure examples of this type are rare for they tend to be complicated by the presence of private schools, frequently of a religious character, not under the direct control of the state. Leaving aside such circumstances, the type case in Europe is France, where all education is controlled from a central Ministry of Education in Paris. This system is the most common form of control in the developing countries of Asia, the Arab states and Africa, with Indonesia as the outstanding example.

(b) *A federation of states, within each of which education is politically controlled in much the same way as in type (a).* In such systems, however, it is common for the federal government to have some responsibilities for education, either in terms of finance or in terms of certain groups of institutions. The best examples of systems of this kind are Australia and Nigeria.

(c) *Nation states with separate governments but with some political control of education at the local level.* Pure examples of this kind are hard to specify, with the most important exception of a number of eastern European countries such as Roumania and Poland. Even among these, as in mixed types (for example the United Kingdom or Sweden), there tends to be some central political control, particularly of general policy. Local administration, therefore, is not really independent of central government control.

(d) *Federal systems with a great deal of local control of education.* This is the familiar North American pattern, which has been exported more or less successfully to a number of Latin American countries.

Most systems are mixed, although the ingredients of the mixture vary greatly. In a unitary state or a federation, the central government usually controls finance, teacher training, and to some extent, school building, while local authorities accept responsibility for the day-to-day running of the system: teacher employment, school plant planning, and equipment. These systems usually have a central Ministry of Education together with some form of local school board, as in England and Wales. There are a number of arrangements whose rationales are rather more difficult to understand, which sometimes leads to acute administrative problems.

The situation in Thailand is worth describing in some detail, as it illustrates some key issues of administration in major cities. Politically speaking, there is a central government, residing in Bangkok, and the country is divided into seventy-one provinces, each controlled by a governor appointed by the Ministry of the Interior of the central government. There are also a number of municipalities, more or less identical with the capital town of each province, although some provinces have more than one municipality and some have

none. There is also a central Ministry of Education that is responsible for all education except primary education, which is controlled by the Ministry of the Interior, university education, controlled by the Prime Minister himself, and some special schools controlled by the relevant ministries, for example Agriculture or Health.

In the City of Bangkok, then, leaving aside the private primary and secondary schools, there are three major ministries involved in the administration of education, the Ministry of the Interior alone working directly through the municipalities. Urban planning around the school system is almost inconceivable under such conditions, particularly since the school land is owned by the relevant ministry and not by the municipality. One further point is important: in Thailand, as in the United States, the politically responsible person is also the administrative head. The Minister for Education runs the ministry and is responsible to the cabinet. The British "administrative head", as distinct from an elected politician, is a permanent civil servant who continues in office irrespective of the government, and is roughly equivalent to the Thai and United States offices of Under-Secretary. The direct administrative responsibility of a minister is an important factor in considering administrative control of school systems because of its restriction on continuity of policy.

The patterns of administrative control are even more diverse than those of the political systems on which they depend. Probably the most common systems on a world basis are those in which the central Ministry of Education of a federation is both the political and the administrative centre. Next most common is probably the extreme opposite—those whose system is based on a local neighbourhood unit, as in the United States or Canada. Given the

FIGURE 5-1
Paradigm of Classification of Control for Educational Purposes

| | Medium-sized (2 million +) | | Larger-sized (4 million +) | |
	Developed	Developing	Developed	Developing
Central control	Sydney	Singapore	Paris	Bombay
Local control	Toronto	Bangkok (mixed)	London	Tokyo (mixed)

great diversity among the geographical and cultural factors, as well as demographic problems, with which each government must contend, arguments would be raised in defence of almost any workable system of education.

Our immediate concern, however, is with education in the metropolis, and at first sight it would appear that there is much to be said for a locally-based, political and administrative system. It seems reasonable that local interests will be most closely involved, that children will be looked on as children and not merely as pupils, and that adult goals will be most closely linked with immediate social goals, where funds are locally raised, where teachers are employed or even trained by the local community, where schools are built and the curriculum planned on this basis. This argument holds sway in Britain, the United States and Canada, anyway. However, it may be argued that total planning of education in relation to economic and social needs is difficult to achieve under these circumstances, and it is mainly for this reason that most developing nations have eschewed the pattern of local control.

For purposes of analysis it is useful to postulate three variables by which cities may be classified:

(1) type of control of the educational system;

(2) size of the city;

(3) stage of development of the country itself and of the city within it.

In Figure 5-1 these variables have been used to construct a paradigm that illustrates the essential differences among eight exemplary cities.

A cursory glance at the *type of control* of the educational system within which each of these metropolitan areas operates, shows that special problems are created purely as a function of the structure without beginning to consider the functions of the system. Perhaps the best example of this, apart from the special and difficult problem of finance, is that of teacher training and teacher employment. In most systems where there is central control, teachers are trained by the state and then employed by the state Ministry of Education. There is no "free market", and as a consequence any one school or even group of schools has little or no control over its own staffing. Young people are selected at school level and are offered training, sometimes with pay. On graduation they are allocated to schools and usually shifted from school to school over a period of time either to meet staffing needs or to gain promotion within the system. In systems of this kind it is rare for teachers to spend their entire careers in one school or even in one town or suburb within the system. A highly centralized administration with control over a large area is almost forced to move teachers frequently if it is to provide good promotional possibilities for them and, more important, equality of opportunity for all children in all schools no matter where they are located.

Because it is centralized and because the minister is a politician and, therefore, likely to be motivated for political reasons rather than those of efficiency, such a system cannot afford to have rich districts and poor districts in terms of quality of teaching. Where there is local control, on the other hand, teachers may be employed within the system itself, depending on its size, but more commonly the free market operates. A variation of this is a central training system with local teacher employment as exists, for example, in Japan and Thailand. However, problems of pensions and other teacher rights become acute under local systems or mixed systems of this character.

Within this context, problems created in metropolitan areas by the central control of education are obvious, and they are exacerbated by the political issues. Again, Sydney is a good example. The city government has nothing to do with the planning and control of education as such, although it does have a major responsibility for transport, water and sewage, electricity, land use in general, and other services vital to an operating school system. Control of education is firmly vested in the Minister of State and, through him, in a civil service hierarchy of administrators and teachers. Political control is thus with the central government and its servants, who are necessarily as much concerned with rural areas, towns hundreds of miles away, and with other cities as they are with the problems of metropolitan Sydney. Further, the central government is almost invariably ex-

pected to develop education policies in response to the disparate demands of industry, commerce and the like, rather than from a concerted expression of policy needs from the metropolis per se.

On the other hand, even a city with some control of its own educational destiny has other, equally difficult problems. For example, only in the most affluent countries, and decreasingly so among them, can a city raise enough funds to maintain its own education system without forms of aid from the central government. Indeed, almost the only pure cases of this kind of control that remain are in the modern city states of Hong Kong and Singapore.

The issue of finance is one to which we will return, but for the moment it is enough to point out that central governments as a rule are loathe to give money without some political control over ways in which it is spent. This is most acute for the medium-sized metropolis with a population around the two million mark, particularly if it is a state capital or the only large metropolitan area within the state or country. Many of the problems of developed metropolises like Toronto, Copenhagen or Liverpool spring from the conflict between local and central governments, although, in theory, education in the area is the responsibility of the former. In cities in developing countries these issues are sharpened. If we take Bangkok again as an example it will be recalled that control of primary education is the responsibility of local government, with the central ministry looking after secondary and vocational education and all forms of tertiary institutions. However, almost all the funds come directly from central taxation. In addition, the Ministry of the Interior and the Ministry of Education both exercise considerable influence over what happens in the primary school—the Ministry of the Interior because senior city officials are responsible directly to the Ministry, not to the people of the city, and are often civil servants of the central government; and the Ministry of Education because teacher training, supervision of curriculum and other purely technical matters are its overall responsibility. In fact, the city itself is left doing little more than looking after the day-to-day running of the schools, with a little control over new buildings and sites of primary schools. A similar situation obtains in a number of Latin American cities, particularly in federal states like Argentina and Brazil.

If we turn to very large metropolitan areas to look at the same issue, the problems are much the same but are greatly magnified in scope. Paris, for example, has a conurbation population of nearly 8,000-000 out of a total French population of about 48,000,000. All education in the nation is controlled from the central ministry: teacher training and employment, building construction, equipment, text books, curriculum and so on. While it is true that there are local advisory boards, and some delegation of authority to the communes of the Department to raise and spend money, strict control is maintained over this. One consequence of this arrangement has been a uniformity of schooling. The relative gains and losses in effectiveness that uniformity entails are not under discussion. What is relevant are the implications for schooling in Paris itself. Essentially, the government has to choose between the following alternatives: it can devise a system relevant to Paris and hope that it will work for the rest of the country; or it can try a broad, general pattern that ignores the special needs of the metropolis. Because of political and administrative issues, the theoretical policy decision has been for the second alternative, but the administrative reality comes very close to the first. Clearly, the administrative reality has nothing to do with the people who work in the *hôtel de ville*; it is determined by the civil servants in the Ministry of Education. The rigid, highly efficient character of the French civil service is still the most powerful single force in the education system, and while this civil service is Paris-oriented in a very special sense, it does not plan the metropolis as such.

Of a quite different character are the problems of Tokyo, a developing metropolis with a metropolitan government controlling education within its borders and with little responsibility outside that area. Tokyo, in addition to being very large in terms of population, is also the national capital and the site of the national legislature. Further, a large proportion of its education budget comes from central funds. Teacher training and employment, curriculum, and certain other matters, although the responsibility of the Tokyo Board of Education, are undertaken according to laws and regulations laid down by the central ministry. The ministry also determines general education policy so that, large though Tokyo is, the education administration is in many ways no more than a body with delegated authority. It is by no means comparable with the Board of Education of, for example, New York City. Education in Osaka or Kobe is, with minor variations, no different from that in Tokyo. There is one additional point that does give the metropolis and other Japanese cities more autonomy: land use and, hence, urban planning are to a large extent controlled by the metropolis itself (including decisions on school sites, the architecture of school buildings and, to some extent, the kind of equipment to be installed).

The question of equipment raises a somewhat different issue common to metropolitan areas that are also national or state centres, whatever the nature of the administrative control. Because of its facilities for communication in the broadest sense, the metropolis frequently and increasingly offers specialist services to its hinterland. The obvious examples are educational radio and television. This immediately raises administrative, and hence political, problems whenever there is local control. We shall return to this point.

Thus far, mention of local control has implied control of the entire metropolitan complex by one authority, but this is frequently far from being the case.

Metropolitan areas are frequently fragmented into sets of sub-systems of education, although these are still under government control. This sort of fragmentation is mainly confined to North America, particularly the United States, and in a rather different way to England and Wales. In many countries, there are other kinds of sub-systems that are controlled by private individuals or organizations, sometimes for profit, but more frequently for religious or social reasons. Percentages of children in such sub-systems vary greatly. For example, in the United States about 14 per cent of children are in private primary or secondary schools; the corresponding figure for Australia is over 20 per cent. In the Netherlands at the primary level it is nearly 70 per cent and in England and Wales between 7 and 10 per cent. In most countries there tends to be a higher concentration of these schools in metropolitan areas than in small towns and rural areas. For example, in 1960, 79 per cent of all children in private schools in the State of Illinois were in Chicago. In the same year in Thailand about 30 per cent of all secondary school children were in private schools. About 80 per cent of these were Bangkok, where they comprised more than half the total secondary enrolment.

In some developing cities, particularly in Africa, the percentages are even higher because until recently the only organized secondary systems for indigenous people were run by missionary societies. The situation is changing rapidly but the balance in a number of countries is still with the non-governmental schools. For any city this creates special difficulties, some of which are related to the question of education of minorities, but others are rather more general and fundamental. For example, in northern Europe, North America, Australia and New Zealand, where the majority groups tend to be Christian and non-Roman Catholic, most non-governmental schools tend to be Roman Catholic and controlled by teaching orders or by a local parish system. This parochial system may or may not be integrated with some overall supervision parallel to the government system of control. There tends to be rather more autonomy in non-curricular areas for the private schools, so that small school units tend to be more independent of the overall controlling authority than are the equivalent government schools. In terms of curriculum, on the other hand, the reins are rather tighter on the parochial school.

By and large a parochial system may be considered as a sub-system, particularly within a city, although political control and administration are independent of the government system, because city planning is able to integrate a parish educational system, *mutatis mutandis*, within its overall education programme. State or local financing of such schools is common and the regulations and rules accompanying this aid are a means of ensuring integration without exercising direct administrative control.

The situation is somewhat different for the schools controlled by religious teaching orders, for not only are they independent of the state or city system, but also (for the most part) of the local hierarchy of the Roman Catholic Church. They tend to operate on a national level in education, and their administrative links are quite different. In those countries where control of education as a whole is central this does not create many difficulties, but where administration is local, special problems inevitably arise. Where control is central, relations and planning with extensive teaching orders, especially as regards matters of planning, are relatively simple, but difficulties crop up in relations with the parochial schools that are controlled locally. The opposite situation obtains where the government education system is essentially local—relations with the parochial schools are easy and with the teaching orders relatively difficult. It should be emphasized that these issues are in a real sense administrative and thus, in one way, independent of the issue of minorities. In quite a different sense they are, of course, closely related to the problem of the education of minority groups.

So far this analysis has focussed on the first of the three variables relevant to planning for education in the metropolitan areas: that of the type of control of education system. It is to the other issues that we now turn, although rather more briefly.

City size is the subject of discussion in other chapters and need not be raised here as an issue in itself. It is important to note, however, that educational problems appear to increase in complexity as well as in magnitude as the size of the city increases. James Conant has already been quoted on the explosive potential in groups of educated unemployed and how this potential increases with magnitude, and brief mention has also been made of the greater variety of institutions to be found in the larger metropolitan areas. There are a number of other indicators. The larger the city the more complex are its relations with its hinterland, and a number of these complexities tend to be educational. Mention has already been made of the call on the metropolitan area to provide facilities outside its boundaries, whatever the administrative system. These certainly increase with size if only because greater size generally means greater resources. For example, the enormous range of educational television available in Japan, blanketing the entire country and school system as it does, is run from Tokyo and is workable because of the nature of Tokyo. Yet, Paris, even with 8,000,000 people, does not appear to have the resources to provide a similar system in France, despite the far more heavily centralized school system and the more highly developed social and economic system. On a different level, metropolitan areas are able to provide specialized services such as correspondence schools for persons from centres outside their own boundaries. The nature and extent of such services are largely a function of size and complexity.

Similarly, the larger the city and the more varied the institutions it provides, the more migration there is to it for educational purposes. For example, roughly half the secondary school children in Bangkok in 1963 came from small towns and rural areas. In Tokyo there is an entire city area (Ochanomitzu) almost exclusively devoted to universities and other tertiary institutions, both governmental and private. Some of these institutions are highly specialized and draw a large percentage (estimated at about 30 per cent) of their student body from the remainder of Japan. In the cities of the developing world, where almost all tertiary institutions are located in capital cities, the proportions are even higher and raise administrative problems of great complexity, not the least of which is adequate dormitory accommodation.

Apart from these problems, which concern the relations of the metropolis to outlying areas, there are those of the metropolis itself which are conditioned by its size. There is little need at this point to emphasize the problem of providing books, equipment and other materials for five or six million children in Tokyo, London or New York. Such issues are obvious. More subtle are the administrative problems relating, for example, to special groups. The specific problems of minorities demand more detailed discussion, but at this juncture it is enough to point out that while there may be no case for special programmes for one Helen Keller, in a town of 250,000, there certainly is a case for such programmes when there are as many as a hundred or more children like her in one age group in a very large metropolitan area. These are educational problems of the "who" kind, which become definable and manageable only when the administrative unit becomes large enough to handle them. Thus, not only do new problems arise purely as a function of size, but also the possibilities of solution are directly related to size.

Similarly, new problems emerge related specifically to the *stage of development of the area*. The purely educational problems of Bangkok are different from those of Sydney, those of Calcutta from Chicago, of Cairo from Leningrad, of Singapore from Toronto, although in each pair cited population sizes are roughly equal. Nor are the differences in the scope of problems entirely attributable to historical, cultural and social factors, although these play an important part. More vital are the economic and political aspects of development. It is perhaps well at this stage to specify the sense in which "development" is being used here. At least three different uses of the word "developing" may be distinguished, in addition to the sense in which a great, established metropolis like Rome, Moscow or New York is still "developing". First, there are the metropolises that, although very old and centred in ancient cultures, are undergoing a rapid and profound commercial change; for example, Cairo, Bombay, Rangoon and Bangkok, and indeed most capitals in Asia, Africa, and Latin America. Second, there are a group of relatively new metropolises in these same countries, such as Rawalpindi and Ibadan, where populations have increased three or four times since independence. Third, there are new cities in established industrialized societies, such as Canberra, in Australia, which was little more than a village before World War II and now has some 90,000 people or Alma Ata, in the U.S.S.R., which has been deliberately developed from a remote provincial city of some 230,000 in 1939 to an industrial metropolis approaching 1,000,000 people in 1966; or the whole development of new towns in Britain.

For the moment it would seem preferable to restrict the term "developing" to the standard United Nations terminology: to cities and countries officially classified as "developing" or "emerging" or "nonindustrialized". That is, the term will be restricted to the first and second groups described above, although it will be necessary to make some reference at a later point to the third group because of their importance as planning laboratories. Many of the cities in the first group are indeed old, but in the last twenty years or so, with strong pressures to industrialize and with the impetus of foreign aid of many kinds, they have begun to grow again and to change more than they have in hundreds of years. A most important facet of this change has been the realization that educational development and planning is an integral part of social and economic planning. There has been an upsurge of demand for more education of different types, and for the teachers, buildings, equipment and other facilities that go with such a demand. It has been estimated, for example, that if Damascus were to be equipped with schools and other educational institutions on the same level as a western city of comparable size, all building within the city would have to be devoted to educational purposes for the next twenty years.

The implications of such a statement are staggering when considered in terms of finance and resources, especially when it is realized that for many reasons and for many years to come the main growth in educational building will have to be in the cities of the developing countries. The rate of such growth, if plans are at all realistic, will have to be much faster than anything previously attempted by man. To take an extreme (probably the most extreme) example, in Kabul, Afghanistan, there are about 500,000 people, of whom perhaps 200,000 are of legal school age. The official illiteracy rate in the country is about 90 per cent and in Kabul itself it is at least 80 per cent. Less than 9 per cent of all school-age children in Afghanistan are in school; in Kabul this may be as high as 20 per cent. The official target of the so-called Karachi plan for education in Asia is to get all children into elementary schools for a minimum period of seven years by 1980, but more realistic and recent estimates for Afghanistan suggest that the year 2000 is more probable for this country. At this stage perhaps 30 per cent will go on to some secondary education and 5 per cent to tertiary education. Even such a modest plan will mean building at least 300 schools in Kabul alone before 1980. This is a serious challenge to the planning of the city

itself, which at this stage can best be described as primitive in the extreme. But, unless the schools are developed rapidly there is no chance whatever of economic and social growth even approximating the desired targets.

Such figures could readily be parallelled, although admittedly in a slightly less extreme form, for fifty or more countries, in some of which—India and Indonesia for example—the problems are almost inconceivable in their size. What kind of educational growth pattern can be predicted, much less recommended, for metropolises like Calcutta, Djakarta or Bombay; all with high rates of illiteracy, very few secondary schools, and limited resources? Clearly some kind of growth pattern for education must be conceived, within which the planning of schools and other institutions for the metropolis can be considered. Because the problems of many developing countries are so different in magnitude, they must be treated as different in kind from those of the developed countries, and as a consequence the sort of planning for a metropolis in India or Indonesia must be different from that for a metropolis in the United States or western Europe. This statement is open to challenge, but it is certainly worth discussion.

The role of the metropolis in meeting the demands of the national educational policy must be considered when the decision is made regarding the organization of the administrative structure of the education system. If leadership is to be encouraged from metropolitan areas, it is questionable whether one should argue for a centralized system. Should there be strong political and administrative control from a central Ministry of Education, together with a national teaching force? Are the risks in such a procedure so great for metropolitan development that they outweigh the overall advantages? Such questions are also related to organizational choices concerned with the question of who shall be educated. If there are to be good teachers, there must be a strong secondary school system, which in developing countries is most likely to be concentrated in and around major cities. But few "new" systems of education can afford to develop strong secondary school systems (particularly if they have an important vocational component) together with the essential tertiary institutions, and *at the same time* greater universal compulsory elementary education. Any analysis of documents like the reports of the ministers for education in Asia, Africa, the Arab states or Latin America makes this point abundantly clear. A choice has to be made, and is usually expressed as between compulsory education with a minimal amount of secondary and tertiary education, and slow development of elementary schools linked with development of an elite system for secondary and tertiary education.

However, the choice may be posed another way. Should we develop full-scale educational services as rapidly as possible within the cities of developing countries and go slow in rural areas, or should a somewhat lower level of development be the goal, but with equality of opportunity evenly spread? This choice is frequently complicated by an overwhelming, and natural, popular demand for universal literacy. What should be done? The suggestion has been made more than once that the best answer is to build up full-scale metropolitan systems as rapidly as possible, while accepting the political and economic dangers inherent in slower development of a primary system in rural areas. The choice is not easy but it must be carefully considered by each country for itself, for whatever decision is taken will obviously affect the whole pattern of metropolitan education in the developing countries. It is easy to say that choosing an élite or selective development is undemocratic, but the whole history of education in the developed countries happened this way. Even today in almost every so-called developed country of the world provision of education in metropolitan areas is much greater and usually much better than in the rural hinterlands.

Given the existing kind of economic and social patterns, including those of socialist countries, such a development appears almost inevitable. This trend is strengthened in the developing countries by their need to grow quickly. Whereas established cities in developed countries tend to have established school systems, cities in developing countries have to build up their systems as they themselves grow. Established systems certainly have to change to meet new challenges and new ideas, but it is change within an on-going system whose administrative framework, schools, teachers, building, equipment and so on are for the most part already there. To add one year to what is already an efficiently running twelve-year system causes strains, but they are not nearly as great as trying to create an eleven- or twelve-year system within ten years from scratch. The new secondary school system in Sydney, for example, demanded about a 20 per cent increase in the teaching force over seven years. To give similar provision in Bangkok would mean something like a 600 per cent increase with corresponding demands on other sectors of the education system. Obviously the choice of an administrative system must take into account the type of financing to be employed.

FINANCE

Systems for financing education closely

parallel administrative systems although they are not identical. The divergence is greatest when it comes to allocation of budgets and delegation of responsibility for spending. In the centralized system funds are raised through central tax arrangements and are allocated and spent by the central ministry of education with some relatively minor delegation of responsibility for particular purposes. The type case among developed countries is probably Sweden. All money for education in the government system is raised by public taxation and spent by the Ministry of Education and Ecclesiastical Affairs through the Boards of Education and Vocational Training and their officials, to whom there is some delegation of authority depending on their position in the hierarchy. The annual allocation of funds for education is determined by the central legislature which thus essentially determines policy. There is a precise parallel between fund raising and fund spending, and administrative authority and responsibility are identical. However, there are some exceptions which serve to show the difficulties, if not the dangers, of generalization. Most funds for the universities, for example, are provided by the state, but on the other hand universities have some freedom to raise funds from endowments or research contracts, and they are autonomous in administration. In other countries with a similar form of control, universities also have the freedom to raise funds. Furthermore, there is some state aid to private schools, particularly religious schools, although there are regulations and rules which make them part of the total system. Clearly, just as there is no real "Stockholm" or "Paris" or "Sydney" administrative school system, so the city has no real responsibility for, or control over, the raising or spending of money for educational purposes. Paris, like all the other cities in France, does have some control over purchase of sites and erection of buildings, as do Stockholm and Sydney, so that in this very narrow sense it does play a small financial role in education.

In the decentralized system, predominantly North American, a high proportion of funds is locally raised by some form of rate or tax and is allocated and spent by the local authority which raises the money. This system in any pure form appears to be dying, if indeed it ever existed other than at the primary and perhaps secondary school levels. Theoretically at least, in such a system the city is free to raise funds for its own educational purposes and to spend them as it pleases within an authority structure responsible to the city. An important difference between this and the central system is that while the city raises money for education, the central or state government raises it for general purposes and then divides it up for a number of uses, one of which is education. The trend, however, is increasingly for local authorities to be assisted by "higher" taxing authorities: the suburban system of the city, the city by the state, and the state (in a federal system) by the national government. The important distinction is being made increasingly in terms of spending rather than taxing power. A critical issue for the local authority under these circumstances is to retain something of its autonomous spending authority while still obtaining funds from other sources.

In what way can the local authority satisfy these needs? If, for the purposes of discussion, we assume a three-tier system that we can call local, provincial, and national, then theoretically it is possible to put fund-raising at each level but all spending at the local level. In practice, of course, this does not happen, whatever the institutional or legal position, and even when education is the legal responsibility of the local authority, both national and provincial governments do tend to "own" and run educational institutions. In Australia, for example, taxation of one form or another appears at all three levels, although except for very special purposes it is spent on education by the national and state governments only. In other systems money can be raised at all three levels and spent at all three. This raises the interesting and intricate issue to which reference was made above. Money raised by national or state governments is seldom, if ever, raised for education as such. It is part of the general revenue and is thus divided up according to a budget. Thus funds for education are related to an overall plan, whether this is on a yearly basis or as part of a five-, ten- or twenty-year programme. At the local level, on the other hand, funds are frequently raised directly for education. This has obvious local advantages, but it makes overall planning within the larger system difficult and may also result in discrimination between "rich" and "poor" districts in the way that has so plagued British and American education, albeit less so of late.

This in turn raises the vexing issue of the nature and purpose of allocations from a higher to a lower authority, an issue which is elaborated upon in greater detail by Ursula Hicks in Chapter 12, and which will, therefore, only be touched upon here as it relates to education. Should such allocations be considered purely as grants-in-aid to ensure equality of opportunity, or should they be designed to raise the level of provision in all schools in order to ensure that some kind of overall plan is accomplished? Should there be conditions for the allocation of money? If so, should these conditions be administrative, economic or educational in character? Should there be any supervision of spending of allocations, and if so, should this be purely of an audit-

ing character, or should it also involve the right of inspection of the educational offering? What control, if any, should there be over curriculum or over building standards? These are all real and important issues in a world in which national and provincial governments are providing a growing proportion of the funds for education at the local level. In the last part of this chapter we shall consider what happens in practice in a number of metropolises, but at this point it would be helpful to look at a general set of principles, both on a theoretical level and in the light of the social and political circumstances existing in particular cases.

An examination of the financing of education in one metropolis that closely illustrates most of the issues just raised would be instructive. The best example is probably Tokyo, which is faced with the difficulty of trying to develop an educational system to meet its own particular needs and fulfil its responsibilities. Tokyo is very large; it has "developed" comparatively recently; it has a very mixed system of administration and financing for its schools; and, it has a series of particular functions to play in relation to its own immediate hinterland and to the nation as a whole. Table 5-1 illustrates the general problem with which the metropolis is faced in terms of the number of schools, classes and pupils in 1965.

In 1966 the official metropolitan population of Tokyo was 11,320,716 people living in 3,238,417 family units. Of this total, rather less than 10 per cent of the population (less than one member of every family unit) was in government schools. This is low compared with the national total, for example, of 98,000,000 people in Papan in 1965, about 20,000,000 were in government operated schools. The Tokyo figures, however, do not include private school stu-

TABLE 5-1
Public Education in Tokyo, 1965

Type of school	Number of schools	Number of classes	Number of pupils
Primary	1,036	18,926	737,169
Junior High	495	8,995	386,732
Senior High (full-time)	140	3,087	154,348
(part-time)	144	1,292	52,885
Higher Technical	2	40	1,600
Kindergartens	108	357	13,060
Schools for the blind	4	60	430
deaf	9	182	1,447
handicapped	8	112	1,300
Other special schools	4	15	1,562
Total	1,950	33,066	1,350,533

Source: Tokyo Metropolitan Board of Education, *Public Education, Tokyo, 1966* (Tokyo, 1966).

dents, which at the secondary school level account for 50 to 60 per cent of the pupils, depending on the type of school and the sex of the children. The comparable national figure is 7 per cent. This is a quite staggering difference and illustrates a special metropolitan problem for many countries outside North America: the tendency for many children from rural areas to attend private institutions in metropolitan centres for their secondary education. At the tertiary level the situation is even more extreme: of about 300,000 students in Tokyo in 1964, nearly 70 per cent were in private institutions. This has to be seen against a background of increase in tertiary enrolments of over 44 per cent in the five years 1959-1964, so that in Tokyo today 37 per cent of each age group in secondary school enters a tertiary institution. Overall, at a conservative estimate, about 25 per cent of the total Tokyo population is attending an educational institution of one kind or another. How is this paid for?

Basically, the administrative structure consists of a national Ministry of Education, prefectural Boards of Education and local Boards at the municipal, town or village level. In Tokyo the local Board is the Metropolitan Board of Education, which is appointed by the Governor of Tokyo. The Board controls all education in the metropolitan area except kindergarten, colleges, private schools, and the acquisition and disposal of educational properties. These four functions are exercised directly by the head of the local government.

The local authority is theoretically responsible for the cost of all schools in the area. However, in practice the position is quite different. In 1964, for example, about 49 per cent of public expenditure on education at the primary and secondary level was borne by the central government. More than half this amount was spent directly for compulsory education, about half the remainder for operating national educational establishments such as the National Institute for Educational Research. The amounts for compulsory education are likely to increase still more since the national government has accepted the responsibility for producing and distributing free textbooks in all compulsory education schools (the first nine grades of the system).

The effect of all this in Tokyo is considerable. In 1965 the total metropolitan expenditure on kindergartens, primary, and secondary schools, and some special institutions like libraries and art galleries was 65,000,000,000 yen or about U.S. $180,000,000. Of this amount the national government contributed directly about U.S. $85,000,000, essentially in the form of half of all salaries, together with special subsidies for school buildings and materials, including textbooks. Furthermore, the prefecture in which Tokyo is situated contributed heavily toward the special education of various kinds of handicapped children who are its direct legal responsibility. In sum, the Metropolitan Board has to raise about 40 per cent of funds for its current expenditure on education. For capital expenditure the position is to some extent reversed—the Board needs to find nearly 70 per cent of the costs. Since 1960, capital costs for new building have been running between 20 and 30 per cent of all expenditure, so that at a rough estimate the maximum the municipality has to find is about half the total public expenditure on education for all levels and types other than the tertiary. It should be noted that the length of compulsory schooling, the nature and types of schools to be operated, and the curriculum are laid down by the national government, so that in a real sense the municipality has to find funds for the operation of a system over whose general policy it has no control.

The system is complex and reasonably complete. Almost every child of compulsory school age is at school; about 80 per cent of compulsory school graduates are enrolled full-time in senior high schools, with a further 5 per cent part-time; 36 per cent go on to the tertiary level. The facilities, equipment and material provided are multifarious and the equality of educational opportunity within the metropolitan area is high, much higher, say, than in New York or London, but perhaps less than in Moscow.

Clearly all this is achieved because of the high level of support from the national government. What price is paid for this assistance, if "price" is indeed the right word? First, and most apparent, is curriculum control. As was noted earlier, the curriculum in Tokyo is not very different from that in other Japanese cities like Osaka or Kobe or, more important from that in remote villages in the mountains of Hokkaido. Apart from a few highly specialized technical institutions, there is no deliberate attempt to relate the school curriculum to the specific needs of the metropolis. The education system is designed to serve Japan, not Tokyo, and even within the metropolitan area it is conceived and planned as such. It is true that this statement ignores the specific responsibility of the Tokyo Board of Education for museums, art galleries, libraries and other cultural centres which do serve in the main the needs of the city. However, important though these are, they are marginal to the total educational effort.

Thus it might well be questioned whether local boards of education are the best kind of framework within which national financial resources for education may be used. The question pertaining to the degree of local autonomy as opposed to the degree of deconcentration of central government authority that is desirable in local boards of administration is yet to be answered. In many ways this is the critical question pertaining to the administration of education. Until it can be answered by any one country and any one metropolis, planning for metropolitan education is in a sense meaningless. This is another way of asking whether, from a purely educational point of view, metropolitan areas can be regarded as truly viable entities. This leaves aside questions about educational institutions as occupiers of valuable city space and the infrastructure a metropolis must supply because these institutions exist. To the first of these questions we now turn.

PHYSICAL PLANT

In the long run, and especially in metropolitan areas, the land on which buildings stand may be a more valuable commodity than the physical plant itself. Traditionally, schools have tended to be one- or two-storey buildings, ideally set in a large ground area offering facilities for recreation and sport within its boundaries. Paradoxically, the cost of such provision becomes greater the more the state becomes involved in educational investment, as national standards tend to be conceived in these traditional terms. As Vickery puts it, "Expansion as each school grows in numbers is still conceived horizontally . . . [so that] in areas of high rise dwellings or in slum areas with people living cheek by jowl in frequently appalling conditions, the school often stands in an island of green park land, unused for over 75 per cent of each week, serene and apparently oblivious to the pressures of land and life that surround it."[4]

---

[4]D. J. Vickery, *Factors Affecting Schools in the Human Habitation at the Millenium* (Bangkok, UNESCO, 1964).

Vickery argues that such a conception of the school is no longer viable, and uses a penetrating analysis of the situation in Asian cities to illustrate his point. Much of the next few paragraphs are drawn directly from his argument. The point has been made earlier that in developing countries, particularly in their metropolitan areas, the educational base is a great deal narrower than in the established cities of Europe, North America, Australasia and Japan. For example, it will be recalled that at least 95 per cent of children in these developed countries are in school and a high proportion continue voluntarily to higher secondary school and to tertiary institutions. In the cities in the developing countries the proportion in elementary schools may be as low as 20 per cent in, say, Khatmandu or Kabul, with an average over Africa, Asia and Latin America of certainly no higher than 60 per cent and probably a good deal less. If plans like the Karachi Plan for Asia or the Addis Ababa Plan for Africa are implemented, then there will be an enormous backlog of building to eliminate before real growth can take place. Hence, the new land-use problem for these cities is somewhat greater than for cities in the developed countries, but against this it can be argued that in established cities land values tend to be higher and there is pressure on the schools to release their property for industrial and business purposes.

Returning to Vickery and the Asian cities, taking reasonable projections of populations, and assuming the minimum growth of schools and facilities, he arrives at the figures given in Table 5-2, which indicate the number of school places required in some major Asian cities in the year 2000. What are the implications of such figures for school buildings and plant? Current thinking in the UNESCO Asian Model and in most national development

TABLE 5-2
Population of Major Cities in Asia and the Number of School Places
Needed, Projected to A.D. 2000.

City	Total population (in thousands)		School places needed (in thousands)	
	A.D. 1961	A.D. 2000	Primary	Secondary
Bangkok	1,708	4,980	1,150	540
Bombay	4,422	9,550	2,200	1,040
Calcutta	2,981	6,450	1,490	700
Colombo	511	1,430	330	155
Delhi	2,227	4,830	1,120	524
Djakarta	2,907	6,650	1,540	720
Kabul	236	755	173	84
Karachi	1,912	4,200	970	456
Katmandu	122	252	58	27
Kuala Lumpur	316	1,040	240	113
Manila	1,139	3,370	870	408
Rangoon	822	1,770	410	194
Saigon	1,250	4,960	1,140	536
Seoul	2,983	8,840	2,040	960
Singapore	1,775	5,860	1,350	635
Taipei	979	3,720	865	404
Teheran	1,839	4,820	1,110	522
Tokyo	8,913	12,100	2,700	1,310
Vientiane	100	256	59	28

Source: D. J. Vickery, *Factors Affecting Schools in the Human Habitation at the Millenium* (Bangkok, UNESCO, 1964), p. 9.

plans for Asian countries assumes about sixteen square metres per primary school child and about nineteen square metres per secondary school child.

Vickery again has computed this need city by city and his findings are summarized in Table 5-3. In Bombay, such demands, if met, could result in about 8 per cent of the city land being used for school buildings on current standards in the year 2000. Figures for some other cities of Asia are even more startling, and it is hard to escape the conclusion that present standards will have to be abandoned. Vickery, too, comes to this conclusion and drives home his

point with an analysis of the situation in the most conservative case—Bangkok, a city with about 2,000,000 people and an area of 430 square kilometres, which will grow to about 4.25 million people on a total area of 938 square kilometres by the year 2000.[5] This will give a density of over 4,000 per square kilometre, of which some 1,680 will be children of primary school age and 1,000 children of secondary school

[5]The slight discrepancy between this calculation and that in Table 5-2 pertaining to the year 2000 is due to Vickery's use of two different sources of data.

TABLE 5-3
Land Areas Needed for Primary and Secondary Schools in the Major Asian
Cities, Projected to A.D. 2000.

City	Area needed for primary schools (sq. km.)	Area needed for secondary schools (sq. km.)
Bangkok	18.4	10.5
Bombay	35.2	19.7
Calcutta	23.8	13.7
Colombo	5.3	8.9
Delhi	17.9	9.9
Djakarta	24.6	14.1
Kabul	2.8	1.6
Karachi	15.5	8.7
Katmandu	0.9	0.5
Kuala Lumpur	3.8	2.8
Manila	13.9	7.7
Rangoon	6.5	3.8
Saigon	17.5	10.4
Seoul	32.6	18.7
Singapore	21.6	14.4
Taipei	13.8	7.7
Teheran	17.8	9.8
Tokyo	43.2	24.9
Vientiane	0.9	0.5

Source: D. J. Vickery, *Factors Affecting Schools in the Human Habitation at the Millenium* (Bangkok, UNESCO, 1964), p. 9.

age. At present there is a primary school for every 2 square kilometres of land, and on an average each school has 820 children. The choice by 2000 will be between doubling the size of existing primary schools to 3,360 places each on present sites or of doubling the present number of schools and providing 1,680 places in every school. The situation is similar, though less severe, for secondary schools. Furthermore, it must be noted that these figures were projected on a growth rate of 0.9 per cent, which is low for Asia. Thus the situation may well be far more severe for many Asian cities.

A similar situation can be expected in developing cities in other regions, particularly in Latin America, where metropolitan densities of population follow much the same pattern as those in Asia. In some ways the problems of metropolitan areas of the developed countries are more difficult; for although they already have a substantial investment in school sites and buildings, their growth rates are not very much lower than those in developing countries, and they have problems of replacement as well as those of providing new schools.

Increasing densities of population and pressures on land use may well imply, as is already appearing in New York and London, that the pattern of games for children will have to change rapidly. Small-area, rapid-turnover games are likely to replace the large-area, slow-turnover games. Another solution is for a more intensive use of public open space and facilities by school children. In Sydney, for instance, very few government schools today have their own playing areas, and children travel for half a day each week to playing fields that are used at the weekends by the general public.

Thus far we have been concerned with the relative size of school sites in the metropolitan area, but something should also be said about buildings and equipment and about the kind of use which can be made of them. Most educational planning has been in terms of the traditional classroom and of the traditional division of the school system into primary, secondary and tertiary levels. These traditions are becoming increasingly open to challenge, partly as a consequence of shortages of teachers, buildings and equipment in many countries, partly because of the development of new methods and techniques of instruction and learning, and partly as a result of the research of educators, psychologists, sociologists and others interested in the learning process. New ideas about the learning process are emerging and to meet them it is likely that new kinds of buildings and new uses of buildings will evolve.

Already the effective use of programmed instruction techniques demand special class layout. Closed-circuit television, together with national or provincial programmes to which it can be linked, permits differential development and hence a great deal of movement of children and teachers within the school. In time this may lead to the breakdown of the distinction between primary and secondary levels. Indeed, the revolutionary Swedish school reforms are already pointing to the concept of the "total" school and hence the possibility of doing away with separate buildings and sites for different levels of education. This kind of change is costly in terms of equipment and materials and demands highly centralized service areas but will ultimately be cheaper in terms of buildings and sites. In some of the developed countries, this type of system is reasonably close to realization. The likely trend, however, is for the metropolis and its region to be served by a core of central facilities. This again raises the question of the need to plan for the metropolitan region; an issue which is expanded upon in the concluding section of this volume.

THE EDUCATION OF MINORITIES

For a variety of reasons, minority groups tend to become more clearly identified in the metropolis. This is largely a product of size: it does not deny the existence of members of these groups in other places, but paradoxically enough members of some of these minorities may prefer the metropolitan milieu, where, being among their own kind, they are less conscious of their minority status.

These groups do, however, pose certain educational problems with their special needs and although, for purposes of analysis, the groups may be separated, much overlap may exist. For example, there may be more handicapped children among disadvantaged people, just as there tend to be more disadvantaged children among immigrant families.

A caveat should be offered against regarding these minorities as the most crucial and difficult of problems for metropolitan areas. This is only true, and even then in a limited sense, for some cities of North America. It is true that some of the cities of developing countries, such as Bangkok or Kuala Lumpur, have minority problems (and some, such as Cape Town, Johannesburg and Salisbury, have "majority" problems), but these are different in kind from those of New York or Los Angeles. This is not to deny the urgency or importance of providing for minorities as democratically as possible, but it is to say that the problem must be kept in proportion as part of the larger problem of providing proper education for all the children of the metropolis.

In terms of educational opportunity minorities may be broadly divided into two groups: those who are classified as members of a minority according to socially determined categories and those with a handicap that is not socially imposed. The first set may be classified roughly by (1) ethnic group in a strict sense; (2) religion; (3) social class; and, (4) language. The second set may be similarly divided roughly into three groups: (1) those with physical defects (the blind, the deaf, the cerebral palsied, the crippled); (2) those with a mental handicap (the aphasic, the autistic); and, (3) those with emotional disturbances.

These groups overlap to a considerable extent, the degree and nature of overlap being a function of the particular society in question. Thus, in New York or London the Negro is differentiated by ethnic group and social class but not by religion or language, whereas in Bangkok the Chinese are marked off by language, but not by class or religion and only in a special and limited way by ethnic group. Understandably, members of the second group are put at a special disadvantage if they are also members of a minority group. The blind or deaf child living in a disadvantaged minority has much less chance of adequate education adapted to his handicap than has a child with a similar defect living within the majority group in the same city. Nevertheless, keeping in mind these caveats, it is convenient to consider the two types of minority separately, since they pose somewhat different educational problems.

Socially Defined Minorities

On a world basis probably the most numerous minorities in metropolitan areas are those distinguished by language. In the United States, Australia, Canada, New Zealand and, in a completely different way, parts of Africa, the members of linguistic minorities tend to be immigrants. (The special issue of French Canada needs more detailed analysis than is possible here.) This is not true, at least in the twentieth century, of most cities in developing countries or in Europe. In these countries the linguistic minority has a long and frequently proud history, and the educational problems that emerge are in consequence quite different from those of immigrant minorities. This problem may be seen more clearly in many Asian countries, and in a less pronounced way in Africa, where several languages are frequently spoken within national boundaries.

The extreme case is India with its fourteen major languages spoken by many millions of people. It is apparent that the resulting difficulties are greatest in the cities. Even before independence and Partition in 1947 great cities like Delhi, Bombay and Calcutta had within them substantial minorities of people who spoke languages other than the dominant one. The literally enormous movement of people at the end of that year involved many linguistic as well as social and religious differences. The majority of migrants settled in metropolitan areas. Furthermore, the development of industries, again near the cities, together with the crippling famines of recent years, have brought millions of people to swell the already huge metropolitan areas. In consequence, these metropolitan areas contain unknown but certainly very large numbers of linguistic minorities, sometimes distinguished by religious groupings or caste as well as by language. In such circumstances, what should educational policy be with respect to the language of instruction in the schools?

There are basically three solutions, endlessly permuted and complicated by the administrative, political, religious and other factors. The first and most obvious solution is to have separate school systems for each linguistic group, as in Brussels, thus reinforcing the majority division and creating new problems related to equality of educational and occupational opportunity. The second solution is to provide schools that present all instruction in the majority language, either allowing privately operated institutions to teach in other tongues, or completely outlawing minority languages from the school system. A technique of this kind, while in the long run it may lead to greater communication, clearly increases the difficulties of minorities, placing their children at a considerable cultural and social disadvantage, except where intensive programmes are developed to assist them to learn the language and culture of the majority group. It is essentially the solution used by the United States, Canada, Australia and New Zealand for the great numbers of Europeans who have migrated to their shores. The immigrant problem requires further discussion but for the moment it is sufficient to emphasize that it is frequently allied to the linguistic one.

The third solution is the mixed one: separate linguistic schools or classes at the primary level and majority language instruction in secondary and tertiary institutions. This, too, presents difficulties, particularly in connection with the selection and training of teachers and with ensuring adequate instruction in the majority language during primary school to enable children to cope successfully with secondary work. There are many variants of this technique; perhaps the most completely worked out, if not the most successful in practice, being that used in the Philippines where, within the same school, children begin to learn the local vernacular, shifting slowly to the national language, Tagalog, during the elementary grades and from this to English for most secondary and tertiary work. As a result, in the City of Manila, except for older, recent immigrants from other areas of the Islands, there is no real linguistic minority. The "exceptions" do tend to form a distinct set of minorities, almost all of whom are considerably depressed and disadvantaged economically and socially, and who tend to form the nucleus of Manila's grave problem of violence and crime.

For the metropolis faced with the problems of linguistic minorities, some choice of policy among these three solutions must be made, but once more the decision is not always in the hands of the individual metropolitan area. In a real sense the issue in the long run is one of curriculum, but the decision on the issue has all too frequently to be made on political and administrative grounds. For many metropolitan areas and many countries it is *the* issue that highlights most the difficulties of accommodating metropolitan policy to national policy even when effective control is at the local level. When effective control is at the national level the difficulties become magnified almost to the breaking point. Two quite different examples serve to illustrate the point: Bangkok and Kuala Lumpur.

Of Bangkok's more than two million people, between 45 and 50 per cent are of Chinese ethnic origin, and of these perhaps 25 per cent were born in China itself. Commercially Bangkok is in many ways a Chinese city; most large business owners and managers and the vast majority of smaller businessmen and skilled artisans are Chinese, as are many, if not most, unskilled labourers. In terms of day-to-day business dealings it is much more useful and important to command one or other of the principal Chinese dialects, preferably Teochiu, than Thai.[6] From the purely commercial and industrial viewpoint of the metropolis of Bangkok, therefore, there would be considerable advantage to the community and its people in having Chinese language schools, or at least a mixed system. From a national point of view, however, China is felt as a threat—whether realistic or not—and Chinese is a forbidden school tongue. The only permitted language schools, or at least a mixed and private schools is Thai, with the minor exception of an international school and a school run by the Ministry of National Development, in both of which the language of instruction is English. Thus, national policy overrides the immediate needs of the metropolis.

Kuala Lumpur, with a population of about 500,000, has an entirely different problem. The official language of Malaya within the Malaysian Federation is Malay. Chinese, Tamil and English are recognized, however; the first two because of the large groups of Chinese and Indian people, and English because of its historical, political and commercial importance. Schools in each of these four languages are provided at the elementary level; Malay, Chinese and English at the secondary level; and English and Malay in tertiary institutions, except universities, where English is employed exclusively. Some arrangements are made in Kuala Lumpur for special courses in English or Malay for Tamil and Chinese speakers. Apart from the administrative problems inherent in trying to run such a diverse system in what is not a particularly large city, there is the almost insuperable difficulty of ensuring equality of opportunity at all levels and making equal provision of buildings, teachers and equipment. Consider, for example, the enormous difficulties of teacher training for the Chinese and Indian schools when the medium of instruction in the teacher training colleges is Malay or English.

Such examples could be readily multiplied, especially for developing countries. But the problem also exists in the more developed countries and here again, as in the developing countries, other problems are associated with that of language. Of special significance are the problems of equality of job opportunity (especially in countries where the rural migrant or immigrant is not encouraged to go to the city), and of the child being isolated from its immigrant parents. Again, these are, in large part, a matter of curriculum and special services. The education of these adults is another source of conflict between metropolitan level and senior administrations.

Similar problems exist in relation to other kinds of minorities, particularly when there is an ethnic or religious difference but not one of language. The type case of ethnic minorities is the American Negro. The special problems of Negro education in the major American cities has been exhaustively discussed and debated in many places and there is little that can be added to what has already been said and

---

[6] G. W. Skinner, *Chinese Society in Thailand* (Ithaca, Cornell University Press, 1957).

written. It is worth emphasizing, however, three major themes that emerge from all the research and experiment.

First, and most crucial for education, is the well-established fact that members of such ethnic minorities tend to be under-privileged economically, disadvantaged culturally and socially, and deprived in terms of equality of educational opportunity. Not only do the children of such minorities come to school with much less, in every sense, but also they are offered less while they are there. In general their schools are older and are less well-equipped, their teachers are less adequately trained and poorly paid, and the curriculum is far less suited to their needs and abilities.

Second, and partly as a consequence of the first point, the overall school system has tended to reinforce minority status.

Third, and more on the positive side, education systems can contribute greatly to breaking down the effects of prejudice and segregation if they are carefully designed to this end. Some techniques for this are described later in this chapter, but the point needs to be made that the phenomena and the problems are not unique to the United States nor to the Negro. At one level the problems of slum children in all the great cities of the developed world are very much the same as those of the Negro, with the important exception that there are no ethnic differences to isolate the minority. On another level there are ethnic minority groups in many cities of the developing world, and it is easy to see that their status and future will depend on careful systematic educational and social planning. This is to leave out of consideration the paradoxical case of the ethnic minorities in some South African cities who enjoy a privileged and advantaged situation at the expense of the majority. In cities of this kind the problems are quite different, and it is difficult to envisage any solution other than that implied by a complete reversal of policy.

Two or three additional points must be made with regard to religious minorities which, due to their very nature, choose to be minorities. For present purposes this becomes relevant only if, because of this selection, a separate and distinct education system must be provided. This raises a number of sensitive and difficult issues. If the religious minority wants separate schools for members of the minority, what is the responsibility of the public system? One of the aspects of this problem—administration—has already been looked at, but there are also aspects relating to equality of provision that deserve consideration. In some countries and in some cities public policy has been that there should be no state aid for private schools, thus leaving the problem of provision of sites, buildings, equipment, teachers and so on, to the religious community, which means in essence to the parents of the children involved. In such circumstances, however, curriculum and other standards are generally laid down by the public authority, thus demanding a measure of assurance of equality without providing aid to ensure that this exists. The alternative, where such schools are allowed, is to provide assistance of varying degrees, ranging from "across the board" systems as in London, or for specific and carefully defined purposes, as in Sydney.

For the religious community difficulties are imposed by either solution. In the first instance provision has to be made that the community may not be able to afford; in the second instance aid without strings is rare, and the public authority usually demands right of supervision of various kinds. This is not the place to argue the rights and wrongs of state aid to religious institutions, but the point must be made that the issue has to be faced by education systems, particularly in cities with sizable religious minorities demanding their own school system. The problem becomes even more exacerbated when the religious minorities tend to live in specific areas of the city, such as Glasgow's infamous Gorbals or the Irish sections of Boston or New York, for in cases like this the special problems of slum schools in general are added to those of the religious minority.

In general, the problems of the education of religious minorities tend to be administrative planning problems associated with private schools in general. The difficulties, though these are admittedly overlaid with irrelevant emotional attitudes, are essentially those of integrating a separate system of education into an overall public programme of buildings, sites, equipment, teachers, curriculum and so on. Such problems are most acute in metropolitan areas since, as we have seen, there is as a rule a much higher proportion of private schools in the larger cities than in rural areas. The additional problem, when religious minorities are also linguistic minorities or economically depressed minorities or all three, is essentially that of segregation.

Havighurst contends after careful analysis of both empirical situations and experimental evidence that the effects of segregation involving disadvantage or deprivation can be largely overcome under either of two widespread and fundamentally different sets of conditions: where present trends of metropolitan evolution continue, that is, where there is continuation of economic stratification; or, where serious attempts are made to reverse some of these existing trends. If present trends do continue, he outlines a five-pronged programme for implementing "attempts to provide educational stimulation and op-

portunity to the children of the slum areas":

(1) a multi-track system that separates children into several different groups according to learning ability and social status;

(2) enrichment programmes (within the multi-track system) for working-class children who achieve fairly well;

(3) enrichment programmes for culturally disadvantaged children at the pre-school and kindergarten-primary level;

(4) work study programmes for alienated youth; and

(5) finally, Havighurst contends that "if present trends of educational development continue, different kinds of systems may well have to be evolved from the central city as against suburban children." Somewhat pessimistically he concludes that "the suburbs will have more money to work with than will the central city and their predominantly middle-class character will make them responsive to the proposals for the use of new methods, new kinds of school buildings and new types of school programmes."[7]

As the reader may well surmise, Havighurst's analysis is conceived in the context of the administrative system of the United States and under different conditions not only do some of the trends he sees appear less critical, but also some of his solutions, especially the fifth, become unnecessary. The same limitations may apply to his second set of proposals relating to reversing present trends of city development which are essentially directed toward ensuring that school systems should not be co-

---

[7] R. J. Havighurst, *Education in Metropolitan Areas* (Boston, Allyn & Bacon, Inc., 1966).

terminous with segregation areas. Havighurst sees this as calling for "a fundamental programme of urban renewal that will have the cooperation of suburb with central city." The emphasis must be on the social renewal rather than the physical renewal. Over a twenty-year period he would like to see much greater use of "the mixed school" (a mixture of socio-economic and racial groups where these exist in a community) to bring about a greater integration of the metropolis as a whole.

Turning away from the North American situation, the problem of mainland European cities (western or eastern), or of cities like Sydney or Bangkok, is quite different. In Sydney, where few people live in the central city area, segregation tends to be very much on a suburban pattern. In Bangkok, where there are no suburbs in the Western sense, any segregation is in terms of high and low density areas, which relate in turn to occupational areas as in the European medieval cities and most other present day Asian cities. For these cities, very different though their individual geography may be, the real problem is how to organize anything approaching mixed communities and hence mixed schools. Transport alone is an almost insuperable barrier.

Before leaving the problem, it is well worth looking briefly at two major efforts in relatively old, developed cities to solve the segregation problem within the larger context of metropolitan planning. The two best examples of this, one well on the way and the other still largely in the planning stage, are Moscow and London. In each instance the movement of people implicit in any such programme has been facilitated by the heavy destruction and consequent rebuilding caused by World War II. The Moscow solution calls for sets of neighbourhood blocks, housing in each block being more or less free of the trappings of social status. "Placement" of families within apartments in such blocks was and is relatively easy under the Soviet system of government, and by making family need rather than occupation the criterion for priority in housing a fair randomization of different status groups has been achieved. Leaving aside the disadvantages and dangers involved in such a system, however great they are, it is clear that the system does enable the development of "mixed schools" (in Havighurst's sense) without the transport difficulties it would entail for metropolitan areas like New York. Each neighbourhood block in Moscow has its own elementary school and each set of blocks its own series of secondary schools, all with pupils whose parents represent a wide variety of occupations and status levels.

The London plan is rather different and embraces thirty-two boroughs, each with 140,000 to 350,000 people, and each responsible for health, sanitation, child welfare and libraries. The twenty outer boroughs are also responsible for primary and secondary education, but the twelve inner ones all come under the Greater London Council for educational purposes. Apart from its effect in terms of mixed schools, this plan in effect implies that the ideal unit for efficient control of an education system is somewhere between 150,000 and 400,000 people. A complicating factor, however, is the presence of the central Ministry of Education and the grants that have to be made by it to each local education authority.

It is perhaps unfortunate in international terms that most of the best thinking and planning about metropolitan areas and about education within them should have taken place in a North American or United Kingdom context, where the concept of local authority is so firmly entrenched. Accordingly, the tendency has been to think and plan in terms of "consolidation" or "coordination" or "cooperation" of some kind. As we have seen, the problem of many metropolitan areas in other parts of the world is the reverse of this: not so much that of unifying disparate sections of the metropolis as providing opportunity for the metropolis to assert its identity as a metropolis and not merely as the capital or centre of a nation or state. There is a real need for careful discussion, thought and experiment on metropolitan areas in this framework and on the place of education in such a development.

Handicapped Children

The point has already been made that both the extent of need and the capacity to provide for handicapped children are a function of the size of the city. They are also a function of the city's stage of development, partly because of cost and partly because of the need for highly trained personnel. Even in many of the developed, established cities of the world, provision of education for atypical children has tended to be dependent on private finance, sometimes supplemented by government aid of one kind or another. While there is a growing tendency throughout the world for governments to accept full responsibility for all their children, this is still far from universal and is, moreover, somewhat uneven where it exists. Further, it is generally true that in a metropolis where education is locally controlled there is more probability that there will be provision for the atypical child than where there is central control. The problems are essentially the same as those of providing any education system—schools, teachers, buildings, equipment—with the addition of a complicating factor, that of "identification". This is in part medical or psychological, but, particularly in the developing countries, it is also social. This is especially true of the mentally handicapped whose parents feel some kind of stigma and try to keep the children hidden and isolated.

This problem of identification, from the point of view of the education administration, becomes important only when education becomes compulsory, as an obligation is then imposed not only on parents to get the children to school but also on the administration to ensure that all children do attend and are given opportunities commensurate with the available provision. The establishment of services to ensure proper medical and psychological examination of every child is expensive, however, and calls for the existence of a pool of well-trained personnel. In a country where qualified medical officers are in short supply and psychologists are practically nonexistent, it is impossible to conceive of a thorough diagnostic service. Indeed, a reasonable measure of the development of an education system is the extent to which it does provide thoroughgoing educational, psychological and medical guidance services. In terms of cost, a reasonable estimate is that provision for a handicapped child is somewhere between ten and twenty times as expensive as for a "normal child", depending on the type and degree of handicap. Where a city is struggling to develop a sound system of primary, secondary and tertiary institutions among a population with illiteracy of 50 to 60 per cent or more, it is difficult not to be sympathetic with a view which leaves the handicapped entirely out of public consideration. It is instructive in this connection to look at the magnitude of the problem in terms of the incidence of handicapped children of various kinds. In many ways this is largely a question of definition and of diagnostic standards. For

TABLE 5-4

Percentage of Compulsory Age Group Handicapped and Percentage of Handicapped Special Schools or Classes

Type of handicap	Percentage of compulsory age group handicapped			Percentage of handicapped attending special schools or classes		
				1963	1963	1958
	Tokyo	London	New York	Tokyo	London	New York
Blind and partially sighted	0.07	0.12	0.09	45.1	32.8	47.9
Deaf and hard of hearing	0.13	0.17	0.58	67.0	37.3	14.3
Mentally handicapped	4.25	10.0	2.30	7.3	5.0	26.4
Crippled and spastic	0.34	0.5+	1.00	12.5	18.9	7.1
Physically weak	1.35	1.0+	1.00	2.5	12.6	5.6
Aphasic	?	1.5+	3.50	?	1.6	33.8
Maladjusted	?	1.0	1.00	?	?	7.9

Source: Japan, Ministry of Education, *Educational Standards in Japan* (Tokyo, 1964).

example, the level of I.Q. at which a child is regarded as needing special care is lower in London than in New York, and as a result there are more special classes and greater enrollment in London. Similar difficulties exist with respect to partial deafness or partial blindness. However, given this caveat, Table 5-4 shows percentages of handicapped children in Tokyo, London and New York, three very large cities with excellent diagnostic and guidance services.

Table 5-4 clearly needs careful interpretation, particularly the second percentage break-down. Those who are not "attending special classes or schools" may be either not in school at all, or attending ordinary schools and classes. That is, the degree of handicap may be defined as not warranting "segregation", and modern thought is much in favour of retaining children in ordinary classes or schools wherever this is possible. For example, in terms of the hard-of-hearing this may mean provision of individual hearing aids, at

considerable expense to the state; with the partially sighted it may mean special spectacles and various sight-saving devices within the classroom; and similar appropriate arrangements are possible for members of other specially handicapped groups.

However, if we use as a minimum estimate of need the percentage of the age group who are in special classes or schools, by recomputing the table, we arrive at a figure of 4 or 5 per cent. This is almost certainly far too low even in these three cities where excellent diagnostic services lead to ready identification and where public funds are available for the education of most groups of the handicapped.

In the developing cities of the world the proportions of children in special schools or classes are considerably lower; while a figure of 0.1 per cent, for example, might be an exaggeration for Kabul, the need in these cities is likely to be considerably greater because of deficient pre-natal and post-natal medical services and the existence of many crippling diseases with origins both in infection and malnutrition. Indeed, it has been estimated by UNICEF that in the developing world as a whole the percentage of children physically or mentally handicapped in one way or another may be 10 per cent or more.

Table 5-4, with all its limitations, also serves to illustrate the relative size of the problem for different groups. Take the blind and partially sighted: it is estimated that there are about 2,000 children of compulsory school age in this category in Tokyo. This is obviously a large enough group to warrant special provision of schools and classes with the attendant implications for equipment, buildings, teachers and transport. In Sydney there are about 300 children in the same group, allowing for differences in diagnostic standards. A number of this size may seem marginal, but provision is made on a private basis with some governmental assistance, particularly for children from rural areas. But what happens when city sizes fall below, say, half a million and there is no larger metropolitan area within reasonable distance? If the numbers of blind or partially sighted children fall below, say, forty or fifty, what case can be made for special classes? Equally pertinent, what is the responsibility of the city for its hinterland—not only for education of the handicapped, but also for identification, which is more difficult and expensive? When provision for the handicapped is made by parents and private bodies, this issue does not arise except in a moral sense, but once there is public acceptance of responsibility for the education of handicapped children, then the problem becomes acute. The Japanese solution is to place the responsibility for the provision of schools for the handicapped on the prefectures, which then provide schools within the cities. This arrangement does raise some difficulties of identification as the ordinary schools are established, it will be recalled, by the local boards. There is, however, a reasonably good, if somewhat clumsy, administrative machinery for handling this.

Enough has been said to suggest that provision of education for minority groups within metropolitan areas is fraught with problems, many of them administrative and financial. The demand for such provision increases with city size and development. Indeed, there are some signs, particularly in the United States, that the problems of minorities of various kinds begin to overshadow the more central problem of providing an education system geared to the needs of all the people in the metropolis. Minorities are important, but their members are important as people and not as members of the minority. The more effectively they can be treated as an integral part of a total system, the better for them and for the system.

Approaches to Key Issues

To summarize and lend point to what has been said we may look at the present situation in the eight metropolitan areas classified in Figure 5-1. It will be recalled that three criteria were used: type of control, size of city, and stage of development. The education systems of the metropolitan areas cited as examples have been generally described in passing with the exception of Toronto and Singapore, to which brief attention is given here.

Toronto, a metropolitan area of about 2 million people, has its education managed by the Metropolitan Toronto School Board, which came into existence in 1953 as a result of a consolidation of certain functions of eleven smaller local authorities, each of which, however, retained many

	Province percentage	Metropolitan Board percentage	Local Boards percentage
Operating costs	22.16	34.92	42.9
Capital costs	20.5	22.7	56.8

responsibilities and functions. (As of January, 1967, the number of local boards has been reduced to six.) General policy on compulsory education is laid down by the Department of Education of the Province of Ontario, which also exercises authority over teacher training and school curriculum. Education is financed by local school board taxes and Metropolitan Board taxes, supplemented by capital grants or loans from the provincial government. In 1963, for example, costs were shared among the three groups shown above.

It is instructive to note that, contrary to world trends, the proportion paid from local taxes has slowly risen since the mid-1950's, while the proportion from both provincial and metropolitan sources has fallen. There are several reasons for this, but by far the most important is that in many ways Toronto represents a real effort to develop a municipal policy on education which is integrated with a general plan for metropolitan development in general. The success of this integration depends to a large extent on the degree of cooperation between the local education boards and relevant municipalities within the metropolis, and between the Metropolitan School Board and the Metropolitan Council, particularly since selection and purchasing of school sites and financing of construction is largely in the hands of the Metropolitan Council. The experiment, however, has an importance well beyond the boundaries of Toronto.

Singapore is unique in that, like Hong Kong, it is an artificially created city state,

with a total population of about 2 million people of whom well over 90 per cent live in the metropolis proper. Control over education is vested in a Minister for Education who is an elected member of Parliament, selected from other members to be in the Cabinet. He works through an Under-Secretary and permanent civil service on the British pattern. Finance comes from general consolidated revenue as part of the total state budget. There are no tax funds for education as such.

How do these cities tackle the kinds of problems discussed in earlier pages? Some indications have already been given in the examples we have mentioned. While space does not permit an exhaustive discussion of problem-solving, a comparison of approaches to some key issues will serve to reveal those problems common to most metropolitan areas, and those which are special for some because of size, degree of development or nature of control.

Table 5-5 sets out in generalized and summary form a comparison of some aspects of metropolitan education in the eight examples of metropolitan areas. New York has been added to the list for purposes of comparison, for, although in some ways it resembles London, there are important differences that need some consideration in the present context. It must be emphasized as strongly as possible that this table is designed purely to highlight some of the problems that exist and some of the different ways of handling them (relating to the variables that have been discussed). It does not attempt to give a description of

TABLE 5-5
Comparison of Selected Aspects of Metropolitan Education in Nine Metropolitan Areas

TABLE 5-5 (a)
School Authority and Metropolitan Authority Relations

New York	New York City (metropolitan) government, as such, has no direct responsibility for education; elected local education authorities in boroughs cooperate with metropolitan authorities and state authority.
London	Greater London Council area divided into thirty-two boroughs; twelve central ones combined for educational purposes under one elected authority responsible to the G.L.C.
Paris	No formal links except in relation to advisory boards in communes.
Tokyo	Board of Education appointed by metropolitan authority, but functions within national regulations.
Bombay	No formal links; education controlled by Bombay State; some consultation with metropolitan authority on sites and buildings.
Singapore	State government is essentially city government; total planning at state level.
Sydney	No formal links except in relation to sites.
Bangkok	Municipal government responsible for all elementary education, but not for secondary, vocational or tertiary.
Toronto	Complex set of relations at local level; Metropolitan Toronto School Board elected independently of metropolitan government but works closely with it.

TABLE 5-5 (b)
Finance to Schools

New York	City education tax; federal and state subsidies for specific purposes.
London	Local education tax; Ministry of Education per capita subsidy.
Paris	National budget through Ministry of Education.
Tokyo	Local tax; Ministry of Education grants for teachers' salaries, buildings, and books.
Bombay	State budget.
Singapore	State budget.
Sydney	State budget.
Bangkok	National budget through either Ministry of Education or Ministry of Interior.
Toronto	Local education tax; provincial government grants to metropolitan level for redistribution.

TABLE 5-5 (c)
Metropolitan Responsibility for Some Aspects of Policy in Education

New York	General curriculum determined by New York State, detail by local boards and the school itself; length of school life determined by New York State; type of school and school organization by local authority.
London	General curriculum determined by local authorities (but note major influence of nation-wide secondary-level examination requirements); length of school life laid down by national law; type of school and school organization determined by local authority.

TABLE 5-5 (c), (Continued)

Paris	Curriculum, length of school life, and type of school organization all nationally determined.
Tokyo	Curriculum and length of school life determined by national laws; metropolitan board has minor powers relating to school types and organization.
Bombay	All aspects of policy determined either at national level or state level.
Singapore	All aspects determined by state.
Sydney	All aspects of policy determined by state.
Bangkok	City has some powers relating to types of primary schools; otherwise all policy determined by state.
Toronto	General curriculum determined by Province of Ontario; metropolitan Board and local boards have some freedom on detail; length of school life determined by Province; type of school and school organization determined by local authority within metropolitan pattern.

TABLE 5-5 (d)
Handling of Linquistic Minorities

New York	As part of normal system; some special classes for immigrants; enrichment programmes in impoverished areas.
London	As part of normal system; some special language classes.
Paris	No problem.
Tokyo	No problem.
Bombay	Separate schools for major minorities; Hindi and English compulsory languages at secondary level.

TABLE 5-5 (d), (Continued)

Singapore	Separate schools for minorities; English compulsory at secondary level.
Sydney	As part of normal system; special classes for adults only.
Bangkok	As part of normal system; teaching in foreign languages illegal.
Toronto	As part of normal system; some special classes.

TABLE 5-5 (e)
Provision for the Handicapped

New York	Full range of diagnostic services provided by city; special schools and classes, together with special provision in ordinary classes.
London	As in New York, with rather more dependence on special schools and classes.
Paris	Mixed system; physically "weak" children and various types of mentally handicapped, responsibility of state; full range of state diagnostic services and special classes; physically handicapped children, catered for by special public schools or, more commonly, in private, government-subsidized institutions.
Tokyo	Full range of services run by prefecture within metropolis.
Bombay	A few schools for "retarded and delicate children"; no real diagnostic services; children may be exempt from school because of illness or disability.
Singapore	Legal provision but little actual provision; government homes for blind and crippled and some institutions run by missions.
Sydney	Full range of diagnostic services by state; limited range of classes for mentally handicapped; private blind and deaf school with state subsidy; partially sighted and hard-of-hearing with special aids; physically handicapped in state hospitals or private institutons.

TABLE 5-5 (e), (Continued)

Bangkok	Good diagnostic services for physically handicapped, but limited provision of special classes; limited but growing facilities for mentally handicapped.
Toronto	Full range of diagnostic services provided locally in cooperation with metropolitan authority; wide range of special schools and classes provided by metro board or local authority; some children in ordinary classes with special aids.

the systems in any detail, much less any judgements about the quality of the educational systems provided.

Of the patterns that emerge from Table 5-5 let us consider first the issue of control in education. Where there is complete central control (Sydney, Paris, Singapore, Bombay) there is, as a rule, central financing with the metropolis treated in exactly the same way as other parts of the country, both as to sources of funds and as to their disbursement. Certain special facilities tend to be located within the metropolitan area, but these must serve the hinterland as well as the metropolis itself. There tend to be no special programmes for minorities within the metropolis, but handicapped children from rural areas and small towns can be and often are catered to in special schools. Curriculum is centrally determined and is seldom geared to the special needs of the metropolis except for trade and vocation schools. Although not noted in the charts, it is of interest to mention that teacher training is centrally controlled; teachers tend to be state civil servants and are allocated to teaching posts or administrative posts by the state, the local school having little authority. Administration is usually under a minister elected by the people as a politician and not as a specialist in education, with a permanent state civil service functioning under him. Consequently there is seldom any planning for the metropolis itself in educational terms. This appears to apply to developing, as well as developed, metropolitan areas.

On the other hand, when there is local control almost all of these situations are reversed. As a rule there is local financing, although this is nearly always subsidized to a greater or lesser extent by a central source. There tend to be special local programmes for minorities. There are, however, important exceptions to this, as in Tokyo, where the handicapped are the responsibility of the prefecture, or New York, where there is a federally supported programme for the underprivileged. Again, although it is not shown in the charts, we may note that teacher training may be either centrally or locally controlled, but teacher employment is usually in the hands of the local authority (again there are exceptions, as in Tokyo). Under such circumstances many teachers tend to move freely among systems, while many others remain with the system or, indeed, within the same school for long periods of time. Curriculum, however, is centrally determined in general, though it is usually possible for the local authority, particularly in metropolitan areas, to adapt it to meet local needs. Administration of education is usually in the hands of a professional educator appointed by the local authority. Continuity of board membership is seldom long, and senior administrators tend to change their posts much more frequently than in a centralized system. This varies considerably almost in direct proportion to the size of the system.

In general, local systems are more likely inclined toward integrated planning for the metropolis as a whole. There are dangers, however, in that local boards and local administrators are more subject to immediate pressures of many kinds than are central authorities with their large public service systems.[8] The personal nature of a local system is at once a strength and a weakness. The solution in London and Tokyo is to have large units that, it is claimed, avoid the impersonal vices of state control and state financing on the one hand and the immediacy of local pressure on the other.

The question of control is, as has been emphasized throughout this paper, in many ways the most critical of all problems, for it is reasonable to assume that differences due to stages of development may gradually disappear, and differences that are a function of size tend to be technical in nature and require technical rather than political solutions. Differences related to administration and finance, however, tend to turn into differences of policy rather than emphasis. The issue is critical for the metropolis, but it is also critical for the nation as a whole. It is certainly true that proper metropolitan planning is dependent for much of its success on proper educational planning, and it is difficult, if not impossible, for a metropolis to plan round

an education system over which it has little or no control. On the other hand, no nation, particularly a developing nation, can begin to consider social and economic planning without the integrated consideration of its education system, and in particular the quantity and quality of its manpower output. Diverse sub-systems each following their own goals and policies are *prima facie* inconsistent with overall development.

There is a real dilemma here to which there is no easy answer. The answer for the United States might well be that in diversity lies its strength, and that good local systems, particularly in metropolitan areas, lead to a sound national system for which detailed planning is accordingly unnecessary. That is, any planning that does take place should be at the metropolitan level. The British answer would be different: it would imply that uniformity is undesirable at lower levels, that there must be a maximum of local freedom in curriculum and methods, but that coordination with economic and social planning can be achieved by indirect forms of national control, particularly by finance at the tertiary level. The Canadian answer would seem to be a provincially designed curriculum and teacher training, together with equalizing grants from provincial sources, but attempting to ensure strong metropolitan development through local planning within a provincial framework. The Japanese solution is somewhat similar to the Canadian, with two important additions: the entry of the national, in addition to the provincial, government on the educational scene as a provider of funds and determiner of curriculum; and the existence of school-level institutions within Metropolitan Tokyo that are owned and controlled by national and provincial governments.

Each of these answers is perhaps valid for its own set of conditions, although in

---

[8]See, for example, N. Gross, *Who Runs Our Schools?* (New York, John Wiley & Sons, Inc., 1958).

each there are weaknesses. For metropolitan areas in the developing countries the choice is even more difficult since, as we have suggested, it tends to be overlaid and complicated by other crucial issues relating to priorities in education. Should the emphasis be on metropolitan areas or rural areas? Should there be universal compulsory education or a narrow élite system, a broad-based pyramid with a long needle top or a narrow base tapering slowly? The choice is confounded in direct proportion to degree of development by the lack of administrative personnel. A really underdeveloped country, like Afghanistan or Laos or some of the Arab and African states, simply does not have enough trained people to run a local system. The small number of professionals available in a city like Lagos must be concentrated on national problems and cannot be spread thinly over the entire country in a multiplicity of systems. The effect of this scarcity of qualified personnel on metropolitan development as such is difficult to assess. On the one hand, rural pressure is often so great, as in Bombay State or Nigeria, that metropolitan planning of education must suffer. On the other hand, where the seat of central government is in the metropolis, city planning has often taken precedence over total planning. A good example of this is Bangkok, despite the fact that here educational planning has not been well integrated within the overall system.

The problems of the developing metropolitan areas tend to be rather different in kind from those of the developed ones, partly because of the differences mentioned above and those appearing in Table 5-5, and partly because the education systems often remain impoverished versions of those of the colonial powers when the political and administrative systems change after independence. An obvious exception

to this is Singapore, where social and economic factors take on a different dimension because of the geographic situation. Because it is a city-state it is in the unique situation, for a metropolis, of having education directly financed from a central budget. Singapore combines the major elements of both central and local control. It will be interesting to see by what means and how rapidly it solves its educational problems. One would be tempted to predict a more rapid and integrated social development for Singapore than for any other developing metropolis of comparable size.

For a metropolis like Bombay, however, the outlook is bleak. It is so large that, like London, Moscow or Tokyo, it generates problems of its own, but it must solve these problems within the overall context of Bombay State and of the Indian nation. The solutions open to it are determined outside its political and administrative boundaries because the major function of the metropolis is seen as integrative to the state itself. Similar statements can obviously be made about other great metropolitan areas of India and are true also of conurbations like Cairo and Rio de Janeiro, and smaller ones like Ibadan and Dacca. It may well be that, just as there are stages of development of education systems, as has been postulated by Beeby,[9] and stages of development of cities, as Lewis Mumford argues, so must there be stages of development in the relations between education systems and metropolitan areas. If this is so, to look for an integration of educational development with general metropolitan development outside the total national context is a meaningless exercise until a particular stage of both educational and

---

[9] C. E. Beeby, *The Quality of Education in Developing Countries* (Cambridge, Harvard University Press, 1966).

metropolitan development has been reached.

This is an attitude almost as pessimistic as Havighurst's prediction for the developed metropolitan areas if present trends are not reversed. It is even more pessimistic to suggest that such stages might well follow the pattern of growth of the developed metropolises of the world. It is to be fervently hoped that such pessimism is not justified. The developing cities have enough problems to face in trying to adapt to the demands of industrialization, complicated by population pressures and financial difficulties, without having also to face the problems of underprivileged minorities created by the nature of modern metropolitan areas themselves. The real answers to many of the difficulties are educational, but educational growth must be viewed within the total context of social and economic growth. If the planning of metropolitan development, including education as an essential element, is important for developed, established metropolitan areas, it is crucial for the metropolitan areas of the developing world. For, in nearly every case, upon their ordered growth depends the viability of the nations in which they are situated.

6 Public Health Problems in Metropolitan Areas

Tibor Bakács

The Physiological Aspects of Urbanization

For hundreds of thousands of years, throughout his phylogenetic evolution, man had been a biological entity, perfectly adapted to his natural environment. About ten thousand years ago he left this natural environment for an agglomerative, artificial one: first primitive villages and then, about five thousand years ago, metropolises such as those of Mesopotamia, Egypt and the Roman Empire. Except for these few large cities of the past, however, it is only in the twentieth century that urbanization has become a characteristic trait of human settlements, and it is in part because of the relatively short period of time involved that man has not been successful in adapting to this artificial environment. Another reason, however, is that man is still dependent on his natural environment for sources of energy; yet there are ever-increasing denaturalizing factors harmful to health that have been interposed in the new, artificial environment; factors such as air pollution, water pollution, and physiological stress. Thus, the discord between man and his new artificial environment is increasing, and man's health suffers.

The biological bonds between man and his environment are further threatened by the enormous increase in population that has taken place on a global scale in the last

150 years.[1] The development of the natural sciences has enabled an extension of the duration of human life. One hundred and fifty years ago life expectancy, at birth, was about thirty years, whereas today, in some countries, it is over seventy years. In addition to the quantitative extension of men's lives, a qualitative assault is also being carried out. Active life is considerably prolonged, physical and mental maturity is accelerated, and the symptoms of aging are being retarded. Unfortunately, this situation, representing a positive evolution, is valid for only the one third of the world's population living in technologically or economically advanced countries. In the other countries, although the average life span is less, overpopulation raises grave problems.

From the beginning of our era, the first doubling of human population required 1,500 years. Between 1840 and 1950, the world population was doubled in one century (1.1 billion in 1850 to 2.5 billion in 1950). The doubling of the 1960 figure, that is, 3 billion, requires only forty years. Today, the annual increase in world population is estimated to be about 50 to 60 million persons and, in the year 2000, the world population figure may well be 6 billion.

As is pointed out in Chapter 1, this growth in population is being accompanied by continuing urbanization, which, in turn, is taking place at an increasingly rapid rate around the major urban centres. In 1960, about 20 per cent of the world's population were city-dwellers; by the end of the century the figure is likely to be 60

per cent.[2] The implications of this for some parts of the world are made clearer when we consider that the tropical and subtropical countries of Africa and Asia will double their population every twenty-five or thirty years and that their urban population can, as a consequence, be expected to double in less than fifteen years. With the extremely high birth rate in India, a population of one billion people is quite probable by the year 2000. Of course, urbanization will impose still more serious problems. Calcutta, for example, with a present population of 7 million, will have 36 million in the year 2000, if the present birth rate and rural migration to the metropolis continue unabated.

In the highly industrialized countries of Europe, North America and Australasia, where there has been some success in reducing the infant mortality rate to less than 2 per cent, and where the birth rate has been reduced, the problem of overpopulation is not such a serious one. Still, metropolitanization and many subsequent health problems remain. It is evident that the population increase in these countries does not indicate such accentuated tension as exists in Africa or Asia. In fact, the increase in national income of wealthy countries is greater than the natural birth rate increase; thus, the necessary means to combat the disease and ill health arising from urbanization are provided. But as long as even the wealthy countries earmark insufficient funds for it, the problem of ill health will not be solved.

For developing countries these problems are serious. Certainly, some progress has been made in reducing infant mortality and deaths from acute contagious diseases;

[1]Tibor Bakács, "Some Public Health Problems of Urbanization" (in Hungarian), *Népegészségügy*, Vol. 47, No. 1 (Budapest, 1966), pp. 5-11.

[2]Michel Ragon, "La croissance des villes," *Santé du Monde* (Geneva, Février-Mars, 1966), pp. 6-17.

TABLE 6-1
Urban and Rural Population Percentages in the United Kingdom, Germany, France, and the United States between 1800 and 1950.

Year	United Kingdom Rural	United Kingdom Urban	Germany Rural	Germany Urban	France Rural	France Urban	United States Rural	United States Urban
1800	68	32	—	—	80	20	96	4
1850	50	50	—	—	75	25	88	12
1860	46	54	—	—	72	28	84	16
1870	38	62	64	36	70	30	79	21
1880	32	68	59	41	65	35	72	28
1890	28	72	53	47	62	38	65	35
1900	22	78	46	54	58	42	60	40
1910	22	78	40	60	55	45	54	46
1920	21	79	38	62	53	47	48	52
1930	20	80	33	67	49	51	45	55
1940	20	80	30	70	47	53	43	57
1950	19	81	29	71	44	56	36	64

family planning, however, is rare or non-existent and, consequently, the natural birth rate is still extremely high. Economic development in these countries is still limited. The extremely sharp rise in population is not paralleled by home building, by improvements in the standard of living and by new job opportunities. In view of these conditions, what resources remain for use in the fight against the effect of urbanization on health? Almost none.

One cannot halt urbanization. In fact, accelerated urbanization today is not only necessary—considering the great demand for labour by industry—but unlike the past century, it is now a condition of technological efficiency. In an industrialized country, a rural population of 8 to 12 per cent of the total population with mechanized, agricultural techniques is sufficient to feed the entire population. In highly industrialized countries, such ratios of city to rural dwellers soon will be achieved, thus reaching the point of complete urbanization.

As Table 6-1 shows, the saturation point in urbanization was achieved in the United Kingdom between 1900 and 1930. In the United States, West Germany, France, the U.S.S.R., the Netherlands and Belgium, to name only some of the more developed countries, this condition is expected to be reached by 1970. It is reasonable to assume that, sooner or later, the rest of the world will arrive at the same point.

Thus, the inevitability of urbanization renders the anti-urbanization concepts, so popular at the beginning of this century, unrealistic and unattainable. This rapid acceleration in urbanization has increased the severity of problems of supplying potable water, treating waste water, and eliminating slums. At the same time, completely new problems have appeared, such as those of communications and physiological stress.

Some data collected in 1955 in the United States will illustrate important needs in a few services selected at random. After having been installed in an old sector of a city, a thousand new city-dwellers require, in one year, the following: 36.5 million gallons of potable water; the elimination of approximately 62,050 pounds of waste solids, $65,000 to combat the increased air pollution, one extra hospital bed, 4.8 additional primary school classes, $114,000 extra for school maintenance, 8.8 acres in new school ground and playgrounds, 1.8 additional policemen, 1.5 additional fire fighters, 1,000 more library books and one extra prison cell.

In addition, there are many other needs, such as public wells and provision of their cleaning and maintenance. The volume of refuse to be removed increases, requiring more manual labour.

To meet these requirements, present attempts at coordinated development are insufficient, even in developed countries. In particular, the requirement for new dwellings are not being met. Nor does the demand for urban comforts lessen among already installed city-dwellers, especially in the wealthier countries. Indeed, there are approximately 30 million out-dated urban buildings in the developed countries. As for developing countries, it is probable that 150 million families do not yet have their own dwellings with even the most elementary requirements. In India, an expenditure of $22,000 million would be necessary between 1950 to 1975 to house adequately the inhabitants of towns with more than 100,000 city-dwellers, including new migrants. The sum in question is approximately four times the subsidies that the World Bank could have granted to developing countries during its sixteen years of existence. According to a 1954 estimate, an annual investment of $1,400 million for thirty years would be necessary to catch up on the delay in housing construction in Latin America, (including renewal of outmoded buildings) and to ensure elementary

housing for the ever increasing population. The evolution of developing countries is directed to increasing the rate and extent of industrial development. Although industry induces urbanization, however, it brings with it rapid population growth and the unfavourable hygienic conditions of urban areas. In times, these countries will attain the economic and hygienic level that already exists in developed countries; but we must ensure that this process is not prolonged needlessly. Whereas England, one of the first countries to industrialize, took 200 years in the process, the Soviet Union has taken 50 years. Even this is far too long.

In looking at the present world distribution of metropolitan areas as shown in Chapter 1, one notices that the focal point of urbanization today is gradually shifting toward the tropical regions. Therefore, the rate of development of the metropolises of the developing countries and the consequent harmful effects upon health necessitate serious consideration. The threat to health from urbanization varies according to a country's economic level. In the developed countries, considerable advances have been made in diminishing morbidity and mortality from acute and infectious diseases, and in reducing the danger of epidemics. The more serious problems remaining are of a chronic nature, for example, health problems associated with air, water and soil pollution, and the side effects of physiological stress, such as coronary attacks, which arise from such stimuli as noise and traffic accidents. On the other hand, in the metropolises of the developing countries most of the real problems concern acute illness and contagious diseases that owe their origins to those microbes and parasites found in tropical environments. In these countries and cities, the "health-related" elements of the environment, water, for example, are vehicles for contagious illness—a problem that is infrequently found in Europe and the United States.

Air Pollution and Associated Health Problems

The characteristic "urban climate" in large agglomerations is the product of the combination of natural elements, which are quite harmless, and artificial ones, such as air pollutants. The urban climate exists in built-up city centres especially where there is a dry, warm environment and where lack of air circulation often results in stationary, polluted, toxic air. Indeed, the characteristic dryness and warmth of the urban climate resembles a desert in many ways.

Any city, however, enjoys a certain degree of natural air circulation even if dominant winds do not exercise their usual effect. The warm air of the urban centre, although heavily polluted, is less dense and rises, thus being replaced by cooler, denser, but generally cleaner, air from the periphery of the urban area. This air circulation, with its accompanying purifying effect, provides natural ventilation for cities.

Any evaluation of atmospheric pollution of a city should consider not only natural ventilation, but also such factors as the direction of prevailing winds, the impact of industrial plants and heating of dwellings. Budapest, for instance, is surrounded by large forests and has mountains on the Buda side of the extensive waters of the Danube. The effect of natural ventilation in improving the quality of the air, however, is limited because for over two hundred days per year the products of combustion from large industrial plants are blown toward the centre of the city by the dominant northwest winds. The effect is to increase atmospheric pollution in the central sectors where the air is already considerably polluted by building heating systems.[3]

The extensive layer of smog, so often seen covering any dense metropolis, is a further hazard to our health, even if it is far above us. The penetration of solar rays, including ultra-violet rays, is substantially reduced, and the combination of polluted air and the lack of ultra-violet light exerts a negative and denaturalizing biological effect on the human organism. Moreover, persistent localized illness, particularly in the respiratory organs, may be brought about. In contrast to the polluted air of large cities, natural air is clean, from a biological point of view, and rich in stimuli, and therefore offers a favourable environment. Yet, perfectly pure or sterile air does not exist. Even in a natural state it is somewhat polluted, but the quantity and quality of potentially harmful effects are considerably less than are found in the urban climate and the health hazard is correspondingly reduced. Indeed, the most important aspect of the metropolitan climate is that its harmful effects are felt by hundreds of thousands at one time. Preventing air pollution is therefore of prime importance. However, as with other urban problems, air pollution in the metropolis is the product of many factors acting in concert. It is therefore difficult, if not impossible, to identify any one of these factors with any one of the detrimental effects considered below.

First, we may consider the effect of pollution on the metabolism of the human body. Polluted air enters the metabolic

---

[3]Tibor Bakács, ed., "Environmental Hygiene" (in Hungarian), *Hygiene* (Budapest, Medicina, 1965), Chapter III, pp. 137-360.

cycle in two distinct ways: directly, through breathing and through the intermediary of ionization rays, which, avoiding the circulatory system, influence the metabolism; and, indirectly, in that the umbrella of urban air composed of ash and fog prevents the penetration of solar rays, particularly the ultra-violet rays, thereby inhibiting certain metabolic processes. The deficiency illnesses then begin to appear, the first being vitamin-D deficiency. Substances breathed through the lungs (the quantity of which may be of importance in certain extreme cases) may bring about more serious disorders in the metabolic cycle than will matter introduced in food because the latter avoids the body's detoxifying apparatus.

Impurities in the air may also cause harmful local effects in the respiratory tracts. Trüb and Posch, describing the direct effects on the respiratory tracts, point out that bronchitis in school children in Dusseldorf accounted for 3.53 per cent of all illnesses experienced. In the rural areas of West Germany, the comparable figure was 2.4 per cent. In the same study it was found that the haemoglobin level of the blood was below 80 per cent for 58.5 per cent of the city children, whereas this low count was found in only 20.2 per cent of the rural children. Again, 15.1 per cent of the urban children, as opposed to 7.6 per cent of the rural children, had suffered from rickets.[4]

Hamill informs us that in England, where the air is highly polluted, bronchitis causes a loss of 27 million work-hours each year and about 30,000 deaths per year, surpassing total deaths from tuberculosis and lung cancer.[5] In New York, according to Heimann, 1,500,000 tons of sulphur dioxide enter the air each year from the combustion of 32 million tons of coal.[6] The entry of this sulphuric gas into the lungs brings on a temporary spasm of the smooth muscle in the bronchial wall. Greater concentration provokes serious inflammation of the mucous membrane, which in turn can lead to desquamation, or peeling, of the epithelial layer, thereby preparing the lungs for a subsequent cancer. Another study on air pollution in New York City noted that the rise in atmospheric pollution increases the incidence of attacks of cardiac and respiratory illness significantly.[7]

The presence of other minerals and chemicals in the atmosphere, and their effect upon health, has also been noted by scientists. For instance, traces of beryllium in the atmosphere, can bring about chronic pulmonary granulomatosis; and pneumonia is frequently suffered by the inhabitants of areas proximate to manganese processing plants.[8]

In Japan, Toshio Toyama studied the relation between air pollution and asthma

[4]C. L. Paul Trüb and J. Posch, "Program und erste Untersuchungsergebnisse eines Arbeitskreises des Offentlichen Gesundheitsdienstes in Regierungsbezirk Düsseldorf zur Feststellung von Gesundheitsschäden beim Menschen durch industrielle Luftverunreinigungen," *Zentralblatt für Bakteriologie, Parasitenkunde, Infektionskrankheiten und Hygiene*, I, Abteilung, Originale, Vol. 176, No. 3/6 (Stuttgart, 1959), pp. 207-30.

[5]Peter V. V. Hamill, "Atmospheric Pollution, the Problem," *Archives of Environmental Health*, Vol. 1, No. 3 (Chicago, 1960), pp. 241-47.
[6]Harry Heimann, "Effects of Air Pollution on Human Health," *Air Pollution*, WHO Monograph Series, No. 46 (Geneva, 1961), pp. 159-220.
[7]Leonard Greenburg, Franklyn Field, Joseph I. Reed and Carl L. Erhardt, "Air Pollution and Morbidity in New York City," *Journal of the American Medical Association*, Vol. 182, No. 2 (Chicago, 1962), pp. 161-64.
[8]Heimann, *loc. cit.*

in the cities of Tokyo, Kawasaki and Yokohama, and discovered a significant correlation between bronchitis and the dust content of the air, and between pneumonia and certain cardio-vascular diseases. According to Toyama, asthma in Tokyo and Yokohama is an allergic disease occurring mainly in the summer and directly related to meteorological factors. He deduced, however, that the bronchial symptoms appearing in spring and winter can be accounted for by atmospheric pollution.[9]

Randolph, working in Chicago, showed that allergic persons, who are without symptoms when living outside the city, are immediately affected by symptoms upon entering the city. This was particularly noticeable in the western and northern sectors of Chicago where large plants, foundries, refineries and numerous manufacturers of chemical products and paints are to be found. The allergic manifestations took on many different forms: irritation of mucous membranes, neuralgia, over-excitability, hyperactivity, rhinitis, bronchitis and even the classic symptoms of asthma (fatigue and painful muscular symptoms).[10]

The great increase in automobile use and the development of the diesel-engine are responsible for an increasing amount of carcinogenic substances in the atmosphere. Experimental animals have developed skin carcinoma and lung cancer from contact with or breathing of the combustion products of gasoline, coal, benspyrene and methyl-cholanthrene. It is thought that a linear correlation exists between the increase in lung cancer cases and the increase in motor traffic in the last thirty years. Figure 6-1 shows the rapid increase in cases of cancer of the lung and respiratory tracts in the United States and England in the years 1930 to 1955.

This correlation appears particularly obvious in the northern and western sectors of London where traffic is very heavy and the atmosphere is more than usually polluted; there, mortality from lung cancer is high.[11] Table 6-2 shows some further correlations by Lawther between the size of city and mortality from lung cancer and bronchitis. The situation appears to be similar in Hungary where the incidence of carcinoma of the respiratory organs in city dwellers was double that found in rural dwellers in 1960.[12]

Smog, which is the result of an unfortunate combination of meteorological conditions and the increase in atmospheric pollution, may bring about an intrinsic imbalance within the human body. The respiratory tracts, in particular, are affected and death may result. Many persons may be affected simultaneously. In London, in 1952, it is estimated that about 4,000 deaths were caused by smog in a period of five days, for the most part respiratory tract and cardiac cases. In December, 1962, London once again recorded 700 deaths for which smog was responsible. Here again, the disaster affected the sick who were suffering from bad cases of bronchitis. Francis Taylor, working in Los Angeles, has indicated that nitrogen oxides in the

[9]Toshio Toyama, "Air Pollution and Its Health Effects in Japan," *Archives of Environmental Health*, Vol. 8, No. 1 (Chicago, 1964), pp. 153-73.

[10]Theron G. Randolph, "Human Ecology and Susceptibility to the Human Environment, Part III: Air Pollution," *Annals of Allergy*, Vol. 19, No. 6 (St. Paul, 1961), pp. 657-77.

[11]P. J. Lawther, "Cancerogne in der stadtischen Luft," *Zentralblatt für Bakteriologie, Parasitenkunde, Infektionskrankheiten und Hygiene*, I, Abteilung, Originale, Vol. 176, No. 3/6 (Stuttgart, 1959), pp. 187-93.

[12]Unpublished data from the National Institute of Cancer Research, Budapest.

Figure 6-1
Mortality in Males Caused by Respiratory Tract Tumors, 1930-1955
(United States and England)

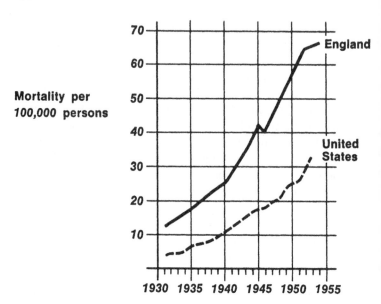

TABLE 6-2

Mortality from Lung Cancer and Bronchitis in England and Wales by Size of Settlement, (1950-1953)

Size of Settlement	Lung Cancer		Bronchitis	
	Men	Women	Men	Women
	(as a percentage of national average .100)			
Large city	126	121	131	126
Cities:				
over 100,000	112	101	107	101
from 50 to 100,000	93	88	90	86
less than 50,000	84	86	85	84
Rural	64	77	62	72

Source: P. J. Lawther, "Cancerogene in der stadtischen Luft," *Zentralblatt fur Bakteriologie, Parasitenkunde, Infektionskrankheiten und Hygiene*, I, Abteilung, Originale, Vol. 176, No. 3/6 (Stuttgart, 1959), pp. 187-93.

atmosphere are primarily responsible for this sad result. Automobiles emit in the order of 65 to 85 per cent of these oxides, and the rest originate from products of heating plants.[13]

In addition to the harmful effects on health, atmospheric pollution damages plants, animals and buildings. Sulphur dioxide is the main cause of this damage. It will eat through cement, concrete, limestone and iron. Architectural decorations and stone mouldings on buildings, monuments and statues suffer. Vegetation is almost completely destroyed in the areas surrounding certain factories emitting sulphur dioxide.

Fluorides have a similarly harmful effect. In Czechoslovakia, a detailed study was made of the destruction of vegetation by fluorides. It was noted that in the region of aluminium blast furnaces in Malacka, the vegetation had been ruined, a further consequence of which was the death of livestock. In Germany, the annual damage caused by hydrofluoric acid is estimated at D.M. 24 million—a very high tax on agriculture. Similarly, the deterioration of vegetation in the environs of Los Angeles alone represents a loss of approximately $5 million in value per year. A calculation for the United States as a whole, on the basis of conditions in Los Angeles, presents a loss of $3 billion per year.[14] In Hungary, the majority of public buildings and monuments are of limestone, and deterioration due to the effect of H_2SO_3 is widespread. Chimneys of the blast furnaces of the Lenin Steel Works at Miskolc daily emit 9.5 tons

of solids into the air over this city. The power plant in the City of Pecs consumes from 280 to 300 railways cars of fuel daily, approximately half of which returns and falls on the city in the form of cinders and ashes.[15] Although the Hungarian government pays considerable compensation for vegetation losses incurred in the environs of aluminium blast furnaces, there are, needless to say, at the same time many harmful effects upon health for which there can be no compensation.

Health Problems Associated with Water Supply, Waste Water Removal and Soil Pollution

A brief description of the role of water in the metabolic process serves to underline the importance of a good water supply, quantitatively and qualitatively, and the implications for the human body of polluted water. The human organism is essentially an aqueous solution in which the processes of metabolism are carried out. To satisfy the requirements of this organism, certain quantities and qualities of solvents must be introduced to maintain a given concentration. Although this process may be successful for relatively short periods of time, with increased age the concentration increases to the "detriment" of the solvent. (The intrauterine embryo at an early stage contains 95 per cent water, the infant 74 per cent, and an adult only 67 per cent.) It is this reducing tendency that brings man to his death. For the building of the body, the human organism has an enormous demand for mineral substances, especially calcium for bones. In the grown adult, this demand is excessive, the amount of dry matter increases and the capacity for

[13]Francis R. Taylor, "The Elimination of Oxides of Nitrogen from Automobile Exhausts," American Medical Association, *Archives of Industrial Health*, Vol. 21, No. 4 (Chicago, 1960), p. 386.

[14]Ralph C. Graber, "Atmospheric Pollution," *Archives of Environmental Health*, Vol. 1, No. 3 (Chicago, 1960), pp. 234-40.

[15]Lajos Szabó, "Air Pollution in Pècs" (in Hungarian), *Népegészségügy*, Vol. 43, No. 2 (Budapest, 1962), pp. 46-48.

concentration and secretion by the kidneys decreases, thus making the secretion of the waste products of metabolism more and more difficult. It is a process of self-intoxication against which the organism defends itself by trying to wash away useless substances from the tissues with the help of increasing quantities of water. Even sound kidneys require a great deal of water for the removal of the waste products of metabolism; malfunctioning kidneys require even more. For the kidneys to remove waste from the blood stream, several hundred litres of water must pass through the kidneys each day. The end product, urine, is produced through successive concentration, and at the time of excretion the volume of water does not surpass one and a half litres. Thus, to economize upon its own water reserves, the organism recirculates its internal water supply. The need, however, is for a plentiful supply of good quality water, and where this requirement cannot be guaranteed life is impossible. The intrinsic balance of the organism is upset, for instance, where there may be sufficient water, but the organism cannot use it for the process of metabolism. The harmful effects of a malfunctioning metabolism accompany disturbances in the capacity for adaptation in the aged organism; they include, for instance, water retention troubles and the accumulation of waste substances in the nervous system and in the organs. Thus the significance of a good water supply, in terms of both quantity and quality, is obvious.

In view of the above-mentioned requirements and the need to put an end to large-scale epidemics, the end of the nineteenth century and the beginning of the twentieth century saw a rapid expansion of water services in many large cities. Throughout the world, the construction of water works began. The supply of high quality, potable water immediately increased. However, as metropolitan areas become larger, and as man's environment becomes more and more artificial, the provision of good quality water, in sufficient quantity, becomes more difficult. At the present time, in spite of accelerated technical progress, an insufficient supply of potable water again threatens the healthy development of metropolitan areas. Indeed, aside from the need to impose rigorous quality controls on the water supply, the local exploitation of water, in most cases, cannot be further intensified.

The available reserve of fresh water in the world is insufficient as measured against demand. Figure 6-2 represents the "exterior" biological cycle of water and, at the same time, possible sources of supply. Barely 3 per cent of the total world water reserve—sea water, icebergs, and so on—is fresh water. In addition, the location of fresh water reserves does not correspond to the location of the industrialized regions of the world. Consequently, water supply for highly industrialized regions has now reached the point of crisis in western Europe (for example, in the Netherlands, Belgium and West Germany) and on the western and eastern shores of the United States. In these regions the quantity of underground water, the traditional source of potable water, has decreased rapidly because the metropolises consume and impoverish the superficial layers of water underlying them. A metropolis with one million inhabitants, for example, requires at the present time an area of about 750 square kilometres if underground water sources are used to meet household and industrial needs. Again, this assumes an annual atmospheric precipitation of 1,000 mm. In Hungary, with an average precipitation of only 700 mm. per year, a larger area is required to meet such needs.

Figure 6-2
Outline of "Exterior" Biological Circulation of Water

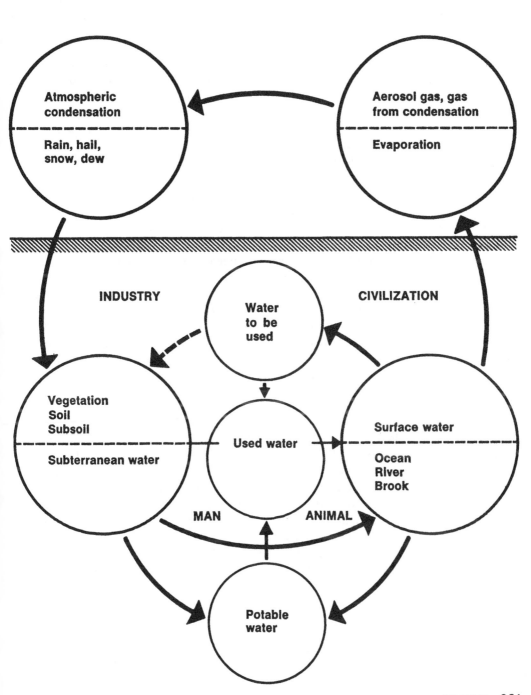

Excessive exploitation has lowered the level of underground reservoirs, making these sources less accessible and, at the same time, has lowered river levels, which are naturally closely related to underground water levels. In western Europe, in certain industrial regions, river levels have decreased by two metres in as little as ten years' time, thus diminishing the possibilities for the river as a permanent, unlimited source of water.

Meeting the water needs of 8 or 9 million inhabitants of the Ruhr area has only been possible—at least for the time being —with the help of the River Ruhr. A great deal of care is taken to protect the purity of the river. The River Emscher, running along the length of the industrial area and parallel to the Ruhr, is used as a canal for waste waters. The River Lippe, traversing the northern region of the Ruhr area, although used for waste disposal, is also used for irrigation purposes. Naturally, a major fall in river levels would threaten this entire regional water supply and waste removal system.

Additionally, the greater consumption of water in large metropolises has resulted in a similar increase of waste water. The tremendous volume of waste water, often untreated, has resulted in considerable damage to the recipient rivers, lakes, and those water reservoirs immediately below the surface. In the environs of large industrial centres, parts of rivers and lakes cease to be "living water" and lose their biological self-purification qualities. The Danube, for example, is seriously affected by the discharge of wastes, although it still retains the characteristics of living water. Budapest produces approximately one million cubic metres of waste water per day (60 per cent of which is of industrial origin and 40 per cent of residential origin), for the most part treated mechanically. The Danube has to digest this tremendous flow and complete the treatment process through self-purification. It is estimated, however, that this process is only completed 60 kilometres downstream from Budapest.[16]

The increased pollution of underground supplies of water for cities and the consequently severe reduction in quality and quantity of supply is forcing industrialized countries to resort to the use of surface water sources. Yet with the pollution of these sources, the problem becomes further magnified. In the industrial regions of the United States, over 90 per cent of the water supply is from surface sources; in England and France, the figure is 85 per cent; in West Germany it is 57 per cent; and, in Czechoslovakia about 26 per cent. This tendency is increasing throughout the world.

In spite of efforts to stem these trends, in the largest industrial centres the demand for water and the great increase in effluent has outstripped the (static) supply of potable water.[17] The consequence of this condition is often a chronic shortage of potable water in many parts of the world. Water consumption may be subject to restrictions. Often there is a cyclic supply, resulting in unsatisfactory sanitary conditions. In fact, while the distribution is cut off, a vacuum is formed in the pipes which may result in the intake of polluted water through joints that are not water tight. Again, restrictions on water supply often decrease personal hygiene standards.

In addition to an increased quantitative demand there is also the need to retain a

---

[16]Bakács, *loc. cit.*

[17]Tibor Bakács, "Some Hygienic Problems of Water Supply in Hungary Compared to the International Situation" (in Hungarian), *Annual Report of the State Institute of Hygiene for the Year 1964* (Budapest, 1966), pp. 195-218.

high quality of supply. The above-mentioned conditions threaten the quality of water. Waste water loaded with toxic agents, such as detergents and pesticides, finds its way into the lakes and rivers, where it endangers future surface exploitation. The increasing presence of these substances in water supplies is slowly bringing previous progress made in the provision of piped drinking water to a halt. The quality of drinking water in many areas is declining, very often because of the need to add disinfectants, such as chlorine, to make it potable.

To meet the increasing demand, new techniques must be developed. Surface exploitation has followed the severe depletion of underground sources. Re-circulation, large-scale regional water works, and the use of sea-water have been tried. None of the techniques has proved entirely satisfactory thus far. C. H. N. Durfor and E. Becker have pointed out the difficulties of producing water for the 100 largest cities of the United States, the most industrialized country in the world.[18] The hundred cities analyzed have 60 million inhabitants—34 per cent of the total United States population, and 48 per cent of American city-dwellers. Water distribution services supply 9,650 million gallons of water daily of which 1,200 million gallons are distributed among the 8 million city-dwellers in New York. Water produced is partially of ground water origin, partially surface. Among the hundred cities examined, twenty rely upon underground water sources alone, fourteen cities use both sources of water and sixty-six cities use surface water sources only. In view of the increasing difficulty of supplying water to large urban areas, water will shortly rank among the most sought after of resources. Among the new production methods, the most unnatural procedure is the recuperation and purification of waste water which, once regenerated, is returned to the water cycle shown in Figure 6-2. At present, this process not only raises problems of an aesthetic and technical nature, but it is also not yet completely satisfactory. In addition to detergents and pesticides already mentioned, there are certain other substances often present in the water than can have ill effects on health and that are more difficult to remove.[19]

An idea of the household uses of water in urban areas in the United States can be gained from the following figures from Akron, Ohio: flush toilets, 41 per cent; baths, 37 per cent; kitchen, 6 per cent; drinking water, 5 per cent; washing machines, 4 per cent; house cleaning, 3 per cent; watering the garden, 3 per cent; washing the automobile, 1 per cent. Another problem associated with the supply of water is the industrial use of potable water. Thus there are cities where daily water consumption per person, including industrial consumption, reaches or even surpasses 1,000 litres (for example, Boulder City, U.S.A.—1,900 litres per day; Buffalo, U.S.A.—1,200 litres per day), and this is not the upper limit.

Table 6-3 illustrates the amount of water required to produce one ton of selected industrial products. To economize on the use of potable water, a possible solution has been sought. In Paris, a network parallel to that of the potable water supply was constructed for industrial use. Industrial

[18]C. H. N. Durfor and E. Becker, *Public Water Supplies of the 100 Largest Cities of the United States 1962*, U.S. Geological Survey Water Supply, Paper No. 1812 (Washington, D.C., 1964), p. 364.

[19]Reinhard Demoll, "Forderung an das Trinkwasser," *Gasund Wasserfach*, Vol. 101, No. 22 (Munich, 1960), pp. 549-51.

TABLE 6-3
Water Required To Produce One Ton of Selected Industrial Products

Product (one ton)	Water Required (in litres)
Soap	2,000
Coal	6,000
Sugar	9,000
Canned Fruit	20,000
Cotton	200,000
Steel	270,000
Synthetic Silk Fibres	600,000
Explosives	800,000
Paper	1,000,000
Synthetic Rubber	2,500,000

water is less costly and is also used for other than industrial purposes. For instance, the fire-fighting and watering network is also linked to this same distribution. Although this solution reduces the difficulty of producing water, careful handling is necessary to avoid risks to health. Thus, for industries near the edge of a river or a lake such a solution is certainly feasible. Certain restrictions would obviously be required for some industries; for instance, it would not be possible to supply both kinds of water to food and dairy industries, where the use of potable quality water is essential.

Perhaps it is due to the present demands of industry that such great strides have been made in the quantitative supply of water through the construction of new water distribution networks in many parts of the world. The percentage of national populations served by such national networks is often high (for example, 96 per cent in Switzerland; 85 per cent in the Netherlands; 92 per cent in Italy; and 50 per cent in Czechoslovakia).

However, unlike the quantitative improvements that have been made, there has

been no substantial improvement in quality since around the turn of the century. In fact, the increasing industrial demand for water has forced suppliers to use methods that, while ensuring the flow required, have threatened health standards and introduced the risks of epidemics. Errors have been made in the location of extraction points for water and in the treatment of waste water that have affected the quality of potable water. This twentieth century decline in standards was felt first in the United States very much as a result of the rapid growth of industry in that country.[20] A similar situation is to be found in Russia. V. M. Jdanov, in discussing the problem in that country points out that the only way in which water-borne intestinal diseases can be eliminated, or even reduced considerably, is through massive expenditure on

[20] Arthur E. Gorman and Abel Wolman, "Waterborne Outbreaks in the United States and Canada, and Their Significance," *Journal of the American Water Works Association*, Vol. 31, No. 2 (New York, 1939), pp. 225-373; and Rolf Eliassen and Robert H. Cummings, "Anaylsis of Water-Borne Outbreaks," *Journal of the American Water Works Association*, Vol. 40, No. 5 (New York, 1948), pp. 509-28.

improvements to installations relating to communal hygiene, the most important being water supply systems.[21] Indeed, it would appear that almost every country undergoing urbanization and industrialization is similarly confronted with these problems. Hungary, certainly, provides yet another example of a developed country which has not yet been able to eliminate this major threat to health.[22]

If in the developed and highly industrialized countries, where the demand for water is ever increasing, the problems continue to multiply instead of decrease, what may be said of the developing countries, where the growth capacity of the national economy is lower, where a medical service is almost non-existent because of a shortage of staff, and where the role of water as a carrier of epidemics is still very great due to a lack of quality control? In these countries, a far greater percentage of city-dwellers lives in poverty-stricken areas, prey to popular superstition and, in many instances, the parasites common to a tropical environment. Consequently, average life expectancy is generally little more than forty years.

Bernd Dieterich and John Henderson, in a study of seventy-five developing countries with more than half of the world's inhabitants, found that only 33 per cent of the city-dwellers in the study area, or 10 per cent of the total population, were supplied with piped water. Even where water was supplied it was not of suitable quality or quantity and in general, the supply was also found to be irregular. With 45 per cent

of city-dwellers and 70 per cent of the total population in regions where it is impossible to lay water pipes, the problems will not be easy to solve. Untreated water, drawn from a river or some other contaminated source is often all that is available. In southeast Asia and south central Asia, for example, two-thirds of the city-dwellers do not have any piped water service, and this is in addition to the fact that population growth is probably higher here than anywhere else in the world. By the year 1977, it is expected that these regions will experience an increase of 207 million city-dwellers over the present urban population, making a total of 450 million city-dwellers, most of whom will very likely be without piped water, as well as many other urban services.[23]

Table 6-4 serves to illustrate the present deficiencies in urban water supply in developing countries. There is limited technical capability, materials and money to alleviate these deficiencies. The necessary construction alone would require from $400 to $500 million annually for fifteen successive years, or a total sum of $6,600 million. Although this sum appears enormous, it is the only feasible way to reduce to a minimum the enteric diseases, so widespread in these countries. In Japan, the introduction of good quality potable water to thirty rural regions resulted in a 71.5 per cent reduction in enteric infection, a 51.7 per cent reduction in infant mortality, and a 63 per cent reduction in trachoma. Similarly, in the state of Uttar-Pradesh, in India, improvements to the water supply reduced mortalities resulting from cholera by 74.1 per cent, from typhoid fever by

[21]V. M. Jdanov, "The Control of Enteric Diseases in the U.S.S.R. in Recent Years" (in Russian), *Sovetskoe Zdravoohranenie*, Vol. 17, No. 10 (Moscow, 1958), pp. 3-7.

[22]Laszlo Kubinyi, "Waterborne Epidemics in Hungary During the Last Decade" (in Hungarian), *Egeszsegtudomany*, Vol. 4, No. 4 (Budapest, 1960), pp. 348-54.

[23]Bernd H. Dieterich and John M. Henderson, "Urban Water Supply Conditions and Needs in Seventy-Five Developing Countries," *Public Health Papers*, WHO, Monograph Series, No. 23 (Geneva, 1963).

TABLE 6-4
Urban Water Supply in Seventy-Five Developing Countries (1962)

| Region | Urban population supplied | | | | | | City-dwellers without supply | |
| | From house connections | | From public outlets | | Total | | | |
	% of Region	% of Total	% of Region	% of Total	% of Region	% of Total	% of Region	% of Total
North Africa	57	10	20	4	77	7	23	3
Africa, south of Sahara	13	2	38	10	51	6	49	8
Sub-total, Africa	34	12	30	14	64	13	36	11
Central America and Caribbean	55	15	30	10	85	13	15	3
Tropical South America	59	29	27	17	86	24	14	6
Temperate South America	67	13	22	6	89	10	11	2
Sub-total, Latin America	60	57	27	33	87	47	13	11
Southwest Asia	39	10	36	11	75	10	25	5
South Central Asia	14	12	20	24	34	17	66	48
Southeast Asia	15	6	26	13	41	9	59	19
East Asia	19	3	30	5	49	4	51	6
Sub-total, Asia	18	31	25	53	43	40	57	78
TOTAL	33	100	26	100	59	100	41	100

Source: Bernd H. Dieterich and John M. Henderson, "Urban Water Supply Conditions and Needs in Seventy-Five Developing Countries," Public Health Papers, No. 23 (Geneva, WHO, 1963).

63.6 per cent, from dysentery by 23.1 per cent and diarrhea by 42.7 per cent.

Our wealth in potable water is measured in terms of our underground and surface water reserves. Apart from water to be used in the future, waste water from large metropolises is entering the soil in increasing volume and with increasing toxicity. Fortunately, soil has a great capacity for self-purification, without which the impact of urbanization on human health would be catastrophic. The billions of micro-organisms per volumetric unit of soil are capable of breaking down and neutralizing many of the substances harmful to human health. Their capacity to do this is not, however, unlimited. Soil in urban areas suffers a great deal from the infiltration of untreated urban wastes. It tends to lose its capacity to break down impurities which, in turn, threaten underground water supplies. Especially harmful are synthetic products such as the chemicals that are used to increase crop production, but which pose health hazards through their initial destruction of micro-organism in the soil and also through direct access to the human organism.

In Switzerland, with the general use of chemical agents in agriculture, the purity of surface water has been affected. It is estimated that 0.3 to 0.5 kilogrammes of the phosphorus and 45 kilogrammes of the nitrogen that have been spread over one hectare of agricultural land, penetrate the soil, rivers and lakes in some areas. In 1962, in the United States, 150 million kilogrammes of pesticides were spread over an area of 36 million hectares. Unfortunately the gravity of this fact is not reflected in these figures: a catastrophe may well be started through the rapid spread of the use of pesticides and through the occurrence of certain hydrological conditions. Already, studies in the United States and in the United Kingdom have revealed that adult human fat contains 19 milligrammes of pesticides per kilogramme. Whereas in West Germany the figure is only 2 milligrammes per kilogramme. Although chemicals certainly have their place in our life, the dangers must be recognized. In a few years it may be too late.

The Problem of Urban Refuse

The removal of urban refuse, its neutralization and final disposal are priority requirements in urban areas. In most refuse, particularly that from the kitchen, there exists the threat of the survival of pathogens of typhoid fever, paratyphoid, dysentery, tuberculosis and anthrax. These species can, in poorly treated refuse, remain in a virulent state for days, weeks or even months. Table 6-5 emphasizes the need for careful disposal of refuse. Luckily, in the metropolises of industrialized countries with a highly mechanized system of refuse collection this threat has been almost eliminated.

The secondary problems of the neutralization and disposal of refuse can be solved by either the incineration or the composting of refuse (biothermal neutralization). A third solution, the filling in of tiled trenches or pits, should be regarded solely as a temporary measure. This last solution, however, is used extensively. It is the least costly, it permits the development of the land at a later date, and the existence of disposal sites spread within the city reduces transportation costs. However, the available land is soon exhausted. For example, it is estimated that Budapest, a metropolis of 2 million people, can use this filling system for another fifteen years only. Thus, another solution will soon have to be found for long-term requirements. Of the two alternatives presently available, incineration and composting, the former

TABLE 6-5
Duration of Vitality of Pathogens in Various Types of Refuse

Refuse	Pathogen	Duration of vitality of pathogens (in days)
Kitchen refuse	Typhoid fever	4
	Paratyphoid fever	24
	Dysentery	5
Refuse swept from rooms	Typhoid fever	42
	Paratyphoid fever	107
	Dysentery	24
	Anthrax	80
Spit	Tuberculosis	120-200

Source: V. A. Gorbov and A. N. Marzeev, *Environmental Sanitation* (in Russian) (Moscow, Medgiz, 1951).

seems to be the more hygienic in that infectious and harmful components are destroyed. Incinerators can also be used to burn the dehydrated wastes produced in the purification of used water. Secondary benefits of this operation are the production of thermal and electric energy. The disadvantage of this solution is that oven construction is expensive, and therefore incineration is economical only in cities whose population surpasses 600,000.

On the other hand, the use of the composting system is declining. Household refuse tends to be poor in compost matter except in those tropical and sub-tropical countries where there is no great reliance on canned goods (although the introduction of pre-cooked and pre-packed meals has resulted in a further reduction in compostable matter). Industrial waste, which represents a sizable proportion of urban refuse, is even less suitable than household refuse for composting. It should be noted too that costs of a modern compost plant approach those of an incinerator.

However, although the change in refuse content has made composting less satisfactory, the increasing amount of plastics and other similar chemical products has posed problems for the incineration process. These materials not only require higher incineration temperatures under pressure (thus adding to the costs), but also release chlorides which result in a greater corrosion of the ovens.

In the final analysis, it must be admitted that the expensive process of neutralizing urban refuse in the wealthy industrial countries is nearly perfect. Another problem of more recent significance, and only mentioned here, is that atomic energy exploitation is raising the problem of the removal and decontamination of radioactive waste. This, however, is a separate problem that cannot be solved by any conventional means.

In developing countries, the neutralization of urban wastes is a health problem of the first order. In the metropolises of most tropical countries, the disposal of rotting

TABLE 6-6
Noise Generated by Vehicles (in decibels)

Vehicle	Warsaw	Prague	Moscow
Trams	90	76-91	90
Trucks	80	88-90	90
Automobiles	60	76-88	76
Motorcycles	90	86-93	86

Source: Karlwilhelm Horn, "Der Grosstadtlarm und seine Bedeutung fur das Zentralnerven-system," *Das Deutsche Gesundheitswesen*, Vol. 15, No. 31 (Berlin, 1960), pp. 1617-22.

household refuse is not the only problem to be solved. The open market places are covered not only with refuse, but also with dead rodents and beasts. In these countries, "the number one employees" of the Highways Department are still the vultures, until technology replaces them.

Sources of Physical and Mental Stress in the Metropolis

While the problems dealt with thus far have always been characteristic of urbanized areas and have only taken on new dimensions in the metropolis, there are other factors affecting human health which owe their origin to recent technological developments and the accelerated way of life, both of which are manifested in the modern metropolis. Brief mention should be made here of the impact on man's health of the increase in traffic and other stress factors in the urban environment.

Any increase in traffic on urban streets results in an increase in noise, traffic accidents and, as we have seen, carcinogenic substances. This increase in traffic is common to all parts of the world but it is the inhabitants of the major metropolitan areas who are suffering most from its detrimental effects. Street noise, while having many other sources, such as pneumatic drills, transformers and so on, is one of the greatest sources of annoyance for the city-dweller. Noise generated from urban expressways, during peak travel periods, may rise above 35 to 40 decibels and may bring about chronic damage to hearing. Other effects are stomach ailments, disturbed intestinal secretions and serious alterations to the nervous system.[24] As Table 6-6 indicates, the noise generated by vehicles often far exceeds 35 to 40 decibels.

Krilova and others have shown that lasting noise, in addition to affecting hearing, results in hypertension.[25] Lehmann has indicated that psychic troubles are experienced with subjection to noise above 30 phons and that noise in excess of 90 phons affects the development of hearing and that in excess of 120 phons results in ganglionic and cellular alterations to the nervous system.[26] It must be stressed, even though it is apparent, that these conditions presently exist in all parts of major metropolises and at all times of the day. Figure 6-3, which refers to Tanács Boulevard, in Budapest,

[24]Karlwilhelm Horn, "Der Grosstadtlärm und seine Bedeutung für das Zentral nerven-system," *Das Deutsche Gesundheitswesen*, Vol. 15, No. 31 (Berlin, 1960), pp. 1617-22.
[25]*Ibid.* (Krilova cited).
[26]*Ibid.* (Lehmann cited).

Figure 6-3
Correlation Between the Average Noise Level and Day Traffic on Tanacs Boulevard, Budapest

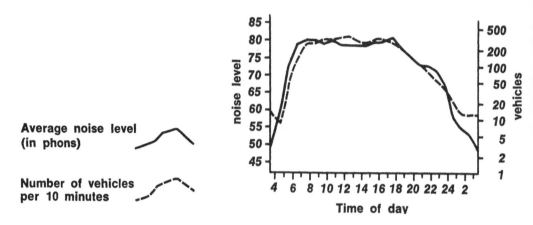

Average noise level
(in phons)

Number of vehicles
per 10 minutes

shows that the noise level remains well above 50 phons for twenty-four hours of the day.

One final observation on noise: on the periphery of the metropolis, it is not trucks that are the major source of noise, but aircraft. Since residential areas tend to be found on the periphery of metropolitan areas, the direct impact of aircraft noise on the health of the metropolitan inhabitants is of great significance and, more often than not, is the main reason for such strong local opposition to expansion of existing airport facilities or the construction of new airports. With the increasing size of jet aircraft it becomes more imperative that far more effort be expended on reducing airplane noise, that is, if we are to be able to take advantage of this form of transport without suffering further mental or physical stress.

Indeed the same comments apply to the other sources of physical and mental stress. To a limited extent the human organism is able to protect itself from these external stimuli, but when it is not, pathological symptoms appear. These symptoms are extensive and include the following: gas-

tritis and enteritis; stomach and duodenal ulcers; coronary, cardiac and vascular troubles and hypertension; metabolic and hormonal disorders (for example, diabetes and thyroid hyperactivity); and, nervous tension.

The last of these, nervous tension, is a common problem among urban dwellers who, unlike rural inhabitants, spend little time in the fresh air, have little physical exercise, live at a much faster pace and are constantly surrounded by noise. Psychosomatic disturbances frequently accompany insomnia, although very often this stems from the consumption of sleeping pills and other products purported to calm nerves. The use of sleeping pills is commonplace. For example, in Germany 98 per cent of all families have sleeping pills in the house, and in England doctors give 20 million prescriptions per year for sleeping pills.[27]

Mental illnesses are also found in greater numbers in the metropolis. In addition to the physical-biological unity that exists

---

[27]*Ibid.*

between man and his environment there is a so-called "psychic climate", rich in environmental stimuli. Mental illness may result if an imbalance develops between man and his environment. As mentioned previously, man is able to counteract any imbalance to a limited extent, but his ability to do this certainly is by no means unlimited. Leo Srole and Thomas Langner have observed in their study of mental health in Manhattan that in one midtown district of New York City, five persons in a thousand are bed-ridden in mental hospitals and 788 out of 100,000 have been out-patients in psychiatric clinics.[28] A similar or worse situation prevails, not only in other parts of the United States, but in all other urbanized countries. This is probably borne out by the 124 per cent increase in criminal cases in cities over 100,000 population in West Germany between 1950 to 1958. For towns of 50,000 or less the increase was only 28 per cent.

The relation between social conditions and health is the concern of the social hygienist. There is a close correlation between poor health and such social factors as low income and poor housing. A further side effect of the poor social conditions is the restriction upon the availability and breadth of social insurance, itself having a direct bearing on the state of health of the individual. The correlation that can be drawn between average per capita income and life expectancy in both developed and developing countries, while needing qualification, is very significant. For example, in 1960, the United States achieved an average, annual, per capital income of $554 and a life expectancy of sixty-two years for males; in Mexico the comparable figures were $61 and thirty-seven years; and in India, $34 and twenty-seven years.[29] Also, generally speaking, within any one country, groups enjoying more advantageous social conditions likewise enjoy a marked increase in life expectancy. Finally, poor housing and the concomitant social ills also have inhibiting effects upon child growth and other aspects of human health. In no city has the desired ideal of one room per inhabitant of every household been achieved, and it is difficult to see how it ever will.

Conclusions

While it is difficult, and perhaps unwise, to make specific suggestions for solving the problems described above, it is possible to make some general observations that apply to all countries alike.

Of primary importance in the elimination of air, water and soil pollution is a programme that embraces multilateral protection for all these elements. Such a programme calls for a two-pronged attack: on the one hand legislation is required, and on the other there is a greater need for research into more efficient and practical measures of control. Research must establish acceptable standards relating to clean air, water and soil. It should be the responsibility of factories and other sources of pollution to observe these standards. Legislation should be enacted and enforced that will ensure close supervision of the location of industry and of persons responsible for installing heating systems and other equipment causing pollution. In the event of violation of regulations, severe

[28]Leo Srole and Thomas S. Langner, *Mental Health in the Metropolis, The Midtown Manhattan Study* (New York, McGraw-Hill, Inc., 1962), p. 428.

[29]C. E. A. Winslow, *The Cost of Sickness and the Price of Health* (Geneva, WHO, 1951), p. 106.

penalties should be imposed, if possible, equivalent to the value of the total damage caused. To maintain a check on the level of pollution, mechanical devices should be set up to record the level of pollution and relay such information to central recording stations where any infractions of the law would be immediately detected.

As it should be the responsibility of industrial plants to eliminate emission of pollutants, the factories should gear production techniques to ensure that regulations are observed and, most important, to ensure that the harmful products are neutralized within the plant. If the latter action is not taken, the harmful agents will enter the environment, where their neutralization is a great deal more difficult if not impossible.

Some steps have been made in the attempt to tackle pollution. For example, in the Ruhr industrial region in West Germany, a disaster plan has been established for such times at which air pollution extends beyond what is regarded as the acceptable upper limit. It has to be made clear, however, that industrial establishments are not the only source of air pollution. The automobile is increasingly responsible. In the Ruhr area, it is estimated that factories account for only one third of the total pollution, whereas automobiles alone account for another third. Thus, there is the added necessity of filtering automobile exhaust fumes.

To protect potable water reserves and the soil, recourse must be had to modern methods of purifying used water. Tremendous sums are being used to install technical facilities for the biological purification of waste water, while the old mechanical and chemical methods of purification are being abandoned as inadequate. In spite of the efforts being made in highly industrialized areas, its rivers are still severely polluted. For example, the Rhine is unable to carry out the process of biological self-purification until it reaches the sea. In some metropolitan areas, the situation is such that further urban extension will be inhibited by the lack of potable water. In these regions, underground waters or surface waters are no longer sufficient to meet the demands on drinking water. Thus, the possibility of desalinating sea water is of increasing interest, although proposed methods are extremely costly.

In small countries, such as Hungary, there is a further problem associated with the protection of water reserves. Protection of the waters of the Danube, Hungary's largest river, requires the cooperation of the eight countries that it crosses prior to reaching the sea. The same obviously applies to other rivers crossing other countries.

The unfavourable impact on human health of various external elements previously described can be mitigated in large part through better planning. Three specific lines of action are the improvement of the poverty-stricken slum areas in cities, the establishment of satellite towns, and the improvement of traffic systems. To improve the living conditions of urban dwellers in poverty-stricken areas, there is an immediate need for modern housing, public parks, and other facilities offering numerous recreational and leisure pursuits. While the elimination of slums may, in the long run, be achieved spontaneously through the material improvement of the inhabitants, this process is far too slow. Yet any plans for action must necessarily be geared to the financial capacity of the national economy. Many metropolitan areas are witnessing the so-called "flight to the suburbs". Although this is due in part to the congestion that exists in city centres along with the noise and other undesirable residential features, another cause is that old residential buildings are being replaced by large office buildings. Some cities have been attempting to combat this outward movement although their efforts are directed more to the problem of cleaning up the central areas through attractive commercial developments than to providing residential buildings in the centre of the city. The establishment of promenades, excluding vehicular traffic, is a comparitively new concept which makes for a far more attractive metropolis. However, even in places like Stockholm, where some of these developments are found, there is an over-emphasis on commercial activity.

The concept of the satellite town is essentially one providing residential units in an area of green space. These towns, apart from the residential units, generally provide rudimentary cultural and basic service facilities. They are linked with the central metropolis by rail and highways to permit commuting over a long distance within a very short period of time. These essentially residential towns are a revised version of the original "garden cities" in that they are designed to be part of the metropolis rather than separate entities.

The improvement of the metropolitan traffic system although it has made a great contribution to the improvement of health will not be discussed here as it has been the subject of much more detailed examination by John Kain in Chapter 2.

To tackle the problems that have been raised in this chapter there is a greater need for research based on the concept of the metropolitan region as an ecological entity. It is the responsibility of governments to follow up the findings of such research and to ensure the maintenance of a healthy environment. Government action in metropolitan areas, especially the larger agglomerations that exist in North America and Europe and that extend over state or national boundaries, requires a regional approach to the planning and development of urban services, including the provisions called for in this chapter regarding the improvement and maintenance of human health. These are not tasks that can be handled by purely local authorities. Regional offices must be established for the purposes of survey, planning, the protection of the atmosphere, water and soil, and the development of open spaces and urban settlement. These measures with others called for in other chapters will enable us to make progress toward realizing the United Nations Code of Human Rights, part of which states: "all men are entitled to be assured of a living standard which ensures health and well-being for himself and family".

7 Welfare Services in Metropolitan Areas

Albert Rose

The Identification of Welfare Services

In any descriptive or comparative analysis of welfare services the first major problem is one of definition. Although welfare literally means well-being, the term "welfare services" has no widely accepted definition. In some jurisdictions the variety of programmes included in the concept encompasses services in the fields of health, education and labour relations.[1] In the more developed countries, there is likely to be a narrower definition of the term and a sharper separation between welfare services, health services, educational services and other programmes of assistance.

The confusion arises partly because in some countries, such as Canada and the United States, health and welfare services are the domain of one government department and partly because some programmes, such as mental health and child welfare, involve both services. Additionally, there has been some confusion as to the limits of welfare services within the sphere of "social services". This probably stems from the traditional free interchange of the terms in common parlance despite the fact that the term "social services" is

---

[1]John S. Morgan, ed., *Welfare and Wisdom* (Toronto, University of Toronto Press, 1966), pp. 3-4.

becoming more all-embracing, as is borne out by the following statement in a recent British publication:

> Britain's social services cover a wide range of provisions to promote the health and well-being of the people and to improve the surroundings in which they live. . . . To give a full picture of the provisions that exist to promote social welfare . . . [must include] not only the activities most commonly referred to as "the social services"—social security, health services, the care of the old, the disabled and children lacking families, together with education, housing and town and country planning—but also the facilities to promote good working conditions and helping people to get work, the treatment of offenders against the law and the provision of legal aid and advice to people without the financial means to defend themselves in court or obtain justice.[2]

It is clear, however, that no analysis of welfare services in metropolitan areas could possibly encompass all the activities considered by some authorities to fall within the concept of the "social services". The test must be the nature of the body, governmental or otherwise, bearing primary as opposed to secondary responsibility for the administration of the service. Thus, treatment of the subject here excludes the fields of the health services in their broadest sense, the educational services, the field of labour relations, and town and regional planning programmes, which are primarily the responsibility of non-welfare bodies. The analysis that follows is concerned essentially with those services provided for individuals and families to sustain their

functioning in society as relatively independent, self-supporting members. While they are not treated here, the author recognizes that those who are not financially self-supporting constitute a major group of beneficiaries of the welfare programmes designed to provide them with sufficient income to maintain a minimum adequate standard of living. Most of these income-maintenance services have become the right of a substantial proportion of the people of many countries as a consequence of the passage of national, state or provincial, or local legislations.

The more "personal" services required to enable individuals and families to function in society in a socially acceptable manner are often offered initially, at least in Western nations, by voluntary organizations whose staff members, trained or untrained in social work, are financed by means of voluntarily provided funds. Most of the welfare services in the Western world were originally developed and made available through the charitable impulses and voluntary associations of people attempting to meet human need and suffering. Only as these services demonstrated their basic worth, and as the process of identification and diagnosis of human need improved, has the public authority been induced to assume legislative and financial responsibility.

Looked at another way, this may be seen as essentially a matter of a government establishing its priorities as a result of more or less clearly defined pressures, with welfare services generally falling into the latter group and taking a low priority. Yet this problem of definition of needs and priorities is a very real one. Among the two thirds of the world's population, whose annual per capita income is $300 or less, specialization of function and sharpness of definition constitute luxuries that can be ill

---

[2]*Social Services in Britain*, United Kingdom, Central Office of Information, Reference Division (London, May, 1966), p. 1.

afforded. The mass of human need is too formidable, the urgency of the situation is too critical, to permit professional persons to demarcate too clearly their areas of jurisdiction or to distinguish too finely their sets of responsibilities. Thus, in the developing countries the basic requirements, whether visualized by the health professional or the welfare professional, may be the need to provide supplies of pure water or facilities for sewage disposal.[3]

In the Western industrialized nations the welfare services may be developed and expanded upon a firm foundation of basic material provisions that the fortunate citizens, one third or less of the world's population, may take for granted. Upon a secure foundation of basic physical provisions, which go far toward the assurance of health and sanitation, a system of relatively sophisticated welfare services can be created and developed. The social worker in Bangkok, the official of the housing authority in Hong Kong and the psychiatric social worker in a child guidance centre in New York City can scarcely be compared, except from the point of view of their identification and dedication to human well-being. An analysis of welfare services in metropolitan areas must be considered against the backdrop of such differences between urban areas in the developed and developing nations of the world. There are, of course, factors in common, but it would be misleading to over-emphasize the commonality of concerns without a strong reminder of the gross differences that make it possible for a male child born in the United States and Canada today to have a life expectancy of over sixty years, while his male cousin born in India or in South East Asia on the same date may have a life expectancy of less than thirty years.

Basic Principles in the Provision of Welfare Services

ACCESSIBILITY

> Nearly all the services now in being were pioneered by voluntary organizations, especially the churches, and many voluntary services still surround and supplement those publicly and statutorily provided. The two types are not competitive but complementary and merge into each other. Public authorities often work through voluntary authorities specially adapted to serve individual needs and officials cooperate with the workers of the many social service societies.[4]

This exposition of the genesis of modern British welfare services could equally well apply to most urbanized Western societies. However, from the time of the passage of the first major Poor Law in England in 1603, any governmental responsibility for assistance to the poor, more specifically the destitute without means, was assigned to local government, whether it was a town, city or rural parish. This pattern of local responsibility in the field of public assistance, that is, the payment of relief to the needy in the form of cash or kind, was taken to the eastern seaboard of the United States, the Atlantic provinces of Canada and to Australia and New Zealand by the British who settled these areas. The Elizabethan Poor Law became the foundation for public welfare services in many areas of the world and survived, with modifications, well into the twentieth century.

---

3 "Social Aspects of Urban Renewal and Redevelopment," *Commission Reports, VI*, International Conference on Social Work (New York, International Council on Social Welfare, September, 1966), p. 21.

4 *Social Services in Britain*, United Kingdom, Central Office of Information (London May, 1966), pp. 1, 91-94.

The origin of welfare services in religious institutions or voluntary associations did not, however, ensure that the basic material or counselling services that did exist would be available to all needy residents of the appropriate locality. In fact, programmes offered under religious auspices were available only to members of a particular religious faith; and even for those there might well have been a stipulation that service would be provided only to the devoted or "true believers". Indeed, even today the mission providing "a meal and a message" is not uncommon. Similar stipulations applied to those seeking help from other voluntary organizations. In short, there were the "deserving poor" and "the undeserving poor". The poor but honest person or family, the unfortunate victim of bad luck rather than the miserable perpetrator of an immoral act, was entitled to help. Those judged undeserving included the so-called drunkard, the "fallen woman", and other men, women and children who did not behave precisely as the dominant class serving on the governing comitees of voluntary organizations believed they should.

The evolution of the social services during the past century has involved an incessant drive by lay and professional persons within voluntary and governmental organizations to ensure that the welfare services are accessible as far as possible on a universal basis throughout the particular governmental jurisdiction or geographic area served. The past fifty years of rapid urbanization have given urgent significance to the basic principle that every individual and family should have access to services required to meet their fundamental human needs. Systems of social security involving social insurance and public assistance programmes were introduced in Germany in the 1870's and 1880's and in other Western

European nations soon after. What is now described as "the welfare state" had proceeded for some distance in many nations, including the United Kingdom, before the outbreak of World War 1.[5] During the 1920's these nations continued to improve schemes of unemployment insurance, old age pensions, and similar social security programmes.

In the United States, however, when the seriousness of the economic depression began to be felt by 1930 or 1931, it was realized that in only two of the then forty-eight states were there statutory provisions for unemployment insurance and in only eighteen states were there systems providing for old age pensions. Even in these few jurisdictions the programmes were quite new and not sufficiently developed to be of much assistance. The New Deal was the first real assumption of social responsibility at the national level. The passage of the Social Security Act in 1935 brought the United States for the first time into the mainstream of public assumption of responsibility for the basic welfare services designed to provide sustenance and well-being for all citizens. Although no similar legislation has yet been passed in Canada, a series of federal acts, beginning with the Unemployment Insurance Act of 1940, have been placed on the statute books. In both these affluent nations the tendency has been, since the end of World War II, to expand and broaden specific programmes to make services accessible to wider groups of people in need.

The rapid expansion of urban areas, and in particular the physical and social restructuring of the modern metropolis, have made the principle of universal access more

---

[5]William Beveridge, *Social Insurance and Allied Services* (New York, The Macmillan Company, 1942), p. 5.

difficult to uphold. Traditionally, the welfare services were located to provide help to people who lived, for the most part, within the heart of the urban area. Even within the central city of a large metropolitan area the welfare services were located within a relatively confined geographical area. As the metropolis expanded, there tended to be a dispersion of the needs. In the 1920's this did not take place on such a scale that it made visits to those requiring assistance impossible.

During the past twenty years, however, the pace of urban expansion has far outstripped the capacity of the welfare services to ensure accessibility through appropriate decentralization of offices and facilities. The failure of the welfare services to decentralize in metropolitan areas was, to some extent, the consequence of inadequate resources, not merely financial but also in terms of staff and even transportation. This, however, is not the only explanation. When the outward movement of urban families into the vast suburban areas occurred, it was assumed, particularly in the Western developed nations, that the residents of the new localities were much less in need of welfare services than those who remained in the urban centre. This may have been true of basic material assistance but it certainly did not apply to the great range and variety of counselling services. For example, no welfare service, branch office or otherwise, was located within a mile or more of Watts, in Los Angeles, the site of a notorious riot and serious destruction of life and property in the summer of 1965. Within the Watts area, more than half the adult males were unemployed and a substantial number of households consisted of female heads of families with dependent children supported almost entirely by a federal-state programme called Aid to Needy Children.

In almost every metropolitan area there is an awareness that welfare services must be made universally accessible, not merely geographically but without reference to such factors as race, creed, nationality, ethnic origin, citizenship or place of birth. Unfortunately, resources are still insufficient to ensure that everyone in need will receive the most appropriate assistance. All too often the decision to provide a service throughout a metropolitan area is but a paper decision. More often the real choice facing welfare administrators is between strengthening a service provided to those who reside in a limited portion of the metropolis and the dilution of the service over a wider area. Only in the more developed nations can it be said that the basic provisions of the public welfare services are universally available, and even this assumes that the needy have sufficient knowledge and understanding of the services available and also know how to apply for them. Richard Titmuss has pointed out that even under British conditions, which are often assumed to approximate a well-developed welfare state, those in the higher social and economic classes have a better chance of obtaining the services to which they are entitled than have those in a relatively lower social and economic position —those for whom, in fact, these services were originally conceived.[6] Frequently the latter do not know how to secure the services to which they are entitled. The concept of universal access, therefore, is by no means a simple one.

STANDARDS

In many parts of the modern world (and this statement is not restricted to the

[6]Richard M. Titmuss, Essays on 'The Welfare State', 5th impression (London, Unwin University Books, 1964), pp. 34-35, 226-27.

developing nations) the standard of welfare services is the barest minimum that can be provided to sustain life. Two or three centuries ago the needy (whatever the cause of this need—physical and mental illness, disability, accident, injury at birth, mental defectiveness), were usually lumped together within the confines of a formal, common space, described in British terms as an almshouse. Indeed, in many parts of Western Europe, even in modest-sized villages, a home for the aged may still be found. In North America the equivalent institution was known as the county home or the "poor house". Persons seeking material assistance were expected to enter an institution of this type. In the late nineteenth century, there was a move to provide "relief" outside the institution, either in kind or in cash, though the latter type was relatively rare.

It has been only during the last quarter-century in most jurisdictions that those requiring public welfare assistance have been given a sum of money, usually by means of a bank cheque, to expend as they wish with the help and advice of a counsellor; and primarily within the last quarter-century, the amount of such assistance has come to be predicted on the basis of a serious attempt to construct and price an appropriate standard of living. The problem seems to lie in trying to find the resources sufficient to meet the needs at acceptable standards. In metropolitan areas of the developed and developing nations alike those in need of assistance live at levels far short of what might be provided if the available resources were deployed differently. This, of course, relates to the question of priorities that was raised earlier.

Standards of service must be conceived in both quantitative and qualitative terms. The ideal situation might be one in which a sufficient quantity of services in great variety were available to meet any forecast of personal and familial needs within a foreseeable time period. In particular, this ideal would require that the workload of any professional person be no more than can be adequately carried. The public welfare services, as well as certain fundamental voluntary or quasi-public programmes, such as those of child and family welfare, have been plagued for at least the past thirty-five years by inordinate demands upon professional staff. The obvious consequence is that many persons and families in the modern metropolis are badly or cursorily served at best, or completely neglected at worst. It is not uncommon, at least in the United States and Canada, for individuals or families requiring personal counselling through the urban community welfare services to have to wait three to six months (or even longer) before they can expect to see a professional social worker, psychologist or psychiatrist. The deterioration of relationships and mental and physical health, which must be a major consequence of such delays, is impossible to measure.

The quality of welfare services depends partially, of course, upon the sufficiency of staff and financial resources available, and in recent years staff training, both within formal educational programmes and on the job, has acquired far greater significance. The worldwide shortage of fully trained professional personnel, graduates of post-baccalaureate programmes in social welfare, psychology, nursing, public health and psychiatry, has drawn attention to the whole question of the differential skills and capacities required for different tasks within the welfare services. If the components of an appropriate service can be clearly delineated and the ensuing tasks be clearly identified, it is then conceivable that appropriate levels and types of training for workers with rather different educational backgrounds can be provided for as well. These matters have been the subject of only a modest amount of research as yet, and a great deal remains to be done.[7] Yet the very question of quality of service depends upon a sound and integrated approach to the whole question of manpower for the welfare services.[8]

INTEGRATION OF SERVICES

The degree to which a modern metropolis is well and appropriately served by the network of welfare services depends largely upon the extent to which the principles of accessibility and standards are satisfied. Both these factors are influenced in an important way, however, by the manner in which the welfare services are related to each other within the urban area. The most important considerations here include the issues of specialized versus multi-functional

---

[7]*Report of the Working Party of Social Workers in the local Authority Health and Welfare Services*, United Kingdom, Ministry of Health and Department of Health for Scotland (London, HMSO, 1959). *Training for Social Work: Third International Survey* (New York, United Nations, 1958). *Utilization of Social Work Staff with Different Levels of Education*, United States Department of Health, Education, and Welfare, Bureau of Family Resources, Welfare Administration (Washington, D.C., 1965).

[8]*Report to the Liaison Committee of the Regents of the University of California and the State Board of Education*, Advisory Committee on Social Welfare Education (1961). F. R. MacKinnon, "Types and Levels of Service in the Social Welfare Field and Related Educational Requirements," paper presented to the Conference on Manpower Needs in the Field of Social Welfare (Ottawa, The Association of Universities and Colleges in Canada, November 29, 1966), p. 32 (mimeographed). *See also* John S. Morgan, "Welfare, Manpower, and the Universities," presented to the same Conference (Ottawa, November 28, 1966), p. 49 (mimeographed).

services, duplication of services and "social welfare planning".

As a general principle it can be stated that there is a functional relationship between specialization and affluence. This relationship will not be equally evident within all metropolitan areas of the more developed and affluent nations because the situation does vary from place to place and each metropolitan area is not equally well endowed or affluent. However, in most major metropolitan centres in the developed nations there has evolved a substantial variety of welfare services with highly specialized functions. A person in need, however, is often confronted with the same difficulty faced by a physically sick person, namely, that he may be referred or may refer himself from one specialist to another in the search for help with his specific, perhaps undiagnosed, problem. Moreover, the different requirements of the members of a family may call for services from various welfare services in the community. Yet it is now clearly recognized that a family should most often be treated as a unit. New programmes involving techniques known as "family therapy" are attracting much attention in large urban centres where a variety of professional persons are available to join in interprofessional teams. If this new approach is to realize its objectives, the multi-functional welfare service, with a variety of professional persons with different skills and perhaps specialized experiences, practising within one framework and, hopefully, within one building, must come into being.

In many metropolitan areas several voluntary services with similar aims and objectives have developed over the past century. In almost every metropolitan area there are community leaders, particularly those active in fund-raising and social-planning activities, who claim that the multiplicity of welfare organizations has led inevitably to a moderate degree of duplication of service. During the past twenty-five years, particularly in North America, serious efforts have been made to identify and eliminate the more obvious cases of overlap. There is much support for the view that agencies attempting to meet similar needs, but perhaps for different sexes or different age groups in the population (for example, the Big Brother and the Big Sister Associations, the Young Men's and the Young Women's Christian Associations) should consider seriously the possibility of federation or amalgamation.

The expansion of the modern metropolis and the creation of vast new suburban areas during the past quarter-century have brought these considerations sharply to the attention of the general public. The new suburbs cannot afford to encourage the duplication of central city facilities throughout the entire metropolitan region.[9] While duplication does not appear necessary on either social or spatial grounds the services within the central city are overtaxed and are quite unable to provide for the needs of new populations in somewhat distant localities. All these factors have combined in the expanded metropolis to underline the need for multifunctional services strategically located throughout the metropolitan area to serve as many needs as possible of the residents, without reference to age, sex, religious affiliation or nature of the problem.[10]

In the metropolis the responsibility for working toward effective integration is usually assigned to an organization described as a council of social services or a social planning council. The discharge of this responsibility consists of a detailed examination of the needs of the residents of the metropolis as expressed by professional persons, community leaders and the clients themselves, and the careful assessment of the social welfare resources available. In this process of identifying both the demand for and supply of services, the general public and the various political jurisdictions should be informed of the evident disparities between supply and demand with respect to geographical location, nature of service and quality and quantity of service available.[11] It must be a basic principle in all metropolitan areas that this process of integration should be pursued with all deliberate speed, not merely to rationalize the existing service system but to ensure that those who are primarily responsible for financing and those who are primarily the recipients of service are served equally well.

Basic Problems

PLANNING FOR PHYSICAL FACILITIES IN A CHANGING COMMUNITY

As we have seen, social service facilities have traditionally been located in or near the central city in the metropolis. It is doubtful that those responsible for city planning were influenced in their ideas to any important degree by the location of the welfare services in the community prior to the onset of the Great Depression of the 1930's. The emerging profession of urban sociology from the early years of the twentieth century was strongly influenced by the

---

[9]Albert Rose, "The Social Services in the Modern Metropolis," *Social Service Review*, Vol. XXXVII, No. 4 (December, 1963), pp. 375-88.
[10]*Ibid.*, pp. 384-85.

[11]*A Study of the Needs and Resources for Community Supported Welfare, Health and Recreation Services in Metropolitan Toronto*, Social Planning Council of Metropolitan Toronto (Toronto, 1963).

theoretical formulation and empirical research that emanated primarily from the University of Chicago school of sociologists.[12] This school saw most of the major metropolitan areas in the United States as relatively similar in displaying concentric rings of social problems and incidence of "welfare cases". At the centre of this series was found the first settlement of population from which those who succeeded in improving their social and economic position moved outward to perhaps a second ring of settlement, and were succeeded in the inner ring by a new wave of newcomers. This process was repeated as successive waves of migrants entered the urban core and proceeded to make their way into new areas of settlement, leaving behind those who did not succeed. In the meantime, of course, the descendants of earlier generations had moved farther and farther away from the original ring of settlement.[13] The incidence of physical disease, of crime and juvenile delinquency and of families requiring financial assistance was much greater in the innermost rings than in those farther out and continued to decline the farther one moved from the urban core.

If this theoretical formulation, for which a great deal of empirical data was provided, can be accepted as a reasonably good approximation of the situation from, say, 1850 to 1930, the location of the welfare facilities was apparently appropriate and propitious. As a corollary, it was assumed that these services need not follow the movement of population outward from the inner ring and, in fact, most services were relatively stable in their earliest locations. By the 1920's and the 1930's, however, some welfare organizations had decentralized to the extent that branch offices were in operation in certain districts of the central city and sometimes within the suburban areas that had developed during the first quarter of the century. Local departments of welfare that came into being after 1920, and particularly after 1930, could not operate in most metropolitan areas from the centrally located municipal offices. District offices were rented or constructed, and to some extent the needy were assisted in making application for help by the proximity of these facilities.

The conditions of the 1930's served to drive home to social scientists and to architects, who often become the first planning consultants to local governments, that individuals and families in need had a vast array of common requirements, extending well beyond the need for basic material provisions. These needs had been provided through varied forms of interpersonal relationships often developed through formal or informal group affiliations.

This recognition has led many planners, if not the majority, to develop their projections after due and careful reference to the public authorities and to voluntary organizations, particularly as they are represented by social planning councils. In this endeavour, however, the physical planners have been disappointed in their contacts with social planners who have not generally given enough attention to forward planning, not merely for their own requirements within existing areas but particularly for those which will develop as the metropolis spreads over many square miles of previously undeveloped or farm areas.[14] This

[12]Robert E. Park and Ernest W. Burgess, *The City* (Chicago, University of Chicago Press, 1925).

[13]John Kosa, *Land of Choice: The Hungarians in Canada* (Toronto, University of Toronto Press, 1957).

[14]Albert Rose, "Co-ordination between Physical and Social Planning in a Metropolitan Area," *Social Service Review*, Vol. XXXII, No. 4 (December, 1958), pp. 374-86.

criticism applies particularly to those welfare services that rely upon voluntary funds.

It is apparent, then, that in the years of fundamental reliance upon the neighbourhood centre or settlement house for a variety of services the physical facility played a central role in the neighbourhood. In fact, it might be argued that each neighbourhood had three central institutions of consequence: the church, the school and the settlement house. The lives of most children and adults could be plotted within the triangulation of these institutions.

In the North American expansion of urban centres in the mid-nineteenth century, the central area served as the main reception centre for newcomers and did so at least every fifteen or twenty years for new groups. The outward subsequent movement of early groups of immigrants made it necessary for traditional social service facilities to adapt their functions with reference to changed cultural circumstances. Despite the different cultural factors, introduced by such groups as Irish Roman Catholics and Jews, the large numbers involved and the fact that the immigrants faced many common problems, made it quite possible for the welfare services to adapt to their needs and serve these groups simultaneously.

In North America, and to a degree in Western Europe, the second wave of urban expansion, which led ultimately to the recognition of the modern metropolis, began in the mid-1930's as poverty-stricken rural people began to move to the cities in search of not only employment but also a satisfactory system of social and health services. Soon after the outbreak of World War II these newcomers were joined by other rural dwellers and migrants from small towns and cities who flocked to the larger urban centres in response to pleas for factory workers needed to produce munitions and other requirements of the war effort. Yet, even this expansion in the 1930's and early 1940's was minor in comparison to the migration of the post World War II period, which was also quite different from that of the previous century. In the United States the post-war migrants were mainly southern Negroes and poor southern whites, although many Puerto Ricans came to New York City and other eastern metropolitan areas, while Mexicans came to the cities of the Southwest and to California, and Filippinos migrated in great numbers before their nation gained independence. These new groups of migrants were fundamentally different from the Europeans in background, motivation, labour-force characteristics and cultural attributes. These migrants of the past twenty years or so have been more confirmed in their poverty, less skilled, composed of larger families and without the cultural background that motivated European emigrants of the past two centuries. This is not the place to explain these phenomena, but discrimination on the basis of race and colour sum up quite well one important aspect of the matter.

In Canada, Australia and New Zealand, most immigrants of the post-war era have resembled more closely the migrants of the nineteenth and early twentieth centuries. While these nations have avoided some of the more serious problems of the American metropolis, the problem of colour discrimination has not been entirely absent. It has always been evident that Australians have feared immigration from Japan; Canadians have feared immigration from Japan and China; and, during the past ten or fifteen years, many people in the United Kingdom have opposed the increased immigration of British West Indians, Indians and Pakistanis. Nevertheless, it is in the

United States that the most difficult social adjustment problems must be faced in the modern metropolis.

In the light of the changed circumstances of metropolitan population growth and spatial expansion, particularly in the United States, the physical facilities, which are an important part of welfare services in the community, have assumed a more important role than in the previous thirty or forty years. As the older parts of the central cities of metropolitan areas have tended to deteriorate in such countries as Canada, the United States and the United Kingdom, the physical processes of urban renewal (encompassing slum clearance, urban redevelopment, rehabilitation of housing and conservation of neighbourhoods) have come squarely to the attention of public officials and private citizens. The fact is, and it must be clear from the previous discussion, that the poor and the needy for the most part sit directly in the path of the urban renewal process. Planned programmes of physical renewal of the heart of the modern metropolis impinge more strongly upon the poor family, the large family, the immigrant family, the single elderly person and the disadvantaged child or adult than upon any other social or economic grouping in the society. A great many socio-economic considerations are involved in helping people to adjust to the demands of physical change in the community, but the social welfare facilities have a role to play that is not unlike that of the neighbourhood house or community centre of the fifty years or more prior to 1920.[15] Any examination of the history of major urban renewal projects in the United States and Canada—in Boston, Philadelphia and Toronto, for example—will make it evident that the needs of individuals and families in urban renewal areas can best be met by a multi-functional agency located in the midst of, or adjacent to, the neighbourhood undergoing physical change. Where neighbourhood centres or settlement houses have not existed previously, several communities have found it necessary to create a modern social welfare facility to provide the modern counterpart and the services of the traditional settlement house.[16]

CONCENTRATION OF THE NEEDY

The distribution of human need is one of the most fascinating and important aspects of the spatial expansion of modern metropolitan areas. Within any modern metropolis, there are generally two patterns of socio-economic demography that exist together. It is rare for one pattern to exist alone, although it is common for one to be dominant.

The first pattern is typical of developed nations. Population growth and the consequent development of suburban communities has resulted in a movement outward from the central city of many individuals and families who hitherto constituted the solid support of welfare services, whether through their tax-paying capacity or their voluntary contributions to charitable funds. In general, the newly-developed areas are characterized by relatively young, self-supporting families, with more than the average number of children for a metropolitan household, with parents of above-average education and with aspirations of a university education for their own children. These families rarely require the services

[15]Herbert Gans, "Redefining the Settlement's Function for the War on Poverty," *Social Work*, Vol. IX, No. 4 (October, 1964), pp. 3-12.

[16]*A Proposal for a Community Development Program*, Action for Boston Community Development (Boston, 1961), Part 2, pp. 1-8.

of the social welfare agencies. Few single adult persons, young or old, and far fewer elderly couples live in the suburbs.

In this first pattern the needy tend to be concentrated in the urban core. The residents of the central city are likely to be older, poorer families, with parents of relatively little formal education and little capacity for steady or highly remunerative employment, or single adults. In particular, the proportion of single or widowed persons over the age of sixty-five may be as much as 50 per cent greater than in the rapidly expanding suburban areas. The older, dependent families in the centre are not small, however; they may be as large as, or even larger than, the new affluent families on the outskirts. The incidence of human need is, without doubt, far greater in the central city (and particularly in the older crowded neighbourhoods) than in the newer parts of the urban metropolis.[17] It is in the central city that the totality of the social impact of the modern industrial urban society is revealed. This has become even more apparent during the past quarter-century, with new and more disadvantaged groups of immigrants moving to metropolitan areas at a time when the economies of such areas have known an affluence never before attained. The contrast between the physical and social misery of individuals and families who have not shared in the expansion of income and the material affluence of the majority of families whose living standards have risen considerably, accentuates the seriousness of the basic anomalies.

The second major pattern is to be found where fringe communities of the needy, sometimes of great size, have been created on the outskirts of previously urbanized areas. This phenomenon is particularly evident in South America—in the metropolitan areas of Rio de Janeiro, Caracas and Lima, for example—where vast slums have been created by migrants who have moved from villages and small towns to the edge of the metropolis in search of some elusive improvement in their standard of living. Whatever they were seeking, they have without doubt been disappointed. Nevertheless, the attraction of the city holds them, despite the inadequacies of their physical and social circumstances. Without water and sewage facilities, without such utilities as light, telephone and transportation services, these families are among the most ill-served in the entire world. Yet, they continue to increase in numbers with the result that any solution to the situation is now almost beyond the capacity of the urban centre.[18]

As indicated previously, wherever one of these two patterns predominates, the other is often found in a minor form. Thus, in some of the metropolitan areas in the more highly developed nations there do exist shack-towns or fringe communities are literally sores on the urban landscape. Often, however, they are beyond the jurisdiction of established governmental organizations and well beyond the capacity and

[17]F. M. Miller, "Social Class, Mental Illness, and American Psychiatry: An Expository Review," and Thomas Langner, "A Survey of Mental Disease in an Urban Population: Prevalence by Race and Income," in Frank Riessman, Jerome Cohen and Arthur Pearl, eds., *Mental Health of the Poor: New Treatment Approaches for Low Income People* (New York, The Free Press, 1964), pp. 16-37, 39-48. *See also* Ronald Freedman, "Cityward Migration, Urban Ecology, and Social Theory," in Ernest W. Burgess and Donald J. Bogue, eds., *Contributions to Urban Sociology* (Chicago, University of Chicago Press, 1964), pp. 178-200.

[18]*Social Aspects of Housing and Urban Development*, United Nations, Economic and Social Council, Social Commission (New York, 1965), pp. 26-42.

range of voluntary welfare services. Similarly, in those metropolitan areas in which the concentration of human need is found in vast shack-towns, there are also poor and needy people in the urban core, and if any are to receive welfare assistance, it is likely to be these central city dwellers.

CRIME AND DELINQUENCY

The old familiar saying that crime knows no boundaries is an important truism in the metropolis. Historically, this fact accounts for the creation of a metropolitan-wide police force as being one of the first regional services for large cities. In London, England, the Metropolitan Police Force has a history of nearly 125 years; that of New York City, some 70 years.

This early attention to crime detection and the apprehension of criminals throughout an entire region is relatively simple to understand. With urbanization and industrialization, the accumulation of property in the hands of the rising merchant and commercial class has increased greatly the number of relatively well-to-do people residing in close proximity to the great mass of poor people. This is not to suggest that crime is simply the appropriation of the property of the well-to-do by the poor, but in fact the potential rewards of theft and murder have increased significantly with the rise of urban society. There is general agreement that the quarter-century since the end of World War II has been marked in many nations by an exponential increase in the rate of crime and delinquency. This argument needs very careful qualification, however, because its application is much more evident in North American society, especially the United States, and much less evident in other Western countries.

The apparent geometric expansion in the number of criminal acts may be partly statistical in that far better records are kept than were in the past. Again, the increase may be due to better techniques of detection and apprehension and to better programmes of training and the relatively higher educational accomplishment of those who enter police work. Despite all these possible ameliorating explanations, there is undoubtedly a positive association between the incidence of crime and delinquency and the spread of urbanization. The residence, if any, of those who perpetrate offences and the locations in which those offences take place have relatively little in common. Social planning councils or similar bodies in many nations have plotted residences of adult and juvenile offenders revealing their concentration in the older, more deteriorated neighbourhoods within the heart of the central city. This is not difficult to explain. For many persons whose tendencies are antisocial or asocial, the dense populations and crowded neighbourhoods in the urban core offer an opportunity for anonymity and a setting in which few questions are asked and few standards demanded. Although many recorded offences occur within such neighbourhoods, particularly with the frequent contact of persons with like tendencies and within neighbourhoods where there is much unemployment, miserable housing accommodation, overcrowding and poverty, the general public has been far more concerned with the spread of criminal acts throughout the metropolis.

In many areas the detection and apprehension of criminals have been severely impeded by the multiplicity of local jurisdictions with separate police forces. In Metropolitan Toronto in 1950, for example, there were thirteen municipal police forces, each operating radio-equipped vehicles on different wave lengths.[19] The failure to integrate the system of communications is the most obvious symptom of lack of cooperation and weakness in an important area of public protection. When the general public realizes that offences will occur within any municipal jurisdiction within the metropolitan area, and that criminals may fail to be apprehended because they move freely across municipal boundaries (while systems of detection and apprehension do not move so freely), there will soon be a demand for centralization of services. In Metropolitan Toronto the time for such action came soon after the inauguration of a system of metropolitan government on January 1, 1954. Within three years the illogic of the divided system of police protection led to the creation of the Metropolitan Board of Police Commissioners and the Metropolitan Toronto Police Force.

A special word of attention must be devoted to the increase in what is called "juvenile delinquency" in metropolitan areas all over the world. After due allowance for the improvement in statistical reporting and for the improvement in detection and apprehension, the fact remains that there has been a substantial increase in the number and seriousness of crimes committed by juveniles. The significance of this fundamental fact rests in its relation to the general inadequacy of the child welfare services and, within that overall rubric, to the protective services in particular. The pressures within modern urban society which lead to family breakdown are among the basic causes of juvenile delinquency. Yet the metropolis has no clear system of social welfare services designed to strengthen the family, to prevent family breakdown and, to whatever extent possible, to forestall the development of criminal tendencies among juveniles (usually defined as those under sixteen years of age). All the weaknesses of the welfare services in the modern metropolis can be drawn out and pinpointed when any thorough exploration of the spread of delinquency is undertaken.

INEQUITIES IN WELFARE SERVICES

Many of the basic problems of the welfare services can be encompassed within one fundamental anomaly, namely, that it matters a great deal where a person or a family resides within a large urban centre.[20] So long as most of the major metropolitan areas of the world are governed and served by a multiplicity of jurisdictions—political, educational and social—there will be fundamental and significant inequities within a whole host of services provided for the residents of such communities.

Some of these inequities are clear and understandable to any examiner. For example, if one community is served by a modern sewage disposal system and an adjacent municipality relies upon septic tanks placed within soil not sufficiently receptive to waste products, there is a clear disparity between the levels of service provided. Inequities within educational and social services are much harder to measure because no individual is apparently entirely deprived of a basic or minimum standard of service. In many metropolitan communities, however, one municipality may offer a tremendous range of opportunities for its children, both in its current educational programmes and in its preparation for

[19]First Report, Section 1, Civic Advisory Council of Toronto, Committee on Metropolitan Problems (Toronto, 1949), p. 38.

[20]Albert Rose, "The Social Services in the Modern Metropolis," Social Service Review, Vol. XXXVII, No. 4 (December, 1963), pp. 362-83.

future roles in the society, whereas in a nearby municipality the standard of educational service may be barely adequate in physical facilities, the numbers of teachers, the quality of the educational preparation of teachers, the variety of programmes offered, library facilities or recreation facilities.

Inequities in welfare services are equally evident but they are even more difficult to comprehend because of the variety of human needs and welfare programmes. First, the basic provision of financial assistance to the destitute, to the unemployable, or to any person or family in need may vary greatly from place to place within the same metropolitan area. These differences in level of benefit, however, are not necessarily related to the basic financial resources of the municipality. The provisions may vary because of the general lack of acceptance by the public, within one municipality as against another, of the need to assist those in poorer circumstances. The administrators may be much more harsh in one locality than in another in interpreting the regulations laid down in national or state legislation. The voluntary services are, of course, unequally distributed throughout the metropolis and, together with the differences in the interpretation and administration of social security programmes, where a person resides may make a tremendous difference when he suffers a catastrophe leading him to apply for service.

These inequities are further compounded by the increased mobility of individuals and families who change residences frequently within the metropolitan area and from one large urban centre to another. When a family is in need of the assistance of one or more of the welfare services it may find itself in serious difficulty, should it choose to move (perhaps unknowingly) across a municipal or local boundary. Since many relatively poor people live in inadequate housing accommodation from which they may be evicted or from which they may move frequently, the possibility of service today and no service tomorrow is not uncommon.

In Western industrial societies much distress is occasioned to individuals and families by the presence of local boundaries and the failure to integrate the welfare services on a metropolitan-wide basis. Differences among the public welfare services from one municipality to another are easily understood, although not by the clients. The difficulties facing the voluntarily supported services are less clearly realized by many people. A voluntary service may provide assistance—in some aspect of child welfare, for example—which is considered so important within the central city that the local government provides a grant of public funds to support that aspect of the service. The impact of this grant alone makes it difficult for the agency to provide that service to the residents of adjacent municipalities whose local governments do not pay for such assistance.

As more attention is devoted in metropolitan areas to the human problems and the development of social welfare services to help meet these needs, far more attention must be paid to the basic inequities in social service provisions. The advocacy of the democratic rights of local areas to develop their own institutions and their own service patterns, and the acceptance of the notion that the maintenance of diversity within the modern metropolis is desirable, must be balanced against the human suffering that may be occasioned as a consequence of strict adherence to these traditional but perhaps irrelevant principles.

Some Solutions

Solutions to the delivery of social services to meet human needs tend to fall into four classes despite different approaches to the problem and the differences in income, nature and standards of service and the incidence of social welfare problems from one nation to another and from one metropolis to another. The most obvious class of attempted solutions falls within the rubric of governmental structure, that is, an attempt to solve problems by changing the formal nature of the political machinery that must, in the last analysis, make the decisions and provide the services. A second group of solutions consists essentially of changes within the organization of the social services themselves. Many experimental solutions rest upon the hypothesis that a more efficient administrative structure will go far toward meeting the needs of disadvantaged urban people.[21]

Many analysts have contended, however, that structural rearrangements, whether at the political or at the administrative level, cannot meet the real requirement of a more efficient system of welfare services—namely, the delivery of greater quantities of service of high quality to every individual or family who requires help. It is maintained that such requirements are better met with solutions that are in the nature of social technology, that is, they attempt to provide services within new programmatic frameworks. Ultimately, of course, all attempted innovations in the provision of welfare services can succeed only if the fundamental problem of finance is solved, and this merits important attention if any of the other approaches are to attain any real measure of success.

Finally, it must be emphasized that structural, administrative, technical and financial solutions inevitably overlap; no major change can be viewed as unitary in nature.

NEW STRUCTURES OF GOVERNMENTAL ORGANIZATION

The emergence of public and voluntary social services within the expanding metropolis during the past half-century has coincided with the titanic struggle for survival of local government. In some measure this statement might be made with respect to any of the major services provided by or through local governments. The pace of urban expansion in its physical form has placed such pressure upon the administrative structure and financial base of local government throughout the world that serious confrontations between the elected officials of local and senior levels of government have occurred with increasing frequency during the past quarter-century. Searching examinations of the responsibilities of municipal government in the case of the welfare services, have resulted in a demand that the senior levels of government develop new programmes to meet the more fundamental needs of all citizens, and assume substantial financial responsibility for those services that are left at the local level.[22] It became clear during the 1920's and 1930's that in most nations of the world local government could not possibly cope administratively and financially with the enormous demands for welfare assistance. Although new administrative organizations were created hurriedly and although financial grants were made by

---

[21]Peter M. Blau and W. Richard Scott, *Formal Organizations* (San Francisco, Chandler Publishing Co., 1962).

[22]John S. Morgan, "The Contribution of the Municipality to the Administration of Public Welfare," *Canadian Public Administration*, Vol. VII, No. 2 (June, 1964), pp. 137-38.

national or state governments to local welfare departments, the results achieved during the inter-war period gave no assurance that local governments could cope satisfactorily with their traditional responsibilities for assistance to the poor and disadvantaged.

The only long-term solution lay in the development of national, state or intergovernmental programmes of social security, encompassing income maintenance services and designed to bear the brunt of all major risks faced by citizens in urban industrial society. A comprehensive programme of social security for all individuals and families in a nation could well be developed by virtue of the powers of national or federal governments in the fields of taxation and monetary policy and through the combination of the techniques of social insurance (in such programmes as unemployment, hospital and health insurance), the techniques of universal transfer payments (employed in such programmes as old age security allowances and family allowances in various countries), and the techniques of public assistance. If all persons within the total population could be assured of basic minimum standards of living in the event of unemployment, disability, widowhood and retirement (to name a few of the more obvious risks), the burden upon the local welfare services would not only be sharply reduced but could be clearly delineated.

This clarification of the responsibilities of various levels of government in the welfare field has proceeded steadily during recent decades. It is now generally held that the national or federal government must develop the overall programmes and maintain the administrative services required to meet mass need at an average level of benefit. Nevertheless, in many countries there is a second level of government, the state or provincial, with certain responsibilities within the rubric of "property and civil rights" that must be exercised. These responsibilities usually include assistance for certain categories of disadvantaged persons such as widowed or deserted mothers with dependent children, unmarried mothers, neglected children and those destitute persons who either fall between the various nation-wide programmes or have suffered such irregularity in their job performance during the usual years of productivity that they have qualified for little or none of the benefits available to most citizens. These state or provincial welfare programmes are often financed by intergovernmental agreement, but they are sometimes wholly within the purview of the second level of government.

A basic assumption underlying this distribution of responsibilities is that as local government is "closest" to the people it should seek, usually in cooperation with voluntary agencies, to provide those services which are more personal and thus "closest" to the people. Again, local government has usually been required to undertake partial responsibility for the administration of the services provided by higher levels to the citizens of modern urban centres.[23] These administrative services most often include assessment of eligibility for benefit, assessment of the needs and living standards of those applying for assistance, and sometimes an assurance that recipients continue to maintain eligibility, that is, continue to demonstrate need for help. As far as the public welfare services are concerned, these functions have usually been performed without the intervention of trained professionals and

---

[23]Amy Leigh, "The Contribution of the Municipality to the Administration of Public Welfare," *Canadian Public Administration*, Vol. VII, No. 2 (June, 1964), pp. 150-57.

consequently the treatment of "personal problems" has necessarily been left to the "private agencies". In the first decade after World War II, however, these arrangements and allocations of responsibilities among various levels of government were by no means settled. The financial burden upon local government in metropolitan areas, if they were to provide welfare services adequate in quantity and quality for their residents, was relatively severe, particularly as metropolitan areas began to expand rapidly.

The second noteworthy change in governmental structure has taken the form of a shift from local to more senior governments of the burden of financial responsibility, whether or not the constitutional responsibilities were to be reallocated or reinterpreted in the light of modern urban conditions. It is not an exaggeration to state that a major element in the relation between local government and other levels of government in many countries is the continuous political manoeuvring to shift the burden of providing the social services (including health, education and welfare) from the taxpayer at the local level to the taxpayer at the state or provincial and national levels. The fact that this taxpayer is one and the same person, whatever level of government taxed him to provide these services, was not always considered germane to the argument. Elected and appointed officials in local governments within metropolitan areas have continued to insist, more and more strongly since 1945, that whereas physical services are an appropriate charge upon local taxpayers who are, after all, taxed upon an assessment of their real property holdings, the personal, human or social services should not be seen as a benefit of the local property owner per se and should, therefore, be a charge upon the wider political unit.

Within the metropolis the major structural defect, as far as welfare services are concerned, is the division of jurisdiction. There has been much pressure to develop new governmental structures for the welfare services through the creation of "welfare units". The transfer of responsibility for welfare services from small municipal units to larger administrative structures, such as the county form of government in the United Kingdom and the United States, is by no means new. In Canada, a Welfare Units Act was enacted by the Province of Ontario more than a decade ago to encourage the provision of welfare services over larger areas (perhaps two or three counties) and within areas not yet heavily urbanized. There are also many examples of metropolitan-wide jurisdiction in the field of welfare designed to remove the burden from the small municipal units and to ensure a more adequate level of service.[24] The Public Aid Commission of Cook County, Illinois, which embraces not merely the City of Chicago but a much larger metropolitan area, is a case in point.

The experience within Metropolitan Toronto merits a brief examination in this regard. Since the majority of the needy were concentrated in the central city, the City of Toronto was the first municipality to create a Department of Public Welfare (in 1931). During the decade after World War II, the department expanded as a division of local government with a substantial staff and budget. Its tasks were to carry out the requirements of provincial legislation and to administer intergovernmental programmes at the local level. During these

---

[24]Arthur Hillman, "Urbanization and the Organization of Welfare Activities in the Metropolitan Community in Chicago," in Ernest W. Burgess and Donald J. Bogue, eds., *Contributions to Urban Sociology* (Chicago, University of Chicago Press, 1964), pp. 245-55.

years the city was responsible for half the cost of general welfare assistance and the provincial government, until 1958, paid the other half. Under conditions of a rising cost level and substantial migration, the budget of the Department of Public Welfare of the City of Toronto grew during years of unprecedented prosperity throughout the economy.

As a consequence of an enlightened administration and the demonstrated efficacy of experimental programmes provided under voluntary auspices, the City of Toronto developed a wide range of welfare services extending far beyond income maintenance programmes. In some measure the city provided services not required by provincial legislation, and in most respects the range and quality of public welfare services far exceeded the standard available within the other municipalities in the metropolitan area. By the late 1950's and the early 1960's, however, the city felt that it could not continue to provide these services without additional financial assistance. In 1958 the federal government entered the field and the distribution of the costs of general assistance (until the end of 1964) were 50 per cent federal, 30 per cent provincial and 20 per cent local. Nevertheless, the responsibilities assumed by the central city far exceeded the combined services and costs incurred in all of the then twelve suburban municipalities.

The city continued its demands for further assistance from the provincial government and, in 1965, the Municipality of Metropolitan Toronto was ordered by the province to take over the local municipalities' 20 per cent share of the cost of general welfare assistance. The pressure for a more adequate metropolitan solution was not confined to the hard-pressed city but existed in a different form throughout the remainder of the metropolitan area. Finally, with effect from January 1, 1967, the government of the Province of Ontario permitted the Municipality of Metropolitan Toronto to create a Department of Welfare to assume all local governmental responsibilities throughout the metropolis.

ADMINISTRATIVE SOLUTIONS

Administrative solutions to the problem of delivering social services in metropolitan areas inevitably overlap with structural rearrangements and with attempts to solve financial dilemmas. Since the administration of welfare services is considered to be either a responsibility of the local municipal government or, at the very least, to be best carried out within the service structure at the local level because of the alleged "nearness" of the civic administration to the people, the first type of overlap is almost completely unavoidable. Proposals to improve the provision of welfare services almost always include elements designed to improve administrative efficiency and to develop a system whereby the service reaches the client in need of help as directly and as quickly as possible. Such proposals necessarily involve a clear delineation of the constitutional and other responsibilities, pertinent to welfare, of the various levels of government involved. Such delineation of responsibility must also be accompanied by a sound programme of financial responsibility.

Administrative solutions have been mainly of two kinds since the end of World War II. First, there have been many experimental programmes, demonstration and research projects, and new techniques introduced without research or experimentation, all designed to provide a more effective welfare staff to deal with the needs of individual or family applicants for welfare assistance.[25] Inexperienced field workers tend to be more interested in offering financial aid to applicants rather than in rehabilitating such applicants to a state of self-support than do the more highly skilled, more formally educated social service workers. Many studies in the United States, notably in San Mateo County, California,[26] and in Chicago, have confirmed the validity of this basic assumption. Impressive results are obtained whenever highly skilled and trained personnel, without unduly heavy workloads, are substituted for the more common employment of overworked, less skilled personnel. Needy clients are assisted more efficiently; they receive the basic financial assistance required but, more important, they are encouraged to make use of or enrol in other programmes designed to assist them to achieve a state of economic self-support. Total financial outlays have been substantially reduced and many, if not most, of the applicants have been assisted to find employment or to enter retraining programmes to improve their economic capacity. Unfortunately, in the larger metropolitan areas there is a noticeable lack of such educated and skilled practitioners and appropriate supervisors.

The other common solution attempted within this field is the decentralization of services throughout the metropolitan area.[28] As before, the basic objective is to achieve a productive and efficient system of social welfare services. Such decentralization of the public welfare services in the central city within the metropolis has been somewhat easier to achieve than has been possible for the voluntary services. In most metropolitan areas the central city has been largely built up and heavily populated for many years and it was not uncommon for the department of public health or the department of welfare of a local government to map out various service districts within the city for the purpose of more efficient administration. At the same time, however, this unilateral pursuit of administrative efficiency by each department creates one of the main difficulties for the civic administration—the great variety of service districts with overlapping boundaries. Not only is this an administrative inconvenience from the point of view of the elected officials but it also inhibits the efficient use of data. Regretably, all too often office space and personnel engaged in the administration of public welfare are still located within the city in a manner that makes it impossible for needy persons to complete applications, to participate in interviews designed to establish their eligibility for assistance and to receive such assistance in cash or in kind at a reasonable distance from their places of residence.

Decentralization rarely accompanies rapid urban growth. The new political units created during the past thirty or forty years have been mainly concerned with providing

[25]Donald B. Glabe, Leo J. Feider and Harry O. Page, "Reorientation for Treatment and Control," *Public Welfare*, Vol. XVI, Special supplement (April, 1958), pp. i-xxiii; *To Prevent and to Restore*, New York State Department of Social Welfare (New York, 1962); Edwin J. Thomas, et al., *In-Service Training and Reduced Workloads* (New York, Russell Sage Foundation, 1960).

[26]Douglas McGregor, *The Human Side of Enterprise* (New York, McGraw-Hill, Inc., 1960).

[27]James M. Gripton, "A Study of Work Orientations, Job Attitudes and Supervisory Styles of Public Assistant Personnel." A doctoral dissertation in preparation, University of Toronto School of Social Work, 1967, pp. 50-68.

[28]*Report upon a Proposed Realignment of the Jurisdiction of the Children's Aid Societies Operating within the Metropolitan Toronto Area*, Muncipality of Metropolitan Toronto (Toronto, 1956).

the more common "physical" services—supply of pure water, facilities for sewage disposal, street lighting, streets and sidewalks, educational facilities, parks and playgrounds. It has been argued already that one reason why welfare services, like health services, have not been widely provided in suburban areas is because it was assumed that they were not widely required. When it became apparent that this was not the case and that needs were substantial and also widely dispersed, the problem of decentralization within the newer suburban municipalities was brought to the forefront.

In the Municipality of Metropolitan Toronto, for example, until 1965, the administration of public welfare services was a relatively minor activity in almost all the then twelve suburban municipalities and was often handled by persons nominally on the rolls of such civic departments as the office of the Clerk, the office of the Commissioner of Works and the like. Most of these municipalities expressed pride in their modest expenditures on welfare services by comparison with those of the central city,[29] yet it can be safely assumed that a great deal of human need remained unmet in these areas by virtue of the inaccessibility of services. By 1966, when the Province of Ontario ordered the amalgamation of all welfare services into one Metropolitan Department of Welfare, it was learned that total welfare expenditures on behalf of approximately two thirds of the population of the metropolis (those living outside the City of Toronto) amounted to one third, while expenditures within the city amounted to two thirds of the grand total within the metropolitan municipality. This

finding discredited the common assumption that there was relatively little human need beyond the boundaries of the city. It is now the task of the Metropolitan Commissioner of Welfare to ensure that those services required throughout the area are, indeed, provided as resources become available.

Within the voluntary welfare services the matter of administrative decentralization has been even more difficult. Since the traditional, privately financed programmes were usually created to assist either the disadvantaged or destitute members of the urban population, or to aid newcomers to become acclimatized to the requirements of an urban society, these services were normally located within the heart of the central city. Their capacity for relocation or extension through a system of branches was severely limited both by their traditional orientation and by a severe scarcity of financial resources. In August, 1955, the Municipality of Metropolitan Toronto commissioned a firm of management consultants to examine the responsibilities of the two Children's Aid Societies then operating within the City of Toronto: the Children's Aid Society of Toronto and the Catholic Children's Aid Society of Toronto. The report of this investigation was issued on June 15, 1956, and recommended that both of these societies be expanded to serve the entire metropolitan area.[30] These recommendations were accepted in principle but the task of decentralization has not been fully accomplished more than a decade later.

In summary, administrative solutions to the problems of welfare services in metropolitan areas hold some promise, but they are relatively limited in both scope and

[29]*First Report*, Section 1, Civic Advisory Council of Toronto, Committee on Metropolitan Problems (Toronto, 1949), pp. 49-54.

[30]*Report upon a Proposed Realignment of the Children's Aid Societies*, Municipality of Metropolitan Toronto (Toronto, 1949).

importance. The great majority of the residents of metropolitan areas who require the services of welfare agencies will have to be helped by publicly provided programmes. These programmes will be staffed by substantial numbers of workers who, with every good will to serve human need, have neither the skill nor the time to deal with the detailed and highly personal problems of their clients. As a consequence, many individuals or families in the modern metropolis will remain more or less permanently in receipt of welfare assistance and the only approach to their human needs will be a financial one, which itself will be inadequate to provide them with more than a minimum standard of living.

SOCIAL TECHNOLOGY

Technical solutions for providing welfare services to the residents of the world's metropolitan areas can be visualized only in the form of new programmes of service. It is reasonable to suggest that the provision of welfare services has some tendency to proceed in cycles of fashion. In North America, for example, a series of studies published in the early 1950's provided conclusive evidence that a small proportion of the urban population absorbed a substantial proportion of all available public and voluntarily supported welfare services in the community.[31] This rather startling though not novel finding led almost directly to a great concern with a unit of need described as "the multi-problem family", on which considerable research and practising professional effort was expended during the 1950's. This has proved to be but a passing fashion, however, since attention turned

---

[31]Bradley Buell and Associates, *Community Planning for Human Services* (New York, Columbia University Press, 1952).

rather quickly, after the presidential election of 1960, toward the various questions involved in the so-called "war on poverty". During the 1960's an important proportion of social welfare manpower, research funds and published literature has been channelled into tackling the problems arising from the conditions of absolute poverty amid the affluence of the wealthiest nation in history.

In a world in which at least one half (and perhaps as much as three quarters) of the more than 3 billion inhabitants do not enjoy a reasonably adequate material, let alone a non-material, standard of living, the provision of welfare services to assist in the alleviation of poverty is absolutely fundamental. For this reason the discussion of social technology which follows is primarily related to the role of welfare services in the alleviation of poverty within the world's metropolitan areas.

Poverty in most of the teeming cities within the developing nations is a matter of basics—food, clothing, shelter, pure water, sewage disposal and public health services. The welfare services in such metropolitan areas can scarcely be distinguished from the overall attempt of the society to provide employment, income and basic services necessary for a human existence. It does not seem productive under these circumstances to attempt to explore technical solutions to the provision of welfare services in the midst of widespread poverty. Rather, the discussion will have to be confined to a consideration of that one third or so of the world's urban population in which there is sufficient affluence that only a portion may be considered to be relatively poor. It has become a matter of political importance for the most affluent nation, the United States, to demonstrate that it can meet the problems of poverty within its midst. Many of the most noteworthy products of social technology have appeared in the past six or seven years and it may be too soon to offer anything more than a preliminary judgment. Nevertheless, the war on poverty in both the United States and Canada illustrates, as no other programme can, the very serious difficulties that the welfare services face in attempting to meet human need in the metropolis.

Generally speaking, the urban poor have many basic characteristics in common, whether in France, West Germany, the United Kingdom, Sweden, Canada or the United States. Oscar Ornati has identified a series of poverty-linked characteristics for the United States that serve, with little modification, to point up the conditions shared by the poor residents of metropolitan areas throughout that part of the world which is not, comparatively speaking, poor. These poverty-linked characteristics are the following: non-white; female; sixty-five years of age or over; fourteen to twenty-four years of age; rural farm residence; southern residence; no work experience; part-time work experience; families with six or more children under eighteen years of age; less than eight years of schooling.[32] In this major study the research team examined (in 1960 census material) the relation of these social and demographic characteristics to three low-income levels. The lowest income level representing extreme poverty was that under $500, the middle level was that under $2,000, and the highest, that under $4,500. All characteristics showed a strong association with low income.

In the United States the realization that in the midst of a generally affluent society a significant proportion of the population was judged to be in poverty was a most distressing revelation. The publication of Michael Harrington's book, *The Other America*, in 1962, emphasized the conditions under which significant groups in the population lived. In the middle years of the Kennedy administration a number of social and economic forces moved toward a concerted attack upon the problem of poverty, which President Johnson enunciated as the "war on poverty". In 1964 the establishment of the Office of Economic Opportunity within the administrative department of the President of the United States, gave formal legislative and budgetary sanction to a major effort to bring affluence to the entire population of the United States. Since 1964 several billions of dollars have been spent and many new, innovative and experimental demonstration and research projects have been launched under the aegis of the Office of Economic Opportunity.

A further examination of poverty-linked characteristics makes it clear, however, that this concerted attack, even in America, is not a simple undertaking. Poverty is not just a matter of low income or the absence of income. It involves age, status, social class, race, geographical location and, in particular, the attitudes or state of mind of those persons or families considered to be poor. One of the most difficult problems is to determine what the welfare strategy ought to be or, in other terms, where the attack should be made. For poverty is also, in the view of many social scientists, a matter of culture, and poverty is perpetuated from one generation to another by virtue of the repetition of the basic social and economic experience of the parents and children from one generation to another. If poverty does involve economic and social disabilities as well as a cultural

---

[32]Oscar Ornati, *Poverty Amid Affluence*. A Report on a research project carried out at the New School for Social Research, New York (New York, Twentieth Century Fund, 1966), p. 38.

phenomenon, no one attack upon the problem will necessarily be successful.[33]

In the view of some theorists, major emphasis must be placed upon the children of the poor, who must be provided with an opportunity (through such programmes as "Operation Headstart") to gain the necessary educational preparation in the pre-school period. They could then enter the customary educational programme with as much preparation and background as children of families who were neither poor nor so disadvantaged from a cultural point of view. Operation Headstart has been considered relatively successful to date and increasing amounts of money will be spent upon its expansion during the coming year or two. Nevertheless, the ultimate consequences of such preparation will not be known for many years to come, and many students have observed that the child who has been assisted by this programme continues to remain for the majority of his years of schooling within the circumstances of his culturally deprived family setting.

Other strategists have argued that the main emphasis should be placed on young people who require encouragement and financial support to complete secondary education and, if possible, university education. Some programmes, including the HARYOU programme in the Harlem district of New York City, have been initiated to encourage young Negro students to complete high school or in some other way to gain the education and skills required to grasp more significant opportunities in the labour market. As Herbert Gans and others have pointed out,[34] the main problem facing these young people is discrimination in employment, and it has been argued that palliative programmes, in the form of technical solutions to the disabilities of young people from poor families, are doomed to failure because the major requirement is a fundamental change in the structure of society itself.

A third approach recommended by many students of the subject is based upon the argument that it is the head of the family who requires the most important consideration in that a major change in his income-earning capacity would do a great deal to upgrade the socio-economic conditions of his entire family. It is reasoned that the failure of the male head of the family destroys the normal set of role relations and expectations within the family and creates an image of a weak and inadequate father for the children, particularly the male children who tend to duplicate the performance of their father. Much effort and money have been expended to induce poorly educated fathers with little occupational qualification to enter retraining or vocational rehabilitation programmes designed to upgrade their qualifications, capacity for employment, and thus their income. In Canada several hundred millions of dollars have been expended in this way but the results to date have been disappointing. There is much evidence that disadvantaged male heads of families are ill-equipped to enter retraining or rehabilitation programmes and that a substantial majority fail to complete them. This failure adds further to the discouragement of the persons involved. Thus far there is no real evidence

[33]Albert Rose, "Strategies for Implementing Social Change," *The Role of Government in Promoting Social Change*, Proceedings of a conference at Arden House, Harriman, New York, November 18-21, 1965 (New York Columbia University School of Social Work, 1965).

[34]Herbert J. Gans, "Urban Poverty, Social Planning and the Uses of Sociology," in P. Lazarsfeld, W. Sewell and H. Wilensky, eds., *The Uses of Sociology* (New York, Basic Books, Inc., Publishers, 1967).

that programmes designed to elevate the economic status of family heads have produced satisfactory results.

Perhaps the most notable American attempt to alter the social and economic circumstances of a group of poor people is known as Mobilization for Youth (MFY) in New York City.[35] This major community action agency was a prototype for many of the later developments, not merely because of its early inception and relevant experience but because it was conceived, developed, financed and implemented in the midst of a multitude of favourable portents. It was launched by President Kennedy on May 31, 1962, with the assistance of the first demonstration award under the Juvenile Delinquency and Youth Offenses Control Act, to mount a systematic, three-year attack on juvenile delinquency. Public and private grants totalling $13,200,000 were assembled. Its conception was rooted in a sound theoretical formulation expounded in a 617-page *Proposal for the Prevention and Control of Delinquency by Expanding Opportunities*.[36] The "opportunity theory" of Cloward and Ohlin is too well known to require elaboration. It is sufficient to emphasize that the breadth and variety of efforts in the fields of work, education, community development, services to individuals, families and groups, legal services and research have never been matched. The example and experience of MFY have been of enormous value in both the United States and Canada.

Nevertheless, from the point of view of strategy, the great difficulties experienced in organizing low-income families and individuals to develop and eventually to assume political power are of prime importance. The "involvement of the poor" has not only been difficult to justify on the basis of past success in social reform, but the social and political action in partnership with clients in the 1960's has been rough indeed. The community action agencies quickly learned that their plans of political protest were strongly opposed by organizations and agencies with established reputation and influence. Thus, in 1964, Mobilization for Youth came under severe attack in the press and local government. The agency was forced to submit to an investigation, to closer supervision of its activities and to the virtual emasculation of its militant programmes.

The difficult problem of transferring some power to the representatives of the poor, so that they can play a participatory role in the development of programmes designed to alleviate poverty, has been a contentious issue in many American metropolitan centres. The effort has taken on several different aspects, including the following: the provision of legal aid services; the development of multi-functional social planning services within neighbourhoods designated for urban renewal operations; and the assignment of a new class of "community worker" to be present or resident in poor neighbourhoods. There is developing, then, in the midst of North American metropolitan areas, a general practitioner in welfare services akin in same degree to the general practitioner in medical services. However, the community worker in North America is not yet so clearly recognized as is the French *assistant du quartier*, whose responsibility is to serve a group of families (possibly as many as five hundred at any one time), offer personal counselling,

---

[35]*Progress Report*, Mobilization for Youth Inc. (New York, 1963). See especially "Introduction" and "Action Programs".

[36]*A Proposal for the Prevention and Control of Delinquency by Expanding Opportunities*, 2nd ed., Mobilization for Youth Inc. (New York, 1962).

referral and follow-up services.[37] Within any particular time period this worker in France may meet or serve only a modest proportion of the families on his list, but it is known to all these families that he is available and is ready to discuss their problems when they require his services. There is also a specific time period each week (*permanence*) known to everyone during which the worker is in his office and available to anyone without appointment. The emergence of the social welfare generalist within the modern metropolitan area may be one of the most significant developments of the past two decades.

FINANCIAL SOLUTIONS

The fundamental problem of providing and delivering welfare services adequate in quality and quantity to meet the needs of the residents of a modern metropolis is not solved merely through the input of additional or even vast quantities of financial resources. Some aspects within the overall problem are certainly amenable to the injection of far greater amounts of money—in particular, personnel education and training—but the fundamental "human needs" to be met within the world's metropolitan areas cannot be solved in this way. Aside from the improbability that most governmental jurisdictions at the present time are able to devote much greater absolute and relative amounts of resources to welfare, the essential difficulty lies in the basic socio-economic characteristics of the needy people themselves. It is possible that the allocation of much larger amounts of general welfare assistance to disadvantaged persons and families would provide them

with a more adequate standard of living than is now afforded. This solution, however, is certainly not designed to prevent the continued incidence of poverty as the affluent society and the modern metropolis continue to expand hand in hand.

It is axiomatic, of course, that in many nations a much larger proportion of the national product could be devoted to welfare services with important consequences in the alleviation of distress, but it is equally evident that the developing nations do not have sufficient resources to enable them to move equally and simultaneously on such fronts as the development of an educational system, the development of basic economic resources and employment opportunities, and the development of social services within the society. If total resources are as limited as they appear to be, the welfare services (as conceived in Western countries) might merit a relatively low priority and would have to be built into the general social and economic development of the nation as a whole. The experience with financial aid programmes of the developed nations provides no assurance for the developing nations that these funds will rebound, directly or in the short run, to the benefit of the most needy in the society. The developing nations may be faced with the unpleasant alternative of writing off their present poor, if they are to build a reasonably adequate future for the generations to come. In such a context welfare services can be little more than a palliative.

The most significant problem in the welfare field is the development of a programme of preventive services designed not to alleviate but to forestall the incidence of the almost insoluble social problems facing urban centres at the present time. The field of prevention in the welfare context is one that has been scarcely explored. For more than three hundred years' attention has

---

[37]W. A. Friedlander, *Individualism and Social Welfare: An Analysis of the System of Social Security and Social Welfare in France* (New York, The Free Press, 1962), pp. 139-45.

been concentrated on the provision of the basic physical requirements of poor people and to the development of social responsibility for the destitute members of the society.

The time has come to reconsider the allocation of resources within the welfare field in the direction of prevention and away from an uncoordinated system of income maintenance programmes which are, in any event, beyond the capacity of local government to provide. What is needed in modern metropolitan areas is a programme in the field of welfare services akin to the experience in the field of public health and sanitation during the last century, which has done so much to eliminate past scourges and to prolong the physical life of man. In the Western World preventive public health measures have reached their finest development, yet, even in the West the question may well be asked, "What does it matter, or what benefit is it for the society that its members maintain long physical life without personal and social fulfilment?"

To date, there is little objective evidence that efforts to prevent the incidence of social breakdown can be successful. The question may well be asked, "Is prevention possible?" In Western countries there are many scholars in the social welfare field who believe that prevention is not possible. In a world of swirling movement of population, from non-urban to urban conditions, from relatively poor to relatively affluent nations, there is not much reason for optimism on this score.

8

Servicing the Metropolis: A Commentary on World Opinion

Simon R. Miles

While many of the issues raised by the authors of the preceding essays were either strongly supported or rejected by the metropolitan area study groups and seminar participants, there were others on which opinion was divided. This chapter draws upon these reactions. The nature of these reactions, of itself, provides an interesting comment on the degree to which it is possible or acceptable to develop both universals for the provision of any one service, and a more comprehensive approach to the servicing of the metropolis. In most instances, some comment was made on the degree to which situations in different metropolitan areas around the world are comparable, and this is noted where relevant. The purpose of the comments that follow is, however, to add to and qualify what has been said in the preceding essays, with a view to identifying the different ways in which we can improve life in the contemporary metropolis. From this, and from the similar comments appearing in Chapter 14, we shall be able to advance to the establishment of an integrated approach to the comprehensive planning and administrative process for the metropolitan area.

Transportation

Much of the comment on metropolitan area transportation relates to two basic

243

postulates of John Kain's essay: that existing transportation infrastructure and technology is not being exploited fully; and, that the institutional machinery that exists for the provision of transportation services is, generally speaking, extremely ineffective. Although Kain has focused his attention on the former issue, the review groups and conference discussions not only offered constructive criticism of his proposals but also had much to say on the improvement of the institutional structure.

There appears to be no real agreement as to how useful an exercise it is to extrapolate from American experience in the field of metropolitan transportation to metropolitan areas in developing countries. Probably, much depends upon how specific one wishes to be. It is recognized that no one solution will necessarily have universal application whether it applies to the design of a transportation system or to an institutional framework for the planning and arrangement of transportation services. Local conditions must prevail. In some ways, these local conditions are becoming more alike. And yet, in other ways the differences become greater, with the growing economic production gap between the developing and more developed nations. Kain sees that the degree of similarity of transportation problems is likely to vary according to the similarity of national incomes, and this is reflected in the varying interest accorded Kain's proposals. For example, in the developing countries, there is much greater interest in improving the use of existing road systems than introducing new modes of transportation or initiating major programmes of land-use relocation.[1]

---

[1] One extremely simple illustration of this point is the difference that will be produced in the flow of traffic if all the pot-holes are filled. The editor has witnessed such a situation in Manila: the pot-holes being filled just prior to the local elections.

MOBILITY: NOT AN END IN ITSELF

There are two basic objectives of metropolitan transportation. The first is mobility or the efficient movement of goods or people and the second is that of assisting in the attainment of the broad social, economic and cultural goals of the metropolis. While the first of these objectives is recognized as the positive contribution of a transportation system to the attainment of the second, mobility cannot be pursued as an end in itself without considerable likelihood of incurring some deleterious effect whether it be economic, social, physical or aesthetic. These negative effects of a transportation system are often far more visible than are the positive advantages derived from increased mobility. For example, the overhead expressway that cuts through the centre of Boston is regarded by many as being not only unacceptable aesthetically but also an effective barrier to the spreading of, in this instance, the Italian community. The economic advantages of the expressway, while obvious, are not immediately measurable, and one cannot help but suspect that they would be similar if the expressway were below ground. Thus, it is the quality of life in the metropolis that is of greater concern than mobility *per se*.

ROAD VERSUS RAIL

In most metropolitan areas mobility, or lack of it, is a problem mainly of rush-hour, or peak-hour, periods. In some cities, such as Bangkok, the problem may be intensified and extended throughout the day by the variety of vehicles and lack of road space. For many different reasons, the solution to the pressure on transportation infrastructure is generally seen as being a rail-based rapid transit system. This seems to apply to both developing and developed countries. In Bangkok, for example, a rail-based transit system is now being developed. The study group there regarded Kain's proposals for limited access bus transit routes as calling for far greater human and financial resources than Bangkok could draw upon. Certainly, if some of the other techniques that Kain proposes are to be implemented, extensive demands must be made on "know how" and other skills, and the point is well made. It might be questioned, however, whether a rail-transit system does not also make considerable demands of this nature, even though, as in the case of Bangkok, the transit system is above ground.

The San Francisco study group, while not rejecting Kain's proposals on the basis of resources required, felt that road-based transit systems were not the most efficient and economical, nor aesthetically and socially acceptable ones. At the same time, it was recognized that road transit systems are necessary and should be used in conjunction with rail transit systems. Indeed, this compromise is normally accepted if adequate attention is to be given to all elements of a transportation system, from the origin of a journey to its destination. The San Francisco Bay Area Transportation Study Commission is designing a five-mile, rail-based transit system that by 1972 will be able to accommodate 30,000 passengers per hour in each direction on each route. By comparison, an eight-lane highway with two reversible lanes to accommodate major traffic flows would accommodate only 7,000 to 8,000 cars per hour. The purpose of the Bay Area Rapid Transit District (BARTD) system is primarily to provide relief for peak-hour traffic. It is recognized that many passengers will have to make use of road systems to reach transit systems, yet in spite of the disadvantages of the vehicle changes involved it is expected that the system will carry 50 to 60 per cent of peak-hour travellers.

By comparison with the rail-based system, bus transit does not have nearly the same passenger capacity. The BARTD rail system will permit operation of a ten-car train every ninety seconds in one direction at a speed of 50 m.p.h., with stops averaging every 2½ miles. The ninety-second interval is the time necessary for a train to decelerate, unload and load passengers and accelerate. Buses running on an exclusive right of way at comparable top speeds of 80 m.p.h., under comparable safety standards, would require sixty-five or seventy-five seconds to perform the same process and an interval of at least forty seconds. On this basis ninety buses could be operated per hour in a single lane. With seventy-five-seat buses, the capacity of the lane would be 6,750 seats. To haul 30,000 seated passengers per hour at seventy-five passengers per bus, 400 buses would have to be operated on one lane with an interval of only nine seconds. This is obviously an impossibility.

It is on these grounds that the San Francisco study group questions some of the cost arguments proposed by Kain. Furthermore they argue that it is not enough to consider the direct dollar costs alone. Indirect costs must also be considered, and in costing a transportation system attention also must be given to the following:

a. The need to minimize the social and economic costs of the space required for transit systems which are involved in: (1) parking facilities; (2) the dislocation of families; (3) the elimination of urban land from the property tax rolls;

b. The need to minimize the negative impact of transportation systems on urban aesthetics and the physical environment; for example, (1) adherence to certain standards of architecture and design; (2) the problem of air pollution;

c. The need to encourage, through selection of the type and location of major transportation facilities, the development of the urban area so as to maximize open space;
d. The problem of maximizing passenger safety;
e. The effect of a transportation system on land values in the area.

It is appropriate at this juncture to mention certain aspects of finance that merit attention, especially as they relate to techniques of reducing pressure on existing facilities. There is little support for the concept of a self-supporting transit system. Subsidization of a transit system can be justified in that its benefits go equally to those who do not use it and those who make frequent use of it. Also, while justice might be introduced if users of highways were charged for this privilege, such a move could be regarded as highly regressive in terms of fiscal equity. A telling observation can be made regarding pricing systems designed to inhibit use of road and/or parking space and at the same time serve as a source of revenue: they are unsuccessful in realizing the former objective to the extent that they are successful in raising revenues. Unfortunately, the latter objective is often pursued at the expense of the former.

THE COORDINATION OF TRANSPORTATION SYSTEMS

More drastic solutions to congestion and to the related problem of urban sprawl call for a complete replanning of the entire urban area. The Athens study group proposed that the congestion in the central city would be greatly relieved if the present monocentric pattern of development were to be replaced with a plan for a polycentric metropolis. This would bring the place of work closer to the residential area, which would not only reduce congestion

but also, in a city such as Athens with four rush hours, make more leisure time available. In this way, further growth of the metropolitan area could be accommodated more easily.

This approach takes account of the real need to relate transportation planning to other aspects of planning. It also relates to the proposals for a decentralized administration of metropolitan government, which will be dealt with in Chapter 9. However, while this issue is taken up more fully in later chapters, both Kain and the study groups have indicated that the fragmented nature of administration and decision-making pertaining to metropolitan transportation is a very real problem. The Milan study group members expressed themselves very forcefully on this issue. Conflict theory finds its application here. The Milan group indicated that while the groups, the field of conflict (transport) and the dynamics of a conflict situation exist, the fourth element—the control group required to regulate the conflict—is missing. In Milan, as elsewhere, there are a great number of management agencies, regional agencies influencing transportation and intra-departmental or agency organs. Additionally, there are many interest groups, some powerful elites and others more ephemeral and less influential organizations. In all there are about forty governmental or quasi-governmental groups and a similar number of non-governmental groups having an interest in transport for the Milan area. The absence of an overall controlling agency has left its mark in the form of urban sprawl stretching far out over the plain of Lombardy. The problem is compounded with sprawl; for, with no economic incentive to provide a transit system, the public must take to private automobiles which in turn increases congestion.

In some cities this control group might take the form of an "umbrella" agency which would coordinate all organs presently playing a part in the provision of metropolitan area transportation. This was seen as a possible solution for the San Francisco Bay Area. In this instance, there was no real need foreseen, by the study group, to completely restructure the existing politico-administrative machinery. There appears to be an acceptance of the existence of these organs, and although some of them could well be eliminated, the general tenor of the San Francisco group on this point was a conservative one.

Utilities

An insufficiency of financial resources and a lack of institutional coordination appear to be the two major obstacles to improvements in the provision of utilities in metropolitan areas. Yet, while money can solve many problems, it cannot bring about the coordination of planning and administration, which bear so heavily on the cost and quality of services.

Reactions from the study groups and the seminar discussions indicate that the problems associated with these two central issues are unlikely to be solved without major reforms of governmental machinery. In principle, there was general agreement with Hanson as to the eminent desirability of a single multi-purpose metropolitan form of government. Far less interest was shown in resorting to any one of the many legal forms discussed by him in Chapter 3.

However, despite the fact that the "status quo" evoked a sense of frustration, the reactions were both positive and realistic. This was particularly well illustrated by the politically sensitive tenor of the review by the Rio de Janeiro study group. Obviously, the problems confronting a fast-growing metropolis in a developing country with heavy demands upon its scarce resources call for a more comprehensive approach to their solution. Yet, the Rio group recognized that political attitudes cannot be changed overnight. The reactions from Budapest, with its planned economy, were oriented toward seeking reductions in costs through technological improvements rather than major adjustments to an administrative machinery that is reasonably satisfactory. The experiences of both Edmonton and Nagoya complement those of Rio de Janeiro and Budapest. Edmonton's utilities are almost all provided by the city, the jurisdiction of which covers virtually the entire built-up area. In Nagoya, on the other hand, although most of the major utilities are publicly owned, there is an obvious attempt to introduce more business-like methods into the provision of services, whenever this is possible.

Paradoxically, despite this contrast in approach, the objectives sought in each metropolis are very similar, and before discussing the issues of resources, planning and administration, it may be salutory to examine what it is we demand of public utilities.

THE SEARCH FOR EFFICIENCY

The basic requirement made of a utility is that it provide a useful and efficient service as defined by the business principles of optimum charges and minimum operational costs. Yet "efficiency" should not be equated with "least cost". A major social objective of a utility is that it should be provided with reasonable equality. A service may become useful only with some additional expenditure and loss in economic cost efficiency; for example, with a frequent bus service operating over an extensive route network. However, while the social responsibility of a utility must be recognized, due compensation must also be made for any operation designed to achieve this objective. Thus, while public authorities that manage or have control over utilities should not regard them as mechanisms for subsidies, indirect taxation or the redistribution of income *per se,* undoubtedly these social side-effects will exist to some extent even with due compensation being made. The dangers of restrictive rate policies on the operating ability of a utility will be referred to later.

Of the various ways in which efficiencies can be achieved, whether they be administrative, technological or resource in their orientation, most have some direct relation to the scale of operation of a utility. It is generally assumed that there are certain economies that can be realized with increased scale although little concrete evidence of this has been produced to date. Increase in scale of activity can be either through an increase in areal activity or in intensity of activity. All study groups agreed that a regional perspective is required for the provision of utilities in the metropolis. This is an absolute necessity for the efficient long-term planning of the expansion of the metropolis, for the present coordination of services and additionally, for the reduction in the delivery costs of most services. An example is the electricity supply in southeast Brazil. As a result of mergers there are now only two private electricity companies in Rio de Janeiro, and these are being linked to a grid that is being formed to serve the whole of southeast Brazil. Savings will be realized by both the private sector and the central government authority responsible for supervising the private companies.

The water supply in Rio de Janeiro, however, is far from satisfactory. Each municipality had its own water supply system in 1967, and although there was some discussion of the possibility of coordinating these into one distribution system in the future, the present situation is one in which some municipalities suffer water shortages even though the mains to supply the City of Rio de Janeiro (with no acute water shortage) cross their territories. The form of coordination most likely to be accepted politically would be a holding company that would control individual, existing enterprises or authorities. A similar arrangement is being contemplated for the joint planning and use of sewers in the metropolitan area. While it is readily admitted by the Rio study group that these administrative arrangements are not the most satisfactory, they do represent an improvement, and they also permit the realization of certain technological efficiencies.

Another means by which efficiencies can be achieved both in cost and in quality of service is the transfer of responsibility to a senior level of government. The contemplated transfer of Edmonton's city-owned telephone system to the provincially-owned telephone company is a good example of this.

Again, there is great scope for bringing about cost reductions through the introduction of technological improvements. Perhaps the most interesting of these is the suggested development, in Budapest, of a network of tunnels to house all major services. While this concept is by no means new, it has not been introduced on any major scale (at least, not to the knowledge of the editor). In Budapest, the only tunnels of this kind that exist at present pass under the Danube to facilitate inspection and repair of the water and electricity installations. The study group referred to a study carried out by the city administration which showed that above a certain density of traffic and utility networks it becomes economically justifiable to provide these common service tunnels. As the density of persons or services in central Budapest is nowhere as high as it is in some other major cities in the world, it is likely that this information would prove to be highly profitable elsewhere.

Another interesting development is in the area of district heating. In 1967, 31,000 flats in Budapest were served by this new system, and it was then expected that about 8,000 additional units would be linked to the system each year thereafter. Comprehensive district heating systems will be the source of energy supply for future large housing and industrial estates in the suburbs. Natural gas is used to drive the turbines to generate the steam and hot water for the district heating systems. Despite the high investment costs, it seems that district heating systems will be the most common form of heat source in the foreseeable future in Budapest.

Illustrations of the losses that can be suffered from an unwise use of resources were given by the Rio study group. The group cites the transfer of 4,000 workers from the old tramway company to the bus company established in 1963. The funds set aside to compensate these workers for the loss of employment was used instead for the purchase of trolley buses. Thus, not only is the new company grossly overstaffed with a ratio of sixteen workers per vehicle (as compared with 6.5 workers per vehicle in private companies), but its vehicles are constantly being brought to a standstill because of the frequency of power failures in the city.

Before leaving the matter of efficiency, it is of interest to note that there was little real agreement among programme participants on whether private or public enterprise could provide services more efficiently. Both cases were stated but without any real supportive evidence. Perhaps the best case for public enterprise is that it is likely to have a better credit rating on the capital market, especially in a developing country. Yet, as Hanson points out, it is in the developing countries that the publicly-run utilities suffer most from inefficient personnel operations, which is partly due to their inability to offer salaries competitive with those offered by private enterprise. Again, the desire to introduce more business-like methods of operation of publicly-run utilities and the attempts being made to attract further private investment in utilities to fulfil the need for capital, can be interpreted as recognition by the public sector of the values of private participation in operating utilities. Obviously, a good case can be made for both sides; private enterprise tends to be more efficient, but the public sector is likely to be more conscious of the social responsibilities of a service. Whatever form is decided upon for the distribution of the service, it will probably require some form of senior government regulation and supervision.

Having discussed the basic qualities of efficiency and social responsibility that are sought in utilities, it is appropriate to consider some of the factors that hinder their realization and ways in which these constraints might be mitigated. Essentially, these constraints are the lack of, or mismanagement of, resources and the lack of satisfactory integrated planning and administration on a metropolitan-wide basis.

EASING THE RESOURCE CONSTRAINTS

Of the two resources of major import to the improvement of services, financial resources are considered to present far more of a problem than human resources. There was little discussion of the nature of the need for these resources. All accepted that technological developments have led to a mounting and sustained need for more capital to build new plants and upgrade existing ones. As long as technological efficiencies can be obtained man will seek to improve standards of living and, in short, he must pay for these improvements.

The lack of capital is the major problem in the provision of utilities: it far transcends related issues of public versus private ownership. The challenge is to make capital available. While a government may be in a stronger position than a private company to borrow money on the open market, especially in a developing country, the credit of that government will depend on its viability and financial strength. Just as a small company is unable to raise sufficient money to provide an efficient service, so is a small and weak municipality. The implications are obvious enough: if government is to accept responsibility for the provision of utilities, then the governmental units have to be strong and efficient in their operations. In short, the pressures of the market-place are pushing for larger, multipurpose governmental units. Certainly this factor was a major force that led to the creation of the metropolitan tier of government in Toronto in 1953. One element of the crisis immediately prior to the appearance of Metro Toronto was the extreme difficulty experienced by some of the suburban municipalities in borrowing money on the municipal bond market. As Frank Smallwood has observed:

> Metro's third major achievement [the other two being the realization of a strong political base and the resolution of crises with specific services]—the provision of the capital financing necessary to meet steadily expanding local demands within Greater Toronto—represented another crucial consideration that led to the original establishment of the new governmental program. The fact that the new Council was given exclusive jurisdiction for assessment and capital

borrowing for itself, for the area's special boards and commissions, and for the thirteen local municipalities provides the most striking evidence of the high priority that Metro's creators placed upon this particular responsibility.[2]

The example of Metro Toronto would appear to be worthy of note for other metropolitan areas. From their statements it would appear that the Nagoya and Rio study groups appreciate the advantages of a metropolitan form of government for these purposes. (The Edmonton group, interestingly enough, did not believe that a municipal authority is any more competitive on the market than is a private company.) However, the Rio group, believing that more capital should be made available in general, faulted the past actions of the senior governments which had imposed severe restrictions on rates charged by privately-run utilities, with the resulting drain of private investment from this sector. It is only since 1957, and more so since 1964, that the situation has shown signs of improvement.

Before leaving the matter of borrowing, it is important to note the credit difficulties confronting urban governments in many developing countries. In such countries, internal resources are usually scarce, and the country may well be experiencing severe inflation. Thus a loan, on favourable terms, acquires added value, and it is because of this that the Rio study group suggested that such loans should be made conditional. In this way the recipient authority (and in the case of some international loans, a recipient country) could be forced to reform the structure by which it provides the service in question and, thereby, improve the effi-

---

[2]Frank Smallwood, *Metro Toronto: A Decade Later* (Toronto, Bureau of Municipal Research, 1963), p. 12.

ciency of the service. The Rio group cited the failure of the Inter-American Development Bank to apply certain additional conditions to a loan to the City of Rio de Janeiro for the development of its water supply system. The loan was already subject to certain conditions that did, in fact, produce a reasonably efficient water system for the central city. If the Bank had been astute, however, it could probably have used this loan to bring about the creation of an efficient metropolitan-wide water service rather than one serving the central city alone.

The problem of the scarcity of financial resources on a national basis is very often insurmountable. Senior government assistance in making large sums of money available for capital works projects in the cities is obviously highly desirable, if not a necessity. However, the limited resources available call for the development of a system of national priorities. Budapest offers some guidance in this regard. The city has a twenty-year development plan as well as a five-year plan that spells out, among other things, the financial resources required in that period. These requirements have to be cleared with various central government ministries before being incorporated into the national economic plan. Thus the national plan allocates "investment credits" to certain sectors of activity and parts of the country. Any "credits" required beyond the number allocated in the plan for, say, waterworks development in Budapest, are met by the transfer, to the Municipal Water Company, of a limited amount of the investment funds also allocated to individual industrial enterprises for their own growth.

Apart from loans, a major source of revenue for utilities is the rates charged. The setting of rates and the impact of rate structures was an issue of major concern to study groups and seminar participants. A basic question is whether utilities should be run to produce a profit, whether they should be merely self-supporting and non-profit making, or whether they should be subsidized. Each of these policies has certain advantages and disadvantages according to the conditions under which they are to be applied.

For purposes of simplicity in accounting, it would be highly desirable for the rates charged to cover operating costs and provide for all the necessary capital costs incurred in plant expansion and upgrading. If this is not the policy to be pursued, then, in the case of a publicly-owned utility, there has to be some form of internal transfer or a grant from a senior government. Also, in the case of the privately-owned utility a government grant would be required to meet not only the deficit incurred but also some margin of profit. While the use of government funds to meet the costs of providing a social service is generally acceptable, the use of funds to provide a profit for private enterprise is political anathema. However, as the setting of rates is either carried out, or supervised, by a senior government, it would appear obvious that the above situations could be avoided by setting rates high enough to cover all costs involved. Unfortunately, this is not always done. In Brazil, for example, the policies pursued by the government until 1964 have resulted in a dearth of private capital in the electrical utilities in Rio de Janeiro. The major problem was that it was only in 1957 that costs were considered in the setting of rates, and it was only in 1964 that the monetary correction of company assets was taken into account. With rapid inflation and frequent devaluation, the net result was that companies were faced with the progressive decapitalization of their assets and, consequently, heavy losses. A similarly poor pricing policy existed with regard to water rates. Although this situation has been remedied in Brazil, the effects are still being felt. Despite this bad experience, there seems to be general agreement that rate setting should be the responsibility of a senior level of government.

In Budapest utility rates are restricted to cover operating costs, as major capital projects are funded by grants from the central government.

Where utilities are privately operated, the degree of government intervention varies. In Nagoya the private utility rates are set by respective central government ministries. For example, private electrical utilities, which do make a profit, have their rates set by the Ministry of International Trade and Industry at a level that allows an 8 per cent return on the value of business assets. The Ministry of Transport permits a 10 per cent return on equity capital of private transport utilities.

Nagoya's public utilities, on the other hand, are able to set their own rates although they have to do so within the limitations of the Municipally Owned Public Utilities Law. This does not mean necessarily that profits are realized. The transportation services are predominantly publicly-owned, and it is estimated that commuter transportation fares are anything from 30 per cent to 70 per cent below actual cost. Any profits that are realized ·by publicly-owned utilities in Nagoya have to go first of all toward meeting any back debts and of the remaining amount, if any, 5 per cent is placed in a sinking fund and the other 95 per cent is disposed of according to the decision of the municipal council concerned.

The Edmonton situation provides some interesting comparisons. Privately-owned utilities operate at a profit although their rates are fixed by the provincial Public

Utilities Board. There appear to be no problems in the financial stability of these companies. The city-owned utilities on the other hand, have their rates set by the city and, in some cases, operate on an even higher profit margin than do the private enterprises. One reason for this is that the utilities are seen as a lucrative source of municipal revenue. For example, although the major source of revenue in Edmonton is the real property tax, in 1965 the contributions of the utilities were equivalent to 27 per cent of the property tax levy. Because of the public resistance to increases in the property tax, Edmonton's utilities are seen as an expandable source of revenue.

Although the study group appreciated that high utility rates were an indirect form of taxation, there was no mention of the fact that if this is to support general municipal services rather than the services for which the rate was levied, then it is a most inequitable indirect tax. Inasmuch as 85 per cent of the profits were disposed of in this way in 1965 (the other 15 per cent being used for future capital expansion by the utilities), there are obvious grounds for questioning this policy. Indeed, it was noted by the group that accord could not be reached on this question. It is of interest to note that while the inequity of this indirect form of taxation cannot be avoided, it can be ameliorated by widening the area of utility provision in order to spread the service to all people and, indirectly, to tax a full range of income groups. Naturally, if the area of jurisdiction of the multi-purpose authority, which has call on the profits of the utility, is also extended, then the benefits of the indirect taxation are similarly spread, thereby reducing the inequity somewhat.

The alternative to having utilities support themselves or even produce a profit is to have them operate at a loss. The only condition under which this appears to be generally acceptable is when we find the utility providing a social service. There was general agreement with Hanson that any losses incurred as a result of operations on this basis should be met by some form of transfer, as already mentioned above. (This assumes, of course, that the utility itself is not expected to provide the subsidy, as this would only lead to the rapid depreciation of the assets of the utility.)

Direct subsidies can, of course, be provided for services of a social nature. One example is the subsidy provided for water services in Budapest, which are met by the Tax Office of the central government; another example is found in Nagoya, where the Japanese government subsidizes major capital water works projects. This is almost the only form of subsidy in Nagoya other than the internal transfer subsidy for operation of the municipal subway system. Indirect subsidies are provided through tax relief to both private and public utilities in both Edmonton and Nagoya.

Another source of funds is a tax on users of competing services. In Rio de Janeiro, one source of revenue for the proposed subway is being sought from private motorists through licences to park in the central city. Sympathy for increasing the tax on the motorist was indicated by the Edmonton group, which thought that the full costs of municipal roads construction and maintenance should be met by road users. Presumably the argument is aimed at the commuter who is not willing or prefers not to use public transportation services. Hopefully, before imposing such a tax, the municipality would be prepared to offer the commuter a useful and efficient public transportation network.

As has been mentioned previously, while the importance of human skills in the provision of utilities was recognized by the study groups and the seminar participants, these problems are not as pressing as those associated with financial resources. While the lack of skilled managerial personnel was recognized as being acute in the developing countries, there was no discussion of the need for training institutions to produce competent personnel. Neither was there discussion of the importance of the attitudes of the politicians who will decide whether or not the acquired skills are beneficial and, indeed, whether or not staff should be supported and encouraged to take "in-service" advanced training courses in public administration and related fields. Perhaps these matters are taken for granted, but in the view of the editor this is doubtful. An exception can be seen in the Budapest situation, where the requirements for labour and managerial personnel, arising from the anticipated utility requirements, are built into the five-year plan for the city. Whether the city then assumes full responsibility to produce the required personnel is not clear. Certainly there is no problem in obtaining labour in Budapest, as "competition" to live in the city is strong. At the present time migration to the city is controlled.

There was, however, some discussion of personnel problems and the implications that ensue for the politico-administrative aspects of utility-provision. In the developing countries especially, the independent public utility is often preferable to the service integrated with municipal administration. There was general agreement that municipal government is often of dubious competence and has a tendency to interfere with the operations of the utility; a weak executive may be forced to use patronage in his personal practices. It was agreed that publicly-owned utility executives should be compensated on a level comparable to opportunities in other sectors. From this it follows that the wage scales may have to be freed from those of the civil service, which are often extremely low.

The other major problem is settling on the amount of discretion given management in the hiring of employees. It is assumed by the management of utility operations that only they are competent to judge the suitability of prospective employees. This is probably correct and calls, again, for managerial freedom from political interference.

On the basis of these arguments the independent-public-agency form of utility is seen as being highly desirable. However, as has been made clear from the discussion of financial resources, these are not the only considerations to be taken into account. The main conclusion of those at the seminar was that, as the independent public utility may also be subject to political interference (and sometimes justifiably so), there seems to be little or no escape from a state of dissatisfaction without an enlightened legislative body.

THE DEMANDS UPON THE POLITICO-
ADMINISTRATIVE MACHINERY

Many of the problems associated with the planning and administration of utilities have been touched upon already in discussing the goals sought and the resource constraints upon the realization of these goals. These issues will be reiterated only to the extent that it is necessary to illustrate their implications for the politico-administrative machinery that has to be developed to facilitate a comprehensive planning process.

One failure of utilities planning is that it is rarely part of a comprehensive process. The separation of a utility from other services for purposes of management very often repeats itself in the organization of planning for a utility. The failure of utilities

planning is that it does not achieve the integration either between like services provided throughout a metropolitan area or between related services throughout the same area. This integration is required, not only to take advantage of the technological economies that can be achieved, but also to have the services in question offer maximum benefits to the user.

Unfortunately, the main weakness does not rest with the number or quality of planning personnel or the techniques being used. If it did, it would be easier to remedy. The major problem rests with the politico-administrative machinery itself. For this to be changed requires a change in attitude on the part of the public and, even more so, among the decision-makers themselves.

The many examples of poor planning that have already emerged are basically due to the lack of comprehensive planning machinery or the lack of interest, by decision-makers, in improving the conditions.

This applies whether we speak of the time dimension, such as the inability to plan ahead because of the lack of financial resources or of the ability to anticipate what limited resources will be available; or, with the space dimension, the inability to guarantee continued transportation services for commuters arriving at main rail termini in a city.

Thus, there has to be physical, social and economic integration of services planning and, with this, the integration of financial planning. Additionally, the machinery has to provide for the effective performance of the management functions of regulation, supervision and operation of the utilities. The way in which all these elements, plus that of adjustment to the machinery through administrative planning, are tied together as part of comprehensive planning is described in Part IV of this volume. The present objective is to illustrate the demands made by utilities upon this machinery.

The great diversity in geographic, political, economic and administrative conditions and the inherent differences among the efficient scales of operation for various utilities make it virtually impossible to determine absolutely which utilities should be operated at the various levels of government. Both Forstall and Jones (in Chapter 1) and Dupré (in Chapter 11) have some further suggestions to make on this point, and these will also be discussed in Chapter 14. To be brief, in assigning functions to any agency at any level of government or to private enterprise it is necessary to have due regard to the three values discussed by Dupré: efficiency, participation (by the public) and liberty (freedom from oppression or undue control). Unfortunately, the pursuit of these three values inevitably involves compromise since the efficiency value will generally lead to the structuring of large units, whereas the values of participation or liberty tend in most instances to call for smaller units. A clue to the long-term solution to this dilemma is contained in Chapter 9. There, Gorynski and Rybicki suggest that the size of a polycentric metropolis and its cells should be determined on the basis of the participation and liberty values and that the efficiency value be regarded as a secondary consideration. The message is that utilities exist to serve people and not that people live to enjoy efficient utilities.

As the area for the optimum economic operation of utilities is often larger than the metropolitan area, and thus of existing municipal or metropolitan governmental jurisdictions, the decision must be made either to try to make the metropolis part of a larger service area or to make it a separate area for purposes of planning and operation, subject to overall regulatory and supervisory coordination by a senior government. If the values of participation and liberty are to be respected, the latter alternative seems to be more desirable.

For some services, such as electricity, the former solution may be more appropriate, for there is little public interest in everyday control of electricity services. This particular example, however, brings up the question of assigning different parts of a service to different levels of government. In Brazil, the generation of electricity is now largely a public responsibility and is provided on a regional basis. The distribution is then caried out by local companies—in this case mainly, but not entirely, privately-owned. As can be imagined, this particular division of labour makes additional demands upon the coordinating and financing body. In the Brazilian instance, the publicly-run electrical services are coordinated through a holding company—*Electrobras*.

As regards the provision of electrical power, the situation in Budapest provides an example of a cross between the two cases referred to above. The generation of electric power is a responsibility of the Hungarian Electric Works, which has six generating stations throughout the country, one of which is the Budapest Municipal Electric Works. The Hungarian Electric Works receives policy direction from the Ministry of Heavy Industry, and the Budapest Municipal Electric Works distributes electrical power to the city according to policy formulated by the central agency. There is also the Permanent Public Utilities Committee, a standing committee of the Budapest City Council that is responsible for city policy directives and works through the Public Utilities and Services Administration to coordinate the separate plans emanating from the city-controlled utilities. These separate utilities operate as trading companies under the control of the Public Utilities and Services Administration. As the electric company is not one of these, its activities have to be coordinated through the five-year city plan approved by the central government. The process of coordination in this instance can be seen as long and arduous. Hence, there is a desire to keep more control at the local level.

The most favourable situation, as remarked earlier, appears to be a multi-purpose, metropolitan-wide authority that is regulated and supervised by a senior government to the extent necessary to serve the values previously mentioned.

In the absence of a metropolitan form of government, committees of utility executives should be responsible for ensuring coordination between the plans of their separate authorities. The objective sought here is to bring into the planning process those responsible for implementing the plans. The success of these committees will depend largely on the spirit with which they are developed. Whether they are compulsory or voluntary will have little effect on the spirit of cooperation among individuals.

The seminar participants were undecided as to whether the provision of utility services by separate metropolitan-wide agencies is conducive to the development of a single multi-purpose metropolitan government. On the one hand, it is argued that the existence of these metropolitan-wide, single-purpose agencies would bring about public recognition of the need for coordination through a multi-purpose government. On the other hand, to the extent that they are successful and are coordinated by existing senior government agencies, they serve to underwrite the longevity of the "status quo".

The implications of public attitudes and the demands made by utilities on governmental machinery in the metropolis will be taken up again in parts III and IV.

Housing

In attempting to synthesize all that was said about housing problems in the world's metropolitan areas, it is tempting to select housing policy as the issue central to these discussions. Many questions were raised relating to the content of a housing policy, the criteria considered in its development, the organs of the public or private sectors responsible for that development and its implementation, and the policy framework for housing planning. While answers to these questions may produce a housing policy, it is not necessarily a housing policy that will produce the houses desired. There were representations from enough metropolitan areas that boasted some form of housing policy, and yet had enormous housing problems, to bear out this point.

In short, the real issue is not a question of "policy or no policy", but the relevance of that policy to economic and social conditions. Housing policy cannot stand alone. It must relate to social and economic policy in general and be viewed in the context of these broader social issues. It might, indeed, be best to refrain from speaking of "housing" policy *per se*. This, however, is impossible. Having offered these caveats, it is now possible to proceed to an examination of the issues that the study groups saw as being of greatest significance in resolving the housing problem that confronts all major urban centres throughout the world.

SOCIAL NEED AND ECONOMIC DEMAND

The universality of the "housing problem" is truly remarkable, yet understandable.

Housing is a highly visible and extremely personal good. For many it provides basic shelter, for some it provides comfort too. The individual cannot avoid being closely associated with it even if he were to wish otherwise. It is also extremely easy for the individual to compare his housing with that of others. Largely because of this visibility, the "housing problem" is often a major issue in political campaigns. Yet, there are differences in the nature of the problem, and these differences relate directly to the difference between social need and economic demand.

The significance of this distinction was enunciated clearly by the Glasgow group in referring to Lionel Needleman who says:

> The social concept of housing "need" has to be distinguished from the economic concept of housing demand. The effective demand for housing relates to the accommodation for which people are able and willing to pay. It takes no account of social desiderata or of personal aspirations that cannot be fulfilled because of lack of money. Housing need, on the other hand, is the extent to which the quantity and quality of existing accommodation falls short of what is required to provide each household or person in the population, irrespective of ability to pay or of particular personal references, with accommodation of a specified minimum standard and above.[3]

If we accept that housing policy is to be a part of a broader social policy, it is most important that the goals of housing policy should be more concerned with meeting social need than with meeting the effective economic demand of the market-place. Once basic social needs have been met in housing, resources should then be directed

---

[3]Lionel Needleman, *The Economics of Housing* (London, Staples Press, 1965), p. 18.

toward basic needs in other areas, such as health and education.

In most countries the "housing problem" is one of providing that specified minimum standard of accommodation alluded to by Needleman. In a few affluent countries, however, this basic need remains for only a small percentage of the population. Yet, because of man's desire to forever improve his lot, there is continuing pressure in these countries for improved housing—the effective demand described by Needleman. An excellent example of this is provided by Toronto, where the housing conditions were acknowledged by the study group to be very good relative to conditions found elsewhere. Despite this, housing is probably regarded as the biggest problem (and certainly the biggest headache) for the local politicians. The reason for this is that there is too great a tendency to cater to effective economic demand, which is unlikely to be satisfied under any circumstances, and virtually no effort given to meeting social need, which in the case of Toronto is not insurmountable.

Despite this criticism of misguided policies, the effective economic demand is obviously significant for those developing and modifying housing policy, for it does indicate the extent to which the public, under varying conditions, is able to acquire housing of a certain standard. By keeping this in mind, policy-makers are better able to ensure that the social need can be satisfied. Determining the extent of the gap (between social need and effective economic demand) is obviously no easy task for either a housing authority or its critic. The desire, however, or social conscience that motivates the government to provide housing to meet this gap, is an effective third force in housing demand. The other two factors are the effective economic demand and the residual demand for poor quality housing by those unable to participate effectively in the market. Indeed, one objective of increased government intervention is to reduce the significance of the demand for sub-standard housing.

HOUSING STANDARDS

It is not particularly significant, at this juncture, to elaborate further upon some of the present standards that form the basis of policy in some metropolitan areas. Sazanami has provided us with many examples of the variations in floor space, density and age of buildings that are acceptable under varying conditions. It is up to the policymakers concerned to decide whether or not their standards reflect the criteria that should be observed. Thus, there was more interest shown by the study groups and in the seminar discussions in the characteristics of these standards and the manner in which they might be realized than in the mere comparing of statistics.[4]

What are the major considerations that

---

[4]However, it should be stressed that there was great interest in improving the availability of statistics, and the presentation of these statistics, on housing conditions. There was a strong plea for making statistics available on a metropolitan-wide basis in addition to the present nation-wide statistics, which are often quite unrepresentative of conditions in metropolitan areas. For example, whereas house building rates across Canada averaged less than 8 units per thousand population in the period 1963-1967, comparable figures for Metropolitan Toronto were 12 units per thousand population. A more important requirement is a more accurate picture of the nature and extent of the need. Also, some statistics presently in use are lacking in sophistication. A simple example is that of using persons per dwelling as an indication of density. The Glasgow group indicated a strong preference for the formula proposed by the World Health Organization (WHO); that of persons of the same household per room. (As a rule of thumb, the WHO regards more than 1.5 persons of the same household per room as being representative of overcrowded conditions.)

should be reflected in housing standards? Obviously, those charged with setting standards must distinguish between social need and economic demand. Only the former should be reflected in any housing standard. However, the degree to which equality can be achieved in housing conditions must also be considered. Varying opinions were expressed on this point, although it is possible to reconcile most of these. Concern ranged from the national to the local scene. For those countries experiencing substantial emigration or immigration in addition to internal migration, there is a need to pay due attention to the equality of housing standards not only throughout the country but also as they relate to standards in other countries.

The Glasgow group, recognizing housing conditions as a significant element in general social conditions, indicated the need for improved standards in order to stem the flow of population to other parts of Britain and also abroad. The Toronto group, representing an area which receives about 17,000 immigrants each year recognized housing as one of the major reasons why immigrants are attracted to a country such as Canada. It was pointed out, however, that although the immigrant loan assistance fund had recently been raised substantially, there is no attention given by the federal government to assisting the provincial and local governments in housing immigrants. Thus, just as Glasgow regrets the loss of population through emigration, so Toronto regrets the burden immigrants impose on local resources.

While the social conditions in Glasgow can be improved through upgrading, among other things, housing conditions in the local area alone, the Toronto situation calls for greater equality of housing conditions throughout Canada. (The fact that Toronto also receives about 17,000 in-migrants from other parts of Canada each year further supports this stand.)

While common international standards are perhaps unrealistic, nation-wide standards are regarded as possible in the more developed countries. It is in the under-developed countries that not only nation-wide but metropolitan-wide common standards are thought so difficult to achieve. The Lima study group indicated that in their case the availability of few resources makes equalization of conditions, through subsidization, a low priority. In developing countries especially, there is a greater need to gear housing policy to the immediate needs of different social groups. While this results in a wider range of standards, the policy should be designed to encourage self-help toward improving family housing conditions. Since substantial aid to each family unit is impossible, housing policy should reflect the fact that it is part of social policy and should enable each family to select its own priorities for the utilization of its few resources. If a family wishes to put more emphasis on education for the children, it should be permitted to do so. In such instances, the governmental authorities should be more concerned with ensuring that low-income families are located near jobs and are provided with the basic public utilities rather than with the number of rooms available.

In the seminar discussions, these viewpoints were reconciled only with the switching of discussions to socio-economic possibilities rather than goals for housing. Observing socio-economic possibilities, however, implies not only a maximizing of social welfare, but also a striving for greater equality as more resources can be made available. For the metropolitan area, the administrative arrangements necessary for achieving these objectives were seen as taking a regional form. There was strong support, therefore, for the regional approach to urban government as proposed by Sazanami. While this is examined in greater detail later in this section, it is also of significance for another aspect of housing standards: that housing conditions reflect not only the living unit but also the environment.

The size of the residential community was commented on by several groups, and there seemed to be general agreement that the scale of settlements being developed in the U.S.S.R. is generally very satisfactory. Perhaps it is more important, however, to stress the importance of access to the communal services (in communities of various sizes) in terms of time, ease and cost of travel rather than call for them implicitly in terms of the population size of the community.

The optimum physical size of towns is obviously related to the optimum size of administrative units. This was the basis of the Athens review, already commented upon in the section on transportation. The Athens group added to the comments on housing, by indicating that communities of 20,000-30,000 population are not only more satisfactory in terms of environmental conditions, but also in terms of government effectiveness in providing services.

While it is obvious from the above that the quantitative and qualitative aspects of housing needs are the basic determinants of housing goals, it is equally important to stress the role of other policy considerations that may result in a modification of these goals. Such considerations pertain mainly to the various aspects of housing supply, whether it be the nature and use of resources available, the planning of housing, or the administrative and decision-making arrangements for housing supply. The study groups and the seminar discussions generated a wealth of information on local problems experienced in these areas of concern. Here we shall examine those observations that contribute most to the establishment of criteria on which a sound policy may be based.

THE OPTIMAL USE OF AVAILABLE RESOURCES

Rehabilitation versus New Construction
As the objective of all housing policies is to provide not only more housing, but housing of improved quality, rehabilitation versus the construction of new houses is one of the first issues confronting decision-makers. Although there is a great need for cost-benefit studies on housing investment in these two sources of supply, it is fairly safe to assume that old housing that cannot be provided with essential facilities is not worth rehabilitating. However, as this element of the housing stock is eliminated, the choice becomes much more significant. In Europe especially, but also in those other parts of the world where a large percentage of the existing housing stock is old and in reasonable condition, timely investment in maintaining existing stock may be a wise move. This was certainly the opinion of the Glasgow group, who tended to favour this approach, where feasible, in conjunction with the further development of new towns to relieve the population densities of the older parts of the city. A similar concern for reducing densities and, at the same time, preserving older buildings for their contribution to the aesthetic quality of the urban environment, was shown by the Leningrad study group in a supplementary comment on housing.

As implied above, however, the feasibility of rehabilitation depends upon the costs involved. Until now housing policies in all countries have been limited largely to a consideration of the direct costs of rehabilitation versus construction. Judging by these costs alone (and seminar participants fully realized that these were not the only costs to be considered), rehabilitation appears an expensive proposition. In Toronto much of the older housing was of poor quality when constructed. The Toronto group cited cost estimates of over $10,000 per unit for one scheme that had been proposed. Yet, although this is in addition to the acquisition cost of the property, it should compare favourably to the cost of purchasing a new house at $25,000 to $30,000.[5]

It is at this juncture that one has to consider the politics of urban renewal. Large, highly visible, clearance schemes seem to be far more "popular" than sporadic rehabilitation in many cities. The possibilities of making government loans available to assist private individuals in carrying out their own rehabilitation did not have much appeal for the Toronto situation. Most of the occupants of the older buildings are aged, often retired persons not wishing to assume new debts, or poor people who are not in a position to assume these added burdens.

For the developing countries, rehabilitation on any extensive scale might be ruled out for two reasons. First, there are relatively few dwellings suitable for rehabilitation and second, many governments do not have the funds available to develop an extensive and equitable programme. Hence, there seems to be greater interest in programmes that encourage individuals or groups to improve their living conditions through self-help. Governmental agencies and others may find that their resources are best deployed in providing basic ser-

[5]The cost of housing is exceptionally high in Toronto, relative to other cities in Canada. It is also increasing at a rapid rate. The average price of houses sold in 1967 was $24,681. In 1968, the figure rose to $27,637 and the average for the month of April, 1969, was over $30,000.

vices, as alluded to above in the instance of Lima, and leadership in organizing such programmes. Much good work in this field has been accomplished in Manila, where university groups play a strong part in developing these programmes. This approach not only makes good use of labour resources available, but develops a sense of pride in accomplishment.

Rehabilitation is unlikely ever to provide an adequate supply of housing, and for the most part governments must look to new housing. The availability of land, the state of the building industry, and the financing and planning of new housing are major issues in this regard.

Regulating the Land Market
Of the means of land-market regulation discussed by Sazanami, that of public land acquisition is probably of greatest interest, if not wholly acceptable. The experience of Belgrade, where settled land of an urban character has been nationalized since 1958, is that public ownership of land is conducive to improved environmental conditions and that it also facilitates the large-scale developments that are necessary to maximize the advantages of the latest technological innovations in the building industry. In addition, the governmental control of most housing development assists in regulating the growth of cities in Yugoslavia.

Regulation of city growth was not seen as a problem by the Toronto group. The timing of the extension of services is an effective tool in this regard. For Toronto the real problem is the spiralling price of land values and, to a lesser extent, the control of land use. Public ownership of land is one solution. However, it would have to be on such a massive scale that it would amount to nationalization which the group saw as being not only something which the present governments would be unprepared

to finance but also something which people would be unwilling to accept from an ideological standpoint.

Just how much land has to be nationalized, and when this has to be done relative to the process of physical urbanization, are questions that need further study. In Budapest, for example, an attempt at a compromise has been made whereby the state acquires only that land required for public purposes. Although the state plays a large part in residential development in order to ensure that sufficient housing units of acceptable quality are available to low-income groups, it does encourage private ownership of housing. Most of the land that is best suited for urban development is privately owned. Speculation is curbed through the imposition of a strongly progressive tax on land owned for purposes other than personal residence. (Individuals are not permitted to rent out residential property for a livelihood.) Another way in which the state effectively kills speculation is by selling serviced, subdivided lots to individuals at subsidized prices on the condition that they be developed, presumably to certain standards set by the state.

The limited availability of financial resources for housing in Lima has been referred to already. State acquisition of land is seen as an important way of reducing the cost of housing. The establishment of a land bank, which had acquired about 20 per cent of total developable land as of 1967, has enabled the government to provide incentive areas of serviced land for occupancy and development by persons with low incomes. These incentive areas permit the realization of self-help programmes referred to earlier.

Obviously, the less land that is acquired publicly, the less control there will be over land prices and the more subsidization will need to be of land made available for housing low-income families. The extent to which this land is available in any one metropolitan area is just as much a reflection of the values of that society as is the extent to which speculation in land values is tolerated. As mentioned above, the prevailing ideology accepted by the people and the decision-makers is the controlling factor in land policy. It is with regard to the ownership of land that it is most significant as it is the cost of land that can form such a high percentage of the purchase price of a housing unit.

Reducing Construction Costs

The other major elements of housing cost, that of construction and that of money, are also affected by political ideologies but to varying degrees. The cost of construction can be heavily influenced by the scale of organization and management of the building industry. In a market economy there is always an ideological tendency to discourage monopolistic practice, whereas in the planned economy every attempt is made to coordinate and reduce duplication in productivity.

The experiences of Toronto and Glasgow, which can probably be taken as representative of the market economy situation, show that there is much room for improvement in present mass-production techniques. First, there are too many factory-based building systems available due to the competitive nature of the industry. Second, the production runs of any one system are not long enough to permit real economies to be realized. This is due to the lack of research on the design of the prototype unit; the uncertainty of the housing market as a result of the irregular flow of mortgage funds (this will be discussed further); and, the variations in requirements set by local authority building codes. There is obviously room for improvement in all these areas, but it was apparent from the seminar discussions that not too much faith can be placed in the ability of the building industry to improve upon the present situation.

From the evidence submitted by the Leningrad and Budapest groups, representative of planned economies, it would appear that considerable savings can be realized through mass-production. In Budapest, given even the modest increases in land prices alluded to earlier and the increased costs of services, the reduction of building costs is seen as one way of achieving considerable savings in the market price of housing. About 50 per cent of new housing construction is carried out by the state and is produced according to mass-production methods. With the introduction of these techniques, it is estimated that the savings in labour, which is one of the problems confronting the local industry, have been about 40 per cent.

In the U.S.S.R. about 65 per cent of the housing construction is carried out by government agencies and the remainder mainly by cooperatives. The components are about 70 per cent mass-produced, and the continuing improvement in this regard has enabled the 1960 housing stock to be increased by 70 per cent as of 1966. The reduction in labour required as a result of the change over from "conventional" to mass-production techniques is comparable with that cited for Budapest. The overall cost reduction per housing unit is about 60 or 70 per cent. These great savings have been achieved by strict control over the number and types of designs being produced. With the improvement in quantitative production, attention now has to be focussed more on improving the qualitative aspects of the design of the units.

For the developing countries the lack of capital required for the establishment of building plants is a major problem. This would be yet another reason for supporting the type of action taken by the Peruvian government to provide serviced land and leave individuals to build their own homes and expand them at such time as they are financially able to do so. At the same time every possible incentive should be given to the construction industry and related trades by way of credit, tax free materials and so on, in order that the industry may more rapidly reach a point at which it can produce sufficient materials at marketable prices and housing units on a mass-production basis. Increased productivity in the building industry cannot be discussed, however, without reference to the development of the national economy in general, and it is appropriate at this point to turn to the economic and financial aspects of housing.

The Cost and Availability of Money

While it is the basic objective of a housing policy to supply more and improved housing, this can be done only within the limitations of the national economy. The influence of the national economy on the cost of housing and, therefore, on its availability to the public will affect all but those few who, under any conceivable circumstances, are able to purchase housing for cash. For the bulk of the public, unable to avoid the strictures of the money market, there are two factors—the cost of money and the availability of money—that strongly influence ability to rent or purchase a housing unit. Both cost and availability of money depend on national ideologies and economy. Although there is not much that can be usefully added to the previous comments on the influence of ideology, it may be useful to mention, as a background to discussing the financial aspects of housing, the interest shown in the relations be-

tween housing investment and the national economy. Sazanami's stress upon the significance of housing investment as related to G.N.P. and to gross national capital formation was thought misleading by the Glasgow group. While these ratios prove to be a satisfactory rule-of-thumb measure of the activity of the housing market relative to its potential in developing countries or in countries with a fast rate of population increase, it is not so in all cases. For example, existing housing stock is fully adequate for a population that is neither growing nor changing in family composition. The Glasgow group listed a number of variables that should be considered in assessing the adequacy of a country's investment in new housing. These variables are:

(a) the existing housing stock including its composition and condition;
(b) the rate of growth of population and of family formation;
(c) the rate of growth of income and its distribution;
(d) the view of the society on socially acceptable housing standards;
(e) the development of housing policy by the state;
(f) the proportion of the state's revenue to the total national income, and an assessment of the possibilities of increasing the role of the public sector;
(g) the proportion of the state's revenue actually devoted to house construction;
(h) the mobility of the population in terms of location;
(i) the costs of construction;
(j) priority in the total investment policy of the community between investments designed to increase the rate of growth and social investment, and priorities within so-

cial investment between housing and other types of investment.

It is only once these and other factors have been analyzed that it is possible to make a judgement about the adequacy of investment in new housing.

Although there was some dispute over how best to determine the size of housing investment, there was absolute concurrence with Sazanami that this investment must be adequate. The relation of housing development, and urban development in general, to the growth of the national economy was seen by all as being a *sine qua non*. However, much depends on the availability and the cost of money. The latter largely reflects the former, but not entirely. For just as interest rates vary from source to source, so may the availability of money be subject to certain conditions that are not reflected in the interest rate.

On the availability of money for the housing sector, Sazanami remarks that in European countries about 45 to 55 per cent of the funds invested in housing are from government sources. From the study groups and the seminar discussions there emerged the general agreement that the governments of non-socialist countries should play a far stronger part in making more money available for the housing market. Housing, as a basic necessity, should not be subject to the changing fortunes of private enterprise. In Toronto, perhaps typical of North America in this respect, only 17 per cent of investment in new housing over the last decade was a direct investment of public funds, and 18 per cent invested indirectly through the insured mortgage programme. Thus, 83 per cent of the funds have been privately financed, and although the insured mortgage programme has encouraged an additional flow of private investment, the problem has been the high cost of this money. Another problem, experienced in Toronto and Glasgow, is that of obtaining mortgages on older properties. Both groups felt that far more could be done by central governments through making loans available for rehabilitation of older houses to standards not necessarily comparable to new housing.

Whatever may be the various means by which government, either by itself or in conjunction with private enterprise, can contribute to an improvement in housing finance, the one problem that must be resolved is the high cost of money in so many countries. It is of no use to make money available through government sources if the interest rates are so high that the loans cannot be amortized over a reasonable period of time (generally about thirty years). This applies to short-term credit made available to the building industry and related industries (which will affect the end price of housing) just as much as it does to mortgage money.

CLOSING THE PURCHASING POWER—
HOUSING COST GAP

As Sazanami points out, the big challenge is to reduce the gap between the average income of a household and the rental cost or purchase price of housing. This brings us back to the influence of national economic policy on housing policy. The most satisfactory solution proposed by Sazanami —one that is attainable only in the long run and, then, only with an expanding economy—is to increase household purchasing power through increasing productivity in sectors other than housing. However, in the discussions on housing in the seminar more interest was shown in the immediate steps that might be taken to reduce the gap between cost of housing and family income. These short-term measures, which normally turn out to be permanent, generally call for some form of government subsidy. Sazanami describes various means of subsidization; Albert Rose also suggests that more attention should be given to income maintenance as an administratively simpler and more comprehensive form of subsidization. Apart from rent control, most of these suggestions, such as public housing, tax relief, rent supplement and low-interest loans, were well-received by the groups, although the problem of political acceptability of such measures was questioned by the Toronto group and that of economic feasibility by the Lima group.

As mentioned, rent control appears to be a controversial issue. For the Lima group, rent control was seen to result in a loss of interest on the part of the landlord in maintaining his building, and also as a deterrent to private investment in low-rental dwelling construction. While Toronto suffers from very high rents relative to North America generally, nothing has been done about introducing rent control, although it is being considered. In Glasgow, rents average about 4.6 per cent of a family income, mainly because of the public ownership of so large a percentage of housing. However, public ownership does not necessarily mean low rents. The Belgrade group indicated that rents represented about 10 per cent of a family income in 1960 but are now between 15 and 20 per cent. This tendency toward economic rent obviously is acceptable, providing other measures are available for subsidizing low-income families. In Budapest rents for low-income families are about 5 to 6 per cent of the family income. This compares with the situation in Leningrad. These rents are further subsidized as they do not include the cost of maintenance and servicing.

Unfortunately, the manner of control—perhaps the most important aspect of rent control—was not discussed fully in the Programme. While rents related to income may be satisfactory for the most part, with an increase in family income, they result in a corresponding increase in rent. This erosion of incentive can be eliminated in large part if a wider margin of increase is permitted before ensuing rent increases. A system in which rents increase only by a certain percentage each time a tenancy changes gives rise to the situation cited by the Lima group.

Public housing appears to be the most acceptable form of governmental intervention in the housing market. The Toronto group decried the current low level of government participation; only 2 per cent of the present housing stock in the Toronto area has been built or administered under public auspices of various kinds. (This is representative of the Canadian national scene.) The group proposed that this 2 per cent be increased to 10 or 15 per cent if the needs of the poor are to be met. Only in this way would those families, presently with an income too high to qualify for public housing but insufficient to secure adequate private housing on an ownership or rental basis, be satisfactorily housed. (In 1967, this "grey area" extended from $6,000 to $9,000 family income per annum.)

The Glasgow group appeared to be satisfied with the extent of government participation (60 per cent of housing constructed between 1945 and 1966) and was more concerned with keeping costs down and improving the planning and administration of housing.

Just as there was interest in encouraging more public sector participation in Toronto, there was similar interest in the private sector, whether individuals or cooperatives, in such places as Belgrade, Budapest and Lima, where there is either too great an acceptance of governmental responsibility or lack of public funds. It was recognized that government has to accept responsibility for assisting with the financing of housing for low-income groups, but, at the same time, policies for these groups have to be linked with policies for middle-income groups. Individuals should be strongly encouraged to invest in their own housing. Apart from all the social benefits this may produce, it frees public funds for other investments.

To encourage private-development ownership, the Lima group was prepared to urge the Peruvian government to exercise its expropriation rights to assemble land for private development where necessary and make available low-interest loans to assist the private purchase of housing. However, inasmuch as low-interest loans were seen as a burden on the economy, these were recommended only for the lower-income groups interested in purchasing houses.

In Yugoslavia cooperatives are associated with the enterprises subject to workers self-government. Thus, the funds of a firm are used to purchase housing for the workers of that firm. In Belgrade, workers cooperatives are assuming a growing role in the ownership and arrangement of housing, although the construction is generally carried out by building firms. Cooperatives were strongly favoured by representatives from metropolitan areas in developing countries. They offer strong credit ratings and, at the same time, provide a useful link between industry and workers. They have proved particularly useful in this respect in Calcutta, where firms often play a significant role in the welfare of their workers. However, their value in terms of quantitative production of housing was not upheld in the United Kingdom or in North America. In Belgrade, the ownership of private family dwellings is being encour-

aged, but the priorities are given to multi-unit apartment developments because of the rapid rate of urbanization in the city.

In Budapest, where, as has already been mentioned, much of the best building land is privately owned, private development is encouraged for personal needs. (Development for renting is not permitted.) The state and cooperatives build most of the housing but sell them under various conditions, mainly according to the income of the purchaser and the size of his initial down-payment. Those able to offer the larger down-payments receive priority in purchasing a unit in an apartment building. Even for higher-income families, state loans at 3 per cent interest amortized over a thirty-year period are available. These units are generally sold at 20 per cent less than true cost. Cooperatives are able to obtain loans of 50 per cent of the cost of building houses which can be amortized over twenty-five years at a 2 per cent interest rate. Loans for apartments are even more favourable. An individual wishing to build his own house is able to live free of property taxes for fifteen years. Middle-income families, if selected by the municipality, may be able to purchase a home, which has been state subsidized to the extent of 25 per cent of its cost, with a 15 per cent down-payment and amortization of the remaining debt, interest free, over a period of thirty years. These measures are probably the most advantageous that exist anywhere for facilitating private home ownership and, while not available to all, do qualify Sazanami's comments on the existence of such funding arrangements.

THE PLANNING AND ADMINISTRATION OF HOUSING

One other area in which substantial improvements could be realized remains to be discussed: the planning and administration of housing in metropolitan areas. There is little point in reiterating the many complaints about the plethora of housing agencies, both private and public, in many metropolitan areas—especially those in the more developed countries. While some of the arguments advanced above for encouraging private investment in housing justify the existence of such organizations, the real problems arise in coordinating housing development with the urban development of the metropolitan region. Closely associated with this overall problem are the problems of varying building codes and standards, the inefficiencies of "mass production on a small scale", and the poor financial position of many small housing agencies.

Housing should be the responsibility of those organs of government or other institutions that are financially able to provide the services expected of them. This, and the scale of the housing problem in most metropolitan areas points to the need for a greater degree of centralization of housing planning and administration.

The Glasgow group arrived at the conclusion that the physical dispersal of urban growth in new towns, which results from attempts to lower central city densities, calls for a form of regional government that would include (in one region) the metropolis and the new towns. This allows not only for the development of large or small satellite communities without burdening the tax resources of the "reception area", but also permits a better regulation and diversion of the flow of rural migrants to the city through the provision of improved community facilities in the surrounding hinterland. While there was general sympathy for the outcome of the Glasgow group's findings, there was also a strong interest shown in retaining the involvement of local government or its

agencies to ensure adequate public participation in the planning of housing. The Athens group, in keeping with their proposal for the polycentric development of the metropolis and complementary decentralization of administration, essentially supported the proposals put forward by Gorynski and Rybicki in Chapter 9. There was further support for this general approach from the Toronto group. However, we shall examine this subject further in parts III and IV.

Education

VALUE PRIORITIES AND THE ACHIEVEMENT OF EDUCATIONAL GOALS

The fulfilment of the social, psychological and cultural functions of what Philp refers to in Chapter 5 of this book as the "socialization process" is the major goal of the formalized education system. As Philp explains, the education system is not the only contributor to the socialization of man— much depends, for example, on the social and physical environment in which the process takes place—but it probably is the most important single contributor. While this much is applicable universally, it is quite apparent that there is obvious variation from one metropolitan area to another (or, more generally, one country to another) in value priorities, and that this is reflected in the functions given greatest attention in the education system. In Greece, for example, the lack of attention given to the fulfilment of the economic aspect of the social function (which, it will be recalled from Philp's essay, also provides for the economic role of the individual) may well lead to an imbalance in manpower resources in the near future. On the other hand, in the view of the Detroit group, an undue emphasis on the individual as a source of manpower, dominates the philosophy of some local education authorities in the Detroit area. It is this same emphasis that the Detroit group saw as being both a product of, and contributor to, the distorted values of urban America.

The problems facing Detroit are repeated in other major metropolitan areas in North America and some of them in other parts of the world. But the problems of budgetary inequities, high in-migration, high intra-city mobility, poverty amid affluence, and racial discrimination, are not simply products of urban growth, which can occur without them. The most fundamental cause of all these problems, in the view of the Detroit group, is the value structure upheld by urban American society. The values of vocational success, competitive achievement, patriotism and even, to some extent, good citizenship, do not equip urban man to deal with the problems he and his fellows have created: of law and order, race relations, poverty and city-suburban tensions. There has to be a readjustment of this value structure if any really long-lasting change is to be brought about in urban communities not just in North America but in most parts of the world.

The values of social fairness, economic justice and international peace were seen by the Detroit group as being far more laudable. Indeed, it is of interest to note how closely this relates to the thinking of the Prague study group which also saw international understanding as the most significant value that might be derived from the education process. What is taught in schools, to be discussed later, bears closely upon this.

It is hard to underestimate the importance of the above in understanding the nature of metropolitan problems and searching for possible solutions to them. Obviously, there is basic agreement on why

we teach: more important perhaps is what, how, where, when and even whom, we teach. Thus, one can move from an examination of the intent or goal of formal education, to that of the qualities to be sought in developing an education system, the influence of resources upon the satisfactory development of the system, its planning and administration, and to the demands made on the comprehensive planning machinery and process desired for the metropolitan region.

THE BASIC QUALITIES OF AN EDUCATION SYSTEM

In discussing the qualities to be sought in a metropolitan education system, one is immediately confronted with the problem of relating the metropolitan system to that of the country as a whole. Despite the special problems pertaining to metropolitan education and its need for special attention, the system cannot be divorced from the rest of the country. Indeed, the two levels must be closely integrated if the characteristics of equality, choice, continuity, flexibility and efficiency are to apply.

Although none of these characteristics can be discussed without reference to others, it is helpful to identify the implications of each one for education planning and administration.

By equality is meant equality of opportunity for the individual to optimize his potential. This means matching the ability of the individual to existing and future opportunity offered by society. In the metropolitan context this implies equipping individuals with the desires and skills necessary to enjoy life in great cities. It does not mean that all will receive the same type of education. Yet just as this implies choice, it also implies some overall control of the choices offered and the number of persons to whom they are offered. As many of the educational institutions in the metropolis serve the entire country, or at least an extensive hinterland, the control of numbers must lie with a central or regional form of government. If this is not so, the situation will arise in which individuals, through no fault of their own, acquire certain skills that assist them but little in taking advantage of the opportunities offered in later life.

An equally important consideration is that the process of matching skills with potential opportunity should not be seen simply as a process of producing professionally qualified personnel. As the Prague group commented so perceptively, there is a grave danger of producing too many qualified people and not enough educated people.

These are the dangers that Philp attemps to highlight when he warns that opportunity must be found in economic and social terms as well as in educational and psychological ones. This much was generally agreed upon. What could not be resolved, however, is the problem of who should be offered the choice. For although Philp is far from arguing for the education of an elite based upon wealth and class, he is arguing nevertheless for a realistic limitation of formal education if the metropolis, and the country, cannot absorb vast numbers of "highly educated" people. The Prague group felt that the principle of this limitation is undesirable. However, the alternative they propose stands up only to the extent that the economy can support it. Theoretically, the economy develops as the education of a generation becomes more sophisticated. Indeed, it develops as a product of this. The problem, of course, lies in relating the education of a generation to the rate of advance of the economy. The Prague group emphasized the importance of general education, especially to the age of fifteen, but also as a general principle for all education programmes. Without this basic general education, it is argued, people lack the ability to understand and communicate or the ability to educate themselves informally later in life. The advances in science and technology will not be utilized to the full if the lay public does not appreciate the advantages that may be forthcoming. Again there will be little or no understanding of the moral aspects of life. Indeed, the Detroit group also stressed the need for a greater emphasis on ethical values in the curriculum.

While it is vital to make available to all good general education, a continuing education programme going beyond that of formal schooling is envisaged as being a recognized part of the system in many countries. Adult education programmes, "in-service" training programmes, retraining programmes and active participation in cultural programmes all ensure to add continually to an individual's knowledge and his ability to better understand his environment and to take advantage of the opportunities it offers. The pace at which this programme is run is obviously limited by the resources available. The question is whether the intellectual appetite of the participants will ever outpace the resources available!

The concept of education as being part of the continuing process of the socialization of individual members of society implies continuous upgrading and change of operating conditions. The process is a continuing experiment to which adjustments have to be made due to the lack of knowledge of what is needed at the outset. This change is basically the reason why flexibility is so important a quality in an education system. Being most easily introduced at the local level, flexibility calls for more local control of education, although there is a similar need for a willingness and ability to change in other parts of the system. Certainly the greatest demand for flexibility is in metropolitan areas where minority groups tend to be concentrated.

Most metropolitan school administrations are still learning how minority groups can best be helped. The Detroit group, in discussing the problems of educating the poor southern in-migrant, decided that these people need to be taught in their own terms if they are to be taught anything at all. With families constantly on the move, either within the city core, or back and forth from the city to part-time jobs in rural areas, mainly in the South, the problem is more than one of a child's education; it becomes one of the socialization of a family. This situation is akin to that confronting many large metropolitan areas in developing countries where the family, or certain of its members, rarely remains long in one place. Flexibility to accommodate these needs, rather than avoiding them, is a basic requirement.

Also, with the metropolis being the one place where special schools are most easily located, its education system has to be able to accommodate their special requirements. Schools preparing people with special skills, whether in the cultural or scientific fields, will generally find far better back-up facilities available in the metropolis. The use of these facilities, such as museums and libraries, has to be integrated with the formal education system.

In discussing the quality of efficiency in relation to most services great interest is shown in cost-benefit studies and other means of measuring productivity. With education the situation is somewhat different, mainly because of the great difficulty in measuring the quality of a product of one system as opposed to that of another. It is almost impossible to have sufficient control over the "productivity pipeline"

which provides a child with his formal education (let alone the informal elements) to be able to make fair comparisons between procedures. Also, every child really is different when he or she enters the system and the socialization process takes an extremely long time to complete!

This should not deter the spirit of experimentation although the interest in keeping costs down should be secondary to that of improving the standards of education within the limits of resources available. For example, Philp's suggestions for reducing the amount of land required for schools in the central parts of metropolitan areas can be challenged on several counts. The findings of research carried out in Prague on the effect of different types of school buildings on the school life of a child indicates that neither buildings of four storeys or one storey are very satisfactory. Thus it might be assumed that multi-storey schools, similar to office buildings, would be even less satisfactory. The present tendency in Prague is to favour two and three-storied schools.

Similarly, the idea of having playing fields far from the school buildings can be questioned. Not only is lengthy travel undesirable for children but the benefits of having space around the buildings is salutory for physical and mental health. Any city has to have open spaces and they should be designed for use. A preferable solution is the more intensive use of school grounds for other community activities. It is this type of decision that is so likely to reflect the values of society.

The administrative inefficiencies of a fractionated education system, such as that which exists in metropolitan Detroit, appear obvious in the extreme cases. For example, the metropolitan Detroit area has 97 school boards each of which has taxing power and 82 of which provide their own schools and have extensive autonomy in hiring and discharging teachers, determining curricula and planning community education programmes. Although this is an extreme case, the task of making administrative savings in an education system is just as difficult as it is in any other service.

THE USE OF RESOURCES

Money

As money is never plentiful enough, our main concern is to make the best use of what is available. The interrelated nature of the metropolitan and national education systems requires centralized decision-making in allocating financial resources to education. This does not necessarily mean that resources have to be raised by the central government and then redistributed, although this might be the simplest procedure. It simply means that a national plan is necessary in order that basic education needs are known, that priorities be established and that these requirements then be integrated with plans for development in other sectors.

Also of concern, however, especially if all funds are not derived from the central government, is that financial resources of education systems be approximately equitable. Although metropolitan Detroit's fractionated system has already been criticized as typical of those needing reform, it does serve to illustrate the problems that arise from heavy reliance upon the property tax. For example, in the Detroit area there are two neighbouring school boards where assessed property value per pupil, in 1966, was $4,810 in one case and $53,156 in the other. Even after senior level grants had been made to the two boards in question, the poorer one was still taxing its citizens twice as heavily and yet having only two thirds of the resources per capita to spend

on its schooling as compared with the wealthier board.

While these inequities could be ironed out within the metropolis through the introduction of one metropolitan-wide school authority, this would not solve the inequities that exist across the country. Since such great stress has been placed upon uniform standards for all educational institutions offering similar education programmes, it is natural that there be central government support to ensure equal availability of funds. Yet, in the United States, if one examines the purpose of the present federal government funds, one finds that they are available only when sought by a local board for specific activities recognized by the federal government as meriting encouragement. These programmes include the development of vocational education programmes, the improvement of science courses, the expansion of library facilities and special instruction for children from low-income families. Thus the equalizing effect is dampened not only by the limited range of programmes covered, but also by the nature of the programmes. For this reason it is not surprising that although senior level grants to school boards in the Detroit area have increased considerably in past years, there is a feeling among school administrators, in that metropolis, that senior level grants should also be made available for capital and current expenditures for regular programmes.

However, the recent introduction of a Michigan State income tax, with much of the revenue going to schools to provide state-wide relief from the property tax, can only be viewed as recognition of the income tax as a more equitable revenue source. Of course, one of the main reasons why these reforms have been slow in appearing is that the local boards have been reluctant to relinquish their control over the education system, and the senior governments have not wished to be accused of intruding upon the bailiwick of the local boards. Naturally, if the senior governments are to provide more of the funds they will want to have some say in how those funds are dispensed.

It is of interest to note that the central financing arrangements applying to other metropolitan areas did not come in for criticism. The only complaint, expressed by the Athens group, was that there might be more local freedom in administering the funds. In the case of Detroit, however, senior government financing of education would bring about not only fiscal equity, but, hopefully, a radical reform of the local administrative system. This matter of central-local relations will be picked up again in Part III.

Teachers

Limited human resources, whether of teachers or administrative personnel, can be attributed for the most part to the availability of financial resources, but not entirely. For, as the Athens group observed, it is the lack of manpower planning in the past that accounts in part for the acute shortage of teachers in Greece today. Teacher-student ratios in public schools in 1961 were 1:42 at the primary level and 1:67 at the secondary level. In Athens itself, the situation is even more aggravated as a result of the heavy in-migration of families from the rural areas. Although private schools take up some of the slack in Athens, especially at the secondary level, where they account for nearly 20 per cent of students in general education courses and 85 per cent of those in technical education courses, it is doubtful whether the shortage of teachers in public schools will be solved for a long time to come.

Other comments on personnel relate to the quality of teachers. One problem facing many large metropolitan areas is the lack of teachers able or willing to teach in schools where the children do not seem to be particularly interested in learning. These schools are generally in the poorer, more dilapidated parts of the city. There may be a high turnover and frequent absenteeism of children and staff and, consequently, little rapport develops. The children may come from broken homes and lack parental control or interest in their education. The lack of teachers qualified to take on the education of these children is a major problem in Detroit, just as it is in many other cities. The special skills required should be compensated for by additional salary. Preferably the teachers should be male, to make up for the lack of parental authority in the home, but they should also have a proper appreciation of the problems confronting the children with which they will be working. The application of a pricing policy was seen as quite justifiable—it should be recognized that the job of teaching in these schools is far more demanding than teaching in a quiet surburban area. Obviously, this policy has implications for the control of the hiring of teachers and setting of salaries throughout the metropolitan area.

TECHNOLOGY AND URBAN FACILITIES

It is appropriate that technological resources and urban facilities be discussed together, for they both relate directly to the process of acquiring knowledge in the most effective and efficient manner. Television, radio, film and correspondence have all been developed to some extent and there is likely to be continuing progress made in their use. Providing that the student is able to have access to teachers for consultation and discussion, these techniques have obvious benefits for transmitting information. These advances are likely to be just as advantageous to the developing countries as they are to the more developed countries. High standards of lecturing become available to the entire area served by the network and efforts can then be concentrated on improving local tutorial techniques. Whether costs incurred in developing a metropolitan or nation-wide system are likely to be matched by savings in personnel and physical plant is difficult to determine. There seems to be insufficient research on this matter to date.

However, these developments, which will facilitate the spread of information, will never affect the city as a centre for learning. The *raison d'être* of the city is access: access to places, to goods and services, to persons and, most important, to information. Personal access to all the sources of information in the many institutions in the city and especially access to the knowledge and skills of other urban inhabitants will never be replaced. This generating, transmitting and receiving of information has always been a key function of the city and its institutions. In the future, proper cognizance of this fact will serve to encourage greater organization of the education system to make communication more effective and to broaden participation in the learning process. The spatial organization of the metropolis at that stage should, as Philp says, "be planned around the complex of schools and other educational institutions." It is appropriate at this juncture to turn to the various aspects of planning the education system in the metropolis and its integration with the community as a whole.

EDUCATION PLANNING

Philp's plea for the integration of education planning with social and economic planning is unquestionable. It is the manner in which this integration takes place and the implications of different policies that give rise to contention. It is on these issues that we shall focus our attention.

Perhaps the most basic question relates to the establishment of priorities within the education system. It is quite apparent that the limited resources of a developing country impose severe strictures on the scope of its education system. It is not as evident that a more developed country with relatively plentiful resources may end up with an imbalanced education system because of a poor sense of priorities. (Indeed, to the extent that the system does not exploit fully the potential of the resources available, the system might be regarded as a poor one.) The poor sense of priorities is a product of the value structure of a society, which is reinforced, in turn, by the contribution of the education system to the socialization process. Without the international exchange of ideas and exposure to other ways of life, this could be a vicious circle. Fortunately, there is much room for change. A major requirement of those responsible for setting priorities is the quality of introspection and preparedness to adopt new approaches if the education system appears to overemphasize any one of the social, psychological or cultural needs of the individual and of society.

If one accepts this argument, it is difficult to justify treating the education system of a metropolis separately from that of the country as a whole. The issue then becomes the extent to which the country as a whole, or its education administrators, should be sensitive to the nature of the problems of the metropolis, which are in fact national problems!

Although there was general agreement that the needs of the metropolis should be considered within the context of a national system of priorities and that economic development, as a precursor to social development, is one of these priorities, such consideration does not necessarily relate to the question of whether priorities should be adjusted. For example, Ibadan faces a genuine dilemma in having to provide uniform education and, at the same time, provide present and future trained technicians with higher skills than are normally provided in the public education system. Is it better for a developing nation to limit educational opportunities at the primary level in order to provide funds for secondary education for a selected few? Similarly, at what stage in development can the emphasis on vocational training be decreased and more attention be given to education of a more general nature? Unfortunately, there seems to be no general solution to this problem. To say that the production of manpower must be closely linked with the long-term development of the economy, that no more skilled personnel should be produced than are essential and that all remaining effort be put into providing a general education for society as a whole rather than selected members, only partly answers the question. For success, the answer assumes sophisticated long-term planning and still relies on a value judgement as to when the emphasis on manpower planning should be eliminated.

Another aspect of education planning relating to the social function of the socialization process is the education of minority groups. There seems to be no obvious agreement about whether conscious efforts should be made to assimilate these groups into the community, or whether they should be allowed to maintain their own identity. The Detroit group was very much in favour of total assimila-

tion on the grounds that two or more different groups are unlikely to exist side by side without conflict. It is doubtful whether the solution proposed by the Detroit group —that of mixing the children of the groups in question by busing them to schools in different parts of the metropolis—is really a satisfactory long-term answer to the problem as it exists in Detroit, or as it appears elsewhere. Certainly busing will help to break down race prejudice; but it also serves to make the child more aware of the racial tension around him. In the short run it may be a way to start on the problem, but in the long term a conscious effort should be made to prevent the formation of communities of poor persons who are readily identified as a group by the colour of their skin.

The value of the neighbourhood school, especially for young children, is not one to be dismissed lightly. The ability to identify with a community is probably more important for a child than it is for an adult. The effects of lengthy commuter travel on an adult's ability to identify with a community and the significance of this for participatory democracy are stressed by Gorynski and Rybicki in Chapter 9. In the planning and functioning of the metropolis, children (as well as adults) must be considered. While this does not entirely answer the problem of how to break down prejudices, it goes part-way toward a solution that relies on "what is taught" as opposed to "the type of person with whom one is taught".

What is taught has been alluded to already in stressing the values of a general education that should provide an ethical base upon which society can operate and from which will emerge the values of economic justice, social fairness and international understanding. Only in this way can the cultural functions of socialization be provided for satisfactorily. Education was also seen to be the chief means by which man attains self-fulfilment. The Athens group anticipated that this psychological function will attain far greater importance in the future and education will consequently be placed at the centre of man's interests. There will be a greater awareness of the value of exploration on the frontiers of knowledge for man's understanding and spiritual well-being. At that point, development will be synonymous with education, and developmental planners will be educational planners.

Already many principles are beginning to emerge that have implications for the physical form and location of educational institutions and the development of the metropolis. We can now turn our attention to these matters. The physical expression of the education system in the metropolis is an outcome of the interaction of the following elements: the student population of schools, the population density and physical extent of their catchment areas, the degree of specialization of each school or type of school by age group and programmes offered, and the provision and use of access routes provided to and from each school. Adjustments made to any one of these elements across a metropolitan area will affect each of the other elements. To this framework one can add other necessary infrastructure such as dormitories, kitchens and so on, although any adjustments to these facilities will not have the same ramifications throughout the entire system.

Beyond being aware of the widespread effects of any adjustment to parts of the education system, our present concern is not with the interaction of these elements *per se*, but with giving expression to the principles that have emerged from the work of the study groups and the seminar discussions.

Excluding the private school network, there are essentially two patterns of schools in a metropolis. First, there is the hierarchy of the most common types of school serving each age group over a given part of the metropolitan area. Second, there are a number of schools specializing either in the subject matter taught, such as art, music or mathematics or in the manner in which it is taught, such as in schools for the deaf.

Little was said on the principles affecting the form and location of the second type of school although the metropolis was seen as the most obvious location for these schools, which may, in some instances, serve the entire country. The location of supportive facilities to be found in the metropolis, and generally in the centre of the metropolis, may have some influence over the location of these schools, although they too will require recreational areas and be subject to the same influencing factors as the first group of schools in this regard.

It is the first group of schools that provides the spine of any education system at the present time. As a group, the schools are generally broken down by age group served and, in later years, by specialization of interest. Although Philp implies that the technological innovations in the education field might result in the merging of institutions serving all age groups to produce the "total" school, it seems that there will be strong opposition to radical changes to some of the elements of the existing system.

The limitation of the physical size of the kindergarten level is regarded as a very important contributing factor in the educational environment of the child. The same applies, although to a lesser extent, to the size of schools for older children. For example, kindergartens in Prague have no more than 120 pupils each; "basic schools", serving children from six to fifteen years of age, range from 270 to 810 pupils each; and "middle schools" (fifteen to eighteen year olds), like their counterparts in Moscow, take 800 to 900 students each. In both Moscow and Prague the tendency is to build a second or third neighbourhood school rather than expand the size of a large school if the neighbourhood expands.

Again, both the Moscow and Prague groups stressed the desirability of having the younger children walk to school, which in turn has implications for the size of the school relative to the population density of the neighbourhood. A case was also made by the Detroit group for expanding the "walk in" catchment area of a school by exploiting the use of the automobile and school bus. This would enable the school to expand and facilitate the mixing of students from different residential areas. However, whether or not continuous expansion of the school is really necessary does not seem to have been fully answered. Judging by the attitudes taken by the Moscow and Prague groups, this exploitation of transportation facilities is unnecessary. Further support for this approach is found in the stance taken by Gorynski and Rybicki, alluded to above. Perhaps attitudes about the use and purpose of highways vary according to the values of society: either they are there so we should use them; or we should use them only at such times as are necessary!

One last point on the spatial planning of service areas relates to the trend observed by the Prague group toward a wider variety of special schools whether for special needs or interests. This has significant implications for access, the sharing of recreational and other facilities and the location of the schools near institutions that may be of particular interest to the students in question.

IMPLICATIONS FOR THE ADMINISTRATIVE AND PLANNING MACHINERY

Perhaps the most noticeable feature of these opinions on education systems is the degree to which centralized control is accepted not only as inevitable but, in many countries, as desirable. The general impression may have been somewhat different if there had been some comment pertaining to an area such as Toronto, which represents Philp's category of the medium-aged metropolitan area, in a developed country with local control of education. Although this type of metropolis would probably exemplify the virtues of local control, there would necessarily be, as Philp observes, constant pressure from senior government for more control. In both Toronto and in London (Philp's example of the large metropolitan area in a developed country, with local control), the case for local control is strengthened only because of the relatively well-organized division of responsibility between two tiers of multi-purpose governmental units, which cooperate reasonably well with one another. As these conditions often do not exist, the need for coordination of education planning and especially control over spending, throws the responsibility back on the shoulders of a senior government.

Philp also remarks that almost no metropolis exists (with the somewhat misleading exceptions of Hong Kong and Singapore) that is able to support its own education system without financial assistance from senior governments. This being so, senior governments are all called upon to assist local schools financially. Once a senior government offers assistance, it is extremely unlikely that its share of the total education expenditure will decrease. More often, it tends to increase. Thus it is not surprising that a central or provincial (state) government likes to hold on to the controls it already has for coordinating education expenditures in the metropolis with those throughout the country or province (state). Similarly, the general agreement that manpower planning calls for a greater integration of education planning with economic and social planning, again supports the case for centralizing the integration of plans of national consequence (educational or otherwise).

It was felt that some local control was necessary over the types of programmes offered to meet the special needs of the metropolitan area and the planning and provision of schools within the context of the local plans. However, while these interests would likely be most efficiently provided by a metropolitan-wide authority there are also strong demands for a smaller authority to ensure that adequate attention is given to children as individuals and to encourage parental and teacher involvement in each child's education.

In some instances, as in Prague, these three levels of authority (central, metropolitan and local district) are operating. In others, one of the tiers is missing—generally the metropolitan-wide authority. In the North American setting the existence of boards of education, separate from the rest of local government, makes coordination of education planning with all the other aspects of planning extremely difficult. Judging by the lack of defence for the school board and by the overwhelming desire for integration of education with other services, the provision of education by multi-purpose authorities appears to be far more acceptable a solution.

If any system appears to be favoured it is that of central financial and manpower planning, which involves the basic structuring of curricula, the setting of standards, the setting of major examinations and the training of teachers; metropolitan-wide responsibility for the location, building and staffing of senior schools and most special schools; and local responsibility for the location, building and staffing of junior schools. Teacher salaries should be equal for equal service, but arrangements should exist to attract good teachers to schools where they are needed most. With the central government being responsible for major examinations, effective control of the basic content of curricula would be provided for. Within this framework, metropolitan and local authorities would have some flexibility over the provision of texts used and special programmes offered. In very large countries, regionalization of central services is desirable. This can be achieved either through provision of most of what would otherwise be central services by decentralized administrations at the provincial level or by having a deconcentrated administrative system run by the central government.

This is not intended to be the basis of a universal formula, but merely a reflection of the general thinking on the subject of the division of responsibilities. It does not necessarily solve all the problems involved. For example, just how to reconcile common standards imposed by national financial grants and controls with the pluralism inherent in the metropolitan resolution of the unique education problems of the metropolis, could not be resolved.

Health and Welfare

The joint consideration of the health and welfare of the inhabitants of metropolitan areas can be criticized as well as supported. For purposes of description, it may be easier to treat the two separately. Although this has been done by Bakács and Rose in chapters 6 and 7 respectively, both authors are aware of the closely related nature of their subjects, not only one to another, but also to the other major problem areas treated in Part II of this volume. As it was a major objective of the

Centennial Programme to highlight the interrelated nature of these problems and the need for a comprehensive approach to their solution, the exercise of considering health and welfare problems and services together may help to further that objective.

Indeed, this approach may serve to highlight the dangers inherent in the over-specialization of interest being built into any policy. This is probably a danger that confronts the developed countries more than it does the developing countries. The greater availability of resources permits a greater specialization in research and case work and, often based upon the findings of the research, the administration of services. The lack of attention that has been given to the significance of the family as a unit in the provision of social services serves to illustrate how this "specialization" can be self-defeating. Some specialization in research is obviously necessary if our overall knowledge in a given field is to be advanced. "Empire building", based upon the findings of the research, is to be decried, however. What is needed most is more interdisciplinary research, which will encourage, hopefully, the integration of "empires" for purposes of following up on the research findings.

Perhaps the most important issues raised by Bakács and Rose are the need for more and better research and the need for improved coordination in the planning and administration of health and welfare services with other services. While support for more and better research is like promoting motherhood, the point being made here is that, in comparison with other service areas considered in this book, there is a lack of very basic information, which is required before action can be taken. These issues will be considered here along with the sought-after service characteristics of quality, availability and accessibility, which both affect, and are affected by the planning and administrative structure and process.

Bakacs' contention that man has failed to adapt himself physiologically and psychologically to his artificial urban environment received a mixed reaction. There was a general feeling that man can and does evolve along with the urbanization of his environment. Increased life expectancy is a part of this evolution. However, although life expectancy is longer in the more urbanized countries, this does not mean that it is a product of urbanization *per se*. It is a product of a more scientific way of life and greater control over environmental elements that militate against survival. These scientific developments can take place only with urbanization, but the urbanization need not manifest itself in the way that it has to date. Man is confronted with many difficulties in adapting to his urban environment, both in terms of the increasing pollution of his sources of energy and in terms of the mental stress of modern urban living. As these difficulties are basically the result of the misuse of technological developments and are of man's making, man should be able to improve his environment by making adjustments to his way of life and especially to his use of technology. This process of adjustment will require not only a physical restructuring of the urban environment, which will be discussed later, but also much more research to guide these developments.

The New York group, while sympathetic to many of the statements made by Bakács, believed that more pointed studies are required before one can judge the health implications of air and water pollutants and noise. While each of these may be undesirable, one must be wary lest this feeling of undesirability be purely aesthetic in origin. Health surveys in Los Angeles had not been able, as of 1967, to demonstrate any pathological effects of smog.

One of the basic problems in carrying out research of this nature is that of establishing effective control systems. There are so many factors influencing the health of the human body that a study of the effects of, say, air pollutants has to be conducted with extreme caution. Hence, the limited value of the evidence of the deaths resulting from oft-cited smog such as that of London in 1952. This is said, however, only to emphasize the need for improved research on the effects of the pollutants and environmental stimuli on physiological and mental health.

The New York group also suggested that as urban populations, especially in more developed countries, are becoming more truly urban in origin, and relying less on the in-migrations of rural persons, it may be of interest to explore the long-range genetic, demographic and behavioural consequences, if any, of this entirely new biological situation. Of even greater importance would be an effort to determine the limits of man's adaptability to the human environment. The limits of transformations of life in a dehumanized, urbanized environment should be defined not in terms of technological and economic possibilities, but in terms of the unchangeable

physiological and mental needs and attributes of the human species.[6] With answers to this last enquiry, decision-makers would be in a much better position to establish standards pertaining to environmental health and use available technologies to attain these standards.

One of the major problems associated with the determination of public need for welfare services is that of the location of the service outlets. Continuing research is required to assist in the location of service facilities to meet needs best in a changing metropolitan environment.

Factors influencing the extent to which these needs are met are the standards that are established, the resources available, and the organization of the planning and administration of the services. These factors, which apply equally to health services, will now be discussed.

THE SOUGHT-AFTER QUALITIES

In establishing service standards, one of the greatest problems is striking a balance between the need for quality and the need for access. For such services as medical operations provided by hospitals, where the service is of value only if it is of an extremely high standard, the constraints of limited resources (that is, doctors, equipment and so on) tend to dictate the geographic limitations of service provision, whether in a metropolitan area or throughout a country. To spread this type of service too thinly may all but eliminate its effectiveness. For other types of service, such as family counselling, greater use can be made of unqualified volunteers to provide greater access. As trained personnel become available they can replace the volunteer help. Madrid's *Obra de Auxilio Social* (Social Aid Establishment) is an example of a semi-public agency that grew

---

[6]Dr. René Dubos, a member of the New York study group has recently published an excellent discourse on the significance of the environment for the development of human potential. *See* René Dubos, *So Human an Animal* (New York, Charles Scribner's Sons, 1968). Also of interest to the reader is René Dubos, *Man Adapting* (New Haven, Yale University Press, 1965).

out of organized, voluntary, public effort and which is now run professionally. It provides children's eating places, kindergartens, student hostels, children's homes, maternity centres and so on, throughout Madrid. Thus, although Rose may be correct in saying that professionals are far more effective than volunteers, many services can be initiated and made accessible through volunteer effort.

Another criterion that has to be observed in setting health and welfare standards is that of equality. Perhaps more than for any other service (including education), health and welfare standards should be uniform throughout a country. It is for this reason that the distinction has been made between availability and accessibility. For within the limitations imposed by physical access, as described, services should be available to all without discrimination. Furthermore, they should be equal in quality.

Indeed, to some extent there is need for international standards, especially in the health field but also, with the growing migratory movements of labour forces, in the welfare field. Inequalities in the provision of services result in either more hardship or migration to those centres where services are available. On a national scale, the movement of rural peoples to the metropolitan areas causes a multitude of pressures on other services. The Athens group cited the superiority of social services in Athens as a major cause of the migration to that city. For example, many rural workers moved to Athens when a pension scheme for Athenian workers of fifty years of age or more was launched. Only the introduction of agricultural pensions helped to ameliorate this situation.

While the general principle emerging from the above is that services should follow need, rather than force the needy to move to the services, there will be a few exceptions to this. One example will serve to illustrate this point. The maternity and infant care institutions in Athens are designed to serve the whole of Greece. This provision is of particular significance for illegitimate children born in rural areas. Given the social mores of the Greek village, these children seem to receive far better care at an institution in Athens that will also arrange for their adoption.

Comprehensiveness, as a quality of metropolitan services, has been alluded to already in our discussion of the need for treating the family as a unit. References have been made, in discussing education, to the importance of adult education programmes to the education of the child and to the importance of the reduction of travel time in the home-to-work trip in giving the family more time together. Thus the setting of health and welfare standards involves much coordination with other services. By way of illustration, positive steps are being taken in Leningrad to reduce the noise generated by traffic through the development of new road surfaces, the modification of the design of motor vehicle engines, the designation of truck routes, and the use of plantings as noise buffers. Similar efforts are being made to cut down the noise generated by certain types of industry, the most satisfactory of which is obviously the introduction of quieter machinery. Some factories, however, must be insulated, while others can be located in rural areas.

Again, the problem of rodents in Madrid is not simply a product of insufficient effort being put into their extermination. Until the storm sewer system is separated from the household foul-water sewer system it seems hardly possible to reduce substantially the number of rodents from the present estimated 3 million. If the two sewer systems were to be separated, there would be no food available in the storm sewers, and the faster flow in what would be the new, narrower, steep-sided, household sewers would make habitation and procreation far more difficult. The banning of household garbage grinders would also be of help in cutting down the rodent population. The more regular flow of sewage would facilitate its treatment and would result in a cleaner River Manzanares. The cost, however, virtually prohibits the introduction of a metropolitan-wide programme except over a long period of time.

Finally, before leaving the discussion of standards, it is necessary to stress the need for the enforcement of legislation, if standards are to be maintained. Again, this requires equality in enforcement and makes further demands on resources. If resources are not available, and the standards cannot be enforced because of this, then there is little point in setting the standards at all.

THE COORDINATION OF SERVICES

Some comment has already been made on the effect of limited resources on health and welfare services and environmental conditions. The shortage of money, trained personnel and supportive services in other fields, such as transportation, are obvious. The same problems of developing priorities apply here as they do in the other serivce areas discussed. In the developing countries, where pressures are greatest, the luxury of a case-work approach to welfare problems is replaced by a mass social welfare approach. Distinctions are made not so much on the basis of whose needs are to be served as on the basis of what services can be provided. In such circumstances, the most basic needs in the broad area of social services may well be, as Rose

remarks, the provision of water supply and sanitation services to ensure a reasonable standard of health. These needs are closely followed by that for the central planning of economic development to ensure wider opportunities for employment and education.

The satisfaction of the demands for the efficient and fair use of resources and for the coordination of services having a bearing upon the well-being of the individual will depend largely on the planning and administration of those services and the structures established to perform these functions. Central government leadership is required in the establishment of standards and in assisting with the provision of conditional grants for services.

Whatever the framework established by the central government, any metropolitan-wide governmental authority can, if large enough, be responsible for detailed planning and implementation. Indeed, for many services the existence of a metropolitan-wide authority for purposes of planning and implementation is highly desirable. For example, if garbage is not incinerated, burying or composting sites may be required far from the city centre. While these sites can be purchased, the large number of authorities seeking sites often gives rise to some problems. Fewer sites, well-administered, are far preferable to many poorly administered ones. Dumping can also be part of a long-term programme in which the sites are eventually used for residential or industrial development. Indeed, the development of incinerators may prove economically feasible if metropolitan-wide coordination of their location and use can be provided. It is rarely satisfactory to leave this service to local government to provide on a cooperative basis. The "nuisance factor" of an incinerator or a dump is such that no municipality wants one located in its own jurisdiction and even less does it wish to expand its capacity to accommodate other users. This is likely to apply whether the service is municipally run or privately operated on a concession basis. For example, until 1967 there were only two dumps and two composting areas to serve Madrid. Since the composting process, operated by private concessionaires, has proved very successful, two more sites are being opened up. Because of the existence of a metropolitan-wide planning authority this can be accomplished with relative ease.

While the example of garbage disposal above illustrates the advantages of a metropolitan-wide authority, there are also many instances in which the physical and administrative deconcentration of services of such a metropolitan authority is also recommended. For example, the recent annexations by Madrid of adjacent municipalities has left the new city with several pockets of need, representing the centres of the previous municipalities. The services, however, are to be found mainly in the central part of the city where the greatest need has always been concentrated. The policy now is to encourage the physical deconcentration of services, not only to these pockets of concentration but also to the new, fast-developing suburbs. Although land is reserved in these areas for the use of health and welfare institutions, the institutions very rarely move in at the same time as residential and industrial development takes place. This process might be speeded up if the coordinated deconcentration of the presently centrally-located institutions were provided for. If health and welfare centres are to be established, then any participating institution naturally wishes to know which other institutions are to be located in the same new centre or centres. The suggestion has been made by the Madrid study group that a special committee or panel for social welfare and health services be established to plan the coordinated relocation of public and private institutions. This committee would report to the Planning and Co-ordination Committee of the Metropolitan Area of Madrid. Although this Planning Committee serves an area much larger than the City of Madrid, it has no staff and performs no functions of its own. Services administered over the metropolitan area are run either by local or senior governments. The Madrid group finds this arrangement unsatisfactory and would prefer to see a staff attached to the Planning Committee.

The Athens group, in keeping with their suggestions for the deconcentration of services in other fields, would like to see an administrative deconcentration and decentralization in the many health and welfare facilities to the communities (of about 30,000 persons each) that they wish to see established throughout Athens. In this way, local government can assume responsibility for administering many of these services, although planning would require metropolitan-level and senior government participation. In addition, the role of the private agency was recognized as being very significant, especially in the early stages of the development of a service. However, the need to professionalize some of these services has already been commented on. This will require further funds and further supervision by government, although there is no reason why the service should not continue to be privately administered.

Some Preliminary Conclusions

From the foregoing commentary a number of guidelines pertaining to the provision of services in a metropolitan area begin to emerge. These relate mainly to the issues raised in the Introductory Note to Part II and it may be helpful to identify these before moving on to Part III where their implications for the development and operation of suitable institutions are examined.

The biggest problems facing most metropolitan areas seem to be the lack of resources available to meet the needs of the metropolis and the lack of suitable governmental structures to provide for the co-ordination of any one service or number of services throughout the metropolitan area.

The lack of resources, although mainly a problem of lack of money, is not entirely so. Very often resources are not being used to optimum effect precisely because the lack of institutional organization and co-ordination inhibits long-term planning and the assignment of priorities for resource allocation. This is well illustrated in the instance of manpower planning, where initial mistakes are hard to rectify except over a long period of time. Manpower planning also serves to illustrate the importance of balancing social and economic objectives in life. Economic development is recognized as a means to social development, and it should be seen only as that. The development of an educated people, not a "qualified" people, is an end objective in this regard; for education vitally affects any value structure, which in turn is the basis of the decisions made in other fields and, sometimes unfortunately, in the education of future generations. The significance of this value structure has been reiterated many times. While we teach vocational success and competitiveness, we have to forego much social fairness, economic justice and international peace. In all societies there are supporters of conflicting values, but it is obvious that for the increasingly urbanized society, and especially for that now found in the metropolis, there must be a review of existing value structures if this conflict is to be reduced to permit fellow men to live in greater proximity.

Other examples of resources not being used to the full are very evident in the technological fields. The mass production of housing, while realizing some economies of scale, could achieve more in this regard if in some countries fewer lines were being produced and there was more modular co-ordination. On the other hand, social (or non-economic) values again tend to become lost, and the aesthetic merits of most units produced to date leave much to be desired.

Pollutants, whether of air, soil or water, and noise are unnecessarily tolerated at a time when, in many countries, resources are available to eliminate them. Only the desire for economic progress as opposed to the development of a liveable environment diverts man from this relatively simple although large-scale task. Again there appears a conflict in values. There will always be conflicting values, of course, and indeed there should be if there is to be innovation and, through this, improvement in life styles. What is questioned here is whether the resolution of conflict has to await the emergence of a crisis situation, as is so often the case. With matters pertaining to the provision of services in metropolitan areas this is a particularly common occurrence.

A crisis is a relative state. What is a crisis to some is not necessarily so to those elsewhere. In the field of metropolitan services, a crisis is a state of gross dissatisfaction with the service in question. This is a result of one or more of the following factors: the lack of consultation or involvement of the consumer in the planning or provision of the service; the assignment of responsibility to a governmental or other institution ill-equipped to perform the service; and, a lack of resources or unsatisfactory distribution of those resources. To be effective in the provision of services in the metropolitan area, the governmental machinery has to be designed to satisfy the demands being made of it. Hence, there are demands for small units of government or governmental agencies that will be "close to the people". Similarly, there are demands for metropolitan-region governments capable of providing services, not only efficiently but also fairly, over a large area. Further, there are demands that services be coordinated throughout a metropolitan area.

For each of the services examined, there have been demands for regional structures to perform some, if not most, of the tasks. There have been more calls made upon the central government, especially for financial assistance and redistribution of national resources, and there have been demands for more local involvement of citizens in the organization and running of services.

Not all of these demands can be satisfied without compromise and, hence, the value of our present experience for the development of metropolitan areas in the future. For while the present offers us little opportunity other than for making modifications to our governmental machinery and deployment of resources, the future offers us the possibilities of improving the economic and social roles of the metropolis within the development of society and of redirecting the growth of metropolitan communities both in a national and local context. These changes, over time, will have beneficial implications for the process by which we make decisions, deploy resources and provide the services, and together offer opportunities for a vastly improved metropolitan environment and way of life. It is on this somewhat optimistic note that we now turn to a consideration of the development of a metropolitan governmental machinery for satisfying all these demands.

Part III

Development of a governmental machinery

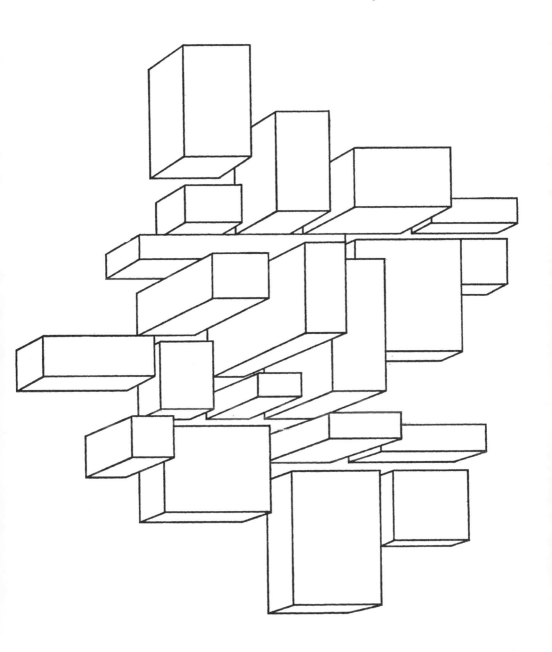

Introductory Note

Is New York governable—or Tokyo, Moscow, London, Paris or any of the other large metropolitan concentrations we are considering here? What of the resource-starved Calcuttas and the booming Seouls of this world? Yes, they each have their governments, but for those who live there this is no answer. The real question is whether or not those governments respond adequately to the multiple demands being made of them; adequately, that is, in terms of the level of tolerance of the local populace.

Here we examine the governmental machinery in terms of its response to those pressures, most of which have been identified, directly or implicitly, in Part II of this book. The metropolitan governmental machinery must satisfy the demands for services to the individual and to the larger community; this must be done both efficiently and equitably; and, at the same time, there must be regard for both the present and the future needs of the society. Obviously, the adequate performance of the service functions and the redistributive functions is not enough. There is also the political function, which involves everything from the provision of means by which individuals and groups can express their demands, to the weighing of these demands against any cost implications and against the priorities of the larger community.

The business of striking a balance between these often conflicting demands is the major occupation of the governmental machinery. Further, as the metropolitan community is only part of a national community, the metropolis cannot be considered in isolation from the nation as a whole. Thus, any machinery designed to resolve the conflict and disparities that exist within the metropolis must be designed also to serve the same functions as the more senior governments. This has been well illustrated in the earlier references to the varying approaches required in the setting of service standards across a national territory in order to achieve the sectoral goals for both the metropolitan and national communities. Some services should be provided according to one standard on a nation-wide basis; others should be provided at different standards.

The satisfaction of demands of different sized communities has obvious implications for the policies that shape the physical size and form of those communities. Although almost all metropolitan areas are growing larger, there is still some choice in the directions that they might take. These directions should be influenced by the needs of the governmental machinery.

All of these issues are raised in the following chapters. The bridge from the treatment of the services in Part II to the examination of the structure of the governmental machinery is provided by Gorynski and Rybicki in their essay on the functional metropolis and systems of government. The functions examined are more than the governmental functions. The authors cover the cultural, social, economic and political concepts of the metropolis as a functional entity. The metropolis is recognized as a fact, not as something that can be dis-

mantled in the same way as it might be studied—that is, piece by piece. The authors' concern is for the growing centralization of control at the national level of government and the wasting away of local authorities—the traditional channels of citizen participation. In response to the pressures of concentration for economic growth and decentralization of decision-making, the authors propose that future growth be oriented toward the polycentric metropolis in which the physical and political communities are of two different scales —local and metropolitan—but with the local unit being far more self-contained as a living unit than, say, the residential suburb in the monocentric metropolis.

Whether monocentric or polycentric, the metropolis represents a territorial community that requires a means by which the separate goals of various interest communities can be reconciled to produce the consensual goals of the metropolis. In Chapter 10, Smallwood focusses attention on the development of a metropolitan political system that will permit effective participation and on the articulation of the would-be consensual goals of that metropolitan community. There are again implications for the physical size of the community, but these are pursued only indirectly to the extent that size affects the functional responsibility of a system and, therefore, citizen interest and participation in the political process.

The attainment of consensual goals is, as the study groups point out, very often a distant goal in itself. The fragmentation of government, the division of responsibility among agencies that lack coordination, and the conflict of interest groups all conspire against the realization of such laudable objectives.

It is against this background that Dupré presents his evaluation of the present effectiveness of intergovernmental relations in serving the values of liberty, participation and efficiency. Although he suggests certain reforms to the governmental machinery if certain values are to be better served, he implies that it will be difficult to serve all values equally by such reform. That these are not the only values that are to be served by the governmental machinery in the metropolis is demonstrated in Chapter 14. The trade-offs between the values served becomes even more complex than Dupré envisages; the ability of the machinery to reflect adequately present-day priorities is that much more remote.

Much of the interest of the study groups was in the financial aspects of intergovernmental relations, a matter pursued further by Hicks in Chapter 12. It is of interest to express the values of liberty, participation and efficiency in financial terms. Liberty is surely equated with autonomous decision-making and, therefore, autonomous financing. Running counter to this idea is that of efficient financial management implying centralized collection and supervision. Participation would call for the taxing of individuals rather than central sources, such as corporations. This last point obviously clashes with the dictates of efficiency. These observations only serve to highlight the significance of financial systems for the political and administrative functions of government. Hicks' treatment

of the principles of metropolitan finance, the potential sources of revenue, and financial administration, are of especial value to any consideration of the governmental machinery. Again, the upholding of the principles of equity has implications for the size of the tax jurisdiction of the metropolis.

But the tax jurisdiction cannot be considered apart from other metropolitan functional areas. Obviously compromise is necessary in satisfying not only metropolitan but nation-wide needs.

Weissmann's essay, in Chapter 13, presents the reader with a more definitive concept of the metropolitan region. The spatial qualities of the region emerge from the functions that the region has to perform for planning purposes. In response to the pressures on the metropolis that he predicts will continue unabated unless a national framework for planning is developed, Weissmann proposes the concept of the city region. The metropolitan region would be part of a national system of regions, and at the same time, an entity with a viability of its own. The emphasis is on coping with the physical growth of the metropolis and the improved economic and social development of the nation. In this way the essay complements that of Gorynski and Rybicki, which stresses the relation between the political and physical development of both the metropolis and the nation.

9

The Functional Metropolis and Systems of Government

Juliusz Gorynski
Zygmunt Rybicki

The creation and growth of metropolitan regions, although permanent and irreversible, may be viewed simply as an outcome of the process of urbanization. Yet today's metropolitan areas are quite different, quantitatively and qualitatively, from urban settlements of the past. Quantitatively, their population is often greater than that of a medium-sized state. Qualitatively, the density resulting from this physical concentration not only gives rise to different principles of social behaviour but also calls for new technical solutions to ensure the proper functioning of the settlement. In turn, these changing physical, social and technical conditions demand new and more adequate organizational and administrative systems for the metropolis and indeed for the nation as a whole. Thus the emerging metropolitan systems are seen as products of the various demands being made upon them and it is the nature of these various pressures being exerted and the response by way of metropolitan governmental and administrative structures that is the subject of examination here.

By "metropolis" is meant here a region with high population density indices in line with infrastructural investments, active community life and exchange within its area, and a low percentage of agricultural employment. Thus this definition disregards the physical structure, whether it be amorphous, polycentric or monocentric, and also the specific forms and systems of government. For these reasons the definition should be seen as somewhat different from the American and German concepts elaborated upon by Forstall and Jones in Chapter 1.

By a "function" is meant the "proper action by which any person, office or structure fulfills its purpose of duty,"[1] and by "functional" metropolis we understand its functions as a whole, regardless of the extent to which they are actually performed. These functions always relate to people, whether they are of a material or nonmaterial character and whether they depend upon economic, social, technical or other conditions. Since the enumeration of all metropolitan functions would require a separate study, it is proposed that the functions be grouped according to the conditions that determine their role in the development process of the metropolis and their impact on the structure of government. Consideration will be given both to external functions (such as the "chief city" function and its part in the education and culture beyond its boundaries) and the internal functions (such as those of housing and transportation). Thus the metropolis is examined in its national as well as its local context.

Also, since a uniform, finite model of government for the metropolis would be purely utopian, we merely propose to offer criteria for the selection and development of optimal systems in accord with the functional dictates of the prevailing conditions at any given time, in any given country.

---

[1] *The New American Webster Dictionary* (New York, New American Library, 1965).

The Functions as Factors of Metropolitan Growth and Restraint

While this chapter focuses attention on the functions that a metropolitan system of government and administration is expected to perform and certain activities that it has to satisfy, it is helpful, in order to maintain a balanced perspective of these forces, to note the influence of geographic factors on the shaping of the metropolitan environment.

GEOGRAPHIC FACTORS

The physical extent of a settlement and the direction of future development are determined largely by geographical location. Classic examples of this are linear or ribbon-like conurbations of considerable length situated between the sea and the mountains. Physical form, however, need not be caused by the negative aspects of natural features; it may also be caused by such positive and attractive conditions as those created by a navigable river. The optimum size of a settlement can be determined fairly accurately through studies of the geographical environment. These studies, in turn, contribute with other planning techniques to the determination of population capacity and also allows initial consideration of the ideal size and shape of governmental units. These ideas can naturally be modified and developed as part of the planning process, but the point being made here is that attention must be given from the beginning to an appropriate system of government in addition to consideration of the physical and other aspects of planning. At present, these matters are either neglected entirely by planners or are discussed only in the final stages of the implementation of urban development plans.

A characteristic feature of the metropolis is the continuing internal and external migration of population. The well-known studies of Chicago, conducted in the 1920's, showed how the physical ageing of various parts of the city and the ensuing conditions of blight led to social change.[2] More recent studies, particularly of American cities, show the considerable influence of automobiles on the increase of internal mobility of the metropolitan population. However, in this case the nature of population migration is different from the trends outlined by E.W. Burgess. The exodus of well-to-do citizens to the suburbs continues as before, but the population total of the central city becomes static or even declines.[3] Houses left by their inhabitants are not down-graded by subdivision into smaller dwellings for low income tenants, but are replaced gradually by offices and shops. The impact of this shift on commuting to and from work is obvious, and it is only very recently that there has been a counteractive tendency to locate (expensive) high-density apartment dwellings in the centre of the city.

These tendencies are, however, characteristic only of highly developed countries. In other countries, the reverse can be observed with a faster rate of development in the central districts of the city than in the suburbs. One example of this trend was the population increase in Warsaw during its period of post-war reconstruction.[4] Again,

it should be noted in passing that conurbations having no monocentric or circular pattern of development exhibit different patterns of migration and population increase.

If we consider the metropolitan area as a whole, migration from the outside is a factor influencing its population increase. Sources of immigration are not only rural districts but also small and medium-size urban centres. It has been stated that in many cases migration takes place in two stages: from the countryside to the small town and then from the small town to the large cities.[5] In this situation, small towns play the role of way stations in the urban adaptation process and should be given facilities to perform these functions adequately.

External migration has considerable impact on urban as well as rural demographic structure. Although increased agricultural efficiency reduces the need for field hands, it is only the most active members, that is the youth, of the agricultural society who migrate to urban areas. This results in the permanent ageing of the rural society and an influx of young single people, or childless couples, to the urban areas. Consequently, the urban birth rate is higher and, although the number of children in urban families is generally smaller than in rural families, the total age structure of the urban and the rural population becomes similar.[6] This is the situation in the more developed countries. Again, it is necessary to note that the situation in developing countries is very different; here, migration is due not so much to the growing efficiency

[2] R. E. Park, E. W. Burgess and R. D. McKenzie, *The City* (Chicago, University of Chicago Press, 1925).

[3] Raymond Vernon, *Metropolis 1985*, New York Metropolitan Region Study (New York, Doubleday & Co., Inc., 1963).

[4] J. Gorynski, "La Région varsovienne," *Cycle d'étude européen sur les aspects sociaux du développement régional* (Geneva, United Nations, 1965), p. 87 ff.

[5] *Ibid.*, p. 139.

[6] See studies carried out by the Warsaw Institute of Housing Research (Instytut Budownictwa Mieszkanioweg), for example, *Prace Instytut Budownictwa Mieszkaniowego*, sheet Nr. 35/XIII (Warsaw, Arkady, 1962), p. 29 ff.

of agricultural production, but simply to rural overpopulation and the expectation of finding some new source of income in the city. Married couples with many children, and sometimes even whole tribes, migrate to city suburbs, all too often finding shelter only in shanty-town settlements. The plight of these people has been described already in Part II of this volume and need not be pursued further here.

Along with "committed" migration, mention should also be made of "less-committed" migration. By this we mean the taking of permanent or temporary employment in a city without a final break being made with the native settlement. This usually means daily commuting to work from neighbouring towns and villages. For cities with large-scale commuting, an enormous difference exists between "day" and "night" populations, which can put considerable demands upon services. Unfortunately, the failure to take account of this difference has often led to major errors in the provision of urban services.

In addition to everyday commuting, other forms of migration associated with a longer urban stay are known. For example, in the cities of many countries, there is a tradition of seasonal employment in construction. Even several years of urban employment do not necessarily result in permanent displacement of the worker's family to the city. In Poland, for instance, a large new group of peasant-workers has been created.[7] The peasant-worker's family is still concerned with agriculture and the income of family members working in construction or industry is often considered as a source of additional funds to be invested in the farm. Although this phenomenon extends the city's influence far beyond the normal commuting area, the large group of peasant-workers resident in the city neither takes part in community life nor feels membership in it. As the prime objective of this group is to increase its financial resources, as opposed to sharing in the urban style of life, it is not surprising that the process of urban acculturation develops slowly.

This increased internal and external mobility has several implications for government. Perhaps the most obvious impact is on the management of mass transportation systems, which become less and less a local affair and demand coordination by regional governing bodies. Far less obvious, maybe, but of great significance, is the diminished propensity of the population to participate in community life and even in local government. For it is only a small percentage of those persons who are migrating that settles permanently in the new community. In small towns or in those parts of metropolitan areas having a continuing rotation in migrant population, lack of identity with the community and an unwillingness to take an interest in its development is often quite noticeable. The increasing discrepancy between the stabilized population and the daytime population, creates new problems in local government, financial policy and the provision of urban services. Yet, as Jean Gottmann writes in a chapter of his book, *Megalopolis*, entitled "Redistribution is an Urban Function",

What happens in Megalopolis must be interpreted as a redistribution of what used to be city functions over a wider territory. The economic and social characteristics of administrative units in this territory and the relationship between them are being modified. The changes must occur in economic and social relationships first; then the legal and governmental framework can be redistributed from time to time to secure as fair as possible a share to all participants. It

---

[7] *Ibid.*, sheet Nr. 23/VIII

is when the services normally provided or regulated by governments become unsatisfactory to the users that a redistribution of governmental burdens comes to pass.[8]

Thus change must be expected, and one of the first signs of change in a growing city are the financial problems resulting from the increase of population and additional services demanded. An analysis contained in the New York Metropolitan Region Study[9] stated that, for a sample sixty-four medium-sized New Jersey municipalities, 83 per cent of the total variance in average expenditure is associated with the size of the community. In an examination of the degree to which operating expenditures for individual types of services varied according to community size, however, there were appreciable differences recorded. Community size was associated with 92 per cent of the variation in education expenditures (except debt), 67 per cent for streets, 59 per cent for general government, 60 per cent for police and fire protection and only 18 per cent for welfare. There is a correlation, however, between the absolute amount of local government expenditure and the per capita value of property. Thus, we may conclude that a large proportion of unstabilized population increases expenditure on local government services without providing proper compensation by way of increased revenue. It is also true that not only the size, but more particularly the demographic structure of

the community has a decisive impact on local government expenditure.

These difficulties decrease if communities extend their boundaries to achieve greater equity. The central city's annexation of a wealthy residential suburb, housing a large number of the central city's commuters, is often a desirable way of increasing the income of the central city. Not surprisingly, such annexation evokes some opposition on such grounds as the threat it represents to active community life in the smaller units, traditionalism and the fear of losing positions in a local government hierarchy. As will be seen, similar problems also arise when trying to improve other metropolitan functions.

THE TECHNICAL SERVICES

The high degree of concentration of population in the metropolis can be maintained only if the technical infrastructure is adequate. Meeting the demands for technical services, in turn, makes certain demands upon the governmental and administrative machinery of the metropolis, and it is to these that we turn now.

Housing

Housing, or in its broadest sense, shelter, is the most basic function of any human settlement and generally accounts for from 40 to 50 per cent of the fixed capital value of a town's building stock.[10] Physically, housing takes on a variety of forms including single-family units, multiple family units and such special housing as hostels, boarding houses, creches, military camps and emergency housing. The accommo-

---

[8]Jean Gottmann, *Megalopolis, the Urbanized Northeastern Seaboard of the United States* (Cambridge, The Massachusetts Institute of Technology Press, 1961), p. 764.

[9]Robert C. Wood with Vladimir V. Almendinger, *1400 Governments*, New York Metropolitan Region Study (Cambridge, Harvard University Press, 1961), p. 36 ff.

[10]J. Gorynski, *Planung, Verkehrstechnische Erschliessung und Versorgung von Neubaugebieten*, 9. Internationale Baufachtagung, Hannover-Messe 1965, Institut für Bauforschung (Hanover, 1965), p. 14.

dation may be owned by a public institution (either state, regional or municipal government) a cooperative, a private or public corporation or a private individual. In most towns a great variety of housing forms are found, although their mix will vary according to the political system, the state of economic development and local tradition.

Because of its extreme importance, housing has always been subject to control by public authorities, no matter what the form of ownership. Elsewhere in this study, Sazanami has elaborated upon the various types of housing and financial policies utilized. Generally, the degree of public intervention is inversely proportional to the state of economic development and the quality of housing. Yet there is no country that does not have a housing problem and does not require some form of public intervention in this field. Indeed, from the administrative viewpoint, the great number of technical, social, and economic problems in the housing field is one of the main reasons for giving the responsibility for implementation and management of housing policy to the authority closest to the local level. This arrangement also facilitates public participation which is particularly active in housing matters. A qualification may be required here regarding the responsibility for the operation of agencies and management for special forms of housing, which might be more suitably placed at a higher level, even that of the central government for major financial matters.

In short, a three-tier governmental structure for housing management appears necessary: the responsibility for most aspects of policy resting at the local level; special housing being the concern of a city-wide or metropolitan authority; and a financing agency provided at the regional or central level. This scheme can be applied under differing political conditions.

Sanitation

Here we are concerned basically with water supply and sewerage systems although it is appreciated that sanitation services also embrace garbage collection and disposal, street cleaning, and heating systems. In many countries, increasing urbanization and industrial development have caused water supply problems that, if left unsolved, become a basic factor limiting further growth. The critical nature of this situation has completely changed the traditional attitude toward water supply. In the past, it has been treated purely as a local concern. With increased population density, however, it has become impossible to draw water and to dispose of sewage within close proximity to one another. The growth of metropolitan areas has resulted in a parallel increase in the distances over which water and sewage must be transported. Simultaneously, there are increased demands for water purification and treatment of sewage. Given the scale of the metropolis, the water supply and often the sewerage system of a metropolis becomes regional or even supra-regional in its coverage.

Similarly, there is a tendency for per capita costs of construction (and often operation) of service systems to be higher on the fringe of the metropolis than in the more central areas. This, coupled with the need to provide for fuller use of technical equipment and better coordination of the needs of the many areas served from central installations, would again point to the transfer of responsibility for such investment to regional or even national organizations.

This should not be taken to mean that all sanitation issues surpass the interest or responsibility of local governments. It simply calls for the re-allocation of responsibilities among different levels of government. The regulation of local services and responsibility for construction and maintenance of local distributing systems, can remain a local concern although there has been a long tradition of inter-municipal unions for this purpose and, maybe because of this, organizations of regional character are now coming to the fore. More serious dilemmas arise regarding the actual administration of such services. Should we decentralize the power of regional government to local municipalities, or should we deconcentrate this power by creating special-purpose field services, responsible directly to regional or national government? This latter solution has been applied for many years on a large scale in the field of electric power and regional gas systems and there is no apparent reason why it could not be applied equally well to sanitation services.

Transportation

Of the many aspects of modern transportation, only some of those influencing the choice of administrative systems in metropolitan areas are examined here.

One significant feature of transportation is that the more economically advanced countries seem to be reaching a point of saturation regarding the percentage of total consumption expended on transportation. In Sweden and the United States, 1960 percentages were 13.5 per cent and 14.4 per cent respectively and estimates for 1980 are 15.0 per cent in both countries.[11] There is, however, a continuing change in

the modes of transportation used. With increasing development of the economy there is a parallel increase in the proportion of trips by private automobile. Trips by other means increase at a much slower rate, while railway travel decreases even in absolute figures. For example, in Sweden, from 1950 to 1963, the GNP increased by 150 per cent while the number of kilometres travelled per person increased by 250 per cent. The number of kilometres travelled per person by automobile increased 550 per cent, kilometres travelled by bus remained unchanged, while railway figures declined 20 per cent.[12] Yet the very popularity of the private automobile has contributed to its ineffectiveness as a means of commuting to and from work as far as time and comfort are concerned. According to a study by Wilbur Smith and Associates, in the United States approximately 34 per cent of all home based trips are for work purposes.[13] The great concentration of these trips during rush hours serves only to accentuate the difficulties of automobile travel. This concentrated use of facilities is also typical of rapid transit systems; in New York, for example, 15 per cent of daily rapid transit is concentrated during rush hours, while in Cleveland the comparable figure is 22.5 per cent.[14]

Keeping pace with transportation demands imposes heavy financial administrative burdens upon governments. According to the report of the Joint Committee on Washington Metropolitan Problems a comparison of costs for handling increased peak hour loads from suburbs served by the

---

11*Les Besoins Croissants d'Espace dans la Région Urbanisée*, Conférence Internationale de la Fédération Internationale pour l'habitation, l'urbanisme et l'aménagement des territoires (Orebro, Sweden, 1956), p. 24.

12*Ibid.*, p. 25.
13Wilbur Smith and Associates, *Future Highways and Urban Growth* (New Haven, 1961), p. 83.
14Lyle C. Fitch and Associates, *Urban Transportation and Public Policy*, (San Francisco, Chandler Publishing Co., 1964), pp. 35 and 238.

Pennsylvania Railroad showed that, for the transportation of 120,000 passengers per hour, investment in highway and parking facilities was seven times as great as that required for rail development. A similar comparison for 80,000 passengers per hour in the New York area showed a discrepancy of 1:16 in favour of rail investment.[15] Edgar Hoover calculates that the addition of every new car to traffic in Pittsburgh results in municipal expenses ten times greater than the price paid for the car.[16] In countries where fewer persons own automobiles the question arises as to why the whole community should carry the burden of investments serving only a motorized minority. On the other hand, in countries where motorization has reached saturation point, costs incurred in traffic regulation and safety procedures should be distributed evenly among all users of public roads.

Many specialists have come to the conclusion that the future will bring a renaissance of public transportation within cities, and recent developments in many cities would seem to substantiate such opinions. Whatever the technical solution —private automobile or public transit— transportation in the metropolis requires metropolitan-wide control, whether for purposes of supervision of safety, for financing, or for planned integration with land use and development in general. As Lyle Fitch has stated:

> If transportation planning is to be a continuing process, and if it is to be integrated with land-use and development planning generally, its most logical location is in the regional planning agencies. For transportation and other planning projects, representation can be given to state, federal, and other interests.[17]

Services

By "services" is meant those material services which satisfy daily needs associated with work, education, leisure and recreation. These include the networks of universities, schools, nurseries, hospitals, wholesale and retail trades, craftsman workshops, restaurants and entertainment centres, parks and sports facilities.

These different facilities have one common feature: in order to give maximum satisfaction, their distribution and organization should meet the postulate of maximum accessibility in terms of time. The majority of these facilities can be graded according to their frequency of use, as is illustrated by the example of the distribution of household goods. Goods for daily consumption are retailed through small outlets distributed evenly throughout the population. Goods used less frequently are retailed through a network of supermarkets and specialist shops, each outlet serving a fairly large area. There are also the wholesaling networks on a local and regional level. Given these complex retail/wholesale arrangements, there are equally complex forms of management, ownership, and principles of financing. Because these systems are so complex, effective coordination and planning are extremely difficult. Also, as these are the services that serve the consumer directly, there is a strong element of public interest in their operation. In principle, therefore, the greatest amount of decentralization is necessary to ensure direct contact with the consumers and thus provide maximum satisfaction.

---

[15]Gottmann, *op. cit.*, p. 681.
[16]Edgar M. Hoover, *Metropolis samochodu* (The Metropolis of the Automobile) (Warsaw: Biuletyn Komitetu Przestrzennego Zagospodarowania PAN 4/23 1963), p. 39.

[17]Fitch, *op. cit.*, p. 81.

ECONOMIC AND SOCIAL FUNCTIONS

In studying the impact of economic functions of cities upon government structure, it is useful to differentiate between external and internal functions.[18] External functions, or the general process of exchanging goods and services with the surrounding hinterland, with any other part of the country and with other countries, are considered "basic", while internal functions, serving the local population, are considered as "local" or "non-basic". It is the interdependence of these functions that is a subject of study of regional scientists. In socialist countries there is particular interest in the interdependence of functions, which, according to Marxist doctrine, generate national income (and which might be approximately compared to basic functions) and the sectors of the economy that either collectively or individually consume this income. The exploration of these interdependencies and the development of models of any regularities that may exist, may assist in our understanding of the effects of these functions on urban growth and development.

It is the basic functions that determine the perspectives of growth and development and constitute a decisive economic driving force in the process of urbanization. In turn, the operation of these functions is automatically ruled by specific laws of development and growth. Here our main interest is in the effects that arise from the territorial concentration of basic economic functions to take advantage of large scale production. The external economies or increased profits that are expected to accrue,

stimulates the formation of a permanently growing industrial agglomeration. This, due to the multiplier effect on local industry speeds the growth of the urban area. As a corollary, the increased demand for tertiary services leads to the development of the technical and social infrastructure, which in turn contributes to making the area more attractive for the location of new basic industries. The interplay between basic and local functions thus has a snowballing effect upon the development. However, such industrial concentration may result in a deterioration in living, social and cultural conditions, in which case it is necessary for local and regional governments to intervene. To prevent the uncontrolled growth of urban agglomerations, certain administrative measures have to be applied. As a rule, however, the effectiveness of such steps is limited mainly because the steps themselves are not comprehensive enough and do not reflect the complex interdependencies that exist between primary, secondary and tertiary activities of both a basic and local nature. Also, the results cannot be expected to materialize overnight. Modern urbanization and industrialization must be based upon long-term comprehensive planning and such plans must be implemented effectively.

The quantitative increase in economic activity, resulting from concentration processes and the development of tertiary occupations, causes permanent changes in the territorial distribution of functions within the metropolis. The shifts can be synthetically reflected by employment statistics for different districts of the metropolis. For example, the nature of the characteristic shifts in employment that have taken place between the central city and the outer ring are reflected in the figures for two American cities given in Table 9-1. Accompanying a general trend toward

---

[18]A comprehensive bibliography on these functions may be found at the end of Chapter 6 in Walter Isard, *Methods of Regional Analysis, and Introduction to Regional Science* (Cambridge, The Massachussetts Institute of Technology Press, 1960).

TABLE 9-1

Percentage Changes in Employment Structure in New York and San Francisco (1948-58)

Activity	Central City/Cities		Suburban Ring	
	New York	San Francisco/Oakland	New York	San Francisco/Oakland
Manufacturing	−5	−5	+87	+27
wholesale	−6	−10	+93	+180
retail	−3	−8	+98	+78
Selected services	+89	+100	+167	+216
Total change	+4	+2	+98	+58

Source: Lyle C. Fitch and Associates, *Urban Transportation and Public Policy* (San Francisco, Chandler Publishing Company, 1964), Table II-I (appendix), p. 253-54.

rapid growth of employment within the ring and moderate growth in the centre, characteristic structural changes take place. In the central city, a distinct concentration of selected services, such as financial and administrative institutions, and printing and publishing firms, is accompanied by the exodus of industry, and wholesale and retail trade, to outer districts. In time, specialized land-use zones begin to emerge. The existence of a master plan and its proper implementation will determine whether or not a subsequent concentration of heterogeneous functions, as well as excessive traffic congestion, will appear in the suburbs. If such a concentration were to arise, it could cause further shifts similar to those presently taking place between the centre and the ring.

Changes in economic functions are followed by respective changes in social structure. As these changes can take place over entire districts of a metropolis this factor should be taken into consideration, along with other aspects of the process of production, when evaluating the location of industries. Increased population is a social effect of the displacement of economic activities from one district of the town or metropolis to new areas. This mobility is either expressed by increased commuting, or by a propensity to change the area of residence as a consequence of change of employment.

The effect of government intervention for purposes of controlling these phenomena varies according to whether the economy is based on private ownership or whether it is a planned economy, where the basic means of production are nationalized. In the latter case, the government can control the location and development of industry by directives. This does not mean, however, that planned economic systems are free of conflicts; these are very likely to arise if the sectors of basic production are controlled by a central government which has not consulted with the local governments in the development of its policy. The essential difference between the planned and the non-planned economy is that conflicts arising in the planned economy can be solved by collective and comprehensive negotiations during the initial planning period with the participation of the different levels of government. Once ratified, the plan becomes a compulsory directive for all implementing organizations. On the other hand, in a market economy there must be an effective way of preventing economic development detrimental to the interests of the people of the metropolis. Given the crucial significance of the location of basic economic production and its territorial concentration, it would appear desirable to entrust the central government with the responsibility for coordinating functions pertaining to these basic economic activities.

CULTURAL AND POLITICAL FUNCTIONS

As with economic functions, a distinction can be made between basic and local cultural and political functions. Although little research has been done on this basis, certain symptoms of a supposed multiplier effect are in evidence. Large scale concentration again appears in this field and it may be that investments in large-scale technical infrastructure for cultural pursuits is equally profitable. The existence of numerous cultural centres bringing together several institutions under a common roof and management would tend to support this theory. By combining theatres and television studios, economies can be achieved in the employment of technical equipment and personnel, while the combination of theatres and cinemas permits the maximum use of buildings.

It seems doubtful, however, whether such concentrations have any similar advantages regarding the quality of cultural activity and its influence on society. Their excessive concentration of such activities in the metropolis limits the active cultural participation to the local population and inevitably lowers the standard of cultural life in smaller urban centres. Even within the metropolis, the law of concentration is relevant. The central districts of the metropolis or even its very core tend to be the focal point of cultural and political activity. Indeed, the difference in cultural and political activity tends to be greater between the centre and the suburbs than for that found between the metropolis as a whole and the small provincial towns.

Essential changes have taken place in this field in recent decades due to the introduction on a far larger scale of mass media, audio-visual information services, cinema, radio, television and the popular press. These have undoubtedly influenced the development of a passive participation in cultural life, although to what extent they have inspired active and creative behaviour is difficult to tell. On the other hand, they have contributed to an even greater concentration of cultural infrastructure in the metropolis with the development of broadcasting stations and studios, film units, major publishing houses and so on.

Finally, it is worth mentioning the position of those cities which are the seats of senior governments. The existence of the capital function leads to new conflicts, tensions and aspirations. If this position brings with it certain privileges, conflicts with neighbouring cities also arise. To solve these problems, many countries have created special forms and systems of local government for capital cities. This has been done by raising the local government to a higher administrative position (Paris, Warsaw), by increasing the personal importance of the mayor through his participation in central government (Moscow), or by forming separate regions, administered centrally and thus excluding local government (Mexico City).

RECONCILING THE FACTORS OF GROWTH AND RESTRAINT

This review of certain functions of the metropolis has revealed a great variety of

Figure 9-1

Tiers of Execution and Administration of Metropolitan Technical Functions

Function	Level of Agency					Impact of Increased Density and Size of Urban Area
	Local	Urban	Metro-politan	Regional	National	
Housing	⊕	+ [a]	⊕ [a]		◎ [b]	←
Sanitation	+	⊕	(+)	(⊕)	(⊞)	→
Transport	+	⊕	(+)	(⊕)	(⊞)	→
Services	(⊕)	⊕	(⊕)	◎ [c]		←

Notes and explanation of signs:

a   Special forms of housing only

b   Financial and general policy

c   Some special forms (for example, hospitals and universities)

+   operating agency

○   administration

◎   top level administration (supervision)

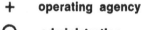   future location of functions

forces and tendencies exerted upon management and administrative systems. Figure 9-1 presents an attempt to illustrate graphically the evaluation and some of the recommendations made pertaining to the infrastructural elements of the urban environment. As can be seen from Figure 9-1, various contradictions arise and perhaps the most discussed of these is the contradiction between public participation and optimum size of administrative areas. This problem can best be presented by means of a diagram, proposed by Isard, shown in Figure 9-2. The EF curve of Figure 9-2 represents the cost of administrative services calculated for areas of various size. From the point of view of effectiveness, point P on a horizontal scale marks the optimum size of the administrative area (OS). The AB curve represents the so-called potential of public participation, which decreases with the increase of time and social distance. Clearly, the structuring of administrative areas on the basis of optimum administrative costs will incur losses in public participation. In practice this problem results in a compromise between the effectiveness of economic outlays and the intensity of public participation. Point C represents the point at which there is considerable gain in public participation and area. The situation represented by OR would be a very democratic unit of government. The compromise would probably lie between OR and OS. However, the choice of the compromise is complicated by the

Figure 9-2
Cost of Administration and Public Participation Potential

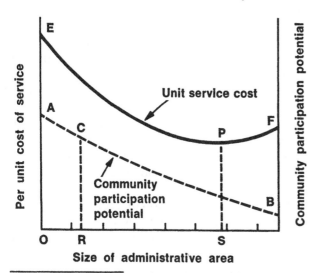

Source: Walter Isard, Methods of Regional Analysis, An Introduction to Regional Science
(Cambridge, Massachusetts Institute of Technology Press, 1960), Figure 5, p. 258.

fact that certain services exhibit different tendencies regarding the choice of a compromise and that these also change over time. For example, modern transportation and sanitation systems are tending toward the regional level, while housing and some other services are showing a greater preference for local administration with increasing urbanization.

If we wish to keep these heterogeneous services under the administration of a common government, we must either retain local multi-purpose administrative units despite the fact that some services would be more efficiently provided on a regional basis or we must create multi-purpose regional governments that, for some services, would have deconcentrated sub-regional administrative units. While this latter suggestion may appear particularly appropriate for metropolitan areas, it is likely to become overly bureaucratic and thus suffer the inevitable decrease in public participation.

Another solution is based on the principle of retaining complete local government responsibility for most functions. This entails removing from local government jurisdiction only those functions whose optimum administrative area surpasses that considered advisable for control by traditional local government. Such functions may be assumed by special regional technical administrations which may form, in turn, their own field services by means of deconcentration or by means of decentralization. In the latter case, local government may perform some technical functions on behalf of the decentralized, regional, technical administration.[19]

Another form of solution is based on the assumption that while planning and investment functions call for a large optimum

_____

[19]S. Humes and E. M. Martin, *The Structure of Local Government Throughout the World*, International Union of Local Authorities Report (The Hague, Martinus Nijhoff, 1961), p. 18.

THE FUNCTIONAL METROPOLIS AND SYSTEMS OF GOVERNMENT   303

area, the actual management of many functions may be more economical over a smaller area. This brings us to the idea of forming inter-municipal public corporations for comprehensive and integrated planning and for the provision of technical infrastructure. Following their initial development, these infastructural investments could be handed over to local government for continued upkeep and operation. In Poland, this concept of an inter-municipal public company has been applied in several instances. Since 1948, such an organization has existed for the investment planning in nationalized housing. This organization has been concerned primarily with the location of housing as it relates to industry and is structured on both a region-wide scale for woiwodships and for large towns. More recently, these organizations have become commonplace and have been extended to cover all communal investments other than agriculture. In Poland, these communal investments would include not only the communal investments in those public services owned and run by local governments in western European countries (such as water supply, public transport and sewers) but also in such areas of local responsibility as retail stores, warehouses, small consumer goods industries, social and educational facilities and so on.

Underlying all these problems is the basic question: is centralization or decentralization of government most favourable for the development of a given area? Decentralization favours both access and increased public participation. On the other hand, it does raise problems of coordination and the creation of favourable conditions for large-scale regional schemes. Such problems are most easily overcome by a single governmental body. Again, a single administrative unit facilitates future planning.

It is impossible to formulate universals, however, and our conclusions should only serve as encouragement for the study of specific circumstances relating to each area at its particular stage of development. The alternatives will vary according to the state of the economy. In a period of expansion, strategic elements of the economy should be concentrated; in a period of stabilized growth, harmonious development of the given area, and any part of this area, can be achieved very satisfactorily through deconcentration and decentralization.

THE INTEGRATION OF METROPOLITAN
DEVELOPMENT POLICIES

Given the above forces of metropolitan growth and restraint and the demands that they make upon the governmental structure of the metropolis and the nation, we shall now examine certain considerations that have to be given to the design of metropolitan development policies if they are to operate successfully within the previously stated limitations. Particular attention must be given once more to governmental structure and the decision-making and administrative roles that different levels of government, with varying jurisdictions, should assume.

Physical Planning and Development Policies

In devising policies for the physical planning and development of metropolitan areas and assigning responsibilities to different levels of government, consideration has to be given to the changing scope of physical planning, the need for its integration with economic and social policies, the desirability of public participation in the planning process and the coordination of governmental bodies responsible for the implementation of the plan with those responsible for producing the plan.

The changing scope of physical planning is due largely to two factors: the mechanization of the building industry and the increased size of urban settlements. With the mass production and modular coordination of the structural elements of building, industrial design has moved into the field of designing individual buildings. Architecture has become more concerned with urban design while town planning, given the current scale of urban areas, has had to assume regional dimensions. National planning, once concerned more with the formulation of economic and social objectives and the significance of the physical resources and terrain for the realization of these objectives, now embraces the planning of a physical framework for the development of urban areas. It is within this framework that the physical planning of metropolitan areas, a combination of both town and regional physical planning, must operate.

The integration of physical planning policies with those of economic and social planning has already been stressed. However, it is of interest to note the differences between the planned and the market economies in this regard. While the early stages of the planning process are similar, the techniques for coordination and implementation vary.

In highly planned economies, physical, social and economic plans are theoretically nothing but different ways of presenting the same plan in terms of space and time. In actual practice there are difficulties in achieving integration due mainly to the different time horizons involved in physical and economic plans. While economic plans cover annual and five-year periods for purposes of implementation, and from fifteen- to twenty-five-year periods for long-term perspective planning, physical plans designed to present the desired final situation may project twenty or even one hundred years ahead, given the average durability of material components of the physical environment. However, as investment plans shaping the future physical character of an area are incorporated into the economic plans, the timing of the implementation of physical plans is in fact governed by economic planning.

In market economies, where almost all investment is in private hands, implementation according to plans is even more difficult. Economic planning is replaced by programming and forecasting, which requires that the implementation of physical plans must be accompanied by efficient implementing policy. The tools of such policy consist of both positive incentives and negative controls. Experience proves that positive measures are much more effective than even the most severe negative measures. While the importance of implementation policy in market economies is emphasized here, it must be stressed that even in planned economic systems planning is not automatically efficient—appropriate policies at both the central and the local levels are again required.

In all countries there is a great deal of public interest in physical planning, particularly in local plans. As the size of the planning area increases, the relative number of discussion participants decreases, and individuals are replaced by institutions and big business organizations. Unrestricted public participation is indispensable in the elaboration of plans for areas the size of an urban district; it is useful up to the level of metropolitan planning; but it becomes fictitious for regional planning. Regional planning calls more for widespread consultation with numerous agencies, institutions and large-scale enterprises.

The assignment of responsibility for physical planning is very much influenced by the desired degree of public participation and the intended relationship with the governmental level responsible for the implementation of the plan. Master plans elaborated by supervisory bodies, even by most competent authors, prove inefficient, unless the executing organ accepts the plan as its own. This means that agencies responsible for the elaboration of urban physical plans should be part of the urban government, and should be obliged to work in close cooperation with representative organs and hold regular public meetings. This principle has to be modified in metropolitan areas where several local governments are involved, and assuming that overall metropolitan physical planning is a prerequisite for harmonious development of the total area and its parts, there are various alternative organizational solutions open. One way is to centralize metropolitan planning in the hands of the most important local government—most likely that of the central city—or create a special state agency. The rigidity of such solutions can be modified by setting up secondary (local) planning agencies, responsible for the detailed planning of smaller areas and contributing to the overall metropolitan planning in a consultative role. A further step toward decentralization is to confer on local (city or town) planning agencies full responsibility for their territory but within the framework of a general outline drawn up by a metropolitan or regional planning agency.

Economic Policies

Economic policies designed to facilitate metropolitan area development must ensure the provision of adequate directives for planned private and public investments, the availability of land, and a variety of conditions for development.

Under normal conditions, and this does not necessarily include the less-developed countries or the newly developed parts of urban areas, metropolitan areas prove very attractive for the investor. All that is required is that this investment be integrated with metropolitan development plans through the use of the tools of financial policy (such as grants and loans), fiscal policy (such as tax reductions), and land policy (such as the controlled allocation of land). While these measures are mainly of a positive nature, negative administrative measures must exist to deter action contrary to the public interests. There are also direct links with social planning in as much as a well-balanced employment profile must be maintained.

The availability of land depends largely on who owns it. In metropolitan areas especially, a great variety of land uses gives rise to a continuing change of uses and the attendant speculation. To ensure that development is not hindered by these conditions, the transfer of land to municipal ownership is probably the most effective measure available. Other public and private investors can obtain user rights on an emphyteutical basis for periods not exceeding the physical durability or technical lifetime of their investments, the usual term being ninety-nine years. After the deliberate, total destruction of Warsaw this solution was seen as a basic prerequisite for its rapid planned reconstruction. Any other systems require the application of complex and comprehensive legal measures such as expropriation, pre-emption, price control, compulsory lease and so on.

The significance of a variety of incentives, arising from differences in local policies, is illustrated by the following quotation from J. Gottmann's *Megalopolis*.

> There is little doubt that the frequency with which the main axis of Megalopolis crosses state lines favoured the extension

of the region along it and the simultaneous growth of many parts of the axial belt. In the seventeenth century the advantages of such partitioning were especially appreciated by the inhabitants for reasons of religion or community politics; in the twentieth century differences in taxation and labor organization have been more important. But, whatever the reason at a given time, it has always been useful to someone to be able to find different administrative regulations by moving just a little distance.[20]

A good example of a conscious attempt to offer different conditions for development, and for living in general, was the policy behind the establishment of some satellite towns for purposes of deterring immigrants from settling in the excessively expanding central city. To this end, the satellite town offered apparently better living conditions and employment in local enterprises. In fact, satellite towns based on this well-balanced concept were quickly transformed into simple dormitory towns due to the lack of coordinated industrial and services development in the immediate environs. As a result, commuting to the central city increased instead of decreased.

The implication for government that can be drawn from these few observations is that the need for achieving efficiency and coordination of economic policy is crucial. The effective horizontal and vertical coordination of economic development policy constitutes one of the basic tasks of all administrative bodies involved and, for this, centralization in the higher echelons of government is highly desirable. Decentralization on the other hand, fosters development through more flexible and positive methods of encouragement although it suffers the limitations of a narrow perspective jurisdictionally. The choice, or

compromise, has to be made between these two ends of the spectrum.

Social and Cultural Policies

The problems of social assistance and welfare policy have been covered already in other chapters. The view taken here is that if governments are to foster adequate social and cultural progress in metropolitan areas, private activities, as a rule, have to complement, but not replace, governmental activities. In this sense, social policy is seen as embracing employment, housing, education, recreation and community life. Its purpose is the internal integration of society, and society's integration with the physical environment, through the optimal use and permanent improvement of the existing material infrastructure.

In spite of the non-economic character of social planning and policy, and the difficulties of expressing this subject in quantitative terms, social planning is closely linked with economic planning and both, together with physical planning, should be considered as a whole. A simple example of this is the importance of the common use of services as a means of community integration. Occupants of new buildings always tend to form a separate group as long as use of different community services persists.[21] Providing separate services for the newcomers can definitely impede integration and freeze the initial situation of social isolation. Purely economic factors may, in such cases, be the cause of socially unsatisfactory conditions.

The integration of economic, social, cultural and physical planning policies will be achieved only with the integration of responsibility for implementation. This calls for a reduction of the number of bodies

---

[20]Gottmann, *op. cit.*, p. 741.

[21]Sylvia Goldblatt, "Integration or Isolation," *Habitat*, Vol. 9, Nos. 2, 3 & 4 (1966).

involved, a greater degree of concentration of functions performed by the administration, and at the same time, the encouragement of active public participation in planning at the local level. It is to the possible reconciliation of these apparently conflicting demands that we now turn.

THE EMERGING SYSTEMS

While the constraints that have been identified thus far would appear to limit severely our choice in the structuring of metropolitan governmental systems, they do, in fact, provide only a framework within which the more detailed structure emerges. This structure is in large part a product of the forces of reconciliation. On what might be regarded as the vertical axis of conflict and/or cooperation, both central and local governments make varying demands. On the horizontal axis similar demands are made for unitary or federated systems of government to facilitate metropolitan-wide integration. In addition to these forces are the internal pressures to which the system is expected to respond in an individual manner.

The Vertical Dimension: Centralization versus Local Control

In most instances, and especially in socialist countries, metropolitan administrations are supervised directly by central government agencies (such as a council of ministers, an individual minister, parliament, or the head of state) or state agencies in the case of federations (as is the case with Leningrad). The greater degree of administrative intervention in the modern state, and the acquisition of responsibility for functions formerly resting within the domain of private enterprise (for example, the organization of industrial production), has brought about a greater integration of these functions and a vast improvement in eco-

nomic, social and cultural conditions. In order that the situation might be further improved it is occasionally necessary to review the assignment of these functions to the local, metropolitan and central governments.

The decentralization of central government responsibilities to local bodies is fairly common in many countries. However, there are also many instances of the devolution of power being only a deconcentration of central administrative responsibilities in that the special bodies that are set up to assume these functions are not democratically elected representative organs of local government. Decentralzation is also limited by the need to maintain administrative uniformity of action and the demands made on administrative organization by large-scale concentration of economic production. Indeed, it is in connection with industrial development that a conflict arises. For, as has already been noted, industrial concentration is the basis of the growth of metropolitan areas and yet the control of industrial location and growth is largely in the hands of the central government. Similarly, the control of transportation systems and education systems is generally beyond the control of the metropolis because of this overriding desire for uniformity of action.

How can this apparent conflict be reconciled? In the socialist countries the principle of democratic centralism combines unity of action in central government with considerable independence of local state organs. In Poland, for instance, the metropolitan-wide Municipal People's Council is directly elected and is responsible to the electors; but, at the same time, it is supervised by the central Parliament (in practice by its Praesidium or another of its bodies). The council is responsible for levying its own taxes, passing its own budget, local legislation and so on. The council also supervises operations of state administrative bodies in the metropolis, such as the post office, even though these are not subordinate to the local council. The council elects its own Executive Committee, which is also supervised by the senior government, in this instance by the Council of Ministers. The Executive Committee directs the administration of the metropolis through departments. However, there are also special committees of council set up for each of these functions, although their roles tend toward general supervision of departmental activities and liaison with the public. The department heads are also supervised by the respective ministers in the central government. Thus there is both democratization and independence of operation at the metropolitan level and also an effective mechanism for vertical coordination of action.

Although the structure of government in the metropolis may vary for purposes of achieving central control and democratic decision-making, there is a strong degree of uniformity in the ways in which public participation is encouraged. The conditions created in the metropolis as a result of urban concentration—such as the cooperation required in the running of large-scale industrial concerns, the advanced division of labour, the common needs arising from common places of living and the rapid communications systems—are all conducive to community action.

This participation is both politically desirable and practically expedient. From the political viewpoint, it is the purpose of democratic government to provide conditions conducive to the cultural, social and material development of man. The interpretation of this principle is based on such practical premises as the political system of the country, the range of administrative

functions assigned the metropolitan government, and the formal and informal machinery designed to facilitate both direct and indirect public participation in metropolitan government. While little comment is needed on the significance of the national political system and the fact that the restriction of the range of metropolitan functions to the satisfaction of basic needs of inhabitants—such as transportation, lighting, heating, water supply and so on—has long been seen as too limiting, it is the development of established systems for public participation that is of greatest concern to us here. Even under the most favourable political and economic conditions, a single system of access does not safeguard the democratic functioning of public authorities. Thus, consideration must be given to both formally institutionalized and legally established systems and to informal systems. In spite of the dominance of indirect representative democracy in the metropolis, the mutual interpenetration of institutions of indirect democracy with the elements of direct democracy often occurs. Particularly favourable conditions exist in large urban centres for the development of elements of direct democracy, and it is to these that we turn first.

Direct Participation
A common form of direct participation is that of the *ad hoc,* or individual, activity of a citizen or a group of citizens, undertaken in the public interest with a view to either improving a given situation (such as a commuting service) or righting an unjust decision. Citizens may express themselves by means of the mass media or a form of intervention. In some countries an arrangement exists for considering public grievances that may be submitted either in writing or orally at a meeting of a local government committee. The decision of the committee

is expected, if reasonable, to be acted upon by the criticized governmental body.

Another form of direct participation is the consultation of citizens. This is generally of a voluntary nature. It may take the form of an explanatory consultation and be conducted by special officials, as between British physical planning authorities and the public, or it may be conducted by authorized citizens. Alternatively, it may take the form of gathering the opinions of temporary or permanent bodies, as in the Netherlands where the system is used for testing draft bills of local government legislation. Consultation is also used to prepare the public for proposed planning solutions. For example, in Paris this would take place before introducing basic changes in traffic regulations.[22]

A third form of direct participation involves citizen groups or special bodies of the local government in making decisions on public issues, which are then implemented by the government or one of its agencies. These indirect representative bodies are generally composed of both citizens and councillors. They may be specialized in nature (such as the organization of private merchants in Poland), or fairly generalized (as is illustrated by the participation of trade union representatives in the allocation of public housing units). In some instances the body may develop public activities themselves as, for example, do the regional hospital councils in the United Kingdom.

A fourth type of direct participation involves the implementation by the community organizations of certain metropolitan functions. The most common are in the fields of child welfare, social welfare, public safety and the supervision of cultural activities. While community organizations ensure the performance of the function, expenditures are covered by the municipal budget.

Indirect Participation

Being far more common and far better known, there is little need to dwell on forms of indirect participation. Since the power and role of the representative governmental bodies are discussed elsewhere, we shall confine our comments to the contacts between the elector and the representative body. In socialist countries the elected councillor is expected to meet with his electors, present reports to them, listen to their opinions and so on during his term office. He is subject to recall by the electorate if it is thought that his activity is not in conformity with the electoral platform instructions. Apart from the link with his constituency, a councillor also has organizational ties such as a party, or social or professional or economic organization that nominated him during the campaign. There are also his associations with block committees and larger groups. The links between the councillor and these committees ensure closer relations between the activities of all government bodies and the local needs of inhabitants. The development of this form of self-governing activity provides a natural basis for stimulating community forces as well as a broad platform for strengthening the principles of the operation of government and public administration in the metropolis.

While these illustrations do not cover all forms of direct or indirect participation in solving the vital problems confronting a community, they do cover the more com-

[22]M. Soysal, *Les relations publiques en matière administrative: II–L'action des administres sur les fonctionnements publiques.* General report of Thirteenth International Congress of Administrative Sciences, 1965.

mon forms.[23] One well-known method that has not been investigated here is the use of the popular vote or referendum. Although this form of direct involvement in decision-making is often used in Yugoslavia and Switzerland, the number and the range of issues settled in this manner is limited.

The democratic implementation of public functions by governmental bodies in metropolitan areas depends on the development of a defined system of activity in which the community is involved. The determination of the form of this involvement must be based upon tradition, the character of the urban area, and the national political and economic structure.

The Horizontal Dimension: Unitary versus Federated Systems

With the increasing intensity of urbanization, man's individual needs become common needs and government is looked to increasingly to meet these demands throughout the urbanized area. In metropolitan areas, various forms of government have evolved. Basically, two rudimentary systems exist: the unitary system, calling for a unified organization of local government and administration; and, the federated system, calling for maximum decentralization of responsibility to local government units within the federation. In practice, neither of the two extreme forms is common and generally a combination of the two is found. In recent years complex metropolitan area reforms have led to yet new combinations as in Toronto, Dade County, Paris and London, and, to some extent, Warsaw. These reforms are needed not only for improved cooperation between central and suburban areas in the metropolis, but also to provide a realistic area of operation for any function and a means for regulating the growth of metropolitan areas.

These latter dictates have given rise, in socialist countries, to a number of studies[24] on the optimum size of urban settlements, and in both socialist and non-socialist countries, there have been many attempts to regulate urban growth.[25] The growth regulation programmes (or deglomeration programmes as they are known) may be of either an active nature, involving the use of incentives to encourage migration from the metropolis to another urban area and the actual relocation of industry in a similar manner, or they may be passive and involve restrictions on development and attempts to reduce commuting. Generally, the passive programmes are more common, as they are less costly and more susceptible to influence by a metropolitan area authority The active programmes, such as the development of new towns in the United Kingdom, generally call for central government participation.

The problem of relating the territorial extent of the governmental unit to a realistic area of operation of services is compounded by the technological advances being made, especially in transportation and communications. Thus, a compromise is generally struck that will attempt first to incorporate within one unit highly urbanized areas that should be encouraged to develop as one and, secondly, to provide

---

[23]See also *Miestnyje sowiety na sowriemiennom etapie* (Moscow, Nauka, 1965), p. 62 ff.

[24]S. Nowakowskiego, *Socjologiczne problemy miasta polskiego, praca zbiorowa pod redakcja* (Warsaw, P.W.N., 1964). This contains a comprehensive bibliography on urban sociological research.

[25]W. Brzezinski, "Plan zagospodarowania przestrzennego. Studium porównawczo prawne na tle prawodawstw kilku panstw europe jskich," *Problemy prawne planowania gospodarczego* (Warsaw, P.W.N., 1964), p. 261 ff.

a means for solving the problems of suburban areas functionally related to the central city, but not otherwise integrated with it politically or administratively. These needs provide the normal rationale for the development of a metropolitan area form of government or the creation of new administrative units. On the other hand, even where changes have been made they may still result in the exclusion of part of the urban area from the unit of metropolitan government, as for example in Buenos Aires and Tokyo. In other areas, such as London and Moscow, very extensive units have been created. This alternative will not be effective in itself, however, unless the unit of government is provided with the power and tools to plan and develop the area as one.

In the Paris region the very extensive District of the Region of Paris has been created as an administrative body to ensure coordination between the eight lower level units of government. It consists of the centrally nominated Prefect and an advisory body, in turn consisting of, among others, representatives of the eight lower units, the chief administrative officers of public services, the prefect of police and the treasurer of the Region. This body debates issues, presented to it by the Prefect, pertaining to the responsibilities of the Region, which include the implementation of central government policy for the economic development of the region, investment planning, physical planning, the provision of transportation and health services, and some control over construction. In addition, the activities of all the members of the advisory body are subject to the supervision of the Prefect.

Another means by which the metropolis, or indeed, simply the central city, may be formally related to its surrounding region is through the creation of common organizational units. Two examples are given here. The first, although rare, is the creation of common administrative organs, subordinate to the metropolitan government and to the governments of the surrounding territorial units. This solution, while successful in smaller units, fails in larger urban areas because of the scale of the problems faced. Polish regulations permit the creation of common administrative organs (in the fields of education, culture, health, agriculture and so on) directed by executive committees of the People's Councils of two territorial units. Although such organs remain administratively connected with one of the executive committees, the implementation of decisions is problematical due to difficulties arising from double subordination.

The second form of the common organizational unit emerges from the practice of organizing common enterprises (for example, transportation, power and water supply, sanitation) or public institutions (specialized hospitals, sanatoria, social welfare centres), and it appears to be much more successful.[26] These units are directed either by a common administrative council or by organs of one of the territorial units. Income or losses are shared according to previously determined principles, proportionally to services rendered. This arrangement satisfies the needs of the suburban areas adjacent to the metropolis without intervention from central government. This solution thus makes possible decentralization of public functions without too great a loss of coordination.

Various other forms of cooperation between metropolitan and suburban govern-

_____

[26]André Martin, "La 'Public Corporation'–forme moderne de l'intervention de l'Etat dans l'économie de la Grande Bretagne," *La Revue Administrative*, No. 93 (Paris, 1963).

ments and public administrative bodies are possible. This cooperation may have an institutional character, as with permanent working or coordinating bodies, or it can be quite informal. The value of such arrangements is that they can be used as means of organizing cooperation between metropolitan organs and those of the state administration on a functional basis. This is particularly useful since the metropolis is often the capital and is therefore expected to combine local and nation-wide responsibilities such as locating government buildings, maintaining public order, and satisfying the needs of foreign representative bodies. Land-use and transportation planning are two areas in which cooperation is especially desirable. For this purpose organs of two types, known as coordinating and auxiliary bodies, are often created.

Coordinating bodies may have the power of decision on issues determined by the cooperating partners. Sometimes a common session of collegial public administrative organs may act as the coordinating body. A coordinating body may operate only in one specific field, such as transportation, in which case the members act according to instructions from their superiors.

Auxiliary bodies, such as research bureaus, play an important role both of a preparatory and an executive nature. This role depends on the problems studied as well as on the necessity for continuing cooperation among people representing cooperating organs. Their part in the elaboration of physical plans for the metropolis and its neighbouring units (regardless of administrative boundaries) makes for a better draft of such plans, and also provides for an exchange of experiences and continuing cooperation among officials of different public services.

These or other methods of cooperation between government and administrative organs can be utilized for solving problems common to the metropolis and its surrounding areas without affecting the jurisdiction of the cooperating organs. The choice of approach depends on the particular conditions and needs of the urban agglomeration, and the degree of success will depend on the functions in question and on the manner in which implementation is handled.

The System as a Response to Internal Needs

Thus far the emerging metropolitan systems of government have been viewed primarily in terms of their response to the needs of the metropolis as an entity or to the demands made upon the metropolis by a supra-metropolitan society. However, within the structure resulting from this interplay of forces there must be sufficient flexibility to permit response to the more local internal demands on the system. In order to round out this review of the relations of functions to government structure in the metropolis, it is now necessary to turn to this matter.

The emergence of the classical structures of territorial self-government in European cities produced, with the existing operating organs of central government administration (of a prefectural type), a dualistic structure of government with both types of organs existing side by side to satisfy various needs of the population.[27] In the large centres, particularly in capitals, there was a tendency toward strengthening the power of the central government organs, as is reflected, for example, in the institution of the Prefecture of the Seine and the Prefecture of Police, both for Paris and its immediate

---

[27]Humes and Martin, *loc. cit.*

neighbourhood. In the inter-war period in Poland, the Government's Commissioner for the Capital City of Warsaw, was instituted to represent central government and self-government powers. Similar tendencies appeared also in Rome and Brussels, and in several places the relics of this dual structure still exist.[28]

Today, a monistic structure is more common and basically this can take one of two forms. The first involves the creation of uniform organs within the metropolitan area with their structures based not on the division between central government and local government, but on the principles of uniformity and integration for the country as a whole as, for example, in Warsaw.[29] The second involves the formation of a particular structure for each individual function, as is the case for example, in Hamburg.[30] In either case, however, supervision is performed by superior organs representing nation-wide interests; the nature of this supervision depends on the general system of government and administration and on the subject of supervision.

More important, however, than the above variations, is the degree to which internal decentralization takes place in the metropolitan system of government and administration. It is out of the confrontation of the two forces, one arising from the uniformity of structure and the other from the citizen demands for the decentralization of powers in metropolitan organs, that there emerges the concept of the metropolis divided into districts, each with its own administrative organs.

The nature of the district organs will vary greatly according to the degree of self-discretion permitted. In Paris, the functions of these bodies are performed by single officials (mayors), nominated by central government, directing their own staff. The power of the district councils is limited, almost exclusively, to some regulatory functions. The position of the mayor is quite different from that of mayors in other communities where they represent territorial self-government and the local community. In Paris, they are officials of the state administration. The limited power of the mayors of the *arrondissements* of Paris is a product of the French system of government and the fact that there are twenty such units, each of which is too small to perform more important functions.

In contrast to their counterparts in Paris, the power of the London boroughs is considerable and their organization is adapted to the principles generally applied throughout Great Britain. Each borough has its own Borough Council consisting of a mayor, aldermen and councillors. The council nominates committees and each of these performs governing functions relating to the responsibilities of the committee in question. The Borough Councils are responsible for public health and sanitary inspection, housing inspection, supervision of trade, sanitary services, street cleaning, upkeep of public parks and gardens, street lighting, supervision of open air markets, public libraries, collection of taxes and other payments, registration of electors, and so on. The decisions of a Borough Council, as well as its Committees, are im-

---

[28]For example, in Stockholm, apart from local government bodies represented by the City Council (of 100 members and 9 City Directors), a Royal Governor is nominated for the general supervision of the overall administration of Stockholm. See Nils Andren, *Modern Swedish Government, Stockholm, Göteborg, Upsala* (Sweden, 1961).

[29]For further details see Zygmunt Rybicki, *Dzialalnosc i organizacja rad narodowych w Polskiej Rzeczypospolitej Ludowej* (Warsaw, 1965).

[30]Roman Schnur, "L'Organisation de l'administration fédérale en République Fédérale Allemande," *La Revue Administrative*, Nos. 105 & 106 (Paris, 1965).

plemented by administrators headed by the town clerk who is responsible to the mayor for the proper functioning of the administration.[31]

Polish legislation provides a more flexible procedure for determining the jurisdictional powers of the district organs (the District People's Councils). Although in all parts of the country the district bodies are responsible for a number of functions and similarly there are certain powers they can never be given, it is assumed that conditions will differ across the country. Consequently, the Municipal People's Councils are given discretion to decide upon which additional powers of jurisdiction can be assigned the district bodies. The assumption of additional responsibilities must be accompanied by the provision of financial and other resources to carry out these responsibilities. In practice, the officials of the Municipal People's Councils tend to favour retention of responsibilities at the metropolitan-wide level, while community politicians or citizen members of commissions favour greater decentralization. Despite this conflict of views, however, informal, social and organizational relations between councillors and administrators are of great significance in facilitating decision-making.

Indeed, decentralization has been very successful in Warsaw. In 1964, the Municipal People's Council made a further move to foster citizen self-government through the creation of more than 100 settlement councils. These councils are elected on a residential area basis. They have no executive apparatus, and all their work is done by volunteer members and citizens. They organize meetings of local inhabitants with councillors and present their motions to district or municipal organs. Not all Settlement Councils make full use of their power, though some have achieved a great deal due to the activity of their members or particular sections of the councils. They are most effective in the improvement of local, social and cultural conditions. They also form an informal, social link between the district bodies and the block committees established some years ago.

Block committees are the smallest units of citizen self-government. The average number in one settlement ranges from six to ten. They cooperate with the administration of the municipal dwelling blocks and supervise (by observing) their work. They also maintain direct contact with the settlement councils. Several block committees have achieved much in providing better living conditions, recreation facilities and improved care of children and old people.

We have described how various forms of citizens' participation provide the basis for the interaction of elements of indirect (representative) and direct democracy. The importance of social supervision over the powers of all representative organs has been also emphasized. This supervision is particularly important in the system of developed municipal services covering not only communal services, but also housing, organization of trade systems, service centres, local industry, public education, health service and social welfare. Proper operation of these functions is an important factor in the process of satisfying the population's daily needs. This goal is achieved through the self-governing activity of citizens as well as through the work of government and public administrative organs of the metropolis.

---

[31]"The City Manager and Chief Administrative Officer," *Public Administration*, Journal of the Royal Institute of Public Administration, Vol. 42 (London, Summer, 1964).

Apart from any specific conclusions that have been made already, more general conclusions can be drawn from our observations of the rapidity of the processes of urbanization and the formation of large metropolitan areas. The controlled growth of large urban agglomerations and the complex nature of the associated technical problems dictate that future development of the metropolis be planned rather than be left to the forces of circumstances. Awareness of the need for such planning and the need for an integrated planning process, embracing simultaneously the physical, technical, social and economic elements are not lacking. The problem, unfortunately, is that this postulate is not always implemented.

The planning area must include not only the actual metropolitan area, but also areas for future expansion. As a corollary, the planning authority may plan for a territory embracing a multitude of administrative units. Hence, it may constitute a regional or state agency or it may be established as a common special-purpose body upon agreement between interested local governments.

The implementation of a plan will depend upon the manner in which it is prepared; upon the relationship of these two stages; and upon the legal, organizational and financial resources made available. The endowment of appropriate authority to governments and administrative organs in the metropolitan area depends on: the country's general political system and structure; the role of the metropolis from the viewpoint of national administration (for example, as capital of the State); and, the degree of decentralization of decision-making on technical and administrative matters. The first two factors are almost independent of the activity and initiative of the citizenry of the metropolis. The third, however, can be modified by the impact of the initative of the metropolitan community without changing the general political structure of the country.

In major urban areas technical problems in the fields of transport, water and energy supply, heating and land use are of particular significance. This increases the importance of the respective administrative organs. The nature of these services and the desirability of an economic and efficient manner of operation dictate that the administration of technical services should be centralized and concentrated at the metropolitan area level. Their administration tends soon to become divorced from the field of general administration, and such specific autonomy and regionalization often spills over the political boundaries of the metropolis. This is particularly true in rapidly developing metropolitan areas where the importance of technical services is paramount and a flexible, dynamic administration for the provision of these services is indispensible. In such instances, the division between the technical and general elements of administration is widened even further.

In spite of tendencies to exaggerate these aspects of technical services, responsibility for, and supervision of, their functioning ought to remain subject to representative organs. There is, however, the additional possibility of decentralizing certain aspects of technical services (for example, by separating energy distribution from production of energy) and allocating decentralized services to sub-units of the metropolis (for example, district governments).

The efficiency of general administration and technical services should be measured in terms of the degree to which citizen needs have been met. The psychological phenomena associated with the community's feelings of co-responsibility for the health of the metropolis must be considered.

One of the basic issues confronting present-day government of our times is the continuing breakdown of the division between rulers and the ruled. This necessitates a search for structural and functional forms of government and administration in metropolitan areas that will develop public participation by involving representative (indirect) democracy with elements of direct democracy.

There are two features of government systems in metropolitan areas that differ markedly from those of smaller centres: the strong links with nation-wide government systems; and the appearance of differentiated and intense social interest for satisfying citizens' needs. In the small and medium-sized town, direct involvement of inhabitants in the functioning of government and administration is quite possible. Inhabitants are aware of the impact of intense public participation on the performance of these bodies. The unsatis-

factory condition existing in most metropolitan systems can be eliminated by establishing organs of self-government for residents, which can contribute to the implementation of tasks defined by law. The efficiency of such collaboration is a function of the range of jurisdictional powers of administrative organs, for satisfying the social, communal and cultural requirements of citizens.

Hopefully, it has been shown that there is no possibility of either automatic transfer of government systems from one territory to another, or the adoption or conservation, in unchanged form, of traditional systems of government developed under different historical conditions.

Finally, the structuring of government systems in metropolitan areas must be based on the structure of nation-wide institutions, the demands of modern technology and metropolitan functions, and historical traditions. The relationships between these and other elements must be considered when studying solutions in other metropolitan areas, when making comparative studies, and when applying foreign experiences to our own situation.

10 Metropolitan Political Systems and the Administrative Process

Frank Smallwood

As Victor Jones and Richard Forstall point out in the introductory chapter of this book, the contemporary metropolis generates "sufficient economic activity to produce highly concentrated and complex human settlements."[1] As a result of this complex interaction between the core city and its surrounding regional hinterland, the contemporary metropolis has coalesced into what William Robson and Luther Gulick have referred to as an "economic and social reality . . . closely tied together . . . by daily human movements and activities."[2] Yet, while the modern metropolis may have achieved an economic—and perhaps even a social—unity, many metropolitan areas have not developed into viable political communities in any meaningful sense of this term. On the contrary, most evidence indicates that the modern metropolis is suffering from a bewildering degree of administrative fragmentation, which makes it extremely difficult to identify and implement consensual goals on a metropolitan-wide basis.

---

[1] See p. 24 in Jones and Forstall's article in this book.

[2] William A. Robson, ed., *Great Cities of the World* (London, George Allen and Unwin, Ltd., 1957), p. 27, and Luther Gulick, *The Metropolitan Problem and American Ideas* (New York, Alfred A. Knopf, Inc., 1962), p. 24.

Despite the fact that this widespread governmental fragmentation has far-reaching ramifications, political as well as administrative, many studies have focussed almost exclusively on what Stefan Dupré refers to as the "efficiency" aspects of metropolitan government. There has been a concerted research fixation on potential administrative breakdowns resulting from fragmented governmental structures, without a concurrent analysis of the potential impact that this fragmentation may have on two other key aspects of the political process cited by Dupré in this volume—the values of liberty and participation. In short, much of the current research in metropolitan government is out of balance. It concentrates too heavily on potential problems of administrative proliferation, without attempting to analyze these potential problems within the context of their impact on political participation and political consent, which are vital to the life of any meaningful community.

The present essay represents an effort to broaden our understanding of both metropolitan political and administrative systems by evaluating the participatory, as well as the purely structural, aspects of these systems. The chapter is organized on the basis of input-output classifications, a research tool which has been developed to provide a more comprehensive overview of the total dynamics of the political process. As used on the following pages, input-output classifications simply imply that any open political system receives various *inputs* in the form of such participatory devices as votes, interest group demands, editorial commentary in the press and so on. In turn, these inputs influence the types of *outputs* the system provides in the form of specific policy decisions, the provision of basic services, and the like. Thus, inputs are the vital factors in any political system

that help shape the outputs the system produces.

Input-output analysis has not been restricted to the evaluation of urban and metropolitan political systems alone. In his study, *The Politics of Developing Areas*, Gabriel Almond suggests that all political systems—whatever their hierarchical level or their geographical area—can be characterized in terms of certain basic, and universal, functions. Some of these functions have a predominately input orientation to the extent that they feed into the system, and some are related to outputs that are produced by the system. Almond lists seven such input-output functions (illustrative examples provided in parentheses).[3]

1. Inputs

a) *Political Socialization and Recruitment* (basic social attitudes toward the system);
b) *Interest Articulation* (informal demonstrations, formal interest groups);
c) *Interest Aggregation* (political coalitions, party systems); and,
d) *Political Communication* (informal discussions, mass media).

2. Outputs

a) *Rule-making* (legislative functions);
b) *Rule-application* (executive functions); and,
c) *Rule-adjudication* (judicial functions).

In a more recent study entitled *A Systems Analysis of Political Life*,[4] David Easton attempts to refine input-output

---

[3]Gabriel A. Almond and James S. Coleman, eds., *The Politics of Developing Areas* (Princeton, Princeton University Press, 1960). Almond's analysis appears on pp. 3-64.
[4]David Easton, *A Systems Analysis of Political Life* (New York, John Wiley & Sons, Inc., 1965).

models by developing such innovative concepts as the "conversion" and "feedback" processes. By emphasizing conversion, Easton is attempting to theorize how inputs are actually translated, or converted, into output allocations on a continuous basis over an extended period of time. According to his analysis, one crucial aspect of the conversion process involves the kind of information that is fed to the governmental authorities responsible for producing outputs. These authorities can only attempt to match outputs with input demands on a continuous basis if there is "a flow of information back to the producers of outputs about the state of the system and its environment and about the effects of their own decisions and actions". Defining feedback as "the property of being able to adjust future conduct to past performances", Easton discusses different types of circular "feedback loops", which consist of four basic components:

1. *Feedback Stimuli*: the initial outputs, of appropriate authorities, that generate specific responses by the producers of inputs.
2. *Feedback Responses*: the specific responses, by the producers of inputs, to the initial outputs.
3. *Information Feedback*: the translation of these specific responses back to the producers of the initial outputs.
4. *Output Reactions*: the conversion of subsequent outputs on the basis of the information feedback.

Easton's analysis of "feedback loops" indicates that a highly complex and dynamic interaction between inputs and outputs characterizes the political process. The present essay does not attempt to encompass the more sophisticated aspects of this "loop" process. Instead, it focusses on a relatively modest use of inputs and outputs as a basic means of classifying some of the key components of metropolitan political and administrative systems in order to suggest the value that input-output techniques can have as an analytical tool, especially in studying the comparative aspects of widely divergent metropolitan areas.

Metropolitan Political Systems: An Analysis of Inputs

One of the major problems involved in any comparative study of metropolitan political systems results from the relative scarcity of data on such systems. In an effort to overcome this problem, a questionnaire was distributed to the metropolitan areas that participated in the Centennial Study and Training Programme on Metropolitan Problems. Respondents from twenty-three areas replied to the questionnaire. Eight of these respondents were from Europe and the Near East (London, Glasgow, Paris, Warsaw, Prague, Belgrade, Milan and Istanbul); four from Canada (Montreal, Toronto, Winnipeg and Vancouver); four from the United States (New York, Philadelphia, Detroit and San Francisco), two from Latin America (Rio and Lima), four from the Far East and Oceania (Tokyo, Seoul, Manila and Sydney) and one from Africa (Ibadan). In addition, more limited political data were received on Conurbation Holland (Amsterdam-Rotterdam-The Hague) and on Madrid, to encompass twenty-five areas in all.

VOTER PARTICIPATION

The questionnaire was designed to provide both quantitative and qualitative data on key political inputs in the different areas. The first input concerned electoral behaviour as reflected in the total number of registered voters and the actual percentage

TABLE 10-1
Urban and Metropolitan Electoral Participation

Metropolitan Area and Election	Year	Total Registered Electorate	Actual % Turn-out
1. Prague (City)	1964	730,069	99.6%
2. Warsaw (City)	1965	883,724	96
3. Milan (City)	1964	1,196,438	92
4. Lima (Metro Area*)	1963	692,724	90.6
5. Belgrade (Chamber)	1963	561,378	87.6
6. Rio (City)	1966	1,497,401	85
7. New York (City)	1965	3,281,689	80.2
8. Sydney (Metro Area*)	1965	1,551,055	74.2
9. Manila (City)	1963	431,541	70.2
10. Tokyo (Metro Gubernatorial)	1967	6,318,000	67.5
11. Paris (City)	1965	1,544,201	65.3
12. Philadelphia (City)	1965	982,000	63.6
13. San Francisco (City)	1965	346,226	61
14. Ibadan (City)	1965	105,562	61
15. Detroit (City)	1965	800,341	58
16. London (Greater London Council)	1964	5,446,756	44.2
17. Istanbul (Metro Area*)	1963	405,133	40
18. Toronto (City)	1966	432,757	36
19. Seoul (City)	1960	1,116,383	36
20. Glasgow (City)	1966	642,023	34.5
21. Winnipeg (Metro)	1964	284,609	33.4
22. Montreal (City)	1966	380,067	32.7
23. Vancouver (City)	1966	253,000	30

* By "Metro Area" is meant metro-wide, but not at a metropolitan, level. No such level of government exists in the three metropolitan areas in question.

of voting turn-outs in the last major elections held in the different areas.[5]

As Table 10-1 indicates, the range of differences in voter-participation rates is quite striking. Turn-outs varied from a low of 30 per cent to a high of 99.6 per cent, with seven cities falling in the 30 to 40 per cent range and four cities falling in the 90 to 100 per cent range. The mean percentage turn-out for all twenty-three areas indicates that an average of 62.5 per cent of the registered electorate voted in the elections cited in Table 10-1. In addition, the respondent from Conurbation Holland indicated that voter turn-out in the Amsterdam-Rotterdam-The Hague area usually averaged from 90 to 95 per cent of the electorate, although he did not provide any statistical data for specific elections. The

[5]It is very important to point out that the voter participation figures that follow are influenced heavily by registration methods. For example, in Toronto, all qualified voters are automatically registered voters whereas in New York City, many of these who have all the needed qualifications have not registered. Thus, the participation figure for Toronto is understandably less than that for New York City. Yet, although comparable figures cannot be made available, useful information on voter participation can be gleaned from the available figures. It is on this basis that the following observations are made.

TABLE 10-2
National and Urban Electoral Participation

National Election	Turn-out	Urban/Metro Election (See Table 10-1)	Turn-out
1. Czechoslovakia (1960)	99.8%	Prague (1964)	99.6%
2. Poland (1965)	96.6	Warsaw (1965)	95
3. Australia (1966)	95.3	Sydney (1965)	74.2
4. Italy (1963)	92.5	Milan (1964)	92
5. Peru (1963)	90	Lima (1963)	90.6
6. Yugoslavia (1963)	90	Belgrade (1966)	87.6
7. France (1965)	85	Paris (1965)	65.3
8. Great Britain (1964)	77	London (1964)	44.2
		Glasgow (1966)	34.5
9. Canada (1963)	77	Toronto (1966)	36
		Winnipeg (1964)	33.4
		Montreal (1966)	32.7
		Vancouver (1966)	30
10. Brazil (1966)	75	Rio (1966)	85
11. Japan (1963)	71	Tokyo (1967)	67.5
12. Philippines (1965)	70.4	Manila (1963)	70.2
13. Korea (1963)	69.8	Seoul (1966)	36
14. Turkey (1965)	68	Istanbul (1963)	40
15. U.S.A. (1964)	63	New York (1965)	80.2
		Philadelphia (1965)	63.6
		San Francisco (1965)	61
		Detroit (1965)	58
16. Nigeria (1964)	26.6	Ibadan (1961)	61

Sources: National turn-outs, from *Keesing's Contemporary Archives* (London, Keesing's Publications Ltd.), Vol. XII (1960-61) and Vol. XV (1967-68), are as follows: Czechoslovakia (National Assembly, p. 17570); Poland (Seym, p. 20837); Australia (House of Representatives, p. 21919); Italy (Chamber of Deputies, p. 19617); Peru (General Election, p. 19516); Yugoslavia (Communal, District, Republic of Federal Assemblies, p. 19680); France (Presidential, p. 21200); United Kingdom (Parliamentary, p. 20349); Canada (Parliamentary, p. 19413); Brazil (Chamber of Deputies, p. 21939); Japan (House of Representatives, p. 19836); Philippines (Presidential, p. 21182); Korea (National Assembly, p. 19822); Turkey (National Assembly, p. 21037); Nigeria (Federal House, p. 20578); Percentage for 1964 Presidential election in the United States from *Statistical Abstract of the United States* (Washington, D.C., Department of Commerce, 1968), p. 371.

respondent from Madrid also failed to provide statistical data on voter participation, but the *New York Times* of November 21, 1966, reported that from 30 to 35 per cent of the voters turned out for the Madrid City Council election of November, 1966.

In an effort to determine how the voter turn-outs reported in Table 10-1 compared to national electoral turn-outs, an analysis was made of voting patterns in national elections that corresponded as closely as possible to the dates of the local elections reported in Table 10-1. Such a comparison is provided in Table 10-2.

It is significant to note in Table 10-2 that electoral turn-outs in the great majority of national contests were either equal to, or substantially higher than, turn-outs in local elections. Actually, the percentage of voters in local elections exceeded national turn-outs by 10 per cent or more in only three of the twenty-three areas reporting (New York, Ibadan and Rio). A number of factors help explain why these three areas experienced this unique distinction. The New York City turn-out for the mayorality election of 1965 was exceptionally heavy because Mayor John Lindsay made extensive use of volunteer, neighbourhood organizations to bring out the vote in support of his successful bid for office. Electoral percentages in Ibadan and Rio, on the other hand, exceed national turn-outs in Table 10-2 due to the fact that the national elections for Nigeria and Brazil cited in this table attracted an unusually low number of voters. The 26 per cent turn-out in the Nigerian election of 1964 was a result of the fact that large numbers of potential voters, especially in Eastern Nigeria, boycotted the Federal election that year in protest over a variety of internal regional disputes that eventually erupted into civil war.[6] In a similar manner, as a result of widespread protests against inflation and other domestic disputes, one quarter of the voters either abstained or cast blank ballots in the Brazilian national election of 1966 and another 25 per cent cast invalid ballots despite the fact that voting is compulsory for all literates in Brazil.[7] Hence, Table 10-2 indicates quite forcefully that the national elections tended to attract a higher percentage of voters than the local contests

unless there were unique factors at work to reverse this trend.

POLITICAL PARTY ACTIVITY

A second major factor covered in the questionnaire related to political party activity in the different areas. As Table 10-3 indicates, nineteen of the twenty-four areas reporting on this aspect of the questionnaire indicated that political parties played an active role in local elections in their respective areas. Of these, twelve indicated that national parties were active in local elections, six reported a mixture of national and local parties, and only one (Montreal) relied on a strictly local (Civic) party. There was no party system at the local level in Vancouver, Toronto, San Francisco or Detroit at the time the questionnaire was distributed (1966-67), although national/ provincial and local party candidates did run in the Toronto borough and Metro elections in December, 1969.[8] In addition, the respondent from Belgrade indicated that functional and territorial groupings, rather than parties, are used as the basis for local elections in Yugoslavia.

It is extremely difficult to make any generalizations about the possible relationship between party activity and electoral participation on the basis of the data that appears in Table 10-3. The only clear-cut relationship that appears to emerge is that voter turn-out is very high in the one-party (that is, Communist) areas of Prague and Warsaw, but as the subsequent discussion indicates, a variety of factors in addition to political party activity may account for this strong showing.

---

[6]*Keesing's Contemporary Archives,* Vol. XV (London, Keesing's Publications Limited, 1965-66), p. 20578.
[7]*Ibid.,* Vol. XVI (1967-68), p. 21939.

[8]Bruce M. Russett, ed., *World Handbook of Political and Social Indicators* (New Haven, Yale University Press, 1964), pp. 84-86. The other nations cited as areas where voting was legally required were Belgium, Greece, Venezuela and Argentina.

TABLE 10-3
Political Party Activity

Metropolitan Area	Voter Turn-out (per cent)	Parties Used	Number of Parties
1. Prague	99.6%	National	One
2. Warsaw	96	National	One
3. Milan	92	Mixed	Many
4. Lima	90.6	National	Many
5. Conurbation Holland	90	National	Many
6. Belgrade	87.6	Non (Functional and Territorial Groups)	
7. Rio	85	National	Two
8. New York	80.2	Mixed	Four Major
9. Sydney	74.2	Mixed	Four Major
10. Manila	70.2	National	Two Major
11. Tokyo	67.5	Mixed	Many
12. Paris	65.3	National	Many
13. Philadelphia	63.6	National	Two Major
14. San Francisco	61	None	Non-Partisan
15. Ibadan	61	National	Three
16. Detroit	58	None	Non-Partisan
17. London	44.2	National	Three Major
18. Istanbul	40	National	Two Major
19. Toronto	36	None	Non-Partisan
20. Seoul	36	National	Four
21. Glasgow	34.5	Mixed	Two Major
22. Winnipeg	33.4	Mixed	Three Major
23. Montreal	32.7	Local	One Major
24. Vancouver	30	None	Non-Partisan

INTEREST GROUP ACTIVITY

A third factor covered in the questionnaire related to interest group activity in the different areas. Respondents were asked to "identify major local interest groups (if any) that you feel play a key role in influencing local political decisions." There may have been some confusion regarding this question since half the respondents either did not provide any answer at all, or else they indicated that there was no interest group activity in their area. Thus, it will simply be noted that the following groups were listed by the eleven respondents replying to this question (with the number of times a group was mentioned indicated in parentheses): chambers of commerce (8), labour or trade unions (7), civic or service organizations (6), builders associations or other commercial groups (5), churches (4), welfare or low income housing groups (3), education associations (3), public employees associations (2), press (2), universities or university students (2), war veterans (2), sports clubs (1).

Even this relatively limited listing indicates that there is at least one striking similarity in interest group activity among the areas reporting. This is to be found in the preponderance of economically-oriented groups named. Both chambers of commerce and labour or trade unions were cited as playing a significant role in local politics by more than half of the eleven respondents who replied to this part of the questionnaire. Similarly, private builders and other commercial organizations were cited by five respondents. Welfare and educational associations, on the other hand, were cited with considerably less frequency.

It is difficult to go beyond the one observation that economically-motivated interest groups appear to exercise significant influence in most of the metropolitan areas reporting. Other metropolitan areas, however, appear to be immune to any meaningful interest group activity at all. The respondent from Toronto, for example, indicated that there were no major interest groups acting at the Metro level. For Toronto this observation is supported by research done by Harold Kaplan who notes that, private interest groups in the Toronto area have been conspicuous by their weak involvement in issues coming before the Metro Council. . . . The only significant source of private influence over Metro Council proceedings is the Toronto press."[9]

---

[9]Harold Kaplan, "The Policy-making Process in Metro Toronto" (paper presented at Annual Meeting of the American Society for Public Administration, Kansas City, Missouri, April, 1965), p. 9.

POLITICAL INPUTS: SUMMARY OBSERVATIONS

Although twenty-five areas represents a very small sample upon which to base any major conclusions, it is obvious that there are some striking variations between the different areas with respect to their patterns of electoral participation, their political party activity and even their utilization of interest groups. This raises the intriguing question of what factors might help explain these basic variations. Obviously, any suggested explanations must be highly tentative. In addition to the difficulties inherent in any cross-cultural comparison, it is impossible to compare results for single elections alone (Table 10-1) and hope to account for all deviant cases. Despite these problems, however, it is possible to speculate on some of the potentially important factors that may have influenced the variations noted.

Indigenous Procedural Influences
Basic procedural differences among the areas may account for some of the variations noted. Voting legislation, for example, can be an extremely important procedural influence. According to the *World Handbook of Political and Social Indicators,* eight nations cited here had some form of compulsory voting requirements in the mid-1960's.[10] Respondents from four of these nations—Italy, Netherlands, Brazil and Australia—all reported high voting turn-outs; for Milan (92 per cent), Conurbation Holland (90 per cent), Rio (85 per cent) and Sydney (74 per cent). The 90 per cent voter participation in a place such as Conurbation Holland undoubtedly reflects the Dutch electoral law which requires that all eligible voters must appear

---

[10]Russett, *loc. cit.*

at polling stations on election day (although they need not cast an actual ballot).

A second procedural factor that could have had an obvious impact on the voting turn-outs reported in Table 10-1 was the fact that the local elections coincided with national elections in four of the areas—Prague (99.6 per cent), Warsaw (96 per cent), Belgrade (87.6 per cent) and Rio (85 per cent). As all four of these areas reported exceptionaly high turn-outs, it would appear that the timing of the elections helps to explain some of the variations in electoral behaviour noted in the different areas.

Differences in governmental structure may also shed some light on divergent patterns of political behaviour in the various areas. In Toronto, for example, Kaplan notes that "the relative absence of pressure [groups] on the Metro Council is partly attributable to the federal structure of government in the Toronto area. . . . The ratepayers' associations are less interested in general policy than its detailed application, and for this reason they act less at the Metro than at the local level."[11]

It is important not to minimize the impact that procedural and structural arrangements can have on the political process in the different areas reporting. In Nigeria, for example, the local electoral system is very different from that in the United States or Canada. In Nigeria, voters are asked whether they prefer to cast their ballots in areas where they currently reside (for example, Ibadan), or whether they would like to continue to vote in their native village elections even though they may have left these villages years ago.

Many voters apparently retain their village loyalties and cast their ballots "back home", even though they currently reside in urban areas. Any such procedure can obviously raise havoc with statistics when one attempts to compare patterns of electoral participation in Ibadan with patterns in a place such as New York City. All of this reinforces the point that extreme caution must be exercised before attempting to generalize too broadly about comparative patterns of political behaviour in a cross-cultural context.

Indigenous Attitudinal Differences
Differences in basic attitudes toward the political process play a crucial role in influencing political behaviour. In one of the more recent studies of voter turn-outs in American elections, four well-known political scientists argue that the intensity, or strength, of electoral preferences can be reinforced by such factors as interest in the political campaign, concern over the outcome, the voters' sense of political efficacy and the voters' sense of civic duty.[12] The high voter turn-out in Prague, for example, is in keeping with the generally high local and national totals that are reported for all of the Communist bloc countries. According to the *World Handbook of Political and Social Indicators,* the following five countries had the highest number of votes in national elections as a percentage of voting-age population: 1958, U.S.S.R. (99.6 per cent); 1957, Bulgaria (99.2 per cent); 1960, Czechoslovakia (98.1 per cent); 1957, Romania (97.9 per cent); and 1958, East Germany (97.3 per cent).[13] Despite

[11]Kaplan, *loc. cit.* Kaplan makes an even more detailed analysis of "Factors Depressing Group Involvement" in Toronto in *Urban Political Systems* (New York, Columbia University Press, 1967), pp. 172-75.

[12]Angus Campbell, Philip E. Converse, Warren E. Miller and Donald E. Stokes, *The American Voter* (New York, John Wiley & Sons, Inc., 1964), pp. 49-64.
[13]Russett, ed., *loc. cit.,* p. 84.

TABLE 10-4
Attitudes Toward Community Affairs

Per Cent Who Say The Ordinary Man Should:	U.S.	U.K.	Germany	Italy	Mexico
"Be active in his community"	51%	39%	22%	10%	26%
"Take part in activities of local government"	21	22	13	5	11
"Try to understand and keep informed about his community"	21	11	24	6	29
"Vote in his local community elections"	40	18	15	2	1

the fact that some American observers are suspicious of these statistics and tend to take them lightly, it appears that general attitudes toward the electoral process in the Communist countries do tend to generate a high degree of voter participation. To a certain extent this high participation receives official encouragement by means of procedural devices that are designed to minimize the difficulties encountered in absentee voting, or by the treatment of significant elections as national holidays. In addition, as already noted, it is a general practice in these countries to hold local and national elections at the same time, which undoubtedly helps to reinforce voter turn-outs at both levels. Finally, the electorate in Communist countries appears to regard the act of voting as a reflection of civic duty. Voting in many of these one-party nations appears to take on the guise of a solidarity ritual in which a vote for the party actually represents a vote for the nation and its ideology. Non-voting or a "no" vote would represent a repudiation of the nation. Despite the worldwide ferment among the young, most citizens of voting age are not apt to express themselves in such a drastically negative fashion. Hence, the high

turn-out in Communist nations may well represent a voluntary commitment, with little coercion involved except for the logical possibility that many members of the electorate may regard voting as an appropriate way to ingratiate themselves with party officials.

In an effort to evaluate variations in attitudes toward local civic participation, Gabriel Almond and Sidney Verba made a sample survey of approximately 5,000 respondents in five countries—the United States, the United Kingdom, Germany, Italy and Mexico—in a recent study entitled *The Civic Culture*.[14] As Table 10-4 indicates, the Almond-Verba study revealed wide variations in attitudes toward community affairs in these five countries. Almond and Verba suggest a number of factors as being potentially relevant to these variations. One deals with structural differences in local governments in the five countries. In Almond and Verba's words, "From the point of view of respondents'

---

[14]Gabriel A. Almond and Sidney Verba, *The Civic Culture* (Boston, Little, Brown and Company, 1963). Table 10-4 is a consolidation of Almond and Verba's Tables V.1 and V.2 appearing in pp. 127 and 129.

TABLE 10-5
Sense of Civic Competence

Nation	Per cent who say they can do something about unjust local regulation	Per cent who say they can do something about unjust national regulation
United States	77%	75%
United Kingdom	78	62
Germany	62	38
Italy	51	28
Mexico	52	38

Source: Gabriel A. Almond and Sydney Verba, *The Civic Culture* (Boston, Little, Brown & Company, 1963), Table VI.1, p. 142.

attitudes toward participation within the local community, there are two types of structural differences that are particularly significant: the degree of local autonomy and the degree to which local structures foster citizen participation." They go on to note "substantial differences" with respect to the extent of the local governments' freedom from external control. This appears to be greater in the United States than in other countries. Great Britain ranks next in local autonomy according to their analysis, while Germany and Mexico rank in the middle in this respect, and Italy displays the least local autonomy of the five nations studied. Another consideration analyzed by Almond and Verba involves educational levels. It was found that the higher the level of education attained, the greater is the apparent interest in local affairs. Still another consideration relates to various aspects of efficacy. As Table 10-5 indicates, the respondents in all five countries appear to believe that they are more effective in influencing decisions on the local level than on the national level.

It is interesting to observe that more than half of Almond and Verba's respondents in all five countries felt they could do something about unjust regulation at the local level. However, less than half of the respondents in Germany, Italy and Mexico felt they could be equally as effective at the national level. This raises the question why Table 10-2 reveals a generally higher turnout in national contests than in local elections. Such a high degree of national participation is difficult to explain if voters feel a need to be intimately associated with small governmental units. To what extent does the size of the political arena actually influence patterns of political participation?

Unfortunately, the evidence gathered from the questionnaires that bears on this crucial issue appears to be quite mixed. On the one hand, a number of the areas listed in Table 10-2 that have large-scale metropolitan governments reveal abnormally low voter turn-outs; among them are London (44.2 per cent), Toronto (36 per cent) and Winnipeg (33.4 per cent). This would tend to suggest that large-scale governmental units may have a negative influence on political participation. On the other hand, it is significant to note that the areas with the largest registered electorates in Table 10-1 reveal a consistently higher percentage of voter turn-out than the areas

TABLE 10-6
Electorate Size and Voter Turn-out

Area	Rank (Size of Reg. Electorate)		% Turn-out (Table 10-1)
Tokyo	1st	(6,318,000)	67.5%
London	2nd	(5,446,756)	44.2
New York	3rd	(3,281,689)	80.2
Sydney	4th	(1,551,055)	74.2
Paris	5th	(1,544,201)	65.3
Rio	6th	(1,497,401)	85
Milan	7th	(1,196,438)	92
Seoul	8th	(1,116,383)	36
Toronto	16th	(432,757)	36
Manila	17th	(431,541)	70.2
Istanbul	18th	(405,1333)	40
Montreal	19th	(380,067)	32.7
San Francisco	20th	(346,226)	61
Vancouver	21st	(253,000)	30
Winnipeg	22nd	(284,609)	33.4
Ibadan	23rd	(105,562)	61

with the smallest registered electorates, as indicated by Table 10-6.

As Table 10-6 shows, six of the eight largest areas (with registered electorates of over 1,000,000) had turn-outs higher than the mean average for all twenty-three areas (*i.e.* 62.5 per cent), whereas seven of the eight smallest areas (with registered electorates under 500,000) had turn-outs below this mean average. This would appear to suggest just the opposite of the previous speculation; namely, that smaller-scale units tend to have a negative influence on political participation!

Perhaps what the above inconsistencies really indicate is that the size of the political arena has a less significant impact on patterns of political participation than other motivational influences. The crucial forces at work here may well be more subtle psychological feelings of access and identity. Just as a feeling of close personal identity may possibly emerge from an association with a small-scale, local governmental unit, so, paradoxically, this same feeling of identity may be found in an allegiance to a larger national governmental entity. A crucial factor here can be the process of political socialization. In many countries, children are encouraged at a very young age to attach a strong sense of personal allegiance to their national political heritage. As a consequence, many adults tend to identify quite closely with their nation, whatever its size. There is little evidence to indicate, however, that any comparable socialization process takes place with respect to the metropolitan area. Instead, the public tend to view the metropolis, and metropolitan government, as something amorphous and impersonal. As Charles Adrian has pointed out, any viable democratic political system must possess legitimacy, an agreement on a set of consensual goals, and a psychological sense of access to the decision-makers on the part of the public.[15] Yet, the metropolitan area often possesses none of these attributes. To the typical citizen, the metropolitan area is often not a meaningful political community, but rather a disjointed aggregate of persons and places. Hence, the electorate may often fail to identify with the metropolis in any meaningful way, and this psychological alienation may lead to political apathy.

A second factor that may tend to activate the national electorate more than the metropolitan electorate relates to the stakes involved in national contests. Although Almond and Verba may be right in suggesting that the respondents in their five-country survey feel less capable of influencing national decisions than local decisions, there is little doubt that the types of choices involved in national elections are invariably of a much greater significance than those found in local elections.

Finally, Doxiadis has suggested a third reason why the electorate may identify more closely with national and even international issues, rather than with local issues, as a result of the impact of technological innovations on the political process. According to Doxiadis,

> . . . today the radio or television set occupies a considerable part of what used to be time for intra-family contacts, chats, and discussions, and people receive news from the other end of the world much more easily than they do

from the other end of their small city. . . . If we subdivide the whole earth into a scale of units of several classes . . . we see that it is becoming easier for us to get news from big distances and more and more difficult from small ones. . . . Since the beginning of his history, man has . . . always tried to create larger social and physical scales. . . . In this process he never eliminated the scales and levels of a lower order—he always enriched his space and never impoverished it. Our era is the first in which the opposite is happening.[16]

If these three basic factors—psychological feelings of access and identity; the importance of perceived stakes in electoral contests; and the impact of new technological innovations on political communication—actually are of significance in influencing political behaviour, it is possible to understand why voter turn-outs in national elections exceed those in local electoral contests. However, the question of political attitudes is exceedingly complex, and a wide variety of other attitudinal factors could also conceivably influence the variations noted in behaviour in both tables 10-1 and 10-2.

Utilization of Alternative Inputs
There are two additional factors that may help to explain some of the variations in political behaviour noted in the different areas. The first involves the degree of emphasis placed on one or more of the three inputs analyzed. Prague, for example, places very heavy reliance on the Communist Party and its ten District National Committees. As a result its electoral turn-out is very high, but the Prague respondent failed to indicate that there were any

---

[15]Charles Adrian, *Public Attitudes and Metropolitan Decision Making* (Pittsburgh, University of Pittsburgh, Institute of Local Government, 1962).

[16]C. A. Doxiadis, "Man and the Space Around Him," *Saturday Review of Literature* (December 14, 1968), p. 21.

major interest groups which played a key role in influencing local political decisions. In Philadephia, on the other hand, parties appear to be less important and voter turnout is considerably lower than in Prague, but the local respondent indicated that a wide variety of interest groups played a "significant role in many areas of decision-making". These groups included the Greater Philadelphia Movement; chambers of commerce; churches; education, labour and civic groups; and the press. The first point, then, is that the various areas can place a different degree of reliance upon any one of the different inputs and this, in turn, can help to explain variations between these areas in the utilization of the other inputs.

In addition, it was noted at the outset that there are many potential inputs other than voter participation, parties and interest groups. As a result, it is quite possible that a given area will minimize the importance of all three of the inputs analyzed in this essay in order to utilize alternative approaches to the decision-making process. To quote once again from Kaplan's analysis of policy-making in Toronto, ". . . issues are most often raised within the governmental sector, generally by appointed officials. If one were to list 'community influentials' in the Toronto area, defining an influential person as someone who plays a major role in public policy-making, that list almost certainly would be dominated by governmental bureaucrats."[17] It would appear that many residents in Toronto prefer to place their confidence in the internal governmental professionals to protect their interests, rather than to protect these interests by means of their own political actions. As a result, voter participation, parties and interest groups all appear to be

relatively unimportant in the Toronto area.

In concluding this aspect of the discussion, two general observations can be made. First, there are some striking variations in political behaviour in the different metropolitan areas analyzed. Second, our knowledge of the metropolitan political process is still so limited that it is difficult to account for these variations with any high degree of precision. This latter point is most significant. In virtually every great metropolitan area action is being taken that is thought to be related to various aspects of the political process, yet our precise knowledge of this process is still very rudimentary. For example, with regard to scale, or size of governmental units, the prevailing belief appears to be that if local urban units are kept small enough this will promote an increasing sense of identity and political activism on the part of local residents. Thus, the basic unit in Soviet city planning is the "commune" and the "microrayon" neighbourhood unit.[18] The new city of Nowa Huta, Poland, is planned upon a basis of small scale self-sufficient residential areas.[19] According to *The New York Times*, the Soviet Union is attempting to freeze the growth of its large cities without much apparent success.[20] A similar lack of success has confronted those attempting to control the growth of London since 1945.

Not only physical restriction but also political and administrative restructuring

---

[17]Kaplan, *loc. cit.*

[18]B. Michael Frolic, "The Soviet City," *The Town Planning Review*, Vol. XXXIV, No. 4 (Liverpool, January, 1964), pp. 285-303.
[19]Ryszard Dzieciolkiewicz, "Nowa Huta, Poland," *Local Government Throughout the World*, IULA, Vol. III, No. 2 (The Hague, March-April, 1964), p. 31.
[20]Raymond H. Anderson, "Soviet Urban Sprawl Defied Official Efforts to Curb the Growth of Cities," *New York Times* (November 13, 1966), p. 122.

is designed to maintain small units. The reorganization of London's metropolitan government led to the establishment of thirty-two new Greater London boroughs which were expected to keep local government close to the people. More recently, another Royal Commission study in Toronto rejected total amalgamation of all the existing municipalities in the metropolitan area in favour of a "four-city" plan which the Ontario Provincial Government later modified into a "six-city" plan. In a similar manner, the Winnipeg metropolitan government consists of a federation of thirteen governmental units; the Dutch city planners are concerned about decentralization and the establishment of meaningful neighbourhood units in the Conurbation Holland area; and the United States Government's "War on Poverty" has stressed the creation of citizen-controlled neighbourhood Community Action Agencies throughout metropolitan and urban areas.

While many of these actions have been influenced by considerations of administrative efficiency, or by local political pressures, many also reflect a more ideological desire to limit the size of metropolitan governmental systems in order to render them "near at hand and easy of access" to use the argument that was advanced by the Royal Commission on Local Government in Greater London.[21] The assumption throughout is that if metropolitan governments become too large they will lose contact with the people and the people will lose interest in them. This may well be correct, but we have very limited knowledge of the extent to which the size of the political arena may actually influence patterns of political behaviour. Only a handful of studies have appeared on this important subject in the United States.[22] In addition, the United Nations and the International Union of Local Authorities are issuing some reports on administrative decentralization and the optimum size of administrative areas, and some interesting theoretical work is being done on this subject in the Netherlands, Poland, Yugoslavia and a few other countries. Yet we still need to know a great deal more about how the size of our governmental units may actually influence the political process. In addition, we need to know more about the entire subject of political socialization as it relates to the initial formulation of our basic political attitudes. Perhaps, if we could launch effective campaigns to encourage the public to think more creatively about common goals for their metropolitan communities in an effort to encourage them to identify more positively with these communities, we would be doing more for the health of metropolitan politics than by splintering these communities into a thousand fragmented local governmental units. Yet, if we are to encourage people to identify with their metropolitan communities, we must consider more than the purely service aspects—the "efficiency and economy" components—of the metropolitan governmental process.

Unfortunately, our obsession with the "efficiency and economy" aspects of urban

---

[21]Royal Commission on Local Government in Greater London (London, HMSO, Command Paper, 1164, October, 1960), pp. 58-59.

[22]Arthur A. Maass, ed., *Area and Power* (New York, The Free Press, 1959); James W. Fesler, *Area and Administration* (Tuscaloosa, University of Alabama Press, 1949); Roscoe Martin, *Grass Roots* (Tuscaloosa, University of Alabama Press, 1957). For analysis of the administration and theoretical aspects of this issue, a number of political studies analyzing the influence of size of place on patterns of party voting have also appeared, such as Leon D. Epstein, *Politics in Wisconsin* (Madison, University of Wisconsin Press, 1958), Chapter IV.

administration has far overshadowed our understanding of the metropolitan political process, especially in the United States. As Robert C. Wood has noted: "Technical specialist after technical specialist [has] moved from one city to another, engaging in painstaking examinations of the current condition of his own narrow proficiency, and solemnly proclaimed that no public health department could work properly unless it had a clientele of at least 50,000 people. . . . Little real attention was paid to the construction of meaningful political communities, or to the question of obtaining political responsibility."[23]

All of which is to note that the field of political input analysis represents a rich area of research that warrants a high priority if we are going to develop the capabilities to control the future development of our great metropolitan areas in an intelligent and a creative fashion.

Metropolitan Administrative Systems: An Analysis of Outputs

At such time as the government policy makers receive political inputs from their various constituencies (these being parties, interest groups, the general public and so on), they evaluate these preferences against their own professional expertise and judgment in order to provide appropriate responses that will meet the needs of their particular metropolitan areas.

In order to obtain information on the types of outputs most commonly provided, the questionnaire was again used. Respondents were asked to list "the five most basic functional services performed by local units of government" in their respective metropolitan areas. Many respondents actually named more than five services and, as a result, a total of 143 different service entries were listed. Although there was fairly common agreement on a group of key services, no less than twenty-four different functional activities were mentioned by one or more of the respondents.

The three most commonly listed outputs (number of times mentioned indicated in parentheses) were: sewage and sanitation (16); education (15); and public health (14).

A second major group of service outputs included: roads and highways (11); traffic and transit (11); parks, recreation and culture (10); police (10); land-use planning (9); water supply (9); and welfare (8).

The above two groups comprise the ten major service outputs listed most frequently by respondents. A third group of services, mentioned less frequently, were: public housing (5); fire protection (4); gas (4); electricity (3); building controls (3); tax collection (2); trade and economic development (2); markets (2); assessment (1); harbour development (1); patents (1); commercial regulation (1); public registry office (1); and cemetaries (1).

The above listing indicates that many areas place a high priority on what the author has referred to in another study as "hard" services—roads, sewers, water supply and the like.[24] In addition, however, considerable attention also appears to be given to "softer" services such as education, public health and, to a lesser extent, welfare. In providing these latter services, governmental authorities become involved with two other supplementary output functions which tend to transcend the actual

---

[23]Robert C. Wood, "A Division of Powers in Metropolitan Areas" in Maass, *op cit.*, pp. 58-60.

[24]Frank Smallwood, *Metro Toronto: A Decade Later* (Toronto, Bureau of Municipal Research, 1963), pp. 35-39.

provision of "hard" goods and services to the public.

The first of these supplementary outputs involves a redistributive function. Essentially, this takes place as a by-product of the provision of services to what are generally regarded as minority, or disadvantaged, groups within the larger community; for example, the elderly, the poor, the sick and the ill-housed. In essence, these redistributive functions are designed to reallocate resources to the disadvantaged. They are often poorly performed, partly because of administrative difficulties, such as the case-by-case treatment often required, and partly because of political conflict. As the groups destined to receive these services possess, by definition, relatively limited resources, they are often unable to exercise the degree of power necessary to secure and maintain the level and quality of services they may desire.

This latter observation highlights a second, supplementary output provided by local governments in the form of the politicial function. It is only in recent years that the political function has begun to receive major attention in the study of local government. This has been given special emphasis in an influential textbook, *City Politics*, by Edward Banfield and James Wilson. According to Banfield and Wilson, local governments provide two major outputs: "One is that of supplying goods and services. . . . This is its 'service function'. The other function—the 'political' one— is that of managing conflict in matters of public importance."[25]

The performance of all three of these functions—the provision of goods and services, the redistribution of resources and

the resolution of conflict—is subjected to an unusually high degree of frustration in many metropolitan areas because, as is recognized by so many others in this volume, the territorial and administrative fragmentation of governmental decision-making and machinery makes it extremely difficult to discharge these functional responsibilities. As this problem of fragmentation can exercise a crucial influence upon governmental outputs in metropolitan areas, it is important to examine some of the remedies that have been proposed to alleviate this problem. Hopefully, these remedies can be of such a nature that they will not only result in better services, but also facilitate citizen identification with, and participation in, the political life of the metropolitan community.

DIMENSIONS OF THE PROBLEM

The problems of fragmentation are particularly severe in the United States, where the existence of a wide variety of local governmental units makes it particularly difficult to provide comprehensive administrative planning and political leadership in metropolitan areas. According to the 1962 Census of Governments, the 212 standard metropolitan areas in the United States at that time contained 18,442 units of local government. (As of July, 1968, there were 233 SMSAs in the United States.)

The fragmentation described in Table 10-7 appears to affect all regions of the United States. According to the 1962 U.S. Census of Governments, eight of America's major metropolitan areas (including special district bodies) possessed 300 or more local units, namely, Chicago (1,060 units); Philadelphia (963 units); Pittsburgh (806 units); New York SMSA (555 units); St. Louis (439 units); San Francisco-Oakland (398 units); Portland, Oregon (374

---

[25]Edward C. Banfield and James Q. Wilson, *City Politics* (Cambridge, Harvard University Press and The Massachusetts Institute of Technology Press, 1963), pp. 1-3.

TABLE 10-7
Local Governmental Fragmentation U.S. Metropolitan Areas

SMSA Population Range (1960 Census)	Number of SMSAs in Group	Local Government in SMSAs (1962)
1,000,000 plus	24	7,227
500,000 to 999,999	29	2,357
300,000 to 499,999	28	2,146
200,000 to 299,999	41	2,141
100,000 to 199,999	68	2,540
50,000 to 99,999	22	531
Totals	212	18,442

Source: U.S. Department of Commerce, Bureau of the Census, *Local Government in Metropolitan Areas, U.S. Census of Governments 1962* (Washington, U.S. Government Printing Office, 1964), p. 2.

units); and Los Angeles (348 units). Actually, the above figures represent something of an understatement of the problem. Whereas only 555 local governments are listed for the New York SMSA, New York's Regional Planning Association counted some 1,467 independent local governmental units in the mid-1960's within the larger geographical configuration which it defined as the greater New York metropolitan region.

Geographical fragmentation of local governments is by no means confined to the United States. The District of the Region of Paris, an area of 5,000 square miles, is divided between 1,305 communes (an average of 3.8 square miles per commune). Other highly fragmented areas in Europe include Conurbation Holland with 70 municipalities in an area of 643 square miles (an average of 7.4 square miles per unit) and Milan with 136 municipalities in an area of 557 square miles, which comprises the area of "The Intermunicipal Plan of Milan Metropolitan Area" (an average of 7.7 square miles per

unit). Other metropolitan areas throughout the world show these same characteristics of geographical fragmentation. For example, Buenos Aires reports 19 municipalities in its 78 square mile metropolitan area (an average of 4.1 square miles per unit); Lima lists 30 units in its 77 square mile urban area (an average of 2.6 square miles per unit); and Calcutta has 36 municipalities in its Metropolitan District (an average of 11 square miles per unit). It is important to note that the figures for areas outside the United States are for multipurpose municipalities only. If all special districts and other non-municipal units were listed (as they are in the United States totals), these figures would even be higher.

The problem of bringing these local units together to develop consensual goals for the metropolitan community as a whole are extremely difficult. The problems of geographical fragmentation are complicated even further by problems of functional fragmentation that result from a proliferation of special-purpose districts and authorities in many areas. For example, a study by the Bureau of Municipal Research, in Toronto, listed 101 different governmental units as playing a role in governing the 240 square mile area of Metropolitan Toronto.[26] Although many of the authorities listed have only a minor role to play in the governmental process, this does not necessarily detract from their contribution to fragmented government.

These are some of the dimensions of the problem. There have been a variety of solutions proposed and utilized, both of a formal and informal nature, that are designed to minimize the consequences of both geographical and functional fragmentation of governmental authority in metropolitan areas.

FORMAL STRUCTURAL SOLUTIONS

One potential solution has involved the partial, or complete, restructuring of local governmental units. John Bollens and Henry Schmandt have referred to partial restructuring as "The Two-Level Approach" and to complete restructuring as "The One-Government Approach".[27]

The Two-Level Approach:
Metropolitan Federation
The most significant of the "two-level" approaches is known as metropolitan federation. Under metropolitan federation, existing units of government are left substantially unchanged, or are subjected to only partial modifications, while an entirely new level of metropolitan-wide government is created in an effort to provide comprehensive planning and coordination of major area-wide services throughout the metropolis. Metropolitan federation has become increasingly more attractive in many countries during the post-war period and there are now a number of areas which utilize this approach, some of the more prominent examples being found in: Tokyo (1943), Toronto (1953), Miami, Florida (1957), Winnipeg (1959), Paris (1961), and London (1963). The first four of the above were similar in that they originally provided for the creation of new, area-wide metropolitan governments without any substantial modification in the existing local municipalities. As of January 1, 1967, however, Metropolitan Toronto consolidated its original thirteen member municipalities into six new metropolitan "boroughs" (although the City of Toronto retained its name as such). The Paris metropolitan plan was based upon a reorganization of the existing government *départements* and the London reform was also based upon the consolidation of existing local authorities into thirty-two new Greater London boroughs (plus the City of London). All of the above plans are identical, however, in that they emphasize the basic characteristics of any governmental federation: a division of functional powers between an area-wide authority and a sub-system of more localized governmental units. In addition, the allocation of powers between the two levels of government is basically similar in most of the areas with the exception of Paris. In Paris, the new area-wide authority known as the District of the Region of Paris is primarily an agent for financing and planning redevelopment. In the other areas, however, the metropolitan governments are granted considerably broader powers, in addition to planning and finance, which usually include basic physical services

---

[26]"The 101 Governments of Metro Toronto," Bureau of Municipal Research, *Civic Affairs* (November, 1968).

[27]John C. Bollens and Henry J. Schmandt, *The Metropolis: Its People, Politics, and Economic Life* (New York, Harper & Row, Publishers, 1965), pp. 400-490.

TABLE 10-8
Federated Metropolitan Governments

Metro Government	Metro Area (Square Miles)	Total number of Multi-Purpose Municipalities	Key Municipalities
1. Greater London Council	640	33	32 Greater London Boroughs plus City of London
2. District of the Region of Paris	4,670	8	City of Paris: Hauts-de-Seine; Val-de-Marne; Seine-Saint-Denis; Val d'Oise; Yvelines; Essonne; Seine-et-Marne
3. Dade County, (Miami) Florida*	2,054	27	City of Miami plus 26 suburban municipalities
4. Municipality of Metropolitan Toronto	240	6	City of Toronto: Boroughs of East York, Etobicoke, North York, York and Scarborough
5. Metropolitan Corporation of Greater Winnipeg	165	13	City of Winnipeg plus 12 suburban municipalities
6. Tokyo Metropolitan Government	1,600	37	23 wards, 11 cities, 3 counties

*Data on Miami Government gathered from E. Sofen, *The Miami Metropolitan Experiment* (Bloomington, Indiana University Press, 1963).

(such as sewage disposal), major arterial highways and a variety of welfare functions. Table 10-8 provides summary data on these metropolitan federations.

The One-Government Approach
Unlike metropolitan federation, the "one-government" approach involves the abolition of existing local governmental units (by means of either annexation or consolidation) in favour of a powerful single governmental authority that covers the entire metropolitan area. Annexation and consolidation were the major tools used in the nineteenth century in both the United States and Canada, which accounts for the original historic growth of many key cities. By means of annexation, for example, Chicago grew from 10 square miles to 190 square miles (by 1900); St. Louis from less than one square mile to more than 60 square miles (by 1900); and Montreal from less than 10 square miles to more than 50 square miles (by 1918). In a similar manner, a number of other American cities have utilized city-county consolidation to enlarge their areas such as Boston (1821), Philadelphia (1854) and New York (1898). During more recent years, however, both annexation and consolidation have become less important as potential solutions to the metropolitan problem in the United States and Canada. Although a large number of individual annexation actions may take place in any given year, the average size of the territories annexed

is generally quite small. The most significant consolidation action in the United States during the past quarter century occurred in Nashville, Tennessee, in 1962 when the local voters adopted a Nashville consolidated metropolitan government for Davidson County. This was an exception to the general rule. In most areas in the United States it is extremely difficult to get the local citizens to accept the concept of the "one-government" approach. This is not necessarily true, however, in cities outside the United States. In August, 1960, for example, 201 square miles of territory were annexed to Moscow in the U.S.S.R.[28]

The Metropolitan Special District
In addition to the "two-level" and the "one-government" approaches, a third structural device that has witnessed increasingly heavy use during recent years is the metropolitan special district. Rather than being given broad coordinating powers for different services, many of these units have been allocated direct responsibility for discharging specialized functional services in their own right. These authorities have proliferated extremely rapidly in the United States to a point where they are now the fastest growing unit of government. Between 1952 and 1962, the number of special districts in the United States increased from 12,340 to 18,323. By 1962, thirty per cent (that is, 5,411) of these special districts were located in the then existing 212 major metropolitan areas.[29]

While special districts may provide immediate relief for particular metropolitan problems, their long-range value is considerably more debatable. In many respects

they involve the long-run substitution of functional fragmentation for geographical fragmentation. This is due to the fact that while they can provide the means for the large-scale geographical administration of particular services, they often create serious new problems of functional coordination between the different districts. To a certain extent, this problem can be alleviated by the establishment of multi-purpose metropolitan districts such as that which has been adopted in the City of Seattle, Washington. Thus far, however, this new Seattle district has limited itself to very specialized functions. As special districts pose potential problems of their own, and as both metropolitan federation and the "one-government" approach often meet stubborn local resistance, increasing emphasis has been placed upon more informal approaches to the problems of metropolitan fragmentation in recent years.

LESS FORMAL APPROACHES
In addition to formal programmes involving structural modifications of governmental units, a number of less formal innovations have been utilized in an effort to alleviate the problems of fragmentation in metropolitan areas.

Inter-Municipal Cooperation
In the United States, one of the most popular of the newer concepts of inter-municipal cooperation is the "metropolitan council" approach which has been utilized in such areas as Washington, D.C. (The Metropolitan Washington Council of Governments), Detroit (The Supervisors Inter-County Committee), New York (The Metropolitan Regional Council) and San Francisco Bay Area (The Association of Bay Area Governments). All of these councils are similar in that they stress voluntary participation by local government

---

[28]See Peter Hall, *The World Cities* (New York, McGraw-Hill Book Company, 1966).
[29]*U.S. Census of Governments, 1962*, (Washington, D.C., U.S. Government Printing Office, 1962), p. 2.

officials. Although the councils do not usually have any formal coercive powers to force cooperative action between their member municipalities, they can serve as a highly useful forum for communication and the exchange of information, and they can provide an organizational base for a wide variety of cooperative projects between different municipalities.

A number of metropolitan areas outside the United States have also been experimenting with cooperative councils, which are often strengthened versions of the American councils. A good example is to be found in the Rijnmond (Rhine Estuary) Authority which was created in 1962 to promote more effective inter-municipal cooperation in the Rotterdam metropolitan area. The Rijnmond Council, which consists of eighty-one members (one third appointed by local municipalities and two thirds elected), is basically responsible for regional planning throughout the Rotterdam area, for promoting inter-municipal cooperation in this area, and for providing technical assistance and advice to the area municipalities. The Council has especially strong powers to designate and plan harbour sites and industrial developments on such sites—activities which are of crucial importance to this major seaport complex.[30]

Another cooperative regional planning group is the Council of Coordination for Development in the Calcutta Metropolitan Area. This body, which consists of the ministers of relevant departments of the government of West Bengal plus local mayors and municipal chairmen, was established in 1962 to serve as the action arm of the Calcutta Metropolitan Planning Organization. It was created to "ensure much needed coordination of the various agencies operating in the (Calcutta) district and thus to help in the preparation of a properly integrated development programme and in its implementation as well."[31]

Similar cooperative councils are being established under the French national "Environmental Planning Programme" in the form of the Regional Economic Development Committees which were instituted in 1965. These committees bring together elected local representatives and other leaders from each of twenty elected local representatives and other leaders from each of twenty national planning regions throughout France in order to assure that local cooperation will be forthcoming to implement national planning goals.[32]

In Germany one of the oldest of the European regional councils—the Ruhr Coal District Settlement Association (SVR) was founded in 1920. This group, administrated by an eighty-eight member Assembly (60 per cent drawn from local municipalities) is responsible for promoting regional planning in the massive Rhine-Ruhr urban industrial complex.[33]

Not all cooperative devices consist of intergovernmental councils and committees. For many years private groups have been influential in promoting cooperative efforts between metropolitan municipalities. Many municipalities in the greater New York area, for example, have been guided by the massive "Regional Plan for New York and its Environs" first published by the Re-

---

[30]H. Van der Weijde, "The Richmond Authority: An Experiment in Intermunicipal Co-operation," *Local Government Throughout the World*, IULA, Vol. II, No. 2 (June, 1963), pp. 23-25.

[31]*First Report*, Metropolitan Calcutta Planning Organization (Calcutta, 1962), pp. 6-7.

[32]Ambassade de France, Service de Presse et d'Information, *Town and Country Environment Planning* (Paris, undated), p. 9.

[33]Hall, *op. cit.*, pp. 122-57.

gional Plan Association in 1929. This plan has been influential, not because of any coercive powers exercised by RPA, but rather because for many years it represented the most comprehensive and intelligent blueprint available to guide planning efforts in the greater New York area. In a similar manner other municipalities have cooperated on various contract service arrangements which have ranged from relatively modest approaches to the elaborate "Lakewood" concept, in which Los Angeles County provides a package of comprehensive services for numerous area municipalities. These are but illustrative of the wide variety of cooperative devices which are available to minimize the problems of jurisdictional proliferation *between* different metropolitan municipalities.

Interdepartmental Coordination
In addition, increasing attention is being paid to problems of lateral coordination involving different departments and agencies *within* individual governmental agencies. As our urban governments have grown in size and complexity, it has become necessary to devise new managerial tools to promote interdepartmental coordination. Historically, heavy reliance has been placed upon traditional administrative techniques originally borrowed from military experience, such as hierarchical organizational arrangements and chain of command. While such traditional techniques can still be quite useful, many modern administrative theorists are concluding that they must be supplemented by newer devices if effective functional coordination is to be assured within increasingly more specialized governmental bureaucracies. Victor Thompson, for example, has argued that traditional "coordination through command" is becoming outmoded because today's commanders must place increas-

ingly heavy reliance upon their subordinates for the knowledge that is necessary to run complex, modern, governmental systems. Thompson has urged that traditional command techniques should be supplemented by "programmed coordination which is built into routine".[34]

Two of the more promising recent innovations which are being utilized in the United States in an effort to promote administrative coordination by means of "programmed routines" involve grant-in-aid programmes and budgetary planning which is backed up by modern computer technology. An increasing number of federal grant-in-aid programmes are attempting to encourage and reward administrative coordination at the local level. The new Department of Housing and Urban Development, for example, is deliberately giving preference in its grant-in-aid programmes to local communities that are cooperating with their metropolitan neighbours on the development of regional planning and related functional programmes.

New budgetary techniques are also being used in an effort to promote more effective lateral interdepartmental coordination within administrative agencies. On the national level, the aggressive approach toward performance budgeting in the Department of Defense indicates the extent to which a computerized budget can be used to increase interdepartmental coordination between such squabbling giants as the United States military services.[35] On the

---

[34]Victor A. Thompson, *Modern Organization: A General Theory* (New York, Alfred A. Knopf, Inc., 1961), p. 179.
[35]See Charles J. Hitch and Roland N. McKean, *The Economics of Defence in the Nuclear Age* (Cambridge, Harvard University Press, 1960), for a discussion of the newer budgetary techniques being utilized in the U.S. Department of Defense. See also, Roland N. McKean, *Efficiency in Government Through Systems*

local level, increasing emphasis is also being placed on both programme and performance budgeting in an effort to coordinate diverse governmental outputs into a more harmonious whole. Rather than listing the more traditional objects of expenditure (that is, personnel, travel, equipment and supplies and so on) on an agency-wide basis the programme budget, as Gladys Kammerer notes:

> ... focuses upon the *functions* to be performed by a given agency or unit of government. It sets forth the work to be undertaken and the objectives sought through that work. ... A *performance budget* carries the program budget one step further into unit costs. This requires the determination of those functions which are susceptible to work load measurement and cost accounting techniques. ... .[36]

Whether one adopts a programme budget or the more sophisticated performance budget, the basic objective is the same; namely, to utilize budgetary tools for purposes of planning and coordinating programme goals, rather than merely listing non-programmatic administrative expenditure items. To cite Gladys Kammerer again:

> Program Planning determines the essence of *what* is to be done: (traditional) budgeting *calculates the cost* of doing these things. ... The very necessity of program planning by line administrators impels them to think in terms of the end-product to be achieved and of the adoption of rational procedures to obtain that product. Attention must, therefore, be given to the organization of men ... and the co-ordination of related activities. A consistent philosophy of administration is implicit in program budgeting.[37]

It is crucial to note that Kammerer emphasizes the need for clarity regarding basic outputs and objectives if either programme or performance budgeting is to be successful. Coordinated decision-making is only possible if alternatives are selected on the basis of a set of consensual goals, and the new techniques such as programme budgeting are largely irrelevant unless there is common agreement on specific programme priorities. As Aaron Wildavsky has pointed out, in the most general sense, "budgeting deals with the purposes of men ... (it) is concerned with the translation of financial resources into human purposes."[38] Although it should be obvious that coordination is not possible unless there is some overall agreement upon common goals, this very basic fact poses difficult challenges for governmental leaders. In large measure, both our metropolitan communities and our internal governmental bureaucracies are made up of a pluralistic mixture of divergent motives and objectives, rather than of any monolithic model of perfect consensus. As a result, the role of leadership becomes crucially important. If coordination is to be forthcoming, our governmental leaders must first strive to secure agreement on a set of common goals to guide the metropolitan community, and then they must employ the newer techniques, such as programme budgeting and the like to achieve these goals.

Analysis (New York, John Wiley & Sons, Inc., 1958), for a discussion of performance budgeting in the management of American water resources development programmes.

[36] Gladys M. Kammerer, *Program Budgeting: An Aid to Understanding* (Gainesville, University of Florida Public Administration Service, 1960), pp. 5-6.

[37] *Ibid.*, p. 10.

[38] Aaron Wildavsky, *The Politics of the Budgetary Process* (Boston, Little, Brown and Company, 1964), pp. v and 1.

Conclusion

It is at this point that the twin concepts of inputs and outputs must merge into a unity. As has already been noted, just as agreement on a set of consensual goals is relevant to promoting the sense of identity that can serve as the catalyst for a more active political community, so this same agreement is essential for the effective coordination of governmental outputs. The job of securing this consensus is complicated by the governmental fragmentation that currently plagues many of our metropolitan areas, but unless and until we experience some type of governmental revolution in these areas, we must look to our political leadership to help us identify and define the larger purposes of our metropolitan systems. While this leadership should have both the power and the territorial jurisdiction to discharge its responsibilities effectively, it is also true that, as Charles A. Beard observed nearly a half century ago, "city government is not an end in itself but an agent designed to serve the purposes of the community."[39] In short, to refer once again to Dupré's essay, in this volume, when we structure our metropolitan political and administrative systems, we must give consideration to the values of participation and liberty, as well as the values of efficiency.

Thus, as we grapple with the increasingly more complex problems of an international metropolitan society, we must attempt to integrate administrative concerns with political concerns in a manner that will enable us to govern our metropolitan areas effectively, yet also within the limitations of democratic controls. Despite its modern metropolitan manifestations, this particular political challenge is nothing new. Perhaps James Madison expressed it as well as it can be expressed in his famous commentary on the United States constitution which appeared in *Federalist Paper 51*, first published in 1788: "If men were angels, no government would be necessary . . . in forming a government which is to be administered by men over men, the great difficulty lies in this: you must first enable the government to control the governed; and in the next place oblige it to control itself."[40]

The challenge which Madison posed to an eighteenth-century, agrarian society is still worthy of the finest minds that we can muster today in this mid-twentieth century era we have come to call the modern metropolitan age.

---

[39]Charles A. Beard, *The Administration and Politics of Tokyo* (New York, The Macmillan Company, 1923), p. 25.

[40]James Madison, "Federalist," reprinted in Peter Woll, ed., *American Government Readings and Cases*, 2nd ed. (Boston, Little, Brown and Company, 1965), p. 19.

11

Intergovernmental Relations and the Metropolitan Area

J. Stefan Dupré

"Truly," states a recent text on metropolitan affairs, "the metropolis is the crossroads of intergovernmental relations."[1] This chapter attempts to explore some questions about these relations, rather than to provide "answers" or "solutions" to intergovernmental "problems". Many of the "answers", which range from "one big city" through metropolitan federation and on to special districts, service contracts, voluntary cooperation and the like, have been elaborated upon already in chapters 9 and 10. Here, we consider in general terms what purposes are served by the existing relations among governments in metropolitan areas. What values, whether of liberty, participation or efficiency do these relations promote in fancy or in fact? Do they tend to advance the ends of particular interests at the expense of others? To what extent do the known "remedies" to intergovernmental "problems" promote certain values or ends as opposed to others or involve trade-offs among these values and ends? These are big questions and, worse still, elusive ones.

---

[1]John C. Bollens and Henry J. Schmandt, *The Metropolis: Its People, Politics, and Economic Life* (New York, Harper & Row, Publishers, 1965), p. 31.

The Development of Intergovernmental Relations in the Metropolitan Setting

THE GENESIS OF INTERGOVERNMENTAL RELATIONS

Intergovernmental relations in urban areas arise for the simple reason that these areas, virtually without exception, are subject to more than one government. Grossly over-simplified, intergovernmental relations are of two basic types—the first between *levels* of government, central and local, the second among *units* of government at the local level. What is lost in the over-simplification is that there may be more than one level of "central" authority, as is the case in such federal nations as Canada and the United States; and again, that there may in fact be a number of levels of "local" government. In the metropolitan area, these phenomena acquire growing importance with the passage of time.

Chronologically, intergovernmental relations between central and local levels are the first to emerge; normally in one of two ways. The first is through the development, protracted or sudden, of a central locus of authority which drains existing local communities of all or part of their autonomy, and upon which these communities in turn become dependent for grants of power. Thus in Europe, the rise of the nation-state overwhelmed medieval corporations, including cities, and placed them at the feet of a sovereign king. In a more restricted sense, a somewhat similar process can be observed in parts of New England, where townships existed prior to the state.[2] The second, and rather more common spring-board of central-local relations, is found where a central authority creates local governments to act as its agents for the performance of delegated functions of its choosing. This is the process in Canada, for example, and in most of the United States. It is again the pattern in France and the Soviet Union, where post-revolutionary regimes created, *de novo*, sets of municipal institutions that reflected their style and values.

In contrast to central-local relations, relations among units of local government arise later. Of course, there are notable exceptions, as can be found in nations that have historically provided a multi-level system of local government through such devices as counties, or in those that at an early stage isolated a particular function, perhaps education, under single-purpose authorities. These caveats notwithstanding, it is fair to say that relations among units of local government are more than anything else a product of the metropolitan phenomenon. Some experts have gone so far as to contend that the two can be equated; Robert Wood points out that the metropolitan dilemma has actually been defined "as the existence of many governments within a common economic and social framework".[3]

The process whereby this has become so is well known. As urban life spilled beyond city boundaries, little if any adjustment in these boundaries took place. Rather, existing rural municipalities were allowed to continue their existence as suburban towns, and, where no previously incorporated municipalities existed, new ones were created. Perhaps the most remarkable thing about this process is its near uni-

---

[2]For an interesting discussion of the evolution of Rhode Island from previously existing towns, see Anwar Syed, *The Political Theory of American Local Government* (New York, Random House, Inc., 1966), pp. 23-26.

[3]Robert C. Wood, "Metropolitan Government, 1975: An Extrapolation of Trends," *American Political Science Review*, Vol. LII (March, 1958), p. 111.

versality. In the words of that redoubtable authority, William A. Robson:

> Bombay was extended in 1950. Sydney absorbed eight adjoining municipalities in 1948. Zurich, Switzerland, absorbed eight adjoining communes in 1934. And there are a few other examples. But, on the whole, this is rare and normally not only a difficult but a virtually impossible operation. The principal tendency in the world is in the opposite direction, by which I mean vast and increasing proliferation of small local authorities in outlying parts of the metropolitan area corresponding, of course, to the outward thrust of population, alike in terms of residence, of industry, of shopping centres, and of other features of urbanization.[4]

Thus, the multiplication of local government units goes hand in hand with the metropolitan phenomenon. The nature of the relations among these units traverses a spectrum ranging from the relatively coordinated (rare) to something that approaches a *bellum omnium contra omnes* (all too common). However, relations of some kind must, and do exist; the metropolitan phenomenon necessarily thrusts municipalities together because their citizens share patterns of life, work, and recreation that transcend local boundaries. Furthermore, whether relatively ordered or disordered, relations among local units tend to be enormously complicated.

THE COMPLEXITY OF
INTERGOVERNMENTAL RELATIONS

In the metropolitan setting one connection between inter-local and central-local relations comes immediately to mind: it is the passivity of the central authority, whether as a matter of conscious policy, cold indifference, or willful neglect, vis-à-vis local institutions, that has given inter-local relations part of their existing flavour. This is an important, but basically negative link. Of greater interest is the fact that central-local relations have become, in a positive sense, inextricably linked with those among local units to produce the bewildering cobweb that is the total network of intergovernmental relations in metropolitan areas today. In this connection, four dimensions are particularly striking.

First, in the sense that the central-local relationship is one in which the central authority has the power to create new local units, we find that numerous local governments have been added to the existing units already thrust into mutual contact by metropolitan development. This stems in part from the incorporation of new municipalities in previously unorganized territory that becomes affected by urban dispersal. Perhaps more often, however, it is the outcome of rather common tendency of the central authority to create special-purpose authities or districts in response to problems besetting certain local services as a consequence of metropolitan growth. Such districts have the virtue of providing one or more services, whether water or sewerage, education or health, expressways or parks, protection or rapid transit, over an area that transcends municipal boundaries and accordingly takes account of metropolitan realities. Under governing bodies which may be locally elected, but which more commonly are elected indirectly or appointed, either by the central authority or the local units or both, special-purpose districts constitute an important and complicating addition to the local units that interact both within the metropolitan region and with the central authority.

---

[4]William A. Robson, "Metropolitan Government: Problems and Solutions," Canadian Public Administration, Vol. IX (March, 1966), pp. 46-47.

Second, we find that to the traditional central-local relationship, where the one has been the legal superior of the other, a new dimension has been added—that of policy and administration. A by-product of the total economic, political and social environment of our times, this dimension has had a particularly profound impact on intergovernmental relations in the metropolitan setting. As the policies, programmes and service standards of local governments have come to have a greater than local, indeed a greater than metropolitan impact, central governments have come increasingly to provide technical assistance and to stipulate standards on a function by function basis. For their part, local governments more than ever before find themselves in a situation that requires them to press for the initiation or alteration of central policies of concern to them. The result is a multiplication of central-local contacts unprecedented in scope. Perhaps some of the most spectacular instances of this process are to be found in federal nations such as the United States and Canada, where constitutional guidelines designate the state level of government as the only central authority vis-à-vis local units. So strong is the thrust of policy and administration that, constitutional law notwithstanding, the federal level of government has been woven into the pattern of intergovernmental relations in the metroplis. The new involvement of the senior level of government in federal nations is one of the most readily visible manifestations of the explosion in central-local relations bred by dictates of policy and administration.

Closely connected to policy and administration but distinguishable for our purposes is the third dimension of contemporary intergovernmental relations in the metropolitan setting—finance. The substantive programmes mounted by central authorities, ranging from education to transportation, often stem directly from the fact that the fiscal resources of local units have proven inadequate in relation to their spending responsibilities. So acute is this predicament that central authorities have also been forced into the breach through such across-the-board devices as general grants, shared taxes and borrowing assistance. The upshot is that local units look to central authorities not only for operational programmes, but also to underwrite their general spending activities. To the extent that local units are dependent on the central authority primarily because of their small size or degree of specialization, the financial dimension of central-local relations has direct repercussions on the intergovernmental setting of metropolitan areas. Inasmuch as financial transactions between central authorities and local governments permit small dependent units to survive, the complexity of intergovernmental relations is allowed to flourish. This is particularly true of those transactions conducted according to the so-called "fiscal need or capacity" principle, whereby local governments receive proportionately higher grant payments in accordance with the extent to which they are judged to be fiscally deficient. Although this principle is well grounded in considerations of equity, it should be made quite clear that fiscal need and capacity payments can create major disincentives to the rationalization of local government in so far as they underwrite units whose financial weakness stems primarily from structural deficiencies.

The fourth and final dimension of central-local relations in the metropolitan setting pertains to the central authority itself, which gradually comes to exhibit such overlapping and uncoordinated jurisdiction as is observable among local units themselves. Even in nations, states or provinces whose administrations are endowed with ministries or departments of local government or municipal affairs, programmes and financial transactions affecting local units emanate from other agencies and departments as well. There has been a decided tendency for the number of central agencies concerned with local government to rise as locally oriented policies and fiscal measures have multiplied. Accordingly, a local unit confronting the central authority has come to occupy a position not dissimilar from that of the metropolitan resident to his local government. The expanding network of central relations with local units breeds new central government actors even as it sustains and increases the number of local units.

Intergovernmental Relations, Values and Interests: Some Concepts

In that the metropolitan phenomenon has spawned intergovernmental relations whose complexity is seemingly self-reinforcing, the outcome is enormous difficulty in public decision-making. Surely, this outcome cannot be considered an end in itself. No one, be he politician, public administrator or political scientist, would defend the governmental situation in metropolitan areas because he considered difficulty a value. Therefore, we must look beyond the immediate outcome if we are to discover what values, if any, are served at the crossroads of intergovernmental relations. The search can begin appropriately by referring to a few classic notions of political science. If these notions have a familiar ring, it is because they constitute the traditional case for local government which is, after all, the denominator common to both central-local and inter-local relations. In no small part the values served by intergovernmental relations in the metropolitan area must be those that arise from the existence of the very governments that make these relations a reality. Three such values will now be examined.[5]

LIBERTY

The first value is classically embodied in the celebrated *De l'esprit des lois* of Charles-Louis de Secondat, Baron de Montesquieu.[6] Much concerned with the liberty of the individual in the face of kingly authority, Montesquieu sees in a host of *corps intermédiaires*, of which municipalities are an important type, a key barrier to the arbitrary use of power. In this sense, local government, as an intermediate layer between the central authority and the individual, promotes the value of liberty. The kind of liberty involved in this case is negative— freedom *from* arbitrary rule.

The contemporary observer may justly pause to wonder whether this particular value deserves a place of primacy in an age of democratic government. Montesquieu himself was prompted to believe that such *corps intermédiaires* as local government served liberty primarily in the context of a monarchical system.[7] But the strength of his basic notion has survived both the passage of time and the evolution of governmental form. Thus, for instance, the 1955 report of an advisory committee to the United States Commission on Intergovernmental Relations, after waxing lyrical on the benefits conferred by local government,

---

[5]For a more detailed and profound discussion of the relation between values and local government, see Arthur Maass, ed., *Area and Power* (New York, The Free Press, 1959). This book, of which I am a contributing author, has had a strong influence on my approach to intergovernmental relations.

[6]Comte de Montesquieu, *De l'esprit des lois*, Gonzague Truc, ed., Vols. I & II (Paris, Editions Garnier Frères, 1949).

[7]*Ibid.*, Vol. I, pp. 19-20.

goes on to say: "More important still is the use of local government to soften the impact of arbitrary State and National Laws and regulations."[8] Where monolithic government remains everywhere a bogeyman, we dare not underestimate the continued vitality of the notion that local government serves the value of liberty understood broadly as freedom from oppression.

PARTICIPATION

Man cannot be understood simply as an individual to be safeguarded from arbitrariness, however. He is emphatically a social animal. It is here that we broach a second value that local government has been said to serve peculiarly well. "Town meetings", writes Alexis de Tocqueville, "are to liberty what primary schools are to science."[9] And why? Because "every individual has an equal share of power and participates equally in the government of the state".[10]

It matters not that the town meeting form of government is now little more than a charming relic of the past. It matters little more that two generations of political sociologists have proven that power is no more equally shared in municipal decision-making than elsewhere, but that a relatively few "notables" determine local public policy.[11] The notion that local government is peculiarly conducive to the value of participation remains a hallowed canon of political theory. As the level of government closest to the individual, the municipality offer him unparalleled opportunity for access to, and participation in, the process of government.

EFFICIENCY

Are the values of liberty and participation ones that must be purchased at the expense of a third value, efficiency? Tocqueville is silent on this point, but he leaves no doubt as to where his priorities lie.

> Granting for an instant, that the villages and counties of the United States would be more usefully governed by a central authority which they had never seen than by functionaries taken from among them; admitting, for the sake of argument, that there would be more security in America, and the resources would be better employed there, if the whole administration centered in a single arm— still the *political* advantages which the Americans derive from the decentralized system would induce me to prefer it to the contrary plan.[12]

For Tocqueville there is no question that democratic values are to be preferred to efficiency. But suppose that local government is also conducive to the efficiency value? On this possibility, few political theorists are more illuminating than John Stuart Mill. As Mill states: "It is but a small portion of the public business of a country which can be well done, or safely attempted, by the central authorities...."[13] and "that, if only on the principle of division of labour, it is indispensable to share them between central and local authorities."[14] Again,

---

[8]Syed, *op. cit.,* p. 93.

[9]Alexis de Tocqueville, *Democracy in America,* Vol. I (Phillips Bradley edition; New York, Vintage Books, 1954), p. 63.

[10]*Ibid.,* Vol. I, p. 67.

[11]See, for example, Robert A. Dahl, *Who Governs?* (New Haven, Yale University Press, 1962).

[12]Tocqueville, *op. cit.,* Vol. I, p. 96.

[13]John Stuart Mill, *On Liberty and Representative Government* (R. B. McCallum edition; Oxford, Basil Blackwell & Mott Ltd., 1947), p. 278.

[14]*Ibid.*

even if the local authorities and public are inferior to the central ones in knowledge of the principles of administration, they have the compensating advantage of a far more direct interest in the result. . . . It is the local public alone which . . . calls the attention of the government to the points in which they may require correction. . . . In the details of management, therefore, the local bodies will generally have the advantage.[15]

If we must acknowledge Mill for the notion that local government can promote the value of efficiency, we are further in his debt for the fact that he stresses the fulfilment of his value in what is explicitly an intergovernmental context. For just as Mill concludes that local bodies will have the advantage in the details of management, so also does he believe that "in comprehension of the principles even of purely local management, the superiority of the central government, when rightly constituted, ought to be prodigious."[16] It follows, according to Mill, that "the authority which is most conversant with principles should be supreme over principles, while that which is most competent in details should have the details left to it. The principal business of the central authority should be to give instruction, of the local authority to apply it. Power may be localised, but knowledge, to be most useful, must be centralized."[17] Parenthetically, we may note that this provides interesting justification for the principle of "democratic centralism" that pervades local government in Eastern Europe. Of course, the extent to which democratic centralism has itself been buffeted by the complexity of the metropolis is another question.

We can cite impeccable authorities who maintain that local government is conducive to the values of liberty, participation and efficiency. To the extent that intergovernmental relations in the metropolitan area are founded upon local government, it is of interest to inquire whether they in turn promote these values. Before undertaking this exercise, however, we must take due note of the fact that recourse to political theorists can also produce notions that attack the very foundations of local government.

THE CHALLENGE OF INTERESTS

In *The Federalist*, James Madison did not set out to attack local government. His overriding concern was to defend the fledgling Constitution of the United States, which proposed a central authority rather stronger than its opponents thought warranted. In the process, however, he hatched a theory that, in hands as unscrupulous as those with which it is unravelled here, must force any disciple of local government to pause in his tracks.

Man, as Madison sees him, is not only a social but a factious animal. Man, animated variously by "a zeal concerning religion, concerning government", by "attachment to different leaders ambitiously contending for pre-eminence and power", or by "the various and unequal distribution of property",[18] is forever forming factions, or interest groups, with those among his fellow men who are similarly inclined or situated. Madison takes an extremely dim view of factions, and for him the capacity of a political system to control the effect of faction becomes a prime consideration. In a democratic system a minority faction can

---

[15]*Ibid.*, pp. 278-87.
[16]*Ibid.*, p. 288.
[17]*Ibid.*, pp. 288-89.

[18]James Madison, "The Federalist, No. 10," *The Federalist* (Modern Library edition; New York, Random House, Inc., 1937), pp. 55-56.

presumably be voted down by the application of majority rule. But how to prevent "a common passion or interest"[19] from capturing a majority? For Madison, the answer lies in a twofold remedy, that is to say representative government and what he chooses to call an "extensive republic".[20]

The extensive republic enhances the probability that those in elective office will be men of virtue and reduces the possibility that government will fall prey to insular self-interest. There remains the likelihood that specific interests will gain some kind of foothold in government. But it is precisely here that the extensive republic has its most outstanding virtue. For the larger the republic, the greater the number of interests encompassed within its boundaries. The outcome will be that the countervailing powers of diverse interests will hold each other in check.

> The smaller the society, the fewer probably will be the distinct parties and interests composing it; the fewer the distinct parties and interests, the more frequently will a majority be found of the same party; and the smaller the number of individuals composing a majority, and the smaller the compass within which they are placed, the more easily will they concert and execute their plans of oppression. Extend the sphere, and you take in a greater variety of parties and interests; you make it less probable that a majority of the whole will have a common motive to invade the rights of other citizens; or if such a common motive exists, it will be more difficult for all who feel it to discover their own strength, and to act in unison with each other.[21]

Stated baldly, Madison's theory gives pause to the most ardent advocate of local government. For the present purpose, it is of no consequence that Madison in fact was eager to concede that local matters might be left to local interests or that the coexistence, in a given state, of local and central authorities, "of unmixed and extensive republics",[22] could provide (shades of Montesquieu) a safeguard against tyranny. Faced with their vulnerability to capture by faction, we must permit Madison to tempt us to say good-bye to the town meeting, to bid farewell to the grass roots. The very values engendered by local government—liberty, participation and efficiency—acquire a hollow ring if these governments are but fronts for special interests. Who is to say that their spokesmen are not "madmen in authority . . . distilling their frenzy from some academic scribbler of a few years back"?[23] It must surely be said of Madison's theory that a better argument for centralization, at least for very large local governments, could scarcely be devised.

Intergovernmental Relations and Values: Some Problems

Underwritten as they are by local government, intergovernmental relations should be conducive to the values of liberty, participation and efficiency. Indeed, if one accepts John Stuart Mill's argument, it may be reasonable to expect these values to be fulfilled more completely in an intergovernmental setting than in one where local government is conceived in isolation. What is the potential, realized or otherwise, of intergovernmental relations in this respect?

[19]Ibid., p. 58.
[20]Ibid., p. 59.
[21]Ibid., pp. 60-61.

[22]Quoted in Samuel P. Huntington, "The Founding Fathers and the Division of Powers," Arthur Maass, ed., op. cit., p. 185.
[23]John Maynard Keynes, The General Theory of Employment, Interest and Money (New York, Harcourt, Brace & World, Inc., 1936), p. 383.

EFFICIENCY AND PUBLIC GOODS

In an examination of the value of efficiency, recent work by scholars in the field of economics provides an appropriate starting point. From their perspective, governments can be looked upon as purveyors of "public goods" or, perhaps more accurately, "non-private goods".[24] It is an important property of public goods that they have "externalities" or "spillover effects". Now, it should be readily evident as a matter of common sense that,

> not all public goods are of the same scale. Scale implies both the geographic domain and the intensity or weight of the externality. A playground creates externalities which are neighborhoodwide in scope, while national defense activities benefit a whole nation—and affect many outside it. Thus for each public good there corresponds some "public".[25]

Depending on its scale, a public good can be conceived as appropriately "packaged" if the boundaries of the unit of government providing that good are such that the externalities of the good are "internalized" to the public served. By way of example, let us take the control of air pollution to be a public good. A single community in a metropolitan area beset by an air pollution problem can take measures to reduce pollution, but it will still be victimized by smog because its area is not sufficient to encompass (internalize) the spatial extent of the evil that its public good, control, is meant to remove. The only area over which air pollution control can be provided with full effectiveness is that whose scale internalizes the object to be controlled and, consequently, benefits from that control.

Given the size of the area required to internalize a public good, the appropriate provider of that good is the government whose territorial jurisdiction coincides with the requisite area. Assuming that no two public goods can be internalized fully over exactly the same area and that the universe of public goods totals n, then n levels of government will have to exist if efficiency is to be maximized.

Indeed, the governmental evolution of many metropolitan areas has been in this direction. If there is validity to the argument just developed, the multiplication of special-purpose districts in particular can be seen in a highly favourable light. The appropriate provider of water and sewerage services to the metropolis is often seen as being the conservation district whose boundaries coincide with the drainage basin; that of transporation, the district corresponding with the journey to work radius; that of playgrounds, a set of infra-metropolitan districts coinciding with the neighbourhood; and so on. But special districts do not emerge alone as an appropriate technique. Thus, for example, the provision of a public good over its appro-

---

[24] A pure public good has been defined by Paul Samuelson as one that "all enjoy in common in the sense that each individual's consumption of such a good leads to no subtraction from any other individual's consumption of that good." In the real world, few goods indeed are marked by this degree of "publicness". It is therefore safer to think of most goods provided by government as being "non-private" in the sense that their objective benefits, though shared, are not available to all in equal amounts. Having acknowledged this technicality, I shall for the sake of simplicity use the term "public goods" as synonymous with "non-private goods" throughout the following discussion. On these points, see Albert Breton, "A Theory of Government Grants," *Canadian Journal of Economics and Political Science,* Vol. XXXI (May, 1965), pp. 176-77.

[25] Vincent Ostrom, Charles M. Tiebout and Robert Warren, "The Organization of Government in Metropolitan Areas: A Theoretical Inquiry," *American Political Science Review,* Vol. LV (December,1961), p. 833.

priate territory can be achieved by voluntary cooperation. On the other hand, public goods will not be provided efficiently unless there exists some mechanism whereby preferences as to the desired quantity and quality of these goods can be registered. It is all very well to say that if each public good is provided over the area in which its beneficiaries are located, "the means are in principle available for assessment of the cost of public control upon the beneficiaries."[26] The means will in fact be available only if the beneficiaries possess the knowledge and inclination to render n verdicts on the goods provided by n levels of government. As this is highly unlikely, we must consider alternatives.

One promising possibility is to have, say, three levels of government—one central, one metropolitan, one local. Public goods can be assigned to each of these levels in accordance with the extent to which the territorial jurisdiction of each comes closest to internalizing the goods in question. If, in terms of inter-local relations, this sounds suspiciously like metropolitan federation, this is because that is exactly what it is. But if we compare existing metropolitan federations to the theoretical ideal, at least two outstanding problems immediately become apparent. The first is that the allocation of functions among central, metropolitan and local levels, especially that between the central and the two local levels, is more often the outcome of historical happenstance than of conscious calculation. The second and perhaps more outstanding problem is that, especially at the lowest level of government, disparities in the territorial jurisdiction of units may preclude a relatively uniform assignment of local public goods, save at the cost of enormous sacrifices in terms of efficiency. Metropolitan "reform" is not simply a matter of erecting special districts or super-governments on the foundation of existing units. In the typical metropolitan area lowest-level governments of similar size, capable of internalizing similar public goods, simply do not exist. In this light, the recent London reforms are of considerable interest, and the changes in Toronto in 1967, though modest, provide another example.

The scientific assignment of public goods to a restricted number of governmental levels, especially if the units at the lowest level are comparable in scale, offers a promising path to the fulfilment of the efficiency value. But it is no panacea. First of all, certain public goods must be tailored precisely to their corresponding areas if they are to be provided efficiently, and in such cases as water supply and sewage disposal, for example, with their ties to the drainage basin, the special district may well be a necessity. Second, however carefully public goods are assigned to their appropriate level of government, there will always be significant spillovers to the extent that the number of levels is restricted. The answer here can lie in the use of conditional grants by the higher level of government. "A higher level government . . . can compute the marginal social utilities and the marginal social costs of the benefits that spill over the frontiers of jurisdiction and . . . can equalize them."[27] Still unanswered, however, will be the problem posed by public goods that can be internalized within a territorial jurisdiction smaller than that of the lowest level of government, for example, playgrounds. Here, failing some device such as neighbourhood government, the situation will be one where, as "the political community contains the whole public and, in addition, people unaffected

[26]*Ibid.*, p. 836.

[27]Breton, *op. cit.*, p. 183.

by a transaction, the unaffected . . . (will) be given a voice when none may be desired. Capricious actions can result."[28] Short of further multiplying the number of governmental levels or creating infra-local special districts, akin to the "block committees" described in Chapter 9, this problem could be insoluble. Whatever the case, the small scale over which certain public goods can be fully internalized seriously damages the argument made by the "one big city" school of metropolitan reformers, at least on efficiency grounds.

Through rationalized intergovernmental arrangements, based on a limited number of governmental levels to which appropriate public goods have been scientifically assigned, and featuring special districts and voluntary cooperation on a restricted basis, together with conditional grants from higher to lower levels, the means are available in principle to fulfill the efficiency value in large measure. The fact is, however, that there appears to be no metropolis that fully approximates the model. Whether or not this is "a good thing" depends upon the extent to which efficiency is not the only value to be fulfilled. The value of participation, for one, may call for somewhat different arrangements.

THE DICTATES OF PARTICIPATION

For Tocqueville, there was no doubt that participation hinged on the power and independence of local government. The citizen is attached to his municipality *"because it is a free and strong community*, of which he is a member, *and which deserves the care spent in managing it. . . .* Without power and independence, a town may contain good subjects, *but it can have no active citizens."*[29] If, for the sake of the argument,

we accept this view, the governmental arrangements dictated by considerations of efficiency, to say nothing of existing metropolitan systems, require rethinking.

For one thing, n levels of government matched to n public goods are out of the question. As argued earlier, this solution is dubious even on efficiency grounds. In terms of participation, it falls apart simply because political responsibility becomes too diffused to engage public attention, save on the assumption that the public's capacity to acquire knowledge and become involved is developed, so to speak, to the nth degree. Meanwhile, the alternative—a restricted number of governmental levels supplemented by special districts and conditional grants—itself requires further thought. With respect to participation, it appears that this scheme's best feature is the restricted number of governmental levels. In particular, if metropolitan and infra-metropolitan affairs are entrusted to two levels of government, each armed with its appropriate mix of public goods, it is reasonable to contend that government will be sufficiently simple to hold public attention and that each level will have sufficient functional importance to attract active participation. There is, of course, the counter-argument which claims that two levels of local government totally defeat participation because the metropolitan level is "too distant" to provide ready access while the lower level is left with "too little to do" to warrant public attention. However, this not only underestimates the capacity of the metropolitan level, as a glamorous and active government, to absorb public interest but also fails to concede the possibility of the lower level units being tailored to provide appropriate public goods with efficiency, thereby assuming an importance fully worthy of citizen interest.

---

[28]Ostrom, et al., *op. cit.,* p. 836.
[29]Tocqueville, *op. cit.,* p. 69, emphasis supplied.

On this point at least, the dictates of efficiency and participation may be said to go hand in hand.

For the rest, however, efficiency and participation begin to part ways. Reasonably full efficiency with few levels of government hinges upon the existence of some special districts and a number of conditional grants. As to special districts, these must necessarily be suspect with respect to participation, and in direct proportion to the extent to which they proliferate. In so far as a special district performs a particularly important function, it may perhaps be tolerated in terms of participation because of the visibility that a function acquires when isolated under its own government. However, entrusting a key function to a special district necessarily undermines the importance of the multi-function level of government that would otherwise provide it, with undoubted sacrifices in the degree of citizen interest in that level.

If there must be special districts, the value of participation underlines the importance of making them responsive to public desires. This is the argument for placing special district government under directly elected boards, as is the case, for example, with the Metropolitan Sanitary District of Greater Chicago. Alternatively, the special district board should be tied as closely as possible, probably through direct appointment, to the level of government whose territorial jurisdiction comes closest to coinciding with that of the district. This at least provides a clear-cut focus whereby public participation can make itself felt on the function provided. How far most special districts (with their boards appointed either by a multiplicity of municipalities or by the central authority and the central authority and the municipalities) find themselves from this practice, is a matter of record. Furthermore, to the ex-

tent that it is the metropolitan level of government whose territorial jurisdiction most closely approximates that of a special district, the value of participation reminds us that special districts are not a substitute for multi-purpose metropolitan government, but are instead a supplementary device to be tolerated only on the grounds of efficiency and tied as closely as possible to the metropolitan level.

If the special-purpose district poses a delicate trade-off between the values of efficiency and participation, the conditional grant device is even more difficult to accommodate. Assuming always that power and independence are the key to public participation, what makes a government powerful and independent? Surely it is the capacity to determine priorities. But few things so distort the priority-setting process at a lower level of government than conditional grants from the higher level. Will there be better roads or better schools if every dollar of road benefits must be matched by a dollar of local taxes, while every dollar of school benefits need be matched by only fifty cents from local resources? It is difficult to imagine a device of greater assistance to, let us say, school interests who are competing for local resources with road interests, than a conditional grant for school purposes. This phenomenon will be examined in greater length shortly. For the moment, suffice it to note that the case for conditional grants must be built almost entirely on grounds of efficiency.

As an alternative to conditional grants, there is the general, or unconditional grant which, not being tied to a specific function, presumably involves a lesser sacrifice in participation because it leaves the local priority-setting process intact. Accordingly, an innovation such as the general grant introduced in the United Kingdom under the Local Government Act of 1958 rightly merits attention. However, a recent study has noted that:

> The introduction of the general grant certainly freed local authorities from some detailed control on general grant services in that it was no longer necessary for them to have small items of expenditure approved for grant purposes. But the area of such freedom cannot be greatly enlarged when so much expenditure flows automatically from decisions already taken or is expenditure which cannot be avoided or which is required as a result of decisions taken centrally.[30]

What this suggests is that if an unconditional grant is substituted for conditional grants while the earlier "conditions" are allowed to reappear through mandatory standards stipulated by the higher authority, little has been added to local discretion. Efficiency continues to be served at the expense of participation.

LIBERTY: REALITY OR ILLUSION?

Now for a word about the value of liberty. If freedom from arbitrariness can be said to be maximized when government is so complex that public decision-making is inordinately difficult, then we can say that the bewildering pattern of intergovernmental relations that characterizes most metropolitan areas fulfils one value. However, to the extent that complexity paralyses decision-making, is not the absence of desired service standards and clear-cut policies itself a form of oppression? Also, to what degree does governmental complexity provide a smoke-screen behind which certain interests can manipulate the decision-making

---

[30]J. A. G. Griffith, *Central Departments and Local Authorities* (Toronto, University of Toronto Press, 1966), p. 504.

process to their advantage, with consequences quite as arbitrary vis-à-vis the general public as any that could be imagined? Clearly, an excursion into the link between intergovernmental relations and interests cannot be postponed.

Intergovernmental Relations and Interests: More Problems

In mounting what by any standards is a telling argument against local government, Madison begins with the assumption that factions are necessarily evil. This assumption is for two reasons. For one thing, interest groups are surely an entirely legitimate, perhaps indispensable, vehicle for public participation in decision-making. For another, there can be little doubt that interests have had a creative, innovating role in political society. But at this juncture we are still in Madison's debt for two very important notions. The first is that interests compete with one another, thereby holding each other in check. The second is that the larger the arena within which interests compete, the more intense is the degree of competition among them. Put another way, a sizable and diverse political arena is likely one where the degree of competition among interests approaches the pure, while a small and homogeneous area is susceptible to monopoly.

THE ENTRENCHMENT OF NEIGHBOURHOOD INTERESTS

If we re-examine the typical metropolitan area with the aid of these notions, we must surely be struck not only by the multiplicity of small political arenas, but also by the relative homogeneity of the interests found in each. As Julius Margolis observes, there has been, "a strong tendency for neighbourhoods of a metropolitan area to become municipalities. . . . The existence of neighbourhoods as cities has the consequence of gathering into one government unit the individuals who have relatively similar preference orderings and who, as a group, tend to have orderings which are distinct from other groups."[31] In brief, the fractionalized pattern of government in the metropolis underwrites political arenas in which particular interests achieve a position of near monopoly. To quote York Willbern:

> Upper-class residential suburbs can and do have a separate governmental jurisdiction and can adopt zoning rules to preserve the "character of the community". Industrial developments, outside the limits of corporate municipalities, can fend off annexation so that they will not have to pay the costs of educating the children of their workers or contend with the welfare problems of those who lose their jobs. A governmental unit with high property valuation and practically no children can exist side by side with another unit with great numbers of children to be educated and practically no property base.[32]

It is noteworthy that, in many instances, the central city has lost much of its earlier heterogeneity and itself acquired neighbourhood characteristics by becoming a haven for a restricted number of groups, often of low income. As it become a neighbourhood, the central city constitutes a political arena marked by increasingly imperfect competition among interests. As such, central cities can become a source of concerted opposition to metropolitan reform quite as often as their suburbs. For example, recent reform proposals in both St. Louis and Cleveland were coldly entertained by city blacks, not least because "they feared they would not be equitably represented in the district government."[33]

The governmental units of the metropolitan area can be viewed as arenas in which competition among interests is highly imperfect, and whose existence the advantaged interests are therefore naturally inclined to defend. Little wonder that "among governmental institutions, the suicide complex is notably absent."[34] Little wonder too, that among the known "remedies" to intergovernmental "problems", almost any observer would rank, in order of increasing difficulty, special districts, a metropolitan government superimposed on existing units, and either "one big city" or the creation of new lower tier units.

THE RESPONSE OF SPECIALIZED INTERESTS

Viewing governments in the metropolis as arenas that confer competitive advantages on certain interests may be illuminating, but it is only a part of the total picture. Various patterns of intergovernmental relations, including technical aid, conditional grants and the like can themselves be seen as the outcome of efforts by competing interests to gain an advantaged position. A particular interest advocating a particular function, better schools or commuter railroads, let us say, may be grossly disadvantaged in the political arena of an individual municipality. However, the same interest may occupy a pre-eminent position in other municipalities, and enjoy relative advantages in the greater arena that is the central

[31]Julius Margolis, *Public Finances: Needs, Sources, and Utilization* (Princeton, Princeton University Press, 1961), pp. 247 and 249.
[32]York Willbern, *The Withering Away of the City* (Bloomington, Indiana University Press, 1966), p. 49.

[33]Scott Greer, *Metropolitics: A Study of Political Culture* (New York, John Wiley & Sons, Inc., 1963), p. 80.
[34]Luther Gulick, *The Metropolitan Problem and American Ideas* (New York, Alfred A. Knopf, Inc., 1962), p. 36.

authority. Under these circumstances, alternative strategies offer themselves. The interest may attempt to move the function to a level of government whose arena provides a competitive edge, or to insulate the function under a special district whose territory will provide a favourable arena, or to overcome the disadvantage suffered locally through mandatory directives or conditional grants from the central authority.

These strategies have been used with telling effect by specialized bureaucratic interests. While not meaning to be pejorative, it seems clear that these interests, born of common professional background, marked by membership that transcends governmental levels, and strengthened by alliances with sympathetic segments of the public, have often been remarkably effective. But to be effective they have had to resort to strategies that enhance the complexity of government. The outcome of these strategies can frequently be beneficial. For example, the trend toward larger school units in the United States and Canada is largely the outcome of a highly professionalized educational bureaucracy striving to realize the goal of self-improvement for its members. Loyalty to the profession is more often than not stronger than loyalty to the local community and yet the profession has often been able to command the respect of outsiders due to the intelligence and activity of its members.[35] Educational specialists have perhaps enjoyed more spectacular and visible successes than their professional counterparts in other fields, but similar stories can be told, whether of planners or social workers, police or traffic engineers.

That the action of bureaucratic interests has meant substantial gains in efficiency there can be no doubt. But these gains are all too often won by insulating particular functions from the political arenas in which unreceptive interests hold sway. The special district, the conditional grant, the central directive—all can be viewed in no small part as protecting the function for which they are designed from incursion by competing claims. Alliances between specialized interests at different levels of government are used to overcome the parochial interests that dominate local units.

It is perhaps in the very success of the devices bred by these alliances that danger lurks. The devices themselves become part and parcel of the governmental structure, and accordingly help, in Robert Dahl's words, "to determine what particular groups are to be given advantages or handicaps in the political struggle."[36] Paradoxically, if not surprisingly, they manage to co-exist with the very units of government whose dominant interests they are designed to overcome because they tend to underwrite the viability of these units. Their very effectiveness reduces the need for more far-reaching reform. As mentioned earlier, such sophisticated variations on the theme as fiscal need and capacity grants can easily go beyond the goal of equity to the maintenance of otherwise unviable governments.

As their programmes take hold, bureaucratic interests in turn can become advocates of the status quo. They breed interlocking directorates that impede change in those rare instances where the central authority undertakes a drastic overhaul. This is a phenomenon documented strikingly well by Frank Smallwood in his study of metropolitan reform in Greater London. There, the opposition of teachers

---

[35]See Willbern, *op. cit.*, pp. 113-14.

[36]Robert A. Dahl, *A Preface to Democratic Theory* (Chicago, University of Chicago Press, 1956), p. 137.

and other education groups finally forced the British Government to leave the London County Council educational system unaffected by the Greater London Act of 1963 for a period of five years. That these groups had their allies at the level of the central authority there can be no doubt. To quote Smallwood: "the Ministry of Education found the Greater London area to be 'an efficient education authority'. As regards the counties, it felt that the London County Council was particularly successful in those parts of their work which require a high degree of professional skill or which raise problems of great complexity."[37]

In Conclusion: Some Questions

All that has been said thus far doubtless promotes the reader to consider how his metropolis measures up to the issues raised. However, before posing the questions that should be asked, it may help to recapitulate.

The fractionalized governments of the metropolis are political arenas whose relative homogeneity endows certain interests with competitive advantages. The result is that any departure from the status quo that endangers these arenas meets ferocious resistance. But within the framework of the status quo, it is readily possible for interests that find themselves disadvantaged in any given set of local arenas to turn to other arenas, especially that of the central authority, where they are not disadvantaged. Specialized bureaucratic interests that benefit from common ties with their counterparts up the intergovernmental ladder, and that can draw support from relatively cosmopolitan groups, are partic-

ularly adept at this. The outcome produces devices like special districts, conditional grants and mandatory directives that insulate certain functions vis-à-vis the interests that dominate local arenas, and enable the proponents of these functions to overcome the competitive disadvantages that beset them at the local level. However, the interests that benefit from these devices in turn become vested, and tend to make far-reaching reform even more remote.

What emerges from all this in terms of the values that local government can aspire to serve? There is a very strong case for saying that these values are far from forgotten. There is a measure of efficiency, tailored to standards developed in relative isolation by specialized interests. There is participation too—albeit a fractionalized kind of participation in political arenas that are anything but comprehensive. As for liberty, if this value is fulfilled by the mere absence of monolithic government, it is achieved in abundance. But on the other hand, what kind of efficiency is realized in the absence of the responsiveness born of widespread access and participation? Does not the value of participation ring hollow when deliberation and consideration are the prerogative of neighbourhood and professional interests? And is not freedom from arbitrariness perhaps better guaranteed if specialist groups are, like the French educationists, firmly tied to a central authority that is perhaps monolithic but nonetheless "extensive" republic? We have now begun to ask some of the questions that this paper was designed to raise. It is possible to construct a set of questions that may prove useful in analyzing individual metropolitan areas by referring in turn to each of the key political values served by intergovernmental relations.

Let us assume first that we wish to pursue the value of efficiency.

---

[37]Frank Smallwood, *Greater London: The Politics of Metropolitan Reform* (Indianapolis, The Bobbs-Merrill Co., Inc., 1965), pp. 185ff.

(a) What functions are now performed by the central authorities, intermediate or metropolitan levels (if any), special districts and local municipalities?

(b) To what extent does each government have a territorial jurisdiction that succeeds in internalizing the public goods assigned to it?[38]

(c) Are the externalities of certain public goods so badly contained at present that movement of these goods to a higher level of government is called for?

(d) If there does not now exist a metropolitan level of government, could such a level in fact be designed so as to internalize a number of important public goods?

(e) In how many instances are incompatibilities in the scale of particular public goods such that special purpose districts are called for?

(f) At the infra-metropolitan level, are the units sufficiently similar in size to permit relatively identical bundles of public goods to be successfully internalized?

(g) Are there an appreciable number of public goods whose appropriate scale is substantially smaller than the territorial jurisdiction of the government or governments now providing them?

(h) Once the best possible match of public goods to levels of government has been made, how many conditional grants will be needed to accommodate the remaining spillovers?

(i) Do unconditional grants accompanied by mandatory directives from a higher level offer any advantage over conditional grants?

Turning now to the value of participation, the following questions emerge:

(a) To what extent does public attention gravitate to a level of government because it is important and relatively autonomous?

(b) Do the existing governments have jurisdiction over a sufficiently large number of functions to make meaningful priority decisions?

(c) How far are decisions as to priorities already predetermined by conditional grants or central directives?

(d) Is the setting such that a second level of government, metropolitan or infra-metropolitan, would have so few functions or be so constrained by higher levels as to be unimportant?

(e) If there are special districts, are they so numerous or their functions so obscure that the public ignores their existence?

(f) Does placing a function under a special district enhance the visibility of that function and increase accountability for it?

(g) Does the public have direct control over special districts through elections? If not, are the governing boards of special districts so appointed that there exists a clear locus of responsibility for their actions?

(h) Assuming that both conditional grants and special districts are to a degree indispensable on grounds of efficiency, do the contrary dictates of participation admit more of one than of the other?

As to the notion of freedom from arbitrary action, it begs at least these questions:

(a) Is this value in fact still served by the existence of more than one level of government?

(b) Is there a degree of governmental complexity beyond which the Scylla of arbitrariness is replaced by the Charybdis of paralysis? Or is complexity itself a virtue in that the greater the number of decision-making bodies, the greater the likelihood that government will at least be under a multitude of elites?

With respect to interests, the questions are no less challenging:

---

[38]For an elaboration on this point, see "Technical Appendix" below.

(a) Are interests embodied in visibly organized groups or in tacit alliances of individuals sharing common occupations, incomes, social class or ethnic backgrounds?

(b) Is the structure of local government such that certain interests dominate in political arenas that are matched to relatively homogeneous neighbourhoods?

(c) Has the central city been losing its heterogeneity and itself been acquiring neighbourhood characteristics?

(d) To what extent do such efficiency devices as conditional grants and mandatory directives bring about local decisions that differ from those that would be made otherwise?

(e) Do these devices tend to underwrite the existence of non-viable governmental units?

(f) What has been the role of professionalized interests in creating these devices?

(g) Do the vested interests of parochial and professional interests reinforce one another and entrench the status quo?

(h) Does the resulting complexity of intergovernmental relations possess the minimal virtue of providing widely dispersed loci of decision-making through which major values are at least achieved in part?

(i) Is the goal of greatly simplified and clarified governmental institutions with which "we might gain a higher level of consideration and deliberation, a fuller and more precise posing of alternative choices, a more accurate appraisal of the probable causes and effects of present conditions and future policies, a more rational conclusion as to the 'best' policy,"[39] but an illusion in a setting where metropolitan problems themselves are neither simple nor clear?

Doubtless, while these questions will provoke useful discussion, and in turn lead

to other questions, they are unlikely to yield answers on which most can agree. We have begun to understand the problems that beset the metropolitan area, and the work of two or three generations has provided us with a list of remedies to some of these problems. But the precise application of these remedies requires diagnostic questions that are still far from adequately formulated, to say nothing of their susceptibility to precise answers. Furthermore, there is the added difficulty that conditions at the crossroads of intergovernmental relations shift so inexorably that we continue to puzzle over old questions when new ones beckon.

Technical Appendix: Apportioning Public Goods

Internalizing a public good for efficiency involves an attempt to confide responsibility for that good to a government whose territorial jurisdiction is such that spillovers are minimized. Unfortunately, opinions as to the scale needed to internalize a public good vary widely because delicate judgement questions are involved, and because it is often necessary to break down a public good, say education, into component goods, for example, primary, secondary and vocational education. These difficulties are not an insuperable barrier to the exercise, however. To help launch discussion of the second question raised in the conclusion, it may be of use to refer to one such exercise, the results of which were published in 1963 by the United States Advisory Commission on Intergovernmental Relations.[40]

Because their spillovers are inconsequential, the Commission cites the follow-

---

[39]Willbern, *op. cit.*, p. 133.

[40]*Performance of Urban Functions: Local and Areawide*, U.S. Advisory Commission on Intergovernmental Relations (Washington, D.C., Government Printing Office, 1963).

ing as functions that can be internalized on a relatively small scale: fire protection; refuse collection; playgrounds and parks; streets and sidewalks; and, health education and maternal and child care services.[41]

On the other hand, the following functions have sufficiently widespread externalities to warrant large-scale—area-wide or perhaps greater than area-wide—internalization in most metropolitan regions. Refuse disposal; police, environmental public health; urban renewal; housing standards and zoning; education; libraries; large parks and recreation areas; welfare; general planning; hospital and medical care facilities planning; transportation, except streets and sidewalks; water supply and sewage disposal; and, air pollution.[42]

The Commission cites yet other functions as ones that have such minor spillovers that they can probably be internalized on a large or small scale. Not surprisingly, there are utility services whose direct benefits can be enjoyed (and paid for) exclusively by the individual consumer. The Commission's list includes: gas and electricity, parking facilities; golf courses, swimming pools and so on; curb, gutter and sidewalk improvements and maintenance; transportation terminals; and, legal services.[43]

If goods are to be assigned to a level of government with efficiency as the principal criterion, it may be necessary to distinguish on occasion between the area needed to internalize the good, and the area over which a good can be provided with maximum economies of scale. Frequently, the two areas coincide. On the occasions when they do not, the problem may be overcome by breaking the public good into components and assigning these to succeedingly larger levels of jurisdiction in such a manner that each level internalizes the spillovers left by the lower levels. Police services are a case in point. With respect to a public good like police, jurisdictional assignments should be geared to components of the good. Thus foot and motor patrols might be local, special investigations and crime laboratories metropolitan, criminal identification files central, and so on.

Once public goods have been allocated to the jurisdiction deemed appropriate on grounds of efficiency, it can be of interest to review the list with respect to the dictates of participation. This exercise was also undertaken by the U.S. Advisory Commission on Intergovernmental Relations, and caused the Commission to reassign two major public goods with area-wide spillovers to the local jurisdiction. Thus, education was reclassified as primarily local because ". . . concerns expressed in large city school systems indicate that effects of bureaucracy associated with large size tend to lessen sensitivity to the public. Sheer size tends to diminish citizen interest and participation, despite the fact that parents' interest in their children's education is largely school-oriented rather than district-oriented, and large districts can have small schools as well as large schools."[44]

Libraries are also seen as more nearly local because ". . . the community library often has its origin in the work of a group of public spirited citizens. . . . Often the library is the center of much of the community's activities."[45]

---

[41]Ibid., pp. 9-15.
[42]Ibid., pp. 10-23.
[43]Ibid., p. 43. The Commission included refuse collection and disposal on this list, but then classified collection as appropriately local, disposal as areawide.

[44]Ibid., p. 12.
[45]Ibid., p. 14.

It is perhaps needless to emphasize that apportioning public goods for maximum participation is the most elusive exercise of all. Here the peculiar environment of different cultures can produce quite different answers. That the U.S. Advisory Commission on Intergovernmental Relations reclassified education as local must be accounted for in no small part by the long tradition of local control in that country. The basic rule of thumb that this author suggests in allocating public goods with respect to participation is that no level of government be left with nothing but insignificant responsibilities and powers.

12 Financing Metropolitan Government

Ursula K. Hicks

The Metropolitan Situation

One of the more important aspects of the general field of local government finance is that of financing metropolitan government. To understand the significance and special characteristics of the problems involved it is useful to examine briefly the features of the metropolis which, from a financial viewpoint, set it apart from the local government scene in general. In the first place a metropolitan area is very large, if not absolutely, at least relatively to the size of other cities in any one country.[1] Secondly, a metropolitan area serves a large "catchment area"; and it is unlikely that there will be other cities of more or less equal size nearby. If there are, preferably they should be included in the one metropolitan conurbation unit; although the political difficulties of this are recognized. More typically the metropolitan catchment area will be larger than could be included conveniently within its political jurisdiction. A third characteristic is that a metropolitan

---

[1]From the point of view of finance, however, the capital city of a federation is not normally in the category we wish to examine, because it is often situated in a separate federal territory where many of the services are provided by the general government, and where there may even be no democratic council that is fully responsible to the inhabitants.

area is generally subject to intense immigration. Nearly all cities in most countries tend to suffer to some extent from "urban drift" as the rural population seeks better jobs and improved amenities than are available in the villages; but the magnetic influence of a metropolis is abnormally strong.

These conditions give rise to special financial problems. On the expenditure side, particular costs will have to be incurred to provide services for commuters—for example, street paving, cleansing and lighting, and transport—in excess of what would be required for local inhabitants. Many of these commuters will probably live outside the tax jurisdiction of the metropolis, thus the problem of a satisfactory boundary policy is acute. Particular costs may be incurred because of "diseconomies of large size". Especially good provision must be made in respect of urban overheads and utilities—such as water, street paving, and lighting—where a large number of people live close together. This is even more true of health services. Moreover, the rate of immigration itself poses special problems of housing, urban development and renewal (for the immigrants are likely, initially at least, to be poor and to need special attention), and special problems of education because the immigrants are likely to be preponderantly young married couples raising families. Indeed many come to the metropolis specifically to seek better educational opportunities; a phenomenon especially true of Africa. All these circumstances give rise to additional outlay over and above what non-metropolitan cities have to face.

On the other hand, size and especially the expansion of the population give rise to improved tax potential. In the first place, property values will be high and rising. It should be possible to derive a great deal of fiscal advantage from this circumstance. Secondly, a large population, including the commuters, ensures a good yield of indirect revenues of the nature of sales taxes, vehicle licences or parking dues. The only (partial) exception to these generalizations is found where the immigrants are extremely poor, as in Calcutta.

Owing, however, to various impediments—legal, constitutional and political—it is difficult to take full and rapid advantage of these fiscal opportunities. Even if they can be fully exploited there is a strong probability that the cost of the necessary development of services up to metropolitan standards (which should be higher than for a non-metropolitan city) will exceed the autonomous revenue sources available. Consequently, there will be a need for assistance from a higher level of government, probably on both current and capital accounts. In a federation the situation in this respect is usually more difficult than under a unitary system, since local government is normally a state subject. In virtually all federations there is a chronic imbalance between state and national financial resources in relation to the services they have to perform. The states themselves are short of funds, and cannot afford to give very much even to the metropolises. Moreover, they may regard them with suspicion (especially over the matter of boundaries) and even with hostility, as having fundamentally different values and interests from those of the rural population. Or again, the state may itself be relatively wealthy but may not wish in principle to impose high taxes on its inhabitants.

Considerations such as these raise a question not so much as to the circumstances and extent to which metropolises should be allowed to extend their boundaries, but as to whether the ultimate solution is not the "city region" throughout the whole country. This appears to be a most acceptable proposition for thickly populated areas (as in Holland, Belgium and parts of England) and it will suffice to note here that this point is pursued at greater length by Weissmann in Chapter 13.

Before we turn to consider the financial problems to which these special metropolitan conditions give rise, we must examine the general principles under which their finances should operate. They differ in degree rather than in kind from those of local jurisdictions, but the degree can be very substantial. In the first place, it is as important in local as in national government that the fiscal structure should be equitable, so that families similarly placed should bear equal burdens and receive equal benefits. This is horizontal equity. Vertical equity, which is concerned with ability to pay and income redistribution, is more difficult to achieve at the local level. Indeed it can be realized fully only on a national basis. Consensus of opinion in regard to vertical equity is not inevitable, nor is it necessary that all local governments should have exactly the same ideology in this respect.

However, equity is not only a matter of principle. It depends heavily on efficient implementation, both of assessment and revenue collection and efficient administration of services. Efficiency must be taken to include first, the ability to exercise optimal choice among ends, and second, between agreed ends the determination of the priorities that will contribute most to the wellbeing and advancement of the city community. Under the heading of efficiency also must be included the problem of the distribution of decision-making between governments. It has been noted that in socialist-type economies there has been a tendency in recent years toward decentralization of decision-making, while in mixed economies the movement has been rather toward greater centralization.

Finally, it is of the essence of responsible local government that there should be autonomous decision-making within the range of ascribed services and within the framework of the nation. This implies that it must be granted the fundamental budgetary rights of choosing the volume and structure of expenditure at the margin, and of choosing the rates and, at least to some extent, the conditions of taxes and other autonomous revenue sources within their competence. These rights are (as we shall discuss below) by no means incompatible with substantial financial assistance from national or state governments, provided that it is given in the right way. If the national government deems it essential to exercise some control over transfers for particular services (either in the interests of good administration or of national policy) it should do so in such a way that the size and structure of the budget remains under local control.

Basic Principles of Metropolitan Finance

PRINCIPLES OF TAXATION

As we have seen already, metropolitan governments have special responsibilities which call for particular financial provision. This does not mean, however, that either the normal principles of public finance or the more specialized ones of local finance do not apply. Adam Smith, writing nearly 200 years ago with remarkable perspicacity, hit the essential targets of a satisfactory tax structure with his "canons" of ability, certainty, convenience and economy.

We have already seen that *ability*, in the sense of vertical equity, must finally be judged in national terms in conjunction with redistributional public expenditure. We may leave it aside here as we shall come to it again in relation to particular metropolitan taxes.

Certainty, concerning the tax bill to be met is of great importance to taxpayers, especially to the business community for whom an unexpected levy, or even the apprehension of one, may seriously hamper entrepreneurial planning. Similarly, an accurate forecast of the revenue that will be available is also of great importance to the metropolitan government which also has to plan in advance for services and for projects extending over a number of years.

Convenience of time and place of payment is hardly less important to both parties. Inconvenient methods of payment not only make it difficult for the taxpayer to fulfil his obligations, but greatly add to the expense of governments in hunting up laggards. A striking example of inconvenient methods of payment damaging the whole image of a tax and leading to the erosion of the base on political grounds has been the British local rate that falls on occupiers of land and buildings. Until 1966 this had been collected by almost all authorities in yearly or half-yearly assessments. The sums required of the lower income groups are often substantial in relation to their incomes. In fact this tax is now the only regular payment that such families have to make on anything greater than a weekly basis.[2]

It is not clear what Adam Smith really wished to include under "economy", but several important and useful meanings can be ascribed to it. First, and most obvious,

it implies the adoption of methods of assessment and collection that are in accordance with good business usage. It is no economy to have tax departments that are either understaffed, or overstaffed, with low-paid inefficient clerks who do not understand what they are doing and have not sufficient incentive or education to work accurately. This is especially relevant for local tax administration, as the competence of local tax departments, even in metropolitan areas, may be less than that of the corresponding national offices. Secondly, and from the viewpoint of the taxpayer, the most economical taxes are those in which the cost of collection is lowest in relation to the revenue that he must make available for public purposes. Thirdly, from the point of view of the whole community, economical taxes are those which exert the smallest possible disturbance in consumption and, hence, in the organization of production. The type of tax that most nearly meets the criterion of economy is a lump sum payment, which (inevitably) removes some of the taxpayer's income, but does not disrupt his preferred personal (family) budgetary structure, and so will not discriminate against any particular type of production. (In economic terms this is a tax that has income but no substitution effects.) The most obvious types of tax that fulfil this condition are moderate taxes on net income or on consumers' goods, the demand for which is highly inelastic. As will be seen later, taxes on property may in certain circumstances also have this effect.

We turn now to the particular characteristics of taxes for the purposes of local finance, and especially of metropolitan finance. The first desideratum is a tax that will yield a *steady revenue*, will not fluctuate violently and, more particularly, that will not decline seriously in times of high unemployment so that either funds for

---

[2]Allen Report, *The Impact of Rates of Households* (Great Britain, HMSO, 1965), Cmd 2582. Local authorities are now encouraged to make provision for installment payments.

important services (such as teachers' salaries) will not be available and the payments fall into arrears, or tax rates will have to be raised just when it is most inconvenient for taxpayers. It is further desirable that a local tax should not accentuate differences between rich and poor areas, especially if there is a national policy of attempting to equalize interlocal financial opportunities. Also, a tax that discriminates between rich and poor areas (such as a progressive local income tax) may prove embarrassing to central revenue needs. A striking example of this was the embarrassment caused to the Federal budgets in Canada and Australia in World War II due to high income tax rates in certain provinces (states). It is also highly desirable that expenditure to be met out of revenue should not fluctuate violently, as this necessitates frequent changes in tax rates even in normal times. This is largely a question of the right choice between finance out of current sources and out of loan, a question which we must discuss in connection with capital finance.

The revenue should be steady, not only in absolute terms, but more important, in *real* terms; that is, the metropolitan government must be able to buy the same "bundle" of goods and services at constant tax rates. It will also want a larger revenue as municipal services expand. Thus a second desideratum is that the revenue should be elastic not rigid, so that as local incomes and wealth rise, tax revenue even at constant rates will expand with them. As this quality belongs *par excellence* to progressive personal income taxes in which even metropolitan governments are not likely to be allowed a large share, this desideratum is not easy to achieve. But as we shall discuss later, something can be done in this direction by the right choice of outlay taxes.

Elasticity of revenue is closely related to

ability to pay, as those who can best afford to pay taxes are those with not only high but also rising incomes. In spite of the restricted field for progressive taxation, this condition can be achieved to some extent. At the other end of the income scale, it should be possible to avoid a revenue system which is highly regressive; that is, a system in which families in the lower income groups pay a substantially larger proportion of their income in taxes than do the more wealthy.

There is, however, another characteristic of local taxes which has received a great deal of attention in the past and is still of some importance—namely, *benefit received*. Many of the services provided by metropolitan governments—police, street maintenance and cleaning, and environmental health—are of direct benefit to the citizens in a much more intimate way than many of those provided by national governments, such as defence and administration. These benefits of local service accrue both to firms and individuals. They may be general, such as traffic control or the suppression of crime, or they may accrue only to particular groups in the community. The groups benefited may be wide, such as all parents who take advantage of public education for their children; or they may be less wide, such as those provided for the sick or elderly; or again they may be very narrow indeed, such as a group of families benefiting from a particular street improvement. The problem is the extent to which compulsory payments corresponding to the benefits received should be exacted from the beneficiaries. There is clearly no general answer to this problem. Some of the special group services (such as geriatrics or those for spastic or retarded children) largely serve groups that could not pay for the services and for which the expenditure is primarily redistributive. In such cases the benefit principle of taxation clearly cannot apply.

This is broadly true also of services such as general education for which increasingly it is becoming the practice to charge no fees. If fees are charged it may happen that the children of poor parents are denied the opportunity of reaching their full potential, and at the same time the community is depriving itself of the full use of their talents in the future. On the other hand, a case can be made for compulsory charges for particular types of education (for example, higher professional or technical education) that will add substantially to the earning capacity of the student himself. In this case there need be less apprehension of income discrimination as it should be possible for a student who has proved himself up to that point to secure a loan. Finally, the small group benefiting from a local improvement, such as street paving, can reasonably be asked to contribute at least to the special expenses. Obviously, other social principles must play a larger part in determining the rates and distribution of metropolitan taxes in the modern world than implied by the nineteenth-century concept of benefit.

There remains, however, the need to find a way to tax *personal benefit* or "unearned increment", which accrues to property owners from the rise in values that takes place as a city develops. Generally speaking, not a great deal of this can be pinned down to specific activities of the city government. Where they can be so identified there is a strong case for a special charge. More generally, the metropolitan tax structure should be geared to secure a *pari passu* expansion in the tax base as property values rise. We shall discuss this below.

Thus far, we have been concerned implicitly with the *formal* incidence of taxes; that is, we have assumed that the economic consequences of a tax, or more accurately tax *change*, affects those on whom the tax is assessed. This may not be true, or only partially true, except with respect to lump sum taxes. In all remaining cases other people also will be affected by some process of shifting or substitution on the part of those on whom the tax is assessed. This process may take many forms.

As far as metropolitan government is concerned, probably the most important processes occur in connection with the property (real estate) tax. Where housing is in short supply, owners will have a quasi-monopoly power and may cause tenants to pay in the form of increased rents when the tax goes up. Thus it becomes, in effect, a tax on occupancy. House owners who are dissatisfied with the tax and who have the means to do so, will move outside the metropolitan jurisdiction, thus forcing the government to raise tax rates on those remaining in order to protect its revenue. This process fosters a tendency for the inner area of cities to fall into serious neglect, giving rise to inappropriate and inefficient usage of property. The ubiquity of the motor car has accelerated this effect, but (in the United Kingdom, at any rate) it is of very long standing. It is common to find poor towns inhabited for the most part by manual workers near wealthy towns even in decaying areas; for example, Oldham and Middleton, in Lancashire, England. In these circumstances there will be no spontaneous incentive for new building within metropolitan jurisdictions, and it is probable that the government will have to give some assistance for redevelopment.

Turning to indirect taxes, insofar as sales or purchase taxes are concerned (taxes assessed at the retail or the wholesale stage respectively), the effective incidence is likely to be on consumers. This will occur unless demand is very elastic (implying that consumers are not resistant to substitution), in which case the effective incidence will be on the manufacturers and sellers in the form of reduced demand for their products, and may be very widely spread. Given the sorts of indirect taxes (taxes on outlay) which are likely to be within the competence even of metropolitan governments, the effective incidence will likely be regressive. The same may be true of a property (real estate) tax shifted onto tenants, as it is the poorer families who are least able to take evasive action. These are some of the reasons that make it difficult to achieve local tax structures which are in accordance with ability to pay in the sense of vertical equity.

TAX AUTONOMY AND TAX SHARING

The final and most fundamental principle of local government finance is that some part of the tax revenue should be *autonomous*. According to the formal definition[3] of federal government the constituent states should have unfettered jurisdiction over *everything* falling within their constitutional competence, including taxes. It is doubtful if this condition ever obtained in the real world, and it certainly does not in present federal circumstances, due to the expansion of the commitments undertaken by national governments. Local governments can put forward no such claim of coordinate rights, but metropolitan governments have a case for a wide area of discretion not only over the rates of individual taxes but also over the choice and range of taxes that they can impose. Such freedom and flexibility is inevitable in view of the importance of the services for which they are responsible. In a federation there is a special need for securing metropolitan autonomy in this sense, for (as suggested

above) the interests of the states within which they are situated are likely to be narrower and probably less sympathetic to metropolitan problems than the interests and attitudes of a national government.

From the point of view of revenue autonomy, the most appropriate local or regional taxes are those where there can be no jurisdictional dispute—*par excellence* those on real property within the area. This avoids all problems of competitive exploitation,[4] which is undesirable if carried to excess because it leads to very uneven tax burdens, as between one area and another when a local tax is piled on top of a tax of a higher government. This condition is satisfied as well, however, by taxes on purchases made within the area—save that the tax may by preference be assessed only on those resident within the city bounds, in order not to drive would-be purchasers away. It is reasonable to suppose that in a metropolitan area the quality and choice of a large range of consumer goods (especially durables) would be far superior to that obtainable on the outskirts. Thus, this danger is relatively small. With respect to food, exactly the opposite may well be true—choice and freshness is better in the rural fringe. But the taxation of most foods is so regressive that it should in any case be avoided.

There is, however, a class of taxes that are useful to metropolitan governments and that are usually in accordance with ability to pay, but over which unfettered autonomy cannot be tolerated. The chief of these are personal income and capital taxes, which may be shared between two or, in a federation, three levels of government. Taxes on net business profits may

[3]K. C. Wheare, *Federal Government*, 4th ed. (Great Britain, Oxford University Press, 1963).

[4]In Australia, however, the states also draw on the property tax, but usually not to a very important extent.

also be shared, but they are always subject to difficulties in determining the value of profits accruing in a particular area to firms operating in a number of areas. This difficulty is virtually non-existent for taxes on gross profits, which we shall examine later. The advantage of all forms of sharing is that the administration of the tax body remains in the (probably more competent) hands of the national government. However, this has to be set against the cost of relinquishing local autonomy.

The sharing process can take several forms. In the first place we must make a distinction as to whether it is the tax *source* that is shared, or the collected *revenue*. The latter form of sharing does not allow for any autonomy on the part of individual governments, although as regards *classes* of governments there may be autonomy in the sense that discussions must (under the constitution) be held at intervals at which the relative shares are discussed. Thus, in India an independent Finance Commission collects evidence from Union and all state governments and makes recommendations on the percentage of income tax to be transferred by the Union government and the amount to be received by each state. A slightly more complicated method of sharing revenue consists of the obligation of the national government to pay the revenue (or some percentage of it) from several taxes into a "Distributable Pool". If desired, the taxes can be chosen as having different locational incidence. In any case, a spread of taxes is more likely to produce a more stable revenue than a single source, especially the progressive income tax which is notoriously fluctuating. Distribution is then made from the Pool to different areas according to a formula taking account of obligations and needs. Although this method of revenue sharing has been applied (I believe) only to states in a feder-

ation, there would seem to be no inherent reason why it should not be applied to metropolitan governments by national or state governments. However, all of these methods of tax sharing provide local revenue that is only partially autonomous, as arrangements have to be made to prevent local authorities from stealing revenue from the national or state budgets.

One other form of revenue sharing is worth notice. It is illustrated by the recent transformation in India of certain state sales taxes into national excises. This operation was undertaken by the Union government mainly because of the low (and consequently inequitable) standard of administration in certain states. The Union government guaranteed to each state the same revenue as that previously derived from its sales tax. The transfer resulted, in some states, in a rise of 200 per cent in revenue collected at unaltered rates. Revenue over and above the guaranteed transfer was at the disposal of the national government, which proceeded to use its powers in what some states regarded as an arbitrary manner. Against the benefit of improved administration, and hence both greater economy and equity, must be set the cost to the states in loss of autonomy and an expanding source of revenue. In fact, owing to subsequent price rises, what the states in India received was not a steady but a declining source of revenue.

Presuming that "free-for-all competitive exploitation" is ruled out, tax (as distinct from revenue) sharing can either take the form of a constant local/regional share placed *on top* (for instance) of the national tax levy, raising the curve of formal incidence right along the income levels, or, placed at the *bottom*, a more or less proportionate local tax taking the first slice of all incomes. It would also be possible for the local tax to be assessed fully progressively with income. Indeed, this refinement was tried with the Swedish local income tax for some years but, interlocally, was found to be intolerably disequalizing.

If the local tax takes the form of an additional percentage on the national tax, full advantage can be taken of central administration. For a bottom level local tax, however, separate assessment would be required. Later we shall discuss in detail these various forms of what are effectively local income taxes. Here, there are only two further points that need to be emphasized. On the one hand, the use of any form of income tax in the local tax structure has the effect both of reducing its regression and of increasing its elasticity as compared with taxes on outlay. Both of these are improvements, on social and economic grounds. On the other hand, the full force of fluctuations in revenue collections stemming from fluctuations in the Gross Domestic Product is transmitted to local governments, which are much less able to bear it than national governments. In this way a shared income tax is less reliable for local purposes than a tax on outlay and very much more reliable than a tax on real property, which, apart from defaults in periods of high unemployment, yields a remarkably steady revenue.

As far as metropolitan governments are concerned, expansion of tax revenue derived from any sort of income tax would be a special benefit, as it is in metropolitan areas that the main wealth of the country is concentrated. Further, it would facilitate the raising of revenue from commuters more than would any sort of sales tax even though, from this point of view, a sales tax is better than a property tax because of its jurisdictional limitation. We shall return to this point.

These various forms of transfer from a higher to a lower level of government can be regarded as types of grants-in-aid. The essential difference between grants and transfers from shared taxes or revenue is that under the grant system the amount which will be available to the local budget is fixed in advance—probably for a short term of years—and so does not fluctuate with the level of economic activity. Hence local budgetary forecasting is facilitated. How important this aspect of transfers will be, as distinct from grants, depends first of all on the relative share of transfers in total current revenue, and secondly on the availability of short-term borrowing to bridge any decline in revenue receipts.

As we have seen, some form of fiscal aid is essential if local governments (or state governments) are to develop adequately the services committed to them. What we have to discuss next is the best form for such grants to take, in view of both national and local interests. Here we are concerned only to map out the ground. We shall study them in detail later.

Principles Relating to Grants
A most fundamental distinction must be made between grants which are allocated to particular services or projects and those which are not so earmarked. Either specific or non-specific grants may be subject to conditions, but on the whole these can be written more easily into specific grants where the conditions can include a stipulation relating to the local tax effort (matching conditions) or to the type and standard of service to be provided with grant aid. Conditions attached non-specific, unallocated grants can also be related to tax effort, but more generally the grant-receiving government has little control over the relevant conditions—such as property values, level of unemployment, or distribution of the population by income or age group.

The effect of a specific grant is to stimulate differentially the grant-aided service. That is to say it is definitely promotional. The conditions written in, however, can contain whatever method of control is thought desirable in the interests of efficient allocation of resources. There is some danger of discriminating against non-grant-aided services, even though the grant-receiving government's financial position may be improved substantially as a result of the specific grant. Thus, a possible distortion is from the preferred allocation of resources by the grant-receiving government, and a consequent reduction in its autonomy of budgeting. For this reason, the total of specific grants in the structure should be kept moderate, of the order, say, of 20 per cent. Finally, unless very careful discrimination in favour of poor authorities is written into the conditions, specific grants tend to be interlocally disequalizing, since only the wealthier authorities can afford to make full use of them.[5]

Unallocated grants do not have a discriminatory effect, as between services, though they may discriminate between areas. Hence they lead to the maintenance of a high degree of budgetary autonomy even when grants form a substantial percentage of total local financial sources. After a certain point, however, local governments are apt to lose their initiative, even with a general grant, relying on grant aid rather than on autonomous revenue sources. Just where this point will come depends on a number of factors, not the least of which is the type of general grant awarded.

---

[5]In the United Kingdom a bad case of this was the failure of the Welsh authorities to take up the grant for tuberculosis prevention in the 1920's, on account of the high real cost of their (quite minor) share, notwithstanding the incidence of the disease was much higher in Wales than in England.

With respect to specific grants, it is even less possible to generalize concerning the optimum size (except that it should always be less than 100 per cent, as this removes all incentive for good management from the grant-receiving government). The crucial point is whether the grant-receiving government or the grant-giving government is more anxious to develop the service in question. British experience has been that very high grants are needed to promote the development of clinics for Venereal Disease, but other health services such as child welfare and maternity may be established with a much lower bait. A classic British illustration of this principle was the grant offered to local authorities, just before the outbreak of World War II, to build air-raid shelters. The local authorities were profoundly uninterested even in rising baits, until 100 per cent was reached. By that time the war was imminent and the resulting shelters were shoddy and doubtfully safer than the open street.

A crucial point in determining both the optimal level and type of grant is the extent to which it is desired as a matter of policy to make the transfer system redistributional as between rich and poor authorities. In principle there are two basic methods of allocating transfers between grant-receiving governments—according to derivation and according to formula. Under the former method each authority receives as near as may be the equivalent of the revenue which has been derived from its jurisdiction. With respect to income taxes (although not with respect to taxes on net profits) this is generally not too difficult. As for the revenue of excises, customs duties and other taxes on outlay levied at the national level, only very approximate results can be expected. The great objection to the derivation principle from the social point of view, however, is that it is interlocally disequalizing—to those that have, more is given. Consequently, wealthy areas (and hence most metropolitan governments) will press for as large a use of the derivation principle of revenue sharing as they can contrive, while poor areas will try to limit its use.

The intention of a formula for the distribution of transfers is to get away from the effects of derivation and thus clear the way for following a constructive policy in a desired direction. This may take many forms, including the promotion of particular types or locations of development and, more commonly, to secure a greater equalization of financial opportunities among local authorities. Locational or redistributional discrimination can be written into the conditions of a specific grant but, probably because the promotional and control conditions which are its main purpose are in themselves likely to call for considerable complication, this is less commonly done than is the case with unallocated grants. Education grants are often an exception. The British education grant discriminated (from 1914) against wealthy authorities by a deduction of a standard tax yield applied to the value of land and buildings in the jurisdiction.

The considerable experience available on the working of differential transfers shows that equalization of opportunity, which is the only form of equalization fully compatible with local budgetary autonomy, can be obtained to any desired degree. Some objective measure (such as per capita incomes or real estate values) of the relative ability of a community as a whole to pay for services may be used, but the formula tends to become very complicated. At the same time the promotional/control aims of transfers need to be allowed sufficient scope. These considerations suggest that just as it is easier to achieve a steady

but reasonably flexible revenue with a tax structure of two or three sources, as opposed to one, at the disposal of local budgets, so a transfer structure of several parts will be more efficient than the use of a single type of transfer.

Transfers may also be geared to assist directly in promoting capital formation. This may be achieved either through special capital grants for selected projects, or by a guaranteed period of additional annual grants designed to cover a substantial percentage of the interest charge on loans incurred. These can also be awarded to particular types of capital formation (such as low income housing) or can be geared more generally in support of development programmes. The General Grant established by the Local Government Act of 1958 in the United Kingdom contained an element of this sort. Grants for capital formation can be even more promotional than those for current account expenditure. They can also stimulate forward planning in order to earn the grant. For this reason, especially, they call for more careful control of outlay—for instance, through insistence on specification of characteristics of buildings and installations in order to qualify for the grant.

From what authority the funds should emanate is the final question to be discussed regarding fiscal transfers. In a unitary country, it is clear that, because of the imbalance between the national budget and local budgets, a substantial part must be derived from the central revenue pool. In a federation, the situation is more complicated because there may well be a dual imbalance: between national and state (provincial) budgets and between state and local budgets. As local government is normally a state matter in federations, the greater part of the funds must clearly be transferred from state budgets. But it is quite conceivable that the state would be unwilling or unable to transfer sufficient funds to secure an adequate support of municipal, especially metropolitan, services. This is especially true if the state constitution has imposed a low ceiling rate of tax on land and buildings, so that even wealthy cities may be in taxation difficulties. In such circumstances the national government may find it desirable to come to the help of cities, either by supporting state governments for the purpose, or by distribution to local authorities (as is frequently done in India), or by inaugurating federal projects within the city under coordinated plans with the city government.

The fact that the funds are ultimately derived, at least in part, from the national revenue does not imply, however, that they need be distributed necessarily by the national government in accordance with its policy. In the Indian case cited above it appears that the states receiving funds for local government exercise considerable discretion in their allocation between areas and to some extent between services. Transfers under the Distributable Pool system are recommended by the Finance Commission, although naturally they must be agreed to by the national government. In both India and Nigeria the Commissions have been appointed *ad hoc* for the quinquennial revenue allocation.

There is experience also of two other systems which are largely independent of the national government and hence of party politics, and which allow for more frequent than quinquennial adjustment to changing conditions. The first and most famous is that applied by the Commonwealth Grants Commission (in Australia), a statutory body, but one that enjoys considerable autonomy. Although its jurisdiction is confined to the poorest states (now only two), its influence has been very wide. It has contributed greatly to the improvement of budgeting and statistics in all states, although it has not yet achieved a uniform system of budgetary accounting. The second system, perhaps best exemplified in Holland, is the establishment of a Municipal Fund into which the Government makes certain payments which are then redistributed according to criteria agreed upon among the participants. The principal objective in establishing this fund seems to have been to increase the independence of municipal finance from national politics.

Of the different methods of allocating revenues that we have explored, only the Municipal Fund is directly concerned with city finance. But there would seem to be no compelling reason why some of them should not be extended to cover at least the larger metropolises—say, those of over a million inhabitants. In another way also, metropolitan governments may find themselves operating in the field of fiscal transfers—namely where a two-tier system of metropolitan government has been established, an arrangement which is convenient for very large metropolitan areas. For instance, the London (rate) Equalization Fund operated successfully over the area of the L.C.C. for many years. It succeeded the still older Metropolitan Common Poor Fund. Although this fiscal transfer has been concerned only with the distribution of autonomous revenue from the (single) local tax, in the case of a wider spread of local taxes it could be given a wider coverage of revenue allocation.

From these general considerations we must now turn to examination of the applicability of particular sources of current revenue to urban and particularly to metropolitan conditions.

Sources of Metropolitan Current Finance

Broadly speaking there are four possible,

regular sources of current finance for metropolitan governments: taxes, transfers from other governments; profits of public enterprises; and receipts from public assets by way of sales, rent or interest. The last two are not likely to be of major importance so we may dispose of them first.

PUBLIC ENTERPRISES AND PUBLIC ASSETS

The enterprises undertaken by city governments are very similar from one country to another, notwithstanding a few original exceptions—a telephone enterprise here, a race course, a mortgage bank or a milk supply there. The normal city enterprises are the public utilities, transport, electricity generation and distribution, gas and water supply, together with markets, parks and amusements. In a small number of countries a substantial share of local revenue is derived from the profits of enterprise, as in parts of South Africa, for example. Where this occurs it will usually be found that the range of permitted local taxes is particularly restricted, so that the realization of profit income becomes a relatively important and prized insurance of autonomous revenue. Some cities make a profit on their transport system: electricity profits are more common; none of the others is likely to bring in much revenue, nor should they be pressed to do so as is made clear by Hanson in Chapter 3.

Both electricity and gas belong to the class of industry with very high capital costs and low and declining running costs as output increases. It would be economically incorrect to charge prices that would restrict demand below the point of cheapest production. Water supply is also in this category; but there are additional (health) reasons why it should be supplied as cheaply as possible, and no net profit should be realized. The same is broadly true of markets, parks, and amusements. Profits on individual items may well be substantial, but taking the services as a whole, net profits are likely to be negligible. Thus it is only in rather odd circumstances that the profits of enterprise can be expected to add substantially to municipal revenues. On the other hand, unless they are intended as a social service—as for instance low-income housing (for which "economic" rents would probably be higher than could be afforded by some families for whom they are intended)—these enterprises should not be operated at a loss. It is economically correct that they should cover their complete costs, including debt service and depreciation, since to cover these costs out of taxation impinges unfairly on those who use little of the service. In the case of gas and electricity such a pricing policy can be arranged easily through a two-part tariff. As for low-income housing, rents can be set in accordance with ability to pay.

Some metropolitan governments own substantial real estate, and from this some revenue can be expected. On the whole, however, more funds are derived from land sales than from renting. Such funds belong to the capital account and by their nature cannot be a regular source of income. It is unlikely that metropolitan governments will keep large reserves in money form; but balances of accounts such as pension funds may bring in some interest. Generally speaking, however, it is possible to use the greater part of all such monies (including sinking and replacement funds) for capital formation, so long as due attention is paid to liquidity at the time at which such monies will be needed. We shall discuss the policy for this at length later in the chapter.

Thus, while these and other minor sources of revenue should not be despised and neglected, they cannot, even in the aggregate, provide a major source of metropolitan finance. The latter must be supplied by taxes, supplemented (insofar as they are available) by grants for current outlay and grants and loans for capital development. Thus, with respect to the current account the most important source of finance is clearly taxes.

TAXATION: LAND AND BUILDINGS

In our previous discussion we uncovered three different types of tax, any or all of which may be available to metropolitan governments and most of which can be fully autonomous: (1) taxes on land and buildings; (2) indirect taxes on outlay, on sales and on the usage of durable goods (vehicle and boat licences, advertisements, and so on); and, (3) direct taxes on incomes and possibly business profits.

Of these, taxes on land and buildings are in almost every country the most important and may well be the sole autonomous local urban tax (as they are in the United Kingdom). This is easily understandable, as in several respects they are uniquely appropriate for local purposes. In the first place, the base is unequivocally localized so that jurisidictional disputes in this respect with other authorities are minimized. Secondly, the revenue is remarkably stable. In certain circumstances land values in particular areas may rise very steeply within a short period; but on the whole real estate values are notoriously sticky in relation to prices in general. Thirdly, abstracting from the costs of valuation (to which we shall return), the tax is extremely cheap to collect. Finally, there is a rather subtle economic advantage if the effective incidence is on the occupancy rather than ownership of buildings (as is likely to be the case with the current, rapid increase of metropolitan populations). Accommodation is such an important element in consumption that a very large revenue can be raised at quite a low rate of tax, and hence with minimum disturbance either to consumption or production.

There are, however, certain disadvantages of taxes on land and buildings. The obverse of stability of revenue is rigidity of revenue. Unquestionably, taxes of this nature have not the flexibility of income or profits taxes. If the right type of tax is chosen, however, and if revaluations are full and regular (at, say, intervals of five years for old property and as soon after construction as possible for new buildings) the tax base should expand *pari passu* with the rise in real estate values. A second disadvantage is that (again, insofar as the effective incidence is on occupiers) the tax is undoubtedly regressive, a larger percentage of small family incomes being spent on accommodation than of large incomes. In these circumstances the effect of the tax is virtually that of an indirect tax on outlay, which is almost always regressive. Investigations in the United Kingdom, however, show that several other taxes (such as those on drink and tobacco) are more regressive.

The crux of an efficient tax on land and buildings lies in the accuracy with which the valuation process is carried out. Valuation is a difficult and expensive exercise, especially where there is a shortage of skilled and experienced valuers, for valuation is a skill that depends very largely on experience. It is a great easement if a higher level of government undertakes valuation of all the units within its bounds. (Since 1948, the national government has been responsible for all valuations for the British local rate.) If valuation is everywhere undertaken by the same authority according to the same methods, the resulting aggregate value of each area will provide a measure of relative taxable capacity, which can be of great assistance when planning a grant structure.

All taxes on land and buildings are alike in calling for regular periodic valuations, but the taxes that can be imposed differ very considerably. The basic distinction is whether they are assessed on annual value (and hence related to rents) or on capital value (and hence related to property sale prices). The two fundamental difficulties of valuation are first that buildings differ so enormously one from another, and secondly that there is never available simultaneously the right sort of evidence for all the units that have to be valued. Valuation consequently has to proceed by *analogizing* from the evidence there is to the evidence that is wanted (and this is where the skill derived from experience comes in). It is desirable to try to minimize the analogizing that has to be done, and hence to choose the base for which the most evidence is available.

The practice of using the occupation base grew up in England from an early date. From the beginning of the nineteenth century, when there was an enormous expansion of urban building, it was normal for town dwellers, especially factory workers, to rent their homes. The annual value base thus fitted well. As a refinement (in England but not in Scotland) it was adjusted to give a "normal" return where the actual rent appeared out of line with that of similar property. In this sort of situation the occupation/rent base is simple to explain to taxpayers, and the tax can be collected very simply and cheaply by the landlord with the rent. This practice has been followed in many of the less developed parts of the British Commonwealth where, because of the sparsity of rent evidence, it is not at all appropriate. Indeed, in India, it is very difficult, if not impossible, under the constitution to use a capital base for a local tax because capital taxes are on the Union (central government) list. A way

round the absence of rent evidence has been sought by making a capital valuation and deflating it to an annual value; but this is far from satisfactory as the deflator is essentially arbitrary.

The weakness of the annual value base is that it is tied to the present usage of the unit (hereditament) and the valuation can only be raised as new rent contracts are drawn up (apart from minor additions for alterations, such as an extra garage or toilet). In contrast, the capital value base, derived from sales evidence, includes the real estate market's view of the property in other foreseeable uses (less the cost of demolition if this is required before a new and larger building could be erected). Where there is an active real estate market —as there is likely to be in metropolitan areas—most of the answers can be found by keeping a careful record of all sales. When, in a particular area, a sale takes place at a greatly enhanced price it will be quite legitimate to raise the valuations of all units in the area, although not to the full extent of the new sales evidence since allowance should be given for the fact that the single sale has something of a scarcity value. Thus as land values rise the tax base expands and at each revaluation some of the additional value, which is not due to the action of the property owners and is consequently known as "unearned increment", accrues to the community.

This is the simplest form of property tax and is used, in principle, almost everywhere on the North American continent although with great differences in detail.[6] But there are two difficulties. First, if there

---

[6]The property tax provides, on average, 88 per cent of local tax revenue in the United States. In the United Kingdom, the rate being the only local tax, it provides 100 per cent, but it appears that in relation to expenditure there is not a great deal of difference.

is not an active real estate market, less reliable valuation methods have to be used, such as measurement (foot super), tempered with considerations of the location of each particular building in relation to others. Such valuations may diverge very much from what they would have been in market conditions. Secondly, as every improvement gives rise to additional tax liability (as is the case with respect to annual value), it is sometimes held that the use of the full capital value base discourages building and modernization. Hence, a method of "derating" the building is sought. This is one aspect of the basis of the Henry George single tax argument. Confining the tax to the full potential value of the site alone, the property owner is free to develop the site to maximum capacity, while at the same time the high tax on the underdeveloped site will impell him to develop in order to acquire some income. In cities at an early stage of development there is likely to be abundant evidence of bare site sales, and valuing by analogy those where buildings have already been erected is not difficult. In built-up areas, where site and building are sold together, one of two procedures is inevitable if it is desirable to obtain a separate value for the site, although neither is very satisfactory. Either recourse must be had to adjusted measurement, as described above, with all its disadvantages, or the whole must be valued and a deduction made for the replacement cost of the building. This latter method is used in Australia to general satisfaction. However, it clearly cannot be fully satisfactory because, if full replacement cost were used, the site could emerge with an almost nil value in times of rapidly rising building costs.[7]

[7]There appears also to be an inertia in writing down site values in declining districts. This may be a deterrent to redevelopment.

Under modern urban conditions it would seem that the indiscriminate encouragement of large buildings in the centre of cities is wrong because it can seriously increase congestion and complicate parking. If this is accepted, there is no case for a tax differential in favour of these buildings, and the attempt to obtain a separate value for site and building is no longer required. But for efficient renewal and development two conditions are necessary: first, that a land use plan has determined broadly the types of buildings that can be erected in particular locations; and secondly, that compulsory purchase can in the last resort be used in order to secure the implementation of the plan. We shall have to return to this question at a later stage. As we have seen, by using full capital value of both site and building as the tax base, a substantial proportion of increasing values will be brought into the tax base at each valuation.

Valuation on the basis of sales where there is a reasonably active real estate market is not a difficult exercise. Nor for that matter is valuation on an occupation base when there is abundant rent evidence. Yet although the tax on land and buildings is an immensely important urban tax it seems frequently to get into trouble and to give rise to considerable discontent. As this is such a common experience it is worth discussing briefly why it should be so. Some of the difficulties are common to any form of tax on land and buildings, but others are peculiar to particular forms. In the first place, for the family these taxes are highly personal. The amount to be paid is obvious. With every rise in the price level the rate tends to rise because local government costs are extremely rigid. The taxpayer knows whom to complain to about this in a way he does not in respect of a national tax. Secondly, housing is a domestic overhead which cannot be so easily adjusted as purchases that may be subjected to a sales tax. The overhead strikes with special severity when there is a sudden fall in family income. The most general case of this is with regard to elderly people. Moreover, they often have to face very high overheads if they continue to occupy the family home when their children have grown up and left the home. The regressive nature of the effective incidence in almost all cases aggravates this.

Objective valuations are clearly difficult, both for occupation values where there is little renting and for site-value valuations where all sites are built up. Valuation weaknesses are much wider than this, however. One trouble is the expense and difficulty of revaluing promptly at regular intervals. Out-of-date valuations quickly become unfair as between one property and another, and the municipality is forced to raise the rates in order to maintain its real income when the price level has been rising. It would seem that councils and their valuers are sometimes afraid of political trouble if a full valuation is made. Hence they attempt a fractional level, which can hardly avoid discriminatory results. These last two points emphasize the importance of having an external and independent valuation, paid for out of central (or state) funds both to save urban government costs and to ensure regularity and uniformity. The only satisfactory method is to take the full value on all property. The rate of tax can then be much lower than on a partial valuation. It should not be too difficult to make this clear to local taxpayers.

These difficulties with the local tax on land and buildings are very real; but they are not inherent in the tax form and all are capable of reform. What seems mainly to be required is courage on the part of the local council. It is urgent that the necessary reforms should be carried out in order that this most important tax can play its proper part in urban development.

There is one more point to be discussed before we leave this sort of tax: the coverage of particular rates of tax. It is common to apply one rate to domestic and another to business property; but much more refinement than this is possible and has indeed been practised in the corresponding British tax—the local rate. For twenty years productive industry was "derated" to the extent of 75 per cent, initially because of the burden of a tax on overheads in a period of depression.[8] Recently, a general reduction of rate liability on low-income families has been introduced in view of the enormous rise in rates, due to increased local authority spending, over the last decade. Further, higher repairs allowances are available to landlords (who collect the rates on the low-income families) to compensate for the greater wear and tear on houses occupied by such families. This easement could be introduced very conveniently into a property tax. The more these special cases are introduced, however, the more the revenue tends to become eroded.

TAXATION: SALES AND PURCHASE

The next form of urban tax to be considered is the family of sales and purchase taxes. These can include taxes on the use of durable goods, such as vehicle licences and parking fees and meters. In the United Kingdom, vehicle licences run at very much higher rates than is usual in North America and are accepted without trouble. They need not be regressive, since they can vary (for example, with the cubic capacity of the

---

[8]Agricultural land (although not farmhouses) has been wholly derated since 1929. The arguments for this are now poor, but in any case it is irrelevant to our present discussion.

engine). However, since the car tax is normally assessed on the domicile of the owner, there may be more difficulty in checking this than in the case of the real estate tax. This difficulty could be surmounted by including cars in the real estate tax base, as they must presumably be annually licenced. Other taxes on vehicles take the form of user charges—such as parking meter charges or, as has been tried in some European towns, a tax on (metered) urban mileage.

Although motoring taxes at the city level are likely to become increasingly remunerative, they suffer from the disadvantage that, being fixed by the year, they are relatively inflexible to revenue needs. This disadvantage is less pronounced with respect to sales taxes, although frequent changes of rates would clearly be upsetting both to business and consumers.

It must be expected, however, that there will already be national and/or state sales taxes, so that there are rather close limits to what even a metropolitan government can hope for in this direction. Moreover, there is a growing tendency for large shopping centres to flee the jurisdiction of the city with its high taxes and locate in isolation on the fringe of the urban area; a further justification for a large metropolitan taxing unit. Apart from such avoidance there are two major difficulties surrounding a sales tax at the local level. First, it has an undoubtedly regressive incidence. Secondly, a thorough inspection is required to make the tax really equitable. This is especially true if it is imposed at the retail level, which is more flexible and in other respects more convenient than at the wholesale level where, as with the British purchase tax, pyramiding is likely to take place.

As regards regression, the State of Indiana is experimenting with a method of (effectively) giving an annual allowance, corresponding to an income tax allowance, against sales tax liability. Recipients would be those families whose incomes are below the income tax exemption limit and who consequently would not qualify for any relief in that direction. This idea would seem to be well worth exploring. The costs of inspection turn very much on the degree of equity aimed at. Indian experience at the state level shows that a very useful amount of revenue can be collected at quite moderate rates, although it is notorious that administration is inadequate. (Reference has already been made to the fact that when certain commodities were removed from state sales tax and transmuted into national excises the general effect was a threefold increase in revenue collected.) Another problem in many developing countries is that a substantial proportion of consumer purchasing takes place in petty street trafficking, which cannot be brought within the range of a sales tax.

An alternative to sales tax with a very similar effective incidence is the tax on value-added. Until now the only authority on the North American continent to impose this tax has been the State of Michigan.[9] But there seems to be no inherent reason why it should not be applicable to metropolitan conditions. This tax is a refined form of the turnover tax, based on gross sales of firms for defined periods (it can be assessed and paid monthly) *less* purchases from other firms. (In some forms, purchases for capital formation are also exempt.) The value-added tax has been used in France for a considerable period. It

[9]J. F. Due, "Value-Added Tax Proposals" in the *U.S. Essays in Honour of C. Ward Macey* (Corvallis, Oregon State University Press, 1966).

appears to have been suggested to Michigan by the big motor manufacturers as a more acceptable levy than higher corporation taxes at a time when the state was suffering from long-term revenue shortages.

The Michigan tax applies to all business organizations, indeed to practically every form of business activity. Although it is assessed on firms (and justified by the benefit principle, which incidentally would be particularly applicable in the case of a metropolitan tax) it is expected that most of the effective incidence in fact will be transferred to consumers in the form of higher prices of goods. This is because in practice the tax on value-added would be similiar to a very widely-based sales tax. It thus would run the danger of being even more regressive; but if the Indiana allowances plan also could be introduced, this could be compensated.

The French method of assessment is simply that all firms are required to forward monthly returns of their sales, together with vouchers covering their purchases from other firms. This would seem to be a less equitable arrangement, but presumably could be geared to assist industries with a high rate of value-added (large wage bill). In any case very small firms are exempt.

The Michigan tax is levied at a single rate, but in France multiple rates have been applied—with at one time as many as fourteen different rates. This would appear to give wide opportunities, not only for relieving the tax on high-employment industries, but also for attracting or repelling industries which the city did, or did not, wish to welcome within its boundaries. It is clear that, regarded as a tax on profits, a tax on value-added would not bear on a firm's overhead and so would be easier than the property tax on firms under depressed conditions. There might be a case, indeed, for a partial substitution of tax on value-added for the tax on land and buildings. Similarly to a sales tax it would also be more flexible, yet, being a tax on gross (not net) income, in so far as it was not shiftable, it would tend to fall indirectly on overhead.

TAXATION: INCOME

If the tax on value-added exempts investment expediture, it becomes in practice a wage bill tax and hence is not far from a payroll tax. In other words it is a species of income tax and we now turn to the scope and applicability of income taxes to metropolitan conditions. As in the case of sales taxes, they clearly can play only a minor role in view of the exactions of higher governments, especially in a federation.

In principle, metropolitan "income" tax could be assessed either on personal income, on corporate net incomes or on both. With the first alternative a modicum of progression would be possible without raising the opposition of higher governments, apprehensive about their own revenue. In regard to corporate profits, progression would be out of place. Size—divorced from rate of return—is no indication of ability to pay: to penalize high rates of return is to penalize good management and initiative. In thirty-eight American states corporate profit (or income) taxes exist side by side with federal corporate taxes. The difficulty of dividing liability between different parts of a firm operating in different locations is settled through one of several possible conventions, none of them (from the nature of the case) completely satisfactory. The danger of tax avoidance by removal of a plant or office to lower tax jurisdiction is always present and would no doubt be worse with a tax at the local level.

The country in which metropolitan governments enjoy the best income taxes is Sweden, where a strong tradition of local income tax survives from the days when the national income tax was in its infancy. Assessment in Scandinavia is by voluntary committee. There is no exemption limit but there is a ceiling to the amount of personal incomes that may be taken by the local authorities. This ceiling is generous—of the order of 18 per cent—within which range a modicum of progression is possible. It is fixed in such a way as not to interfere with the national income tax, which has a high exemption limit and a number of reliefs and allowances.[10] This is an excellent sort of local income tax; but where the national tax is already well established it probably would be necessary to set up a separate assessment administration, which would be costly and inconvenient.

Alternatively, a metropolitan income tax could be approached via a surcharge on a state or national tax, thus enabling use of already established administrative machinery. For this to be possible however, even if agreement could be obtained, the rates and conditions would be laid down already, so that there would be no initiative at the metropolitan level. It is unlikely that any additional progression would be permitted.

By far the simplest form of income tax for metropolitan use would be a straight payroll tax, collected with the national and/or state tax through withholding on all wages and salaries. In this case the metropolitan government (or governments) could probably choose (within limits) its rates and conditions. Such a tax as an addition—or partial substitution—for property tax would have several advantages apart from the economy of using already established machinery. Revenue would be a great deal more flexible and should expand *pari passu* with the rise in the income of the community. (It would fluctuate, as well, causing difficulty in the event that a major firm or industry leave the city—as the stock yards have left Chicago.) Secondly, more directly than any other form of income tax, it would bring commuters firmly within the tax base. Further, the tax base would be widened by including young workers who live at home but are not taxed directly with property taxes. Of course, it can be argued that a payroll tax excluding the incomes of the self-employed and property incomes would be partial and, thus, inequitable; but without access to higher-level income tax returns, any attempt to include the latter would add enormously to the cost of administration. The self-employed are probably not a very large sector, and those with substantial incomes (such as professional people) might be compelled without too much difficulty to file returns. Property incomes would be a more difficult proposition.

How much additional tax revenue would be collected by the methods just discussed would differ greatly from place to place, corresponding largely to the efforts already made and the extent to which property tax assessments were out-of-date and inefficient. But however much the yield of the tax on land and buildings can be improved and additional revenue raised by some of

---

[10]A tax of this nature was discussed in relation to the United Kingdom by a committee of the Royal Institute of Public Administration. It was found to be incompatible with the British system of current withholding in which taxpayers are treated as numbers and tax liability adjusted according to their family circumstances and social security contributions. They file no individual returns, and their identity is not known to the revenue authorities if they fall below the surtax range (unless they own some special form of property that cannot be "coded in").

the other sources just discussed, the expenditure of urban and especially metropolitan governments is bound to go on rising at an accelerating rate. In 1965, in the United States, it was five times what it has been just after the end of World War II; in the United Kingdom, rate demands had increased ten times during the same period.

GRANTS

Even with the most efficient use of resources, the rise in the school population and the demand for higher standards of education, the associated costs of modern traffic problems, the social pressures of integrating immigrant populations, additional costs of extending public utilities as cities grow, and so on, all make it clear that additional grant aid from higher authorities will be required to support local sources. Some of the most appropriate forms of grants are examined here.

As argued above, it is easier to achieve a satisfactory grant system if more than one type is called into service; that is to say, one would look for some grant aid coming on allocated or project terms and another part unallocated, the funds disposable according to the preferences of the metropolitan government. In a federation the larger share of grants would be expected to emanate from a state government rather than from the national budget. Insofar as grants were for the metropolitan area as a whole, it would be preferable that they should be unallocated. It may well be, however, that a state would wish to aid a number of urban areas of varying degrees of wealth, in which case it would be necessary to find some formula that would enable the more backward areas to catch up. A similar problem would arise within a metropolitan area that had a common tax jurisdiction but was organized with subordinate jurisdictions (as with London and New York). What is really wanted is a way in which the more backward areas can be given equality of fiscal opportunity rather than equality of tax rates (between cities).

British experience shows that this end can be attained most simply by making a grant to those areas that have below-average ability to pay. The amount of grant in each case depends upon the difference between the fiscal competence of the area and average ability. Where property tax valuations have been carried out on a uniform basis, a very simple measure of ability to pay is provided. If there is no simple objective index of relative ability available a formula can be constructed, taking into account such measures as degree of unemployment, age structure of the population (giving attention to education and geriatric care), road surface to be maintained, and intercensal change of population (both rapid increase and, more particularly, rapid decrease giving rise to special fiscal difficulties). Such formula items are, of course, no more than different methods of weighting population, and a selection can be made between them, as appropriate. The important point is that, by itself, a measure such as relative property tax yields is useless, as it pays no attention to relative tax effort.

A general formula that gives weight to particular community needs, such as education and child care, acts as a stimulus to the development of such services. But it may be desirable to provide a more direct stimulus to certain services, such as slum clearance and urban renewal, the establishment of additional hospitals or health clinics, or the construction of ring roads. For these services, specific capital or current grants would facilitate more direct promotion and control. The grant-giving authority should assure itself that ample provision will be made for physical maintenance, servicing and eventual renewal of the property—we shall have to discuss this more fully later in relation to loan finance. In addition, as each of these activities has its own particular problems, appropriate conditions of control would have to be written into the grant arrangements. In regard to slum clearance and urban renewal, to give but one example, it has been found by experience (as in Chicago) that if an attempt is made to cover costs on the whole operation and no provision is made for re-housing some of the displaced population at prices they can afford, the result of cleaning up one area is merely to create new slums in the neighbouring ones. Consequently, a grant for demolition and redevelopment should include an obligation to re-house (not necessarily on the spot) a certain proportion of the displaced population in appropriate accommodation. In the United Kingdom it has been found very advisable to appoint trained housing managers to give special attention to families being moved, especially if they include immigrant elements.

Two questions remain to be discussed: what proportion the grant should bear to total outlay on the project and how and when it should be paid over. As a great deal of flexibility is permissible here, it should be possible (if desired) to tailor rates of grant to the relative needs of particular areas for a service and to their ability to pay the local share. To answer the questions however, it is first necessary to make a statistical estimate of the average cost of a "unit" expansion of a service to be granted aid. Such an estimate, which needs to be based on an input/output table, will be required in any event for effective planning. (We shall return to this point later.) The result of the calculation would serve as a standard unit grant at present costs. It can

then be adjusted upwards for a rise in costs and allowance made for special expenses peculiar to the area. It should be possible in some circumstances to adjust the grant downwards, for instance where more economical methods can be used, such as for sewage disposal, where the geographical circumstances allow, or for services where there are striking economies of large scale.

As suggested earlier, the broad percentage of cost that will have to be offered depends on the relative valuation of the service in the eyes of the two authorities concerned: the more interested in expansion the grant-receiving authorities are, the smaller need the bait be. Another consideration is the rate at which a project can be executed and the service come into operation. It is of little use making a grant for a building for which the cooperating labour and equipment will not be available. The rate of expansion in the past of the particular service in question can be a useful guide to this. A grant should always be forward-looking. It is very important for the grant-receiving authority that it should know in advance what is coming to it; consequently it is useful for the grant-giving authority to pledge itself to give a regular sum annually for a defined period (which should be nearer twenty years than five).

At the same time the grant-receiving authority must accept that the continued receipt of grant depends on satisfactory planning and implementation according to the conditions laid down initially. Once a grant-giving authority has pledged itself over a period it should do everything possible to honour the pledge, even to the point of increasing its own tax revenue or raising the funds by borrowing. Only by a firm assurance on this point can a grant-receiving authority plan and implement with optimum allocation of resources, and it is most probable that the grant-giving authority will have broader resources and better credit standing at its disposal. The grant-giving authority, however, should not allow itself to be let in for grants to cover part or the whole of outlay already made. This is to invite the grant-receiving authority to spend carelessly. In the same way open-ended grants, which involve an understanding to finance an unpredictable number of units or expansion of service, should be avoided as far as possible.

Finally, it may be asked to what extent total grants can rise in relation to the total outlay of a local government without endangering its sense of responsibility, its reasonable autonomy and freedom of budgeting. No simple answer on these limits can be given. Much depends on the calibre and experience of the local civil servants and on the accuracy and comprehensiveness of the budgetary methods used. Unallocated grants clearly preserve budgetary autonomy better than specific grants; but the items included in a grant formula need to be strictly beyond the control of the grant-receiving authority so that it cannot alter them to its own advantage.

Specific grants do limit budgetary freedom, not only by distorting the city's "preference map", but through the very considerable amount of supervision inevitable in order that the grant-giving authority may assure itself that implementation is proceeding in an efficient manner. As to complicated projects, the grant-giving authority will probably require regular progress reports detailing both the expenditure incurred so far and the results achieved. If the conditions of the grant have been well devised a light rein will probably be sufficient, as both the need and the incentive for compliance will already have been written in. All this is assuming that metropolitan councils are their own masters, and not subject to mandatory powers by a prefect or burgomaster in the European manner. It is safe to say that better intergovernmental relations will be maintained, and probably spending will be more efficient, the more that persuasion, rather than orders, can be used. This issue, however, is treated at much greater length by Dupré in Chapter 11.

As has been said, there is no single or simple way to establish the "safe" amount for grants in relation to autonomous revenue. Experience suggests that under good conditions grants of over 50 per cent can be within the two limits of autonomy and good management. This statement, however, needs to be qualified according to the distribution of grants between current and capital purposes. Our judgement consequently cannot be complete until we have investigated the sources of capital finance.

Sources of Metropolitan Capital Finance

In the preceding section we discussed problems connected with grants for specific projects and programmes. Most of the additional services that city and especially metropolitan governments are now being asked to undertake entail a considerable amount of capital formation in the course of their development. The greater part of this expansion is being financed from borrowed money. The procedure for borrowing, the type of loans to be sought, the relation between loan finance and current finance (both in general and with respect to particular programmes), and the resulting burden of debt interest, all raise problems of immense importance for the future of metropolitan government and its taxpayers. There is a good deal of flexibility of choice, and the differences in practice in statutory powers (even in the definition of capital investment) are so great, both between countries and between cities in the same country, that it is difficult to generalize about methods and results.

Before we discuss methods of borrowing and their economic results, something should be said as to the relative merits of the financing of capital works from current resources or from loans. The obvious advantage of finance from loan is that the assets can be acquired immediately and the cost can be spread over a long period, depending on the type of loan and the state of the market. Secondly, outlay on capital formation tends to fluctuate with considerable violence from year to year, as one project or another reaches its stage of maximum input and declines. It would be most inconvenient if such fluctuations had to fall directly on the local taxpayer. Against this may be set several arguments in favour of pay-as-you-go. First, the saving in interest charges is very high; as we have seen, debt service eats up a substantial part of current revenue.

There are thus substantial arguments for avoiding borrowing where practicable; practicability depends in the first place upon one's being able to space the time profile of capital expenditure so that it avoids violent fluctuations. This is not as difficult as it sounds, as there are always items in particular programmes—say, school building or urban renewal—that can be accelerated or retarded as required by the fiscal situation. Moreover, relatively small capital items can always be charged to revenue without disturbing annual tax requirements, though the extent to which this can be done differs greatly according to the size of the local budget. Thus the opportunities for metropolitan area authorities to save in this way are relatively good. Yet, in terms of taxpayer discontent and resistance, the price of additional taxes required for such a policy cannot be disregarded. The advantages of a flexible tax structure, which we discussed earlier, are very relevant here.

The two variables that determine the difference in cost are the length of maturity and the rate of interest. It is obviously foolish to issue a very long loan when rates of interest are high, and other things being equal, the shorter the maturity the less the excess cost of borrowing over current payment. On the other hand when interest rates are low, and the project to be financed very long-lived, there is a great convenience in having a loan which can be amortized slowly, say over sixty years.

Other considerations also enter the picture when a government is contemplating the best type of borrowing to attempt. Among these are the expected height of interest rates. If they are high at the time of proposed issue of a loan, it may be better to postpone the issue of a long-term loan indefinitely. This decision, however, turns partly on the estimated willingness of the public to come forward at a particular moment. This may be affected by general considerations influencing all potential lenders, or by considerations affecting particular groups to which local authorities will wish to appeal, such as banks, insurance companies, local firms or the small investor. Further, these potential lenders will have particular preferences with respect to the length or terms of the loan. The city's financial advisers will have to keep a close watch on all these variables in order to take advantage of small differences which arise from time to time and which, in the aggregate, may make a big difference in costs of service.

Borrowing by local authority can be divided into three categories according to maturity: (1) long-term, say from fifteen years upwards, including irredeemables; (2) medium-term, from five to fifteen years; and, (3) short-term (including temporary accommodation), under five years. Each has its advantages and its dangers. If these three alternatives are open to a metropolitan area authority, which is in a better position to use them than less powerful authorities, it will have a wide field for selection and combination in order to secure the optimum debt structure. It must be borne in mind, however, that these are initial maturities. With the passage of time the longer maturities become shorter (apart from irredeemables), so that whenever a new loan is undertaken attention must be paid to the structure of the outstanding debt, as the greater part of potential lenders (especially insitutional lenders) are probably interested in loans of a particular range of maturities. Irredeemables usually are not popular, as is reflected in quotations on the London Stock Exchange, where the irredeemables typically stand lower than those of fixed maturity.

It is necessary to explore the implications of these different loans before a judgement can be reached on the subject of loan finance.

BORROWING: LONG-TERM

It is usual for even metropolitan governments to have some restriction placed on the issue of long-term loans, and the methods used differ substantially from country to country. In federations there may be a general loan authority, such as the Australian Loan Council, which arranges total borrowing and parcels the funds out among borrowers by agreement.

In the newer federations, such as India, it is customary (if not obligatory) for all foreign and a good deal of home borrowing to be undertaken by the national government. Some of the states and local authorities have consequently become heavy debtors to the Union government. In other federations, states impose certain conditions and regulations, or permission must be sought through a local plebiscite. These regulations are imposed to prevent borrowers from acquiring so much long-term debt that they cannot meet the service. The Australian Loan Council, which "nationalized" the state loans, was established precisely for this purpose.

The attraction of long-term borrowing is the large amount that can be raised in a short space of time; this is especially useful to a metropolitan government having extensive needs, for a large issue can usually be made more cheaply per unit borrowed than a small one. Also, from the investors' viewpoint, the size of the issues offer the attraction of high liquidity. For all but the large metropolitan authorities however, this form of borrowing tends to be too expensive unless several can club together to issue a joint loan. Hence the advantage of a metropolitan-wide government, which we come to later.

The typical form of borrowing for North American cities, the debenture in Canada and the bond in the United States, is similar to the British Stock Exchange loan. There are differences, however, due to the different contexts in which they operate. Firstly, there being no central stock exchange with the prestige of London except New York, North American debenture issues are likely to be directed to a greater extent to local investors (apart from Canadian loans floated in the United States, which are of increasing importance). Secondly, the spread of maturities on debentures, while diversified, appears on balance to be shorter[11] than for United Kingdom loans, apart from mortgages. Thirdly, local debentures cannot have quite the status of a British Trustee security backed by the Treasury. Consequently, they may require a somewhat larger rate differential in relation to national government borrowing than do British local loans. (This is being modified by use of provincial guarantees in Canada.) For American cities in the past, local bonds could hardly be regarded as free from risk. This aspect would be reinforced by the greater fluctuation of real estate property revenue receipts than of those from the British local rate. Inefficient collection in some cities may partly account for this difference; but also it would arise from the fact that rate liability in the United Kingdom (which, it will be recalled, is a tax on occupancy not on ownership) is covered by National Assistance payments for those who cannot otherwise temporarily meet their obligations. Similar arrangements exist in some Canadian provinces.

Traditionally, in the United Kingdom the responsible national ministry has exercised some control over term borrowing by local authorities. This was almost wholly in the interests of the local authorities themselves, however, in order to ensure that the debt did not become too extended. The criteria used were rules of thumb with little economic significance. Since 1939, however, Stock Exchange borrowing by local

authorities has also been controlled in the interests of national monetary policy. The Capital Issues Committee (an outlier of the Treasury) has put local authorities strictly in the queue of waiting borrowers. The intent was to keep the rate of interest down (in order to assist post-war reconstructions) and to stabilize the international value of sterling. Consequently, local authorities have been constrained to seek other methods of satisfying their needs, some of which will be examined below.

BORROWING: MEDIUM-TERM

The typical forms of medium-term borrowing are mortgages and bonds. In the United Kingdom bonds have been used predominantly, although the maturities of both forms (now not less than five years on issue) may be identical. Both forms are secured on all the rates, property and revenue of the local authority concerned, even in the case of housing bonds which, uniquely in British practice, are geared to a particular purpose. In North America most bonds are raised on the entire credit and taxing powers of the issuing authority; others on the security of a limited source such as a particular tax, public utility or housing project. Such bonds are normally tied to a corresponding specific application, the thought being that the application will have a special appeal and so attract potential lenders. It is questionable if this brings any real advantage, and the broadly secured or full-faith and credit bond has proved generally more popular.

The use of the mortgage in the United Kingdom has, however, been subject to much change in the last thirty-five years. Before World War II, short-term mortgages of as little as three years were in common use by local authorities to attract local investors. The mortgages could be

---

[11]It would appear that the tendency for shorter maturities in North America compared with the United Kingdom is also true of business borrowing. It can perhaps be related to differences in the organization of banking in the United States, but some other explanations seem required with respect to Canada. It should not be overlooked that state and local government obligations are exempt from federal income tax in the United States. In the United Kingdom they are now (from August, 1967) exempt from stamp duty.

allowed to run on and they were a cheap form of raising money. However, they never brought in large amounts, especially at any one time. After 1945, restrictions imposed resulted in no mortgages of less than seven years being sold and no over-running allowed. There was also increasing competition for the small investor's savings from banks, building society deposits, savings certificates and the like. In 1956, however, with the removal of the seven-year limit, mortgage borrowing immediately increased again, there being a tenfold rise between 1951-2 and 1954-55.

It is doubtful how far the small investor has returned, or will ever return, to his old ways. Apart from the competition of other forms of savings security, he has also made a considerable breakthrough into stock exchange equities via unit trusts, which have expanded enormously over the last decade. Probably the greater part of his savings now go into hire-purchase of consumer durables. In fact the basis of the new mortgage borrowing is substantially different. It is more institutional,[12] and most significantly, a substantial (but not fully identified) amount of it is known to be foreign. Before 1954, a few local authorities were known to have used foreign funds for temporary borrowing. This could not be relied upon, however, as a steady source of funds unless British interest rates were substantially above continental rates. In recent years this has certainly been the case. Owing to exchange rate difficulties and fears, and to uncertainties of potential lenders in their own countries, foreign borrowing by local authorities can be dangerous if carried to excess. The greatly increased needs of local authorities (especially metropolises) for capital funds makes

this a real threat. A sudden withdrawal of funds from the continent, or even a failure to renew mortgages, might put certain local authorities in a very awkward position. (They would be rescued, of course, by the Treasury.) It is also implied that a larger part of local borrowing by-passes central government control. This is reflected in the fact that net borrowing from the central government was negative while "other identified net borrowing" increased tenfold between 1955 and 1964.

In spite of the peculiar influences that have beset local mortgage borrowing in the post-war period, there can be no doubt that mortgages and bonds provide a flexible means of obtaining funds from a wide coverage of potential investors for a range of periods that are very convenient for capital formation.

BORROWING: SHORT-TERM

The need for short-term or temporary borrowing is always present for local authorities. On current account the necessity is to cover a gap between revenue receipts and outlay, since revenue flows in less regularly. It is possible to reduce this gap very considerably by increasing the frequency of tax demands. Toronto did this in 1962 and apparently succeeded in eliminating gaps of this nature. British local authorities, which have now been pressed to offer a choice of weekly rate payments as well as half-yearly demands, should find a great improvement in this direction. Against this must be set a good deal of increased administrative costs in managing more frequent payments. More importantly, such borrowing is required on capital account, partly to bridge the period required to float a long-term loan, partly to fill in gaps in expanded capital programmes. The larger the authority, and the greater its rate of

---

[12] In fact new institutions have arisen to meet the demand.

expansion, the greater is likely to be its need for such accommodation.

As mentioned above, foreign sources, attracted by high interest rates, have sometimes been used for temporary accommodation. The commercial banks, however, have traditionally been the most important avenue for such a purpose. Yet, in the United Kingdom their charges have been high, and the metropolises at least have favoured expansion of borrowing abroad by using either cheaper and more convenient bills or by taking deposits. Certain local authorities have long issued bills (mainly six-month) as a result of private legislation and have found this both cheap and convenient as long as the market rate on bills is below the long-term rate of interest. Such bills were issued by the Discount Houses on the London Money Market and were fully discountable, closely parallel to Treasury bills.

The more recent expansion in issuance of bills, however, has been direct. (The exact amount cannot be checked.) There is little doubt that in recent years British local authorities (largely metropolises) have been borrowing excessively through these new channels, both for mortgages and for bills. Thus they have contributed to inflation at home and possibly to balance of payments difficulties abroad, given the unpredictably volatile behaviour of foreign "hot money". They have been in trouble with the Treasury and the discount houses on account of their competition in the London Money Market. They have also been in trouble with the commercial banks and building societies who resent their competition for deposits. These excesses need never have occurred if the central government had kept a better lookout and had not wished so greatly to expand local investment. The episode—for as such it must be regarded—is of great interest as illustrating new ways in which cities with a good reputation can borrow at lower costs and with a much smaller debt burden than long-term borrowing would have implied. Unless local authorities have additional means of meeting a sudden withdrawal of short-term funds, they might run themselves into considerable danger. One such method is the possibility of using internal funds.

RESERVES

With the expansion of local activities, complementary reserve funds have inevitably grown fast, to such an extent that it becomes important for a large authority to develop a constructive policy for dealing with them. There is much to be said for borrowing such funds for capital formation. The transfer from reserves is a mere matter of bookkeeping. There are no costs nor, necessarily, interest payments. In most countries the largest and most rapidly growing are the pensions funds accumulated on behalf of council employees. In the United Kingdom these represent 67 per cent of total funds available for investment; sinking funds come next with 9.2 per cent of the total. Smaller funds take up the rest.

Broadly speaking, nothing but good can proceed from a policy of employing such funds for investment; but three qualifications should be made. First, pensions funds are in a special position in that the local authority is to some extent committed to take the most profitable course for its employees. This hardly constitutes a case for absorbing them all in financial securities, especially as they do not call for short-term liquidity. Secondly, when the pressure for capital formation slackens temporarily and the return on equities is high, it may pay the local authority to increase its wealth in the long run rather than to continue with capital formation to the greatest possible extent. The answer here would turn largely on development priorities, a subject that we must discuss in the next section of this chapter.

Finally, there is the problem of sinking funds, normally intended for the repayment of maturing debt so that there will be no hangover when replacement of the works has to be undertaken. It is fortuitous if the amount accumulated will serve to finance replacement when it becomes necessary. This depends on the length of the loan relative to the life of the capital acquired by it, on any changes in costs of capital goods, and on interest rates and borrowing opportunities. There is much to be said for abandoning the formal sinking-fund idea and instituting a regular policy of accumulating replacement reserves on a business footing. This would also allow some attention to be paid to obsolescence. An example might be improved techniques for sewage disposal.

Such a policy is facilitated if a consolidated loans pool is instituted, into which all redemption monies and accumulating reserves are paid. These monies are then available to finance any desired investment. This is hardly possible if earmarked loans are raised, such as school or housing bonds. We have already questioned whether the extra appeal of a popular type of capital formation is worth other inconveniences. (In the United Kingdom only the special housing bonds are subject to any restriction.) If separate funds are kept for different services, there is a real possibility that waste will occur as surplus funds from some will not be available for whatever deficit accounts have developed. This is a problem of budgetary structure and practice to be discussed below.

It thus emerges that when funds are accumulating fast, internal finance can be an economical and flexible source of capital finance. Even in the aggregate the amount available can never be very large because of the need for adequate liquidity of the funds. In the United Kingdom, on the average, it appears that about two-thirds of sinking funds are invested internally, but in this, as in all borrowing policy, local practice differs widely. The heavier the programme of capital formation that is being undertaken, the greater will be the proportion of external finance. It is here that a wide spread of financial sources open to borrowers is useful, enabling them to take advantage of different types of potential lenders, at home and abroad, local and national, according to their individual asset preferences and the movement of the different related interest rates. These are opportunities that no metropolis should neglect.

Problems of Budgeting and Control

Given the size and wealth of metropolitan areas relative to the countries in which they are situated and given the fact that the services for which they are responsible—such as education and public health—are the ones that concern the citizens most of all, it is desirable that their budgetary accounting and control be as advanced as that of the nation and, at the same time, open to suggestions and enquiries of citizens.

The annual budget is (or should be) concerned with outlays on established and continuing services, together with such expansions of existing services and such development of new projects as is judged to be feasible for the coming year. In addition, there must be an account of the finances that will be available to implement the expenditure. The key to an efficient budget account is comprehensiveness. It is essential for management and it should be possible for the citizens to see the full financial position of their city and to make an intelligent judgement on it. Individual services no doubt will be the special responsibility of particular commissions or committees, but it is undesirable that these should have a very independent existence. Each will make up its own budget and no doubt urge its own expansion with all its might. There are two very real dangers in this: first, the emergence of ill-balance because one commission is more aggressive than another; and secondly, a tendency to neglect the whole while attending to the growth of some of the parts. Below we discuss methods of budgetary control that should mitigate this tendency.

It has already been argued that a substantial economy in financing development and an improved and more flexible allocation of resources could be obtained by pooling all the reserve funds. If all, or a substantial part, of the reserve funds of the different commissions are included, two improvements may be effected. In the first place, the problem stemming from the inability of a deficit commission to get help from a surplus commission, which may have no urgent use for the funds, would not arise. Secondly, it would be easier for management to keep a watch on the activities of all the commissions if this amount of integration were agreed upon.

Here we are concerned only with the common problems of budgeting in so far as they concern an individual government entity. With some forms of metropolitan organization there are also problems involving the structure of a budget for the whole area and the integration of that budget with the sub-budgets for divisions within the overall area. In the final part of the chapter these will be briefly discussed.

As has been said, efficient budgets are comprehensive budgets. This applies to the budget of any government entity. They should always be two-sided, showing the finance set against the outlay. The use of a double (or divided) budget—current and capital—is much more essential for city or other local government activities than for most national budgets, because a large proportion of their activities is on capital account, concerned with capital formation and maintenance. Public utilities should have their own separate accounts kept according to commercial techniques, with only their surplus of profits (or, *mutatis mutandis*, losses) appearing on the general budget. (For a discussion of the accounting methods for the general budget, see below.)

The annual budget should not only look forward over plans for the coming year, but also backward over the previous years (at least two), whose history will be the foundation for the new exercise. It is probably unnecessary to discuss the *form* of the general budget in detail, since the whole procedure has been thoroughly thrashed out in terms of national accounts, and only minor modifications are required to adapt it for metropolitan use.

We can, therefore, pass to the content of the budget, remembering that it needs to be both a comprehensive record and a comprehensive forecast. That is to say, it is concerned with (1) the maintenance of standards, which often implies a cost rise of existing services—for instance, salaries of teachers and other public servants tend to rise by regular increments, quite apart from any increase in rates, until a stationary position is reached; (2) the expansion of established services along existing lines —more schools, hospitals, libraries, office accommodation, transportation; and (3) the erection of new works—electricity generating stations, waterworks, sewerage, motor roads and additional communications networks. A successful, economically optimal achievement of this total exercise calls for a series of choices, individual but interrelated, five of which can be distinguished and which are described here.

CHOICES IN BUDGETING

The first choice that has to be made concerns the *total size of the budget*, which determines in turn the amount of tax revenue that has to be raised. From the consumption statistics now generally available, it should be possible to make fairly accurate estimates of sales tax revenue, and forecasts of real estate revenue can be even more accurate. The essence of this choice is the benefit to the citizens of the marginal dollar of public expenditure against the marginal dollar of tax which they would have to pay to enjoy it. The choice between tax and loan finance is also relevant here; some capital formation can be carried out by either method of finance, according to the economic situation and, perhaps, in accordance with the policy of the national government. This level of choice, a very real and important one, usually receives far less attention than it deserves. As we saw at the beginning of our discussions, the power to determine the size, not merely the content, of the budget is the first essential freedom of local government.

The second level of choice is concerned with the *relative size of different services* and consequently with the rate of expansion. At this stage services need to be compared as a whole. Is the present allocation of resources on, say, education and roads in the best interests of the citizens? If one were to expand, or expand at a greater rate than the other, there may be a further constraint. Service (or commission) A has in the past conducted its activities more effectively and expanded more rapidly than B. Should A be allowed to expand further in the conviction that it will do a better job than B? Or should B be given special stimulus to expand? Perhaps an examination of why B has not done better is required. At this stage it becomes apparent that some sort of plan is essential so that all relevant services can be compared on a wide basis. Within this general framework each commission can then be given a "notional total" into which it is to fit its estimates.

The third level of choice is *between different parts of the same service* or activities of the same commission. A most interesting illustration of this level of choice occurs within the education service with its wide variety of related activities: primary and secondary schools (including those for handicapped children), junior and senior technical colleges, vocational classes, school meals, school journeys, evening classes and so on.

There are two different aspects of this level of choice that call for particular attention. Consider a new technical college: it is necessary to forecast the time-curve of investment, the demand for cooperating labour and the demand for the service over a number of years. It is to be expected that demands for inputs will rise steadily for a number of years and then flatten out, with at least some of the costs declining from their peak. It should be possible to take advantage of resources not required in the first and last stages of this process for other purposes, especially if labour and equipment are easily transferable within the orbit of the metropolitan government's activities.

Secondly, education is a service that depends largely on the activities of other commissions, particularly roads and transport. A change in road or public transport facilities will change traffic lines, promote the development of new housing sites and increase the demand for new parking facilities. These in turn will call for new schools, a demand which will expand at a gradual but predictable rate. Not only does a decision have to be taken at an early stage concerning the siting of new schools (so that sufficient space may be reserved for play as well as for expansion), but also when to proceed with the building. At an early stage, when there are not many children in the area, it may be cheaper to convey them by bus to established schools. Understandably, this stage is longer for secondary than for primary education.

The fourth level of choice, which needs to be closely integrated with the stage just discussed is *between projects,* with particular consideration to their location. Some of these may be an integral part of a particular service, such as schools, the location of which is influenced by the population to be served. Other projects are more flexible in their location, such as municipal buildings, libraries, swimming pools or sports stadia. This level of choice ilustrates particularly well a factor which is also common, although probably less acute, at the other levels—the effect of special interests. In few instances is choice unfettered by pressures of some kind. Ultimately, decisions will be political and will be the result of compromise, but this is no reason why they should be economically stupid. What is implied is the necessity for a firm economic technique for choosing between projects, which is discussed below.

We still have one more level of choice to examine, and as the decision in this case can be determined to a high degree on economic grounds it will serve as an introduction to the technique of efficient choice. The final level of choice, essentially a technical matter, is *between methods of carrying out a project* that has been accepted for implementation. Let us examine, for example, the erection of a new general hospital to deal with a wide range of patients: a sum of x million dollars is set aside as a budgetary constraint, not neces-

sarily absolute but sufficiently clear to emphasize that the objective is to build, on the scale indicated, the most efficient hospital for the population to be served. Obviously, many of the choices that have to be made with respect to hospital building are of a technical nature and we can do no more than indicate a few of the most important ones. First, there are general questions concerning such things as the height of, and construction materials for, the hospital; the size of wards; amount of circulating space and out-patient accommodation; types of beds; width of doors and floor covering. Next come the types of equipment. Some of this will be general, some highly specialized and probably very expensive. In this age of rapid technological advance the question of obsolescence has to be considered carefully. A choice also has to be made between diagnostic, operative and therapeutic equipment. In this case much depends on the type of patient to be served. Severely ill patients tend to be much more expensive to treat than others. Thus, a hospital that is planned to receive a high weighting of severe cases will inevitably be a high-cost hospital. Much depends also on the length of time it is planned that patients will be kept (for instance, normal maternity cases).

There are such wide differences in practice in these matters that the layman cannot help wondering if some of the decisions do not contain a large arbitrary element. One factor that should be taken into account when making these decisions, especially with reference to the length of stay, is the extent of home visiting and after-care available. There are also the administrative services to be planned, ranging from office organization to kitchen equipment. A most important final factor is the accommodation provided for nurses. Good nurses appear to be in short supply in every country.

Many United Kingdom hospitals are forced to eke out their meagre supply with immigrant personnel who are desperately needed in their own countries. Modern living quarters and good food are a big assistance in attracting and keeping staff of high calibre.

Whatever course is taken on each of these matters affects the amount of finance that will be required, and may affect the type of finance that will be most appropriate, whether tax or loan, and if the latter, whether long- or short-term borrowing. Both balance between departments and the budgetary constraints call for constant vigilance in planning.

Clearly the extensive and varied nature of the development decisions that need to be made by a metropolitan government call for definite principles, the scope of which we must now examine.

BASIC POLICY OBJECTIVES

There are two leading objectives to which the policy of a metropolitan government can be tied: (1) to increase the economic and tax potential of the area so that costs are reduced and real incomes raised, and at existing tax rates create a surplus that will be available for tax reduction or further expansion; and (2) to increase the social welfare of the community by providing more and better common services which will be available to all, but which will not be charged for in accordance with their costs. Indeed, many of them are indivisible in the sense that individual consumption cannot be measured and so cannot be charged for. It is evident that with respect to many projects the two objectives may not be in agreement and they may well involve considerable political differences, the latter objective being more definitely redistributional than the former.

It is hardly desirable to pursue either objective in isolation and to the neglect of the other. This is perhaps especially true of the welfare criterion, as the development of social services is the special responsibility of local government. It is unlikely that any projects within the competence even of metropolitan government will bring about very striking economic advances. But such advances may be very important to the citizens and act as pioneering examples for other local governments to follow. There is thus a dilemma built into choice. Different countries and even different cities will give varying weights to the two elements. If the conflict is sharp it may happen that it is necessary to evaluate a proposed project by both criteria before alloting it a place in the queue.

COST-BENEFIT ANALYSIS

Whatever the level of choice, the criterion on which the decision can be based sensibly is generally known as a cost-benefit analysis. It can be described also as a "discounted cash flow" estimate familiar in entrepreneurial planning, except that both the costs and the benefits are to be weighed from the social point of view. Thus they must include considerations such as "overspill" and community costs and benefits, which would be neither proper nor necessary in the forecasts of private enterprise. Many of these costs and benefits will not have a price that can be used directly in the calculation; sometimes the items can be quantified by the use of "shadow prices" into which a considerable element of judgement enters. At other times not even this is possible. Nevertheless, in order to complete the exercise, a money value according to the best judgement available should be put on each. Even if the result is rather indeterminate, the mere process of drawing up the lists of benefits and costs goes far to ensure that all the relevant factors are taken into consideration.

The first step in the exercise then is to draw up a careful list of all costs that will be incurred if project (or technique) H is selected, with similar lists prepared for project K. Next comes a list of all the benefits that should accrue from the implementation of project H, and likewise for project K. It is important that each entry be accompanied by an estimate of the date at which it is most likely to occur. The cost list should cover all money outlays, including the cost of raising a loan, and the service upon it, because of the wide range of choice in the type of finance to be selected, even if all are not available or appropriate for a particular project. Further should be included the cost of the labour and equipment for both construction and operational service and maintenance when the project is complete. As we have seen, the costs should be reckoned on a social basis, making allowance for those dis-economies that are unavoidably greater when the project is implemented. There is one other cost factor that must be considered always and that may be of great importance. If one project (or technique) is adopted, the resources employed in it cannot be put to any other use. It may be, however, that there is a more important use elsewhere for some of them. If so, in economic terms their *opportunity cost* is high. The opportunity cost of using resources in project K is that they cannot be used anywhere else, say in project M. Yet, it may be that project M cannot be proceeded with for want of them and if project H were selected, fewer of the factors for which opportunity cost is high may be needed, so that both H and M projects will be possible. Considerations of opportunity cost are likely to be particularly relevant in metro-politan planning, as many of the projects are highly competitive in that they make use of similar materials, equipment and skills.

The benefit list should similarly include "overspill" effects, indirectly arising from the new project. The general rule, sensibly interpreted, is as follows. If the project is a public utility that will cover its costs, including debt service and repayment, there will be no additional tax burden. The benefits, especially overspill benefits, may be distributed widely and difficult to trace. Consider the case of a new through road. The main benefit will be the saving of time, and hence cost, in transit. In terms of commercial transport, this is relatively straightforward to estimate; in terms of private vehicles, at least a value can be imputed on the basis of traffic counts.

But this is only the beginning of the effects. As a result of the new road, overcrowding of existing roads will be reduced. Further, land values will change all along the line. Those living near the old road will benefit from a more peaceful life, those living near the new one will suffer from increased noise. The biggest advantage, however, will accrue to the more distant areas that will now be opened to new development with a consequent rise in land values. It should be noted that among these changes there are three separate potential benefits: the income account benefit to those whose transit costs are reduced; the capital account benefit to landowners (some of whom may not live anywhere in the neighbourhood); and, the rise in the tax base (and hence in tax potential at existing rates). If a capital value base is used, this will accrue immediately; if an annual value base is in force, it will accrue only after development has taken place. Against this must be set the cost of connecting public utilities and other public works to the newly accessible areas. Clearly only the major changes can be quantified. Nevertheless, it is useful to list the others for background information. This might conceivably swing the choice between two very evenly balanced projects.

As has been said, each item in the two lists or streams needs to be dated as closely as possible. This is essential, as the basis for comparison is the present value of the two streams. Consequently, each item must be discounted back to the present. A cost that has to be met only five years hence, for example, is not so onerous as one that has to be met next year. In general, real incomes should by then have risen and some of the project benefits may have begun to accrue, either in the form of profits or of tax potential. Again, a benefit maturing in five years' time is less attractive than a similar benefit available next year.

We now reach the crux of the matter. At what rate of interest is the discounting exercise to be carried out? Three quite different types of rates have been suggested, the exclusive use of any of which can give quite different answers. There are, however, two basic principles to be taken into account. First, it is obvious that inasmuch as the rate of interest is to be used for the fashioning of a measuring rod, a government must use the same one for all the discounting it has to do in constructing any one plan or programme. Secondly, a high rate of discount will give relative priority to projects which can be completed quickly and which consequently do not require much borrowing. By the same argument a low rate of discount will favour relatively long gestation projects.

The three rates of interest in question are: (1) the Social Time Preference Rate; (2) the Opportunity Cost Rate; and, (3) the rate (or weighted average of rates) at which funds can be borrowed on the market for the long term. Let us see what is intended by these three rates.

The Social Time Preference Rate is that rate of discount that would give the community the balance it desires (or the government assumes that it desires) between quick-maturing and long-gestation projects. (In choosing, it must be borne in mind that the long-gestation projects are often more productive and so, in the end, a better value; for example, a power station can be built larger and better in four years than in two.) In an advanced country where the community is well supplied with public goods and where the expectation of life is long, the mix between short- and long-gestation projects matters relatively little. The Social Time Preference Rate would be on the order of one or two per cent. Nevertheless, it is worthwhile to check by it in finalizing choice. In poor countries, where the provision of public goods is scanty and the expectation of life is about forty-five years (as in India), the Social Time Preference Rate is very important. The present generation has a right to enjoy a share of the results of its labour and saving, and however anxious the government is to build steel mills and large multi-purpose water projects, it should bear this in mind. Metropolitan governments, however, will not find this rate very useful.

The Opportunity Cost Rate, as emerges from our previous discussion, is the rate of return that could be obtained if the resources to be used in the project in question were either to be invested elsewhere or kept as a reserve (used, for instance, to buy long-term government securities or repay debt). This rate may be crucial for metropolitan government decisions, since a long list of desirable projects always exists with the choice between them often being close.

On the other hand, there may be circumstances, even in advanced countries, in which there is high unemployment and no obvious alternative use for the resources which would be used in the project. In such cases the Opportunity Cost Rate is low. In less-developed countries this is a common occurrence, some factors being in abundant supply (unskilled labour and land), with other factors (managerial skill and specialized equipment) being very short indeed. In such circumstances the use of Opportunity Cost Discount should clarify the choice between projects or techniques that economize in the scarce factors and those that do not.

The market rate of interest for long-term borrowing is the most straightforward and, for metropolitan governments, probably the most generally useful for discounting purposes. Yet there are certain difficulties in its application. In the first place, it is subject to fairly wide, short-term fluctuations and, as we have seen, discounting has to be carried out at the same rate for all projects so that no immediate attention can be paid to these changes. Secondly, even if long-term borrowing rates are very high (as they presently tend to be in most jurisdictions), the use of the actual rate may favour short-term projects in a way that is inconsistent with underlying Social Time Preference. Clearly, a programme should not be based exclusively on current rates. Bearing in mind, however, that the rate chosen for discounting cost and benefit streams is a basis for putting present values on each, it would appear that what is really wanted is a market rate averaged both over a term of years and over the different forms of long-term borrowing likely to be available. The answer would probably be about 4 or 5 per cent, but might well be higher if the danger of inflation is to be taken into account.

With the use of the chosen discount rate, the streams of costs and benefits are reduced to a present value. This can be thought of most easily, perhaps, as a constant annuity. But there is a final check to be made before the projects are put in their order of eligibility for implementation. This check concerns the constraints that may pertain to one or another of the projects. We can list several types of constraints that may have to be considered.

If a metropolitan government exists it is not likely that *physical constraints* will be in evidence. In times of inflationary pressure, however, certain skills or types of equipment may be in short supply. Rather than tie up funds and factors in a project that may be delayed (which causes unnecessary cost), at these times it may be better to postpone implementation until the supply position eases.

So far as metropolitan governments are concerned, any *legal constraints* will be associated mainly with land use and land acquisition. A metropolitan government presumably will have obtained powers of compulsory purchase at a market valuation when this is essential for a particular high-priority project; but the application of this prerogative is time-consuming and expensive. With these two constraints, much depends on careful timing for implementation.

A third constraint relates to the type of *income-distribution effect* a project exhibits. A project proposed for a particular site may have strong redistribution effects. If those are very favourable the council may feel that there is a case for giving a project a higher priority than it would have on pure cost-benefit grounds. This is really a case of upgrading the value of the social overspill benefit. A good example of this would be an urban renewal project, which to be effective is inevitably costly, but which has important moral and security overspill effects. This illustrates the usefulness of setting out the full list of cost and benefit items, in that what is involved in the upgrading becomes easily apparent.

While *administrative and budgetary constraints* should not hamper metropolitan levels of government in the same way as they may affect local governments, there may be severe difficulties in pricing certain items in project plans for which expert advice on valuation will have to be sought. This will increase the cost and will require checking against the possibility of simpler means of reaching the same end.

Many of the obstacles that appear to be administrative constraints may in fact be due to lack of expertise in management. This is a subject to which we must return; but first, having examined all the potential constraints, we must proceed to tie up the investment criteria emerging from the technique that we have discussed. When the discounting exercise has been completed, the projects that have been accepted by policy decisions can be arranged by the degree to which benefits exceed costs. Any project where the present value of benefits exceeds the present value of costs is in principle eligible for inclusion. This criterion can be expressed in several ways, one being the annuity. Thus, the project is acceptable if a constant annuity, with the same present value as the benefits, exceeds a constant annuity with the same present value as the costs.

Yet, to proceed strictly according to the list until all the available funds are used up may not give the best results. At this point the importance of having a good budgetary framework becomes apparent. First, the programme of eligible projects must be checked for consistency with policy decisions concerning the balance between services. Secondly, to follow straight down

the list would only be optimal if all projects were independent of each other. In a large number of instances this will not be so. Schemes may be related in a variety of ways, from complete exclusion (the implementation of one project precluding that of another) to a high degree of complementarity (the implementation of one project producing a very important overspill benefit in relation to another). Hence, the list as it emerges from the cost-benefit computer needs to be examined in relation to other projects and to general policy balance, including the amount of resources that is deemed wise to keep in liquid form. (This sum may change substantially as a programme develops due to changes in the general economic or monetary outlook.)

ORGANIZATION AND ADMINISTRATION

Even the best planning will not yield good results without effective organization and administration. Since most of the problems in this direction lie outside the financial field, our discussion will be limited. The administrative mechanism needs to be directed to two ends in particular: it should be possible to take decisions quickly; and there should be close coordination between all the activities and departments of the metropolitan government, both with respect to civil servants and elected members. There are a number of ways in which these things could be improved in most councils.[13] Ability to attract and keep the right

---

[13]In the United Kingdom a good deal of experimentation in administrative streamlining is currently proceeding, such as in the New Towns that have been created deliberately to disperse population from over-crowded metropolitan areas, especially from London. In the suggestions that follow, some use has been made of this. "Local Government Finance," *Summary of Experiments at Basildon New Town,* Institute of Municipal Treasurers and Accountants (Great Britain, August, 1966).

men at the top is important. Salaries are important, but this is by no means only a matter of finance. There must also be a sense of satisfaction in doing a good job without frustration. This implies a good organizational structure. Organization and Methods experts can assist in keeping the administration abreast of the latest technological advances; but good management goes much deeper than Organization and Methods. For instance, attention must be paid to problems of recruitment, training and pay.

As regards the council itself and its various commissions or committees, modern management techniques allow for a great deal more streamlining than was possible a decade ago. It should be feasible with modern methods of checking to delegate all minor decisions so that the council's time is not wasted over them, and councillors are not faced with voluminous agenda. Two things are necessary if this is to be possible on a large scale. A clear demarcation of responsibility for decisions is required both among departments and at different levels. The position of the elected members can be safeguarded by the right to modify or even reverse decisions not considered to be in line with policy.

Secondly, it is essential to have two strong, but not large, coordinating committees (each with about nine members), one on the executive side and the other concerned purely with financial matters. These would need to meet regularly to receive and review reports of all aspects of the metropolitan government's activities. Even more important, they would be charged with examining the estimates of all commissions for efficiency and consistency with policy. This examination would cover both the estimates for the annual budget exercise and long-term development.

Forecasts of revenue—principally, but not wholly, tax revenue—would need to be made not merely for next year but for the next five or six years in order that they can be brought into line with expenditure plans. The financial executive would also be responsible for overseeing the audit department, whose reports would contain an examination of expenditure set against the estimates, pinpointing inefficiency or possible defalcation. Where, after investigation, it is shown that there has been a misappropriation of funds, there should be no hesitation in giving full publicity to the offence and if necessary, inflicting a full personal surcharge.

ACCOUNTING

Finally, as to accounting techniques, the main problem lies with administrative and other non-remunerative services, as trading services and public utilities (and, for convenience, the roads) should use ordinary commercial or business methods as argued above. For most services the basic form of accounting should be actual cash outgoings and receipts. These can be checked with ease and can form the basis of close estimates for all administrative services. In a few cases of open-ended expenditure (for instance, hospital costs) only stochastic estimates can be made. With experience, these can be reasonably close.

In addition to itemized accuracy, accounts should be able to tell how much a unit of a particular service costs in order to compare its efficiency (performance) with other units. This technique of performance budgeting, which is essentially a different arrangement of the figures to be drawn up, is capable of much further development. As has already been pointed out, the calculation of unit costs is a prerequisite for an efficient grant system. However, it is necessary to emphasize that the practice of performance budgeting, which is not founded on a solid system of cash accounting, has not provided sufficient control of expenditure. Reconciliation accounts are tiresome, but in this case they are worth the trouble.

Current experience will form part of the basis for all these forecasts. Up-to-date income and expenditure analyses are indispensible and can be produced easily by computers. Accurate forecasts on both sides of the budget also depend heavily on an examination of the past. This examination is related to the economic experience, not only of the country as a whole, but also to the area in which the city lies and to its own experience in varying economic conditions. Such historical data are much more useful if they can be broken down finely enough to ascertain which departments were able to go ahead smoothly and which encountered bottlenecks, whether from outside causes or of their own making. In addition, the breakdown should follow the separate experience of individual taxes and other revenue sources. Together these exercises should provide a reasonably firm base for forecasting the practical possibilities for the future.

These methods are applicable primarily to the activities of a single tier, multipurpose metropolitan government. Where there are a number of semi-independent authorities, the situation is more complicated and requires separate discussion, although fundamentally the same type of organization and practice would be required.

Metropolitan Organizations: Some Financial Implications

Any large metropolitan area, by virtue of its size and, generally, its age, is constantly undergoing social, economic and physical change. Whenever change takes place, there tend to be implications for the metropolitan area as a whole. The aging of central-city structures calling for renewal, the general outward movement of population to the suburbs and the overlying movement of rural migrants to the metropolis are all affecting the entire metropolitan area, either directly or indirectly. Because of this there is a need for some form of governmental organization for dealing with the metropolitan area as a whole. Fuller descriptions of various forms of metropolitan government have been provided elsewhere in this book; here the purpose is to elaborate upon the nature of certain financial aspects of metropolitan government organizations. Two major issues need to be considered: first, the question of boundaries or the territorial extent of the area of jurisdiction of a metropolitan authority; and second, the most appropriate sources of finance for metropolitan development.

AREA AND JURISDICTION

A metropolitan area must have a definite boundary if only for administrative purposes. There can be no room for doubt concerning where responsibility for particular services lies, nor from what areas the metropolis has the right to derive its real estate tax.

Some of the financial problems associated with establishing boundaries both for and within the metropolitan area can be illustrated by reference to changes that have occurred in London. The idea of a large area organized as a unity but with some services imputed to particular areas within it, was behind the establishment of the London government system in 1888, with a single supreme London County Council (L.C.C.) and twenty-nine metropolitan boroughs within it. The metropolitan boroughs were the rating authorities, the L.C.C. "precepting" (making specific demands for its needs) on them. The ratepayers also had to find the money for public assistance or Poor Law as it then was, greatly aggravating the inequality of wealth and resources between different boroughs and, as we shall see, causing a series of equalization or redistributional experiments. Technically on the same footing as the metropolitan boroughs were the City of London and the City of Westminster, by far the two richest areas in the country (albeit they contained some of the worst slum areas). Public Assistance, as a cause of inequality, came to an end with the nationalization of the service in 1948; but inequalities in wealth and, hence, in the ability to finance supporting services, remained serious.

As local government responsibilities grew, the new ones fell almost entirely on the L.C.C., which grew in importance and authority (some would say tyranny) far beyond what had been anticipated. Likewise its area, which had never been comprehensive, became progressively less appropriate, with a widening gap between the finance of the areas within and those outside it. (Its lack of adequate comprehensiveness was implicitly realized when the Metropolitan Police District (M.P.D.) was created covering a very much wider area.[14] Finally, in 1965, the L.C.C. was abolished and replaced by the Greater London Council (G.L.C.) as part of an extensive reorganization of London government involving both a considerable extension of the metropolitan area and, at the same time, some degree of decentralization. The area of the G.L.C. corresponds broadly to that of the M.P.D. There are

now thirty-two "London boroughs" with substantially wider powers than those of the former "metropolitan boroughs". But this figure understates the extent of the compulsory amalgamation of weaker authorities and by drawing into the London net areas of surrounding counties and even county boroughs. Normally extensions of metropolitan (or any other county borough) boundaries would rouse fierce opposition from the county losing important rateable value. In this particular case, however, what the neighbouring counties (for instance, Essex and Hertfordshire) lost were by no means all prosperous areas.

Although the G.L.C. area is much larger than that of the L.C.C. it is still not fully comprehensive and demands for a further extension have already been made. This is doubtfully practical or necessary at present. The major emphasis in London's metropolitan reform has been dispersal—overspill—partly on socio-economic, but initially on strategic, grounds as a result of experience of wartime bombing. The objective therefore is to limit the population of the G.L.C. area to 8 million, disposing of the remainder through overspill. Eight substantially "new towns" have been "designated" for London overspill and these are already beginning to provide good additional rateable value for the counties in which they are situated. This is not a complete nor fully satisfactory solution however, because the new town emigrants (and their tax potential) are completely lost to the parent city. In the United Kingdom this basic separation is aggravated by the continuance of the "green belt" policy (which is really a hangover from a less fully motorized world). It is generally impossible to house the "colonists" at all near their former jobs. Indeed one school of thought considers that they should be located beyond commuter range. It is consequently essen-

---

[14]With the "nationalization" of the police in the London area, following an attempt on the life of Queen Victoria.

tial that either their jobs should go with them, or new jobs be created near their new homes.

At first, existing towns and villages that were designated for expansion tended to look askance at the change which they foresaw in the whole character of their environment. No doubt some people moved into more stable surroundings, but so much money has been poured out of national taxpayers' pockets into the new towns that there is now widespread competition to secure designation. This attitude has—certainly in the case of London—been assisted by the enlightened cooperation of the exporting metropolis, which insists that the change-over must be an exercise in partnership.

A continuing partnership, however, would be possible only if some form of regional organization were developed. A small move toward regionalism has been made by the establishment of regional economic planning councils, which appear to be doing good work in producing cooperation between local governments in the regions. The councils, however, are purely advisory bodies and are too big to be regarded as "local" even in the motoring age. Any extension of their powers would meet with united opposition from the local authorities. On the other hand, the establishment of some enlarged authorities in the provinces—for instance a Greater Liverpool or a Greater Tyneside Council—on the lines of the G.L.C., with a new upper tier over the existing authorities would, apart from possible county opposition, appear to be both practicable and useful.

SOURCES OF FINANCE FOR METROPOLITAN DEVELOPMENT

Local Sources

Much of the finance for new works of a metropolitan government will, no doubt,

be derived from bonds or some other form of loan. As the opportunities and costs of acquiring borrowed money vary greatly from time to time and from place to place, and since the various methods likely to be available have been discussed earlier, there is little that need be added here. It is probably the case that a metropolitan government will directly undertake only a part of the borrowing it requires. The main task will be undertaken by a national (or in a federation, a state) government; but, again, circumstances vary greatly. Where there are two tiers of government in a metropolis, it is unlikely that the lower tier will continue to borrow directly once the metropolitan organization has been set up. With centralized borrowing it has proved easier to assure that the cost of debt service does not rise too steeply. On the establishment of a metropolitan form of government it has usually been necessary for the new organization to take over a substantial part of the local debts, just as the Australian Loan Council took over the debts of the states on its establishment.

In the United Kingdom, Canada and the United States, the basic source of finance for development, as for other purposes, is the tax on land and buildings; this is assessed on occupation in the United Kingdom, and on ownership in Canada and the United States. In all three countries (although least in the United Kingdom), there are also some minor sources of autonomous revenue: perhaps sales tax, in some American cities a payroll tax (which is virtually a low rate, more or less proportional, income tax on earners[15]), and now, gener-

---

[15]The payroll tax should be quite practicable in the United Kingdom, but so far the national authorities have set their faces uncompromisingly against it, from an (unfounded) fear that it would queer the pitch for the national income tax.

ally, parking meter fees. The "real estate" tax remains, however, by far the most important autonomous source. Normally this falls on all types of property: residential, commercial, business and manufacturing. The effective tax rate is not necessarily the same on different types of property. For instance, in Britain, the manufacturing industry was assessed on only one quarter of the valuation for twenty-five years from 1929. Residential property, which was rent controlled, was also, in effect, partially derated for a number of years.[16] In London, Winnipeg and Toronto, the lower tier units are the taxing authorities, while the upper, metropolitan-wide tier "precepts" makes specific demands on them for their needs. The lower tier units also raise additional funds for their own purposes.

In principle this is a perfectly workable system, providing that valuation is carried out by a single authority on a uniform basis. In the United Kingdom there should be no difficulty about this, as valuation for rates for the whole country is now the responsibility of the Inland Revenue. In the time of the L.C.C., valuation was carried out by a single body for the whole area with regular quinquennial revaluations. In the Canadian "metros" also, uniform valuation has been established.[17] In New York, however, this uniform valuation is attainable only for the five boroughs of New York City. Outside this area it is hindered by virtue of the large numbers of very small authorities and by differences in state laws.

Uniform valuation provides a basis for a system (or indeed, several varying systems)

of contributions to the metropolitan-wide authority. The total sum to be transferred will be determined by the relative responsibilities for existing services and for development of the two parties to the bargain. Uniform valuation also provides a measure of the relative wealth and poverty of the lower tier governments and, hence, supplies a basis for measures of equalization or redistribution. However, due to both the variation from one unit to another in ratios of residential to other forms of property and the differential tax burdens between the two classes of property, the measure may not accurately define the relative needs and ability to pay of the units involved.

The double problem of aggregate sharing between the metropolitan level and the lower level units and of relative distribution of burden between the units is of course an exact replica of the federal/state problem at the national level in a federation. An accepted solution may be somewhat easier at the metropolitan level, as important executive services fall on the "metro" rather than on the units. Thus, over the decade 1954-64, Metro Toronto's outlay increased fivefold while that of the units barely doubled. In Winnipeg, as of 1964, the system of contribution was changed giving Metro $8.8 million in place of $6.6 million. (We noted earlier a similar rise in importance of the upper tier under the L.C.C.[18]) If the metropolitan level precept becomes very large, to the point where it absorbs nearly the whole of the local tax levy, there is some danger that the units may become atrophied and cease to be active organs of local government. Something of this probably entered into the G.L.C. measure of decentralization.

[16]J. R. Hicks and U. K. Hicks, *The Problem of Valuation for Rates in Great Britain* (Cambridge, Cambridge University Press, 1944).

[17]In the Canadian province of Ontario, the introduction of a uniform valuation has begun across the province, as of January, 1970, with the assumption of provincial responsibility for property assessment.

[18]There are already signs that the G.L.C. has begun to follow the same path.

(Gross) contributions (precepts) from the units to the "metro" are most frequently given on the basis of relative unit per capita assessments. This produces a straightforward proportional tax. But with respect to cash flows in the opposite direction (from the "metro" to the units), it may be necessary to introduce some element of redistribution if equality of opportunity or something approaching equality of services is to be obtained. The most common base for general grant (or subsidy) is equal per capita payments. Unless the highest density of population is in the poorest areas this gives no redistributional effect; indeed, it may give most help to the richer areas with high density modern buildings. In this respect it may be worth glancing at the success in stages of interlocal redistribution policy in London, a city probably unsurpassed in its spread of wealth and poverty.

The first London equalization scheme was embodied in the Metropolitan Common Poor Fund, established in 1867 and administered by a national department, the Poor Law Board. Its area corresponded roughly to that of the L.C.C. established twenty years later. The Fund was responsible for the care of casual paupers and inmates of workhouses (indoor relief) as well as for other social services, such as care of lunatics, boarded-out children and fever hospitals. The finances of the Fund were derived from rates on the participating authorities. A borough whose expenditure on poor relief was more than its share of total London expenditure on the service (on the basis of the ratio of its rateable value in relation to the total rateable value of the whole of London), received a subsidy at the expense of the richer authorities. The system thus took into account both the relative wealth of the boroughs and their needs as measured by their Poor Law expenditure. As time went on the op-erations of the Fund steadily expanded, mainly due to increasing outlay on paupers, until at the time of its abolition it was distributing £6 million annually.[19]

Independent Poor Law authorities came to an end in 1929, and the administration came under the ordinary local authorities. Administration by the L.C.C. only slightly altered the extent of redistribution as compared with the operations of the Common Poor Fund. If anything the dispersion between relative rate poundages increased; this was due mainly to the impact of the Depression. In 1948 a new system of redistribution was introduced along the lines of the Rate Deficiency Grant established all over the country. As even the poorest London boroughs had a per capita rateable value above the national average, it was necessary to devise a separate scheme for London, but the basis was closely parallel to the national plan. Metropolitan borough councils whose rateable value per head was below the London average received a grant (taking account of their expenditure) to bring them up to the London average. The total grant amounted to £7 million, but (as had been the case with the Common Poor Fund, of course) the grant-receiving authorities paid part of their own grant. In fact, only £4 million changed hands, and more than half this was contributed by the two wealthiest authorities: the City of London and the City of Westminster. The final result of this two-way exercise was that nine councils made net contributions while fourteen received net subsidies. The poorer authorities derived their incomes (broadly)

---

[19]From 1894 a further small redistributional measure operated, through the Rate Equalization Fund. This took the form of a levy of 6d in the £ of rateable value, distributed on a per capita basis. The total raised by this was small (initially £75 million) and the additional redistribution weak.

in proportions of approximately one third from their own rates, one third from rate equalization and one third from grants. It is notable that throughout the history of the London equalization schemes the same areas have been net givers on the one hand and net receivers on the other, in spite of noticeable swings of property values and population shifts within individual boroughs. This position has not changed fundamentally in spite of the "nationalization" of public assistance (the great dis-equalizer in the inter-war period), and in spite of the high level of employment in all areas. It is not yet possible to say how far the boundary adjustment and amalgamation of small and poor areas under the London Government Act 1963, which established the G.L.C., will change this pattern.

Grants from Senior Levels of Government
So far as grants from the national government are concerned, it is again useful to compare the London situation with that in North America. The main assistance for London, as well as for the rest of the country, has been for education; first under a specific grant but from 1959 subsumed in the general grant, which also includes assistance for interest on agreed capital spending for schools. The other main type of grant is for slum clearance and the building of low-income housing, either on the site (or other planned interior area) or in an overspill area—the latter, however, without assistance toward moving. These are the two most important types of grant in London.

Within the United States, state grants to cities vary greatly. Only a few states, such as California, seem to be actively interested in the metropolitan problem. Federal grants, although increasing, are still meagre —only $2.1 million federal assistance to domestic municipalities in 1965. These grants are, moreover, heavily geared to rapid transit and road development.

The experience of Metropolitan Toronto and Metropolitan Winnipeg is of particular interest in spite of their short lives. The willing cooperation of the provinces of Ontario and Manitoba has been of great significance, and the federal government has also taken an interest in planning and financial matters. While all urban renewal projects in Canada involve three levels of government, in one Toronto redevelopment area (Alexandra Park) there has been four-level cooperation: federal, provincial, "metro" and city. Achievement in low-income housing under the Metropolitan Toronto Housing Authority was also important, with low-income housing increasing sevenfold in the lifetime of the Authority (1954-1965). However, this large percentage increase is due mainly to the fact that less than 800 units existed prior to 1954. As of April, 1965, the Ontario Housing Corporation has taken over the work of the Metropolitan Toronto Housing Authority, with an increase in grant aid, as we shall presently observe.

The most regularly aided service appears to be education, as is to be expected. The Canadian Federal Government, since 1967, has been meeting 50 per cent of all provincial expenditures on post-secondary education, which has been extremely helpful as the provinces (at least Manitoba) cannot be far off the limit of what they could do for "metros" without pulling back on other duties. To avoid this, the province could, of course, raise more taxes, and this eventuality they may well have to face. For example, the sales tax would likely be more efficient at the provincial level, as it is now in Ontario rather than at the "metro" level.

As regards grants on capital account, Toronto had been receiving 90 per cent

Mortgage Loans from the Canadian Federal Government for its Metropolitan Toronto Housing Authority's low-income housing schemes; but the provincial government decided to take over this work directly. Presumably this change was made partly on grounds of control, as it was accompanied by an increase of the federal mortgage percentage. In addition, low-income rent subsidies in the boroughs are now financed on the basis of 50 per cent federal, 42½ per cent provincial and 7½ per cent metro contributions.

The grants awarded by senior governments to Canadian "metro" development are thus quite generous. So far no system of regularly foreseeable, if not automatic, grants (as is needed for the optimum allocation of resources over time) has been developed. This is due, in part, to the fact that governmental arrangements are still new and experimental, and also to a tightness of tax rates at both the federal and provincial levels.

From this it appears that the Canadian "metros" are in a position to develop both sensibly and relatively cheaply. Unfortunately "metro" development is costly in the short run due to the large amount of long-term capital investment required to meet the backlog of capital plant which was one of the reasons for the creation of the "metros". However, so long as unified financial control can be maintained, in the longer period the increased base of the real estate tax, both for new area development and for redevelopment in the centre, should provide a large amount of additional revenue. It is of first importance that valuations for the tax should be full and regular, based on free market prices. Only thus can

the best value be obtained from the tax. In Toronto the assessment amounted to $2,705 million in 1954, and no more than $4,689 million in 1964, notwithstanding a big increase in population and development. It would seem that the greater part of the increase was due to immigration, and it must be questioned whether sufficient attention had been paid to rising capital values. As we have seen, there are a number of minor levies that can be exploited by "metros" through assessments that are equitable and enforceable, such as parking meters and possibly the tax of urban road use. These should also be used to the full. Some American cities get revenue from local income (payroll) taxes. There is also the possibility of sales taxes. But if, for instance, Manitoba were to introduce a sales tax it would probably want to keep it for itself. All these fringe possibilities aside, the real estate tax remains the basis of "metro" finance.

It is not too early to conclude that the "metro" experiment is a success. It has within it a potential solution of the twin urban objectives of decongestion at the centre and orderly growth within, at the circumference of the "metro" bounds. But the full exploitation of these opportunities will be very expensive, and care will be needed to see that the optimum allocation of resources is obtained throughout. This may imply that more interlocal redistribution may be required than at present, by means of differential grants to poorer areas and less than proportional grants to richer areas. Above all, an adequate development of "metro" or regional forms of local government calls for more taxes at all levels.

13 Planning and Development of the Metropolitan Environment

Ernest Weissmann

In the past two decades we have become so preoccupied with the means of economic growth that we sometimes forget the ends of development. Too often we consider people as mere inputs in the development process.[1] In the struggle to create new capacities, we often lose sight of the wider meaning of development: namely the creation of new qualities of life. Recently, however, the purists' economic approach has begun to change. Improving the human condition is more generally believed to be necessary for both a balanced and a sufficiently rapid economic growth. Nevertheless, as yet the crucial importance of the city as the essential environment in which economic capacities are created (or impeded) and the human qualities of life are enlarged (or frustrated) has not been widely realized. Development theory has not yet fully evaluated the city's role as a major medium of development in our industrial society; nor has development theory recognized the city as the long-neglected link in the many processes and activities of modern civilization and culture with which, for better or worse, we must deal.

---

[1] *The United Nations Development Decade: Proposals for Action* (United Nations publication, Sales No.: 62.II.B.2).

A paralyzing crisis is haunting cities in every corner of our planet, and its implications harbour dangers for the survival of humanity no less frightening than the spectre of atomic war. Unfortunately, neither the origin and consequences nor the urgency of the crisis are yet fully recognized by leaders in the sciences, professions and government. Practical ways of redirecting its spread from cancerous growth to beneficial and balanced development are yet to be conceived, let alone tried out or adopted for action. Affluent countries are slow to act or to use their ample resources; and the poor nations lack sufficient human, technological and capital resources. Squatter towns, social dislocation and individual hardship have become, in the second half of the twentieth century, a common feature of the developing nations' urban explosion and of the physical and social setting in which "marginal man" must subsist. At the same time, less than tolerable urban conditions are causing periodically destructive riots in the slums of affluent countries.

The current urban crisis is complex. It is deeply rooted in the growing interdependence of national economies, which in turn depend upon and are integral parts of world economy and world trade. It has grown into a complex of worldwide problems probably insoluble in terms of national economies and national planning alone. Our society must therefore learn how to share knowledge and to use its resources judiciously, or face degradation. The known and potential resources of every nation will be needed and, of course, should be mobilized and used rationally. The size and urgency of the job at hand require that individual national efforts be supplemented by intensive international, technical and financial cooperation. Rich and poor nations alike need much interdisciplinary re-

search and training; but most of all, we must cease looking to the past and begin to apply progressive technology and the best that science can offer in our time to resolve a problem that "after world peace is probably the most serious single problem faced by modern man."[2]

The two situations, the affluent and the depressed economic settings, differ in substance, in number and intensity, in the rate of decline and obsolescence, and in the kind of measures and resources required to deal with them. However, urgent action is needed everywhere and massive aid in most places. Unless resources and measures adequate to the challenge are devised and put to work rapidly, we may witness a far-reaching disintegration of our industrial society.

Mountains of data on urban and regional development have been, and continue to be, gathered and published, Much of it misses the real issue at hand. Some plans, though based on massive data, are no more than sterile exercises in architecture and civic design. Others, intended to be more "comprehensive", suffer from their authors' professional biases and fail to blend and balance the economic, social, environmental and administrative aspects of urban and regional action programmes. This area of development, in which present action has a far heavier and more sustained impact on the future than in other areas, has seen far less progress. Rare are the plans and programmes that genuinely apply contemporary science and technology to the problems of regional development.

---

[2]*Environmental Health Aspects of Metropolitan Planning and Development*, Technical Reports Series No. 297 (Geneva, World Health Organization, 1965).

When, in the course of its development, a nation recognizes that economic and social conditions are man-made and, therefore, that man can improve them, that country may have changed the course of its history. Through economic and social planning it may obtain a workable balance between its national needs, development potentialities and national interests and, on the other hand, the worldwide opportunities of trade. Through physical planning it can improve the natural environment by creating conditions conducive to man's individual and social progress and cultural development as well as for economic growth.[3]

The greatest need, therefore, is for an integrated approach to the economic, social and physical planning and development of the metropolis within its national setting. Against this background, planned or spontaneous urban development should be viewed as a phenomenon of national development. If we are to accept this philosophy, we must seek a structure that will facilitate, rather than frustrate, integration of national and metropolitan planning and development. It is the firm belief of this author that the application of the concept of the city-region as an environmental unit for planning and development will permit this integration.

Many of the problems that confront the metropolis have been elaborated upon already by others in this book. It may be helpful to elaborate upon the national and international implications for urban development if this relation between city-region and national development is to be clearly understood.

World Urbanization: Dimensions of the Crisis

THE GREAT MIGRATION OF THE TWENTIETH CENTURY

The fear of "overpopulation" is haunting the world, the fear of a "population explosion" in the developing continents where productivity is low and cannot sustain the lives that better health and sanitation are saving. Masses of people are seeking everywhere the better life that modern science and technology have made possible, that industrialization made probable and that education and mass communication have helped to make an aspiration common to all mankind. In our time the city offers conditions for the fulfilment of these aspirations; its seemingly irresistible power draws huge waves of population from the world's rural areas, where insufficient land and conditions of tenure have barely kept them alive in the past and where the agricultural revolution is now making them superfluous. This migration is the largest and the most important movement our society has ever experienced, not only in numbers, intensity and concentration, but also because it is universal and affects decisively the world pattern of distribution of production and people, and because it has unleashed a chain of drastic social changes.[4]

The movement of rural people to the city is not a new phenomenon, but the present explosive growth of cities and metropolitan agglomerations is. Economic and social factors have motivated rural-urban shifts in the past and they continue to motivate them now. There are, however, fundamental differences between the situation

---

[3]Ernest Weissmann, *The Contribution of Physical Planning to Economic Development*, Proceedings of the 1960 World Planning and Housing Congress (San Juan, 1961).

[4]Ernest Weissmann, "The Urban Crisis in the World," *Urban Affairs Quarterly*, Vol. I, No. 1 (New York, Sage Publications, Inc., September, 1965).

then and now. The nineteenth-century industrial revolution in Europe and North America proceeded at a relatively moderate pace. In the main, the market mechanism regulated economic development. As cities grew, a new social structure gradually emerged, though not without shock or hardship for the migrating peasants who were now becoming industrial workers. Industry and people were settling in the most developed areas of growing nations, and world trade favoured the new industrial countries. Now, the headlong rush to the capitals and metropolitan cities of Africa, Asia and Latin America is most intensive, massive and rapid in countries whose natural resources are underdeveloped. Their man-made counterparts, the economic and technological resources and skills, continue to be inadequate. Now, the terms of trade are highly adverse to the developing countries. They must export ever-increasing quantities of primary commodities for a diminishing volume of capital, manufactured goods and other needed imports.[5]

Currently, these differences are becoming even more pronounced. In the highly developed countries, industrialization and general economic development continue, in the main, to match the rate of urbanization and new jobs are being created in step with urban migration. Indeed, many industrial countries now import foreign labour to satisfy their growing manpower needs. In the developing countries, imbalances and

problems multiply under the impact of a massive migration to large cities in numbers sometimes exceeding net population growth. Consequently, most emerging countries already appear to be "over-urbanized" in as much as their industries and related economic activities cannot employ productively all the men and women who migrate to cities or all the young people who reach working age. The growing physical congestion and social tensions created by this concentration of uprooted "unemployables" in slums and squatter towns has reached, in some areas, levels that are bound to distort projected development, delay progress and sometimes reverse hard-earned economic and social gains.

The interaction among economic progress, population growth, urbanization and the human condition is obvious. The world is rapidly changing from an agricultural and rural society to a highly urbanized society. Industry is becoming the chief source of livelihood for progressively larger numbers of people. Rural populations grow at staggering rates, but agricultural technology now produces more food with fewer workers. For this, however, more land is needed per worker. As a consequence, the "push" away from the impoverished village and the "pull" of the city's lure and promise combine to produce a roaring stream of migration. The migrants expect more stable and rewarding employment in the city and hope to find there more adequate health and cultural amenities and better education for their children. However, advances in industry, transport, commerce and most other sectors of urban economies also have resulted in the use of fewer workers to make more and better implements of production and more varied means for a better life. The rural migrant becomes a footloose, un-

---

[5]Recently the United Nations Conference on Trade and Development (UNCTAD) was established in recognition of this "constraint" to the development of a viable world economy. UNCTAD's aim is to secure a fair share of the benefits from the exploitation of the developing countries natural resources for their own progress. This trend is reflected in some of the recent long-term international petroleum and copper arrangements.

employable, urban "marginal man" in the metropolis.

As a result of poorly matched resources, rates of economic growth and population movements, overcrowding and blight are now affecting vast urban areas in most developing countries. These conditions stubbornly defy efforts toward improvement and present us with seemingly insurmountable problems of planning, organization and financing. Too often the already precarious urban/rural balance and the traditional economic, social and cultural institutions tend to collapse. The outward symptoms of breakdown are inadequate shelter and communal services; almost complete absence of sanitation and safe drinking water; filth and squalor in ever-expanding areas; and a growing rate of disease and mortality. At the same time gang activity, juvenile delinquency, crime and vice—the most conspicuous manifestations of personal and social disorganization—have already become part of the social setting of blighted communities in which the "marginal man" lives.

The complexities, the causes and the many ramifications and consequences of urbanization are presented in a nutshell in this news item:

> The human impact of Greece's development in the last decade was emphasized by publication of the first statistics from the general population census. . . . Between 1951 and 1961 the total population of Greece grew to 8,400,000, an increase of 750,000. The greater Athens area, including the port of Piraeus and fifty-five suburbs, absorbed two-thirds of this increase. The population of Athens soared from 1,380,000 to about 1,850,-000. . . . Greece's population increase over the ten years averaged 9 per cent . . . but her five major cities registered increases ranging from 20 to 36 per cent.

> . . . Those figures show the plight of a poor underdeveloped country in which the peasants are running from poverty and under-employment.[6]

The plight of the migrant upon arrival in the city is all too frequently such that the change may appear detrimental to his condition. This is illustrated in the words in the following dispatch:

> Seven thousand "paratroopers" [a colloquial term for squatters] invaded half-a-dozen tracts south of the harbour. The owners requested that the "invaders" be evicted by federal troops but the authorities seem unable to act. Violence is feared since the "paratroopers" are armed. When the men leave for work, women take up their posts with rifles in their arms. . . . Word reached neighbouring communities and factories that home sites can be acquired at 125 pesos. An "office" was set up, funds collected from interested families and plots of land of 100 to 150 square meters were staked out. . . . An avalanche of buyers appeared; and overnight, innumerable shacks and shanties sprung up where the land was barren for many years.[7]

Squatter settlements and shanty towns are mushrooming in cities of all sizes, in all parts of the world. In a number of metropolitan areas they already contain as much as one-quarter to one-half the population. Their existence demonstrates both the ability of the rural migrant to build virtually from nothing and the strength of his will to claim for his family "the right to shelter". To date this dynamic force of the people has been used only in partnership with government for an orderly growth of

---

[6]*New York Times*, (May 21, 1961).
[7]*Ultimas Noticias*, (Mexico D.F. May 5, 1965).

cities. Meanwhile, the newcomer has no choice but to squat illegally because he cannot afford the minimum standards for housing established by law. He who needs it most does not qualify for aid. The wider community does not provide even the most elementary services to the settlement he is obliged to establish overnight.

While this condition of spontaneous urbanization exists in many developing countries, the older industrialized countries have experienced a slackening in urban growth since 1800. In some of the developed countries, urban population now grows by the amount of their total national increase. This seems to happen when four fifths of the population already live in cities with more than 5,000 inhabitants. In such highly urbanized countries about one third of the population resides in cities with more than 100,000 people and about one half in cities with more than 20,000.[8] At this stage urban problems begin to assume a character associated not so much with sheer size and explosive growth, but often with the demands that a new mode of living now makes on the community. Examples of this change are the large-scale emergence of private automobiles, the burgeoning of suburban communities, internal population shifts within metropolitan regions and the consequent need for readjustment and rationalization of physical arrangements in both the city and its region.

In addition to these basic changes in the character of urbanization, there is also a migratory movement from country to country. As a result of the economic boom in Europe in recent years, a massive shift of immigrant labour took place from the impoverished south to the affluent north. A news report described the situation as follows:

> Faced with a growing shortage of native workers the countries of western Europe have had to import some four million foreign workers to keep their booming economies operating at full throttle; Germany plays host to more than one million guest workers; France employs 1,500,000; Britain at least 800,000. Without the Italians and Spaniards, Greeks and Turks and Moroccans, France could not have increased its gross national product by 35 per cent since 1958; Germany could not have enjoyed an annual gross rate of growth in excess of 5 per cent in several recent years; and Switzerland would have been in a desperate shape—for one out of every three persons employed in that small country is foreign. In Belgium the coal mines would virtually cease operating without foreigners who constitute 80 per cent of the work force in some pits. . . . The large-scale migration of labour produces reciprocal advantages as well as grievous social problems. The Southerners go north for higher wages and a chance for a better life once they have returned home with their savings. They agree, at least initially, to take the hard unpleasant jobs discarded by the natives who have been upgraded to better paying, more agreeable work. . . . They are separated from their families for long periods, isolated from the local population and live in virtual ghettos either in the slum sections of big cities, in shanty towns or in bleak barracks maintained by their employers.[9]

_____

[8]*European Seminar on Urban Development Policy and Planning*, ECE/SEM/URB/POL/WPL (Geneva, United Nations, 1962).

[9]Irwin Ross, "European slogan—'Go North Young Man'," *New York Times Magazine* (May 9, 1965). A later survey in the *New York Times* (February 27, 1967) in eighteen countries of Europe gives the following figures for migrant foreign workers: France—2.2 million; West Germany—1.1. million; Switzerland—800,000; Belgium—650,000; The Netherlands—75,000; Sweden—180,000; Austria

Thus, while some foreign workers may realize their dreams, a great number experience dismal conditions from which they benefit little. This is yet another projection of 'adverse terms of trade' into the human condition. Migrant labourers from the less developed countries must accept low levels of living in Europe's affluent countries only because there is a very great discrepancy in the levels of income and standards of decency between the two groups of countries. Rural-urban migration and migration from smaller towns to metropolitan complexes has now become international. The inter-country migrant suffers all the hardships that are traditionally present in internal migratory movements. However, the "host country" has an escape clause with regard to its "guest workers": it can always turn them back out of the country.

The international implications of contemporary urbanization are clearly evident in this unique relationship between the more and the less developed countries. Less than two years later than the above dispatch another dispatch from Rome reporte that ". . . the massive migration by South Europeans to jobs in the north, a vital ingredient of Europe's economic well-being and social stability for a decade, has begun to move in reverse. Recession in West Germany, Britain and the Low Countries, which absorbed many of the more than five million migrant workers, has sent at least 100,000 foreigners home, swelling already large unemployment rolls in their native countries."[10]

—48,000; Denmark—15,000. Of these, 2 million were Italians, 600,000 Spaniards, 400,000 Portuguese, 200,000 Yugoslavs, 250,000 Greeks, and 196,000 Turks. In addition, there were in France 800,000 Algerians and 150,000 Africans from other areas.

[10]Robert C. Doty, "Europe's Migrant Workers Are Facing Job Crisis," *New York Times* (February 27, 1967).

Thus, the condition of life of urban dwellers depends not only on the health of their respective national economies, but also upon that of other economies and their interdependence. The implications for international cooperation in tackling the problems of urbanization are obvious.

TECHNOCRACY OR HUMAN PROGRESS

As more emerging countries adopt industrialization as a national development policy and apply modern technology as a basis for improving agriculture and industry, the pace of urbanization will become more rapid. These new dynamic forces will both spread and intensify the urban crisis in the world. But these same new forces, especially the possibility of harnessing nuclear energy for peaceful purposes, together with mechanization and automation may also help to resolve the crisis. Their far-reaching effect can be no more precisely projected today than were the implications of steam and electric power for industry, or the internal combustion engine for agriculture and transport. Until now this important area of social science remains almost entirely unexplored. The use of nuclear energy in areas remote from other sources of power will facilitate industrial development in many new places. The location of industry will become much more flexible with power available where needed. A greater degree of decentralization of production and settlement will be feasible. Abundant power, intensive mechanization and automation will also bring shorter working hours not only in industry but also in agriculture. Such developments are bound to increase further the demand of rural people for amenities and conveniences available in cities; the process of migration to old or new urban centres will be greatly accelerated. This process can be expected to continue at faster or slower rates, depending on local conditions, until about nine out of every ten people will derive their livelihood from essentially non-agricultural activities. Thus, the prospect of chaotic urban growth and decay, and of declining living conditions, emerges in the midst of rising productivity and potential affluence and in the midst of the "revolution of rising expectations". The prospect of rapid industrialization represents not only a new challenge but also new possibilities for environmental development.

Meanwhile, however, urbanization continues unchecked throughout the world. The developing countries lack the implements, the economic, technical and human resources needed to reduce, through development, the gap between economic and demographic growth; and most industrialized countries appear to be counting on an unlimited elasticity of metropolitan economies and the physical infrastructures and an unlimited capacity of humans to adjust to inconvenience and accept hardship. In the developed market economies, there is a tendency for production facilities to settle near existing consumer markets where sufficient labour and services essential for their success in the short run are also concentrated. In a way, "wealth tends to attract more wealth". But most of the great cities are nearing the point of diminishing returns.

The contradiction between accelerating technology and our ability to use it for human progress has become a concern to scientists, political leaders and philosophers. As Julian Huxley observes:

> The pace of change of human affairs . . . has steadily accelerated, . . . major changes now take place every few years and human individuals have to make several drastic adjustments in the course of their working lives. Where are these breathless changes taking us? Is change synonymous with progress, as many technologists and developers would like us to believe? Is there any main direction to be discerned in present day human life and affairs? No. Change today is disruptive: its trends are diverging. What is more, many of them are self-limiting or even self-destructive—think of the trend to explosive population increase, to overgrown cities, to traffic congestion, to reckless exploitation of resources, to the widening gap between developed and underdeveloped countries, to the destruction of wild life and natural beauty, to cutthroat competition in economic growth, to . . . "private affluence and public squalor", to. . . .[11]

One of technology's most exciting, but also alarming, achievements is the computer, as it pushes the advanced countries into an area of automation. Huxley regards computerized automation as alarming because, coupled with population increase, it tends to split a country into "two nations". In late twentieth century America the two nations would not be the rich and the poor, but the employed and the unemployed: the minority with assured jobs and high incomes; the majority with no jobs and only unemployment pay. The technologically advanced countries, therefore, will have to rethink the whole concept of work and jobs. Two activities that Huxley sees as having to be developed are the teaching and the learning of how to live under these conditions.[12] Our present-day application of technology and the resulting ecological degradation and environmental pollution is, in large part, a product of distorted human values.

---

[11]Julian Huxley, "The Crisis of Man's Destiny," *Playboy*, Vol. 14, No. 1 (January, 1967).
[12]*Ibid.*

Future generations may look back and wonder why we of the twentieth century took so long to see this ecological degradation and to combine the values of the past and the real possibilities of modern science and technology. They may feel that our understanding was clouded by the sudden rush of technological improvement and population increase. "Too quickly man the toiler had to become man the seer. After thousands of years of making societies based on stability he suddenly had to learn to make societies whose very purpose was to accommodate change—while maintaining its heritage of values."[13]

THE IMPLICATIONS FOR INTERNATIONAL DEVELOPMENT

The hard facts of underdevelopment are sobering. According to estimates of the United Nations for the years 1955 to 1957, in a group of one hundred underdeveloped countries with a combined population of 1¼ billion, as many as fifty-two countries with nearly 75 per cent of this total population had a yearly income of less than $100 per person. Another twenty-three countries with one sixth of the total population of the group had a per capita income of more than $100 but less than $200. Only twenty-five countries with less than 10 per cent of all people residing in the developing continents had an annual per capita income of over $200.[14] The comparable amounts for the United States of America, the United Kingdom and France —the top industrial countries—were, in 1957, around $2,500, $1,200 and $1,100 respectively.[15]

How did the factors of population and economic growth affect urban concentration?[16] Until the beginning of the machine age, no country had achieved a state of continuous expansion in its economic progress. The differences in the levels of living among countries, therefore, were not great. The population of the world has since increased more than two and a half times, but world output has multiplied by more than nine times and per capita income about four times. The distribution of population between the present industrialized countries and those that are still underdeveloped (pre-industrial countries) has not altered much over the years. The increase was slightly faster in the now industrialized countries, probably in response to their substantial economic growth. The economic growth of the pre-industrial countries remained low, however, due to their low per capita incomes and, in most instances, to their colonial status. Their economies often continue to stagnate even after independence. The population of the industrial countries, which has grown less than three times has achieved a twentyfold increase in total output and, consequently, a sevenfold increase in per capita output. Their "population explosion" was supported by an "economic explosion", and their urbanization by a sufficient rate of industrialization. The share of agricultural output fell from about 50 per cent of the total world output at the start to about 10

[13]Paul Spreiregen, "The Skin of the Earth," *Way Forum*, special issue on the city, No. 63 (Brussels, World Assembly of Youth, December, 1966).

[14]Paul Hoffman, *One Hundred Countries—One and One-Quarter Billion People*, Committee for International Economic Growth, Albert D. and Mary Lasker Foundation (Washington, D.C., 1960).

[15]These three countries reached the level of $200 per capita income a year as early as 1832, 1837 and 1852 respectively.

[16]This analysis is based on the paper by S. J. Patel, "Economic Distance Between Nations," *The Economic Journal*, Vol. LXXIV (London, Macmillan Journals Limited, March, 1964).

per cent, but its real volume increased about fourfold. The industrial output, on the other hand, increased thirty to forty times over with a fifteenfold per capita rise.

The distribution of world income was considerably altered by these trends. About 1850, the countries that are now industrialized had one fourth of the population and produced one third of the world output. In 1960 they had only a slightly higher share of population, but they produced nearly 80 per cent of world income. Thus, only in the last 100 years have the industrial countries outdistanced the pre-industrial ones in the struggle against poverty, disease and famine. In terms of annual rates this growth seems almost modest when compared with recent achievements. The annual increases were 2.7 per cent for total income; 1.8 per cent for per capita income and 0.9 per cent for population. The fairly small arithmetical difference between the growth rates for population and total income was, of course, magnified in terms of per capita output by a secular, though periodically depressed, development.

How long and how intensive a development will be needed to bridge the gap between the industrial and the underdeveloped countries? A recent analysis showed, for instance, that the current gap between India and the United States in total commodity output could be overcome in eighty years at an annual increase of 3 per cent in Indian per capita output; in 60 years at 4 per cent and in fifty years at 5 per cent. Thus, the transition from abject poverty to relative affluence would require five decades at a per capita growth rate of commodity output of about 5 per cent per annum. Higher rates of growth have been achieved and sustained in a number of socialist countries and in countries with market economies; and the resulting rises in standards of living have, together with

intensive education, brought about a lowering of their rates of population growth by choice. But unless similar rates of economic growth are somehow attained, can we expect that people in the preindustrial countries (who are mostly illiterate and allegedly in a state of apathy) could now be aroused to a very sophisticated understanding of the inter-relationship of their countries' economies and the sizes of their own families? This seems unlikely. Or, should we propose that family planning and birth control be enforced as a substitute for economic growth and education?

The terminology applied by many in discussing the present urban crises and its apparent immediate cause—population growth and migration—conveys the notion of fear, of acute danger and imminent catastrophe. "Over-population" is invoked as the cause of impending world famine when individual developing countries are unable to produce enough food to match their own population growth; and "population explosion" is blamed for the inability of pre-industrial economies to reach levels of self-sustained economic growth. Instead of devising practical ways of attaining and maintaining a sufficient rate of growth in the developing countries,[17] many influential writers blame current population growth for the age-old condition of "poverty-that-breeds-more-poverty". But in the recent past, the same economic, social and political conditions operative in colonial relationships, which enabled most industrial countries to progress, were the very cause for the lack of development in the governed territories. Incomes in the developing countries are consequently low, consumption

---

[17]Say 5 per cent per annum, which was set by the United Nations as a realistic goal for the Development Decade (1961-1970) in light of world resources, productivity and demographic movements.

relatively high and savings very low indeed. To reach a sufficient rate of economic growth will therefore require a long, hard national pull and considerable international aid.

This, then, is the broad setting of the worldwide urban crisis: in the affluent part of the world it is an unwanted concentration of activities, things and people in urban complexes such as the Eastern Seaboard Megalopolis and the West Coast Megalopolis in the United States, or the great urban industrial complexes of Northwest Europe and Japan. And in the developing continents there is a relentless and chaotic agglomeration of people and poverty in amorphous metropolitan areas such as Calcutta, where explosive expansion is less due to the "pull" of industrial and economic growth than it is to massive flight of rural people from misery and hunger. Yet the urban crisis is a manifestation of rapid technological progress. It represents both great promise of abundance and a great challenge to our ability to accept the changes in attitudes and values that will help us to use our immense new productivity for bettering the human condition throughout the world; to begin to close the gap between the affluence of a small part of the world and the increasing poverty of the rest. The growth potential of *our world society as a whole* multiplies with the advances achieved by scientists and technologists in both the highly industrialized and the underdeveloped countries. But the necessary adjustments in our approach to such basic issues as economics, land and government are very slow to come.

These adjustments call for a questioning and revision of our basic concepts of the planning and development of our environment. If this development is to take place it must be through change as part of an organized process, not change for the sake of change. It is in the hope of establishing the basic ground rules for this planning and development process that we turn our attention now to the concept of the city-region.

The City-Region

FORMULATING THE BASIC RULES OF THE PLANNING AND DEVELOPMENT PROCESS

The fantastic growth of world population in recent decades and its heavy concentration in major cities and metropolitan areas in the developed and developing countries alike must be viewed against the background of the social change brought about by economic growth. The relatively inexpensive measures of public health and sanitation undertaken in the developing countries for humanitarian reasons are evidently succeeding in saving lives. Yet, essentially political and economic considerations continue to withhold effective use of the world's accrued human, technological and capital resources for the establishment of a viable world economy, which in turn would enable the developing nations to put to better use their natural wealth and human resources for their own benefit.

Communities as well as traditions are disintegrating in the developing nations under the impact of an expanding world market and the concomitant urbanization. At the same time, in most developed countries the available community services and facilities fail to satisfy the social and cultural needs of their citizens. The productivity of these countries continues to rise, and investment in social development programmes grows at even a faster rate than their productivity; but the quality of life continues to deteriorate. The malformation and malfunctioning of their cities have already caused considerable economic losses. In New York City alone, the 1964 power break, the 1965 transit strike, and the 1966 smog demonstrate that putting the best that science and technology, social engineering, and government can offer to use is at least as urgent for the advanced as for the developing nations.

Formulation of broad lines of development and a varying degree of economic programming have become accepted functions of government in all countries. This is true whether planning is done by a central agency or whether a public authority regulates interest rates for housing loans, for example, or facilitates or restricts, by administrative or economic and fiscal measures, the flow of capital into given investment fields. A growing number of nations are beginning to pay more attention to the social setting in which these economic interventions occur and to anticipate the social consequences of these interventions. Health, education and housing have become, therefore, important components of national planning in many countries. Yet the current "explosion" of urban problems in all parts of the world demonstrates vividly the need to anticipate and to plan as well for the corresponding changes in the physical environment if the full social benefits of economic development are to be obtained. Physical planning, the planning for such environmental changes, may in fact help to reconcile the often conflicting requirements of material production, in the narrow sense, with human welfare. Inadequate environmental development may jeopardize the achievement of the economic goals themselves. There is growing acceptance therefore, of the close interdependence of the mainstreams of development—economic, social, environmental and political—and of the dependence of development planning on political, administrative, and legislative action as means of implementation.

The recognized purpose of economic development is social improvement, but human progress is not an automatic consequence of this development. To obtain it, a nation as a whole must allocate to social development a suitable share of the wealth it produces and plan for an appropriate distribution of the social benefits it can provide. In this context a "plan" is:

(a) *a model of an intended future situation* with respect to: (i) specific economic and social activities; (ii) their location within a geographic area; (iii) the land required; and (iv) the structures, installations and landscape which are to provide the physical environment for these activities; and,

(b) *a programme of action and predetermined co-ordination* of legislative, fiscal, and administrative measures, formulated with a view to achieving the situation represented by the model.[18]

The essential feature of both parts of the definition is the intent to improve the human condition and to raise the level of living through development and change.

The basic elements of planning are: economic and social activity planning; environmental planning; location and land-use planning; and administration. The same elements are common also to urban and rural development planning and to local, regional and national development planning. As any plan invariably consists of more than one element, administrative coordination is a basic requirement for successful implementation. Of fundamental importance, however, is the close integration of socio-economic planning with phys-

---

[18]*United Nations: Metropolitan Planning and Development* (New York ST/TAO/Ser.C/ 64).

ical or environmental planning. A primary objective of balanced development is high total production and a corresponding increase in the levels of living, along with the creation of conditions under which the social values and culture of a society can express themselves freely and benefit humans rather than simply emphasize production *per se*. The effective achievement of this dual objective depends, among other things, on the rational location of the different means of production, consumption and services in an efficient, healthy, convenient, and pleasing environment. To remain effective the environment must be continuously adjusted in order not only to accommodate but also to promote and facilitate the projected economic and social changes and reduce hardship and social dislocation.

REGIONAL DEVELOPMENT

As the concept of development broadens and planning becomes more comprehensive, the questions arise: at what level of government, or at what scale of planning, can the economic, social, and environmental mainstreams of development be most suitably integrated; and at what level can implementation of national plans be conveniently programmed in a rational sequence? Recent experience may provide an answer. Countries that have adopted central planning for their economies have shifted gradually from exclusively national to territorial as well as sectoral planning. This has been done in order to overcome some of the difficulties inherent in the concept of detailed planning from the centre (such as for the locational aspects of development—including full mobilization of local resources, presence of infrastructure and other essential services). In the countries relying on the market mechanism to regulate their economies, planning is essentially local and urban. These latter

countries are becoming increasingly aware of the necessity of expanding the "master plan" at least to cover the area directly influenced by and, in turn, influencing the city's development, that is, the metropolitan region or megalopolis. Thus, starting from opposite planning concepts the "region" is now being accepted by all as a convenient scale or level of development planning, particularly in the different phases of implementation.

As knowledge about and experience of planning and development increase, the inadequacy of such indices as for national average incomes, levels of living and productivity, and so on, becomes apparent. It is now generally accepted that the different geographical areas of a country vary with respect to economic, technological and human resources, productivity, levels of living and their endowment within the essential physical and social infrastructure. Once these regional differences are recognized, national development policy must try to strike a balance between two extreme positions:

(a) to *equalize conditions* in all areas as soon as possible (conceivable even at the expense of total economic growth); and

(b) to *favour areas most likely to grow rapidly* and use the resources so gained to bring about progress everywhere, in the long term.

Such decisions must be guided, of course, by the full knowledge of alternative courses and their effect on the regions and the people concerned. If so conceived, "regional development" could give concrete guidance to industrialization and agricultural reconstruction, and to internal migration in accordance with rational urbanization policies. First of all, the local economies of urban agglomerations could be strengthened by nationalizing the use of external economies and of services and facilities already available; and secondly, to attract rural migration, alternative new urban centres could be created in conjunction with new production centres by developing power and utilities on an infrastructural pattern benefiting from the economy of size and, yet, avoiding congestion. Thus, the concept of *city-region* emerges as the environment in which intensive agriculture, industrial centres, residential communities, cultural and recreational facilities could blend into efficient and pleasing patterns, respecting the individual's, the family's and the community's cycles of daily, periodic and occasional requirements. These local activity centres and their complements serving larger numbers and territories would be linked together by foot or vehicle and set agreeably in the landscape.

City-regions so planned could bridge the gap between national development and the efforts of local communities. At the regional level the many services and facilities furnished in a given area by central, regional and local authorities could cooperate within systems that may attract public and private investment. Furthermore, the region offers a suitable physical framework within which development projects of national significance and those based on local initiative and interest can find their appropriate place. As an analytical and graphic method of planning for economic and social development in a given geographic area for a given period of time, regional planning regulates suitable timing of the execution of specific projects and programmes; designates viable physical locations for these projects and programmes; and establishes rational linkages and interrelationships among them. Like the input/output method, or linear programming, or the critical path method in the economic sphere, a regional plan cannot substitute

itself for the decision-making process, which is essentially a political one. But it can help to conceive a broadly based balance among development projects and programmes by eliminating conflicts in requirements for land, location and timing, and in this way facilitate the choice of the "best" available alternative and promote the execution of the adopted solution.

Two loosely connected development processes seem to occur in many countries: first, a set of large "productive" projects involving few people but designed to create wealth and establish the country's industrial base; and, secondly, development efforts representing, in fact, a struggle for "survival", whether spontaneous or planned, involving the great majority of the people. As an answer to the latter problem, community development programmes have been combined in some cases with agricultural extension to increase food production. In other instances, cottage and small-scale industries have been re-established in rural areas and small towns as partly mechanized undertakings. They were able to produce, at quite an early stage of development, needed consumer goods that the countries' large modern industries would not have manufactured immediately. The rural community has thus been rendered more attractive and the resources of the people have contributed substantially to general development. Rural people have thus begun to participate effectively in influencing the direction and pace of development at the local level.

To facilitate regional development, suitable procedures are needed. In each case, they would be a product of prevailing economic, social, political and cultural conditions. Among other things, the procedures would seek to achieve: (a) pragmatic territorial integration of administrative entities making up the economic base of the metropolitan or the city region; (b) a balanced integration of the different economic and social activities and functions in the region in accordance with assumptions set by specific development policy and strategy; and (c) workable integration of the requisite professional and managerial disciplines involved in development into regional planning teams. In addition, effective communication should exist among these disciplines, and between them and the average citizen.[19] Territorial and administrative integration can be successful only if it is in step with and supported by economic and social integration. It must often proceed only step by slow step, starting with the services and facilities that must be planned and developed on a regional basis, such as power, water and the other utilities; health services and environmental health; transportation systems; or the location of industry or new residential communities within a region. Development must be "controlled" by continuous observation, evaluation and study of the economic and social consequences of different projects or programmes and all their physical manifestations.

The concept of the city-region ultimately implies comprehensive development of extended urban areas efficient economically and desirable socially. Such development will require new departures in professional training and research in general, especially with regard to the integration of the different disciplines involved. Joint research-training projects would tend to bring together the professions concerned and provide an understanding of their respective roles and potential contributions to development. Citizens at large are the ultimate beneficiaries of development, and their participation in the setting and attainment of goals must be secured. The press and other mass media play a most important role in facilitating understanding and acceptance of plans, as well as the individual and group hardships that inevitably accompany development. Patrick Geddes combined the biological and the sociological approach when he concluded that the key to the solution of problems in developing "human communities" was the diagnosis of the complex interaction of people, their environment, and their activities.[20] Clarence Stein and others regard "the region" as the proper environment for the efficient accommodation of a greatly enlarged concentration of people and activities inhabiting megalopolis.[21] The concept of city-region is related to both of these concepts. It differs from both in that it establishes ground rules for a process of productive interplay within the mainstream of development (rather than a community or "regional" plan) as a means of producing a human condition, which is progressively improving as a result of economic, social and environmental development interacting upon each other and as a result of their cumulative contribution to such improvement. The great challenge then is the need to mould a socio-economic, physical environment that is worthy of human achievements and of the potentiali-

---

[19]*Environmental Health Aspects of Metropolitan Planning and Development*, Technical Reports Series No. 297 (Geneva, World Health Organization, 1965).

[20]Jacqueline Tyrwhitt, ed., *Patrick Geddes in India* (London, Lund Humphries, 1947).

[21]Clarence S. Stein, *New Towns for America* (New York, Reinhold, 1957); Derek Senior, ed., *The Regional City* (Chicago, Aldine Publishing Company, 1966); Robert E. Dickinson, *City and Region* (London, Routledge & Kegan Paul Ltd., 1964); and John Friedman, "Regional Planning as a Field of Study," *Journal of the American Institute of Planners*, Vol. XXXIX (August, 1963).

ties of the emerging, highly industrialized, affluent society.

The concept of the city-region offers a rational framework for a practical approach to environmental development. Any development centre (or growth point) can be the starting point for a city-region. The aim of the city-region may well be to create necessary concentrations of activities and people as a basis for development. In other cases the city-region may be a means of loosening the urban agglomerate to make it economically more efficient and/or socially more enjoyable; and it may take any shape that geography, technology and human ingenuity can produce—a string, a necklace, a ring, a star,—any geometric or free form. Its main characteristic, however, must be total flexibility—an ability to adjust to changing development factors, new technology, new economic patterns and rapid "wind" change, and to exploding cultural progress; and, an ability to meet the personal need for privacy, dignity and freedom. Most of all what is suggested here is not a prescription for the city of the future. On the contrary, it is a formulation of the *basic rules of a process* whose aim is to develop the *city-region as an environment*. Such a process of city-building can be started as soon as we decide to act. As we build we can learn, alter and improve.

THE PLANNING AND DEVELOPMENT OF THE CITY REGION AS AN ENVIRONMENT

The Plans
Planning for environmental development usually proceeds in several stages. First, a general *outline plan* for a given geographic area may be formulated by extrapolating its basic development objectives from long-term national development trends and projections for a period of from fifteen to thirty years, within the framework of the planned or anticipated national development. As a land-use plan, it normally designates general locations and land areas for the different functions and activities in the region and suitable zones for the different forms of cultivation and resource development based on capabilities for agriculture, industry and power generation, and on water, soil and climatic surveys. Transportation and utility systems should then be laid out linking the different activities to one another and, as necessary, to the neighbouring regional systems and the existing or planned national networks.

A more detailed *master plan* for the metropolitan region (or city-region) can be drafted for a period of from five to fifteen years. At this stage, precise locations are assigned for such facilities as residential communities, parks, open areas and playgrounds, schools, libraries, cultural centres, hospitals, clinics and other health facilities, industrial complexes and heavy and light industries, agro-industries and service centres. In this phase, feasibility studies should be made for alternative solutions for the supply of power, transport and water, for sewerage and drainage systems, as well as for the more important economic and social projects on which the viability of others depends. Master plans for a city-region should contain special investment programmes for part of the full period based on feasibility studies, and they should also include probable sources of capital.

In the final stage, *detailed action plans* are prepared for specific areas, projects or programmes, tied to given locations to supplement the general outline plan and the master plan. These detailed plans may be designs for residential communities, programmes for environmental health, transport projects or other services and facilities supporting the different economic, social and cultural functions of a city-region. These detailed plans are usually projected in an order of priority, which in turn will depend on the investment resources available and alternatives based on cost/benefit analyses. They may cover current short-term development programmes or capital investment budgets over a period of one to three years.

Changing conditions, technological advances and the evolution of socio-economic values make it essential that both the regional master plan and the general outline plan be reviewed, evaluated and suitably revised in the light of concrete achievements of shorter-term investment programmes. A viable periodicity for this process may be a series of one-to three-year capital-investment budgets for the execution of detailed action plans with the five- to fifteen-year master plan for the city-region, which in turn would be projected within the framework of a fifteen- to thirty-year general outline plan for the region, based on national development trends or plans. Continued review, coupled with an annual revision, would result in the existence at any time of dependable short-term capital investment budgets, a medium-term master plan for the city-region and a longer-term projection of regional development for the larger geographic area. Obviously, the traditional tools of the city planner—zoning and the building regulations and codes—must now be supplemented by tools that are more effective and more complex. Some of them are referred to below, not as substitutes for planning, but as instruments of planning in its contemporary sense—design for development through change.

Land and Water
With the onset of industrialization, problems associated with the use of urban land and water have become most vexing. Particularly in market economies they are becoming more difficult to solve as the concentration of people and activities in the crowded metropolitan areas increases. The two resources are closely related, and the misuse of one often affects the other. The expanding metropolis needs more land and water and must reach for ever more distant locations to meet its higher requirements. The transition from rural to urban living, and from agriculture to industry, carried with it a change in the ways of supplying and using water. Urban communities need rapidly increasing quantities of water for such legitimate purposes as domestic uses, food growing and industry; but the water from the atmosphere is rushed through storm drains straight into the rivers and seas without passing through the soil.

The growing economic burden of storing and transporting water for the metropolitan complexes is multiplied by wasteful use. Costly methods of purification of polluted water, or conversion of sea water, have had to be applied in a number of cases as a result of misuse of a resource assumed to be inexhaustible. In other places, unwise use of underground water caused the sinking of urban land, which then had to be reclaimed and protected from flooding at considerable cost.[22]

The index of urban land values in Japan, for instance, increased by more than 300 per cent between March, 1955, and September, 1960, while the index of wholesale prices rose in the same period only 2.7 per cent.[23] Land prices in metropolitan areas of some other market economies have risen

---

[22]*Urbanization and Regional Development in Japan*, United Nations; reports on Survey Missions in 1960, 1962 and 1964.
[23]*Ibid.*

even more sharply. The scarcity of land is conditioned by geography and topography. It can be relieved to a degree by transport and building technology. Land shortage is being intensified daily by many competing demands, even in countries where urban soil is nationalized. In most market economies, its excessive price (or its market value) is not due to compelling socio-economic exigencies, but to speculation. The objective limitation of land in a given location imposes upon public authorities an inescapable obligation to weigh the social advantages of alternative uses. Based on such analyses, a careful selection of sites for the different essential functions and activities could be made in the light of their potential contributions to the community. The emerging trend in most countries is generally to follow this objective. Indeed, it is now often stressed that the essence of democracy is to prevent selfish private advantage, corporate or individual, from prevailing over the common interest of the community.

Comparative study of urban land problems in different countries demonstrates the close affinity of urban development and urban land. It also reveals the common manifestations of these problems: the simultaneity of their appearance throughout the world, the many common traits of land use and tenure, and the striking similarity in the essential exercise of governmental power of "eminent domain". Governments have the necessary powers to deal with the problem of urban land and they are, generally, applying them. The power to acquire land for a public purpose is an inherent sovereign power; the power to tax it provides broad possibilities for development control and revenue; and the police power over land use is fundamental to attaining and maintaining general welfare.[24]

The frequent reluctance to deal with land in a straightforward fashion is no doubt due to the historical fact that land was man's first means of production beyond the level of mere survival. As such, land has traditionally played and still plays a most important role in all pre-industrial societies. But countries are shifting from agriculture to industry as their primary economic foundation and as this occurs it is essential that land use policies should fall in step with their current and potential role in obtaining the social and economic objectives envisaged by long-term development. Urban land reform is urgently needed to facilitate national development now hampered by obsolete land policies and laws.[25]

A number of metropolitan cities in Europe have used the method of land acquisition in advance of development: a land reserve is created at workable cost, and private speculation at the expense of the community can be controlled. Indeed, this method has allowed the community, rather than the individual or corporate land owner, to benefit from the accrued value of the land due to public investment in such facilities as roads, power and utilities. The

---

[24] Charles Abrams, "Urban Land Problems and Policies," *Housing and Town and Country Planning*, Bulletin 7, Sales No. 53, IV. 22 (New York, United Nations, 1953).

[25] Ernest Weissmann, *Tokyo Memorandum*, Tokyo Metropolitan Government Publication (Tokyo, 1961). The Committee on Housing, Building and Planning of the United Nations, at its 1964 session, stated that "There was a need for *urban land reform* providing for suitable land policies to facilitate the implementation of urban and regional development plans, including such measures as may be necessary to build up land reserves, speed acquisition and prevent speculation." The Committee on Housing, Building and Planning: *Report of the Second Session* (New York, United Nations, 1964), Doc. E/3858 and E/c.6/25.

case of Stockholm is most instructive in this instance.

At the turn of the century, the municipality of Stockholm had the foresight to acquire through due process of law the development rights for all land likely to be affected by the predictable growth of the capital city. According to a master plan (first for the city and then for the metropolitan region) the municipality developed portions of land in a rational pattern paying due regard to economy in layout, to economy of scale of operation, and to modern technology in transport, basic utilities and domestic and industrial power. The plan was, of course, designed to achieve a desirable balance in the location throughout the region of major activities and functions in agriculture, industry, commerce and residence on the one hand, and educational, health, cultural and recreational facilities on the other. The Stockholm region now has excellent services and utilities, at reasonable cost for initial investment and for maintenance, and they are adequate for the expected rise in demand.

Land thus developed is leased or sold for uses established by the regional master plan. The municipality recuperates the investment capital and uses it for further land development; it controls the development of the city and the region according to plan; it guides land development in conformity with set standards; and it polices land use in accordance with Stockholm's plans. But most important of all, the community retains the redevelopment rights over urban land. Future contingencies and urban renewal needs can be satisfied; redevelopment can proceed under these conditions in a rational manner, and the community is spared the burden of lengthy procedures and the high cost of condemnation, acquisition and assembly of large numbers of private lots every time the slightest change is made.

In the process of accelerated urbanization large sections of agricultural land will continue to be transferred from cultivation to industrial and other urban uses. There is, consequently, an urgent need for national policies that could help to guide this transition into desired channels. The policy, to be effective, must of course be supported by the enactment of requisite laws, the establishment of the requisite organization to carry out land operations, and by adequate training of personnel for land planning and development operations at all levels. The physical planner's role in this crucial period is an essential one. It is his job to help the transfer of land from agricultural to urban use to occur in an orderly fashion through suitable land-use planning. In this capacity he must reconcile the rights of the individual with the interest of the community. In addition, by operating on the regional rather than the local scale of planning he can help to obtain a desirable balance in the urban/rural relationship, as well as that between industry and agriculture.

In this respect the world's cities are still an unsettled frontier. "Their forms, their populations and their uses of land have not by any means hardened into a stable mold," states Charles Abrams. "As more land is brought within the urban orbit the form and organization of the metropolis will doubtless change. It would be helpful if we had a few space agencies, appropriately financed, devoting themselves to exploration of how we can make better use of earth space to build better and more comfortable cities."[26]

---

[26]Charles Abrams, "The Use of Land in Cities," *Scientific American*, Vol. 213, No. 3 (New York, September, 1965).

Transport and Utilities

What is true for urban land and water is also true for other urban utilities and services. The obsolescent systems of domestic power, gas, transport and communications are also labouring under the impact of new, massive demands, with poor prospects of closing the technological gap. Yet, the inadequate, obsolete technologies of environmental enginering continue to be used. We are desperately and unsuccessfully trying to satisfy the enormous and highly concentrated needs with the means of another era and far less critical conditions. A large hole in the pavement of any metropolitan city, for instance, will reveal an utter chaos and confusion of pipes, cables and tunnels. Frequent breakdowns, costly repairs, personal and collective inconvenience, and great losses in time and production are the result.[27]

In the field of transportation, we appear to be making progress, at least at the study stage. The contributions of such studies are, as John Dykeman observes:

> (1) the approach to transportation as a comprehensive system of interrelated activities; (2) the recognition of the importance of land uses, demographic and social characteristics and consumer choices in determining urban transport requirements; (3) an appreciation of the role of transportation itself in shaping the development of cities and metropolitan areas; and (4) the acceptance of

the inevitably metropolitan scale of transportation planning. . . .[28]

Unfortunately we do not seem to have advanced much beyond the study stage. Major metropolitan problems are the lack of coordination of urban transportation and the growing traffic congestion. They are frustrating to the authorities and to the people who are exposed to daily discomfort and delays. They are also causing considerable economic waste and loss of production. A mixture of pedestrians and a great variety of vehicles (tramways, trolleys, buses and automobiles; three-wheelers, motorcycles and bicycles; and sometimes rickshaws, bullock carts and donkeys) are usually circulating through a maze of streets too narrow to contain them. The congestion increases nervous tension and air pollution and claims its daily quota of human lives.

It may be sound economic policy to strengthen domestic markets for automobiles as a counterpart to and insurance against the fluctuation in the export market. On the other hand, the predominant concentration of that market in a country's metropolitan areas is unsound both physically and socially. Various measures can help to achieve a more dispersed pattern of automobile distribution. For instance, high registration fees and/or high parking fees in city centres and subcentres may discourage the use of automobiles for commuting. At the same time, provision of large parking areas at low fees, or free of charge, at points where major arteries reach the more congested districts may encourage increased use of public transportation from such points to the centre. Another

---

[27]Several decades ago, Le Corbusier and other urbanists suggested separating the urban functions of transport and utilities and using walks, streets, roads and highways for pedestrians and vehicular traffic and installing utility tunnels interconnecting distribution points for various services in an autonomous and rational pattern. Very few systems have been built in this fashion to date.

[28]John W. Dyckman, "Transportation in Cities," *Scientific American*, Vol. 213, No. 3 (New York, September, 1965).

method of relieving congestion is the prohibition of vehicular traffic in the city centre except for night deliveries. In two cities (Prague, Czechoslovakia, and Ithaca, United States) tests have already been conducted with free public transportation in the city proper as an incentive to reduce the use of private automobiles.

A regular source of trouble is also the fact that in most cases the different means of rapid transit (railways, subways, tramways, trolleycars and motor buses) are not coordinated, either with respect to the service they are to render to individual communities in the city-region or with regard to the service they are to provide to the wider region as a whole. Currently there are overlapping services or gaps in coverage within the city-region as normal features of the service to the market, rather than adequate and equal services to the community. In some cases commuting facilities that have been developed as a part of large-scale land operations often compete rather than complement one another; and the urban dwellers and commuters are the real losers in the squeeze created by profitable supply and legitimate demand. Utilities and transport are essential conditions of survival of our industrial urban society. Their planned and economic operation is vital to the city-region whether they can earn profits or, like agriculture, need to be supported. Of course, more efficient transportation in planned urban environments may greatly reduce this vexing need for subsidies.[29]

Building

Most nations are using between 15 and 25 per cent of their investments in fixed capital formation for residential construction. Another 15 to 20 per cent is absorbed by essential urban services and facilities. Thus, even though a third to a half of the resources goes into these two areas of environmental development, no measurable impact seems to be made on the growing urban crisis. Inasmuch as many of these resources, and a considerable share of resources devoted to all other development sectors, go into structures of various kinds, so building involves from two thirds to three quarters of all investment in fixed capital formation in most countries. Consequently, even a slight reduction in building costs would release the equivalent of hundreds of millions of dollars for other productive activities.

The capacity and efficiency of the construction and building materials industries greatly influence the pace and costs of development, and occupy a pivotal position in national development strategy. Rapid expansion of their capacities should, therefore, be coupled with intensive mechanization. First of all, this would help to reduce costs and also reduce or even obviate the need for imports of many goods and services now paid for in foreign currency. Mass production of materials and components, as well as prefabrication of structural elements and complete structures are essential for this development. However, in the early stages, partial prefabrication combined with self-help, rather than full prefabrication, could be workable.[30]

While there is no other group of industries of similar importance for national development, the level of building technology is generally low and the methods of pro-

[29]Weissmann, *Tokyo Memorandum*, Tokyo Metropolitan Government Publication (Tokyo, 1961).

[30]Ernest Weissmann, "European Housing Shortage," *Housing and Town and Country Planning*, Bulletin 1, Sales No.: 48.IV.7 (New York, United Nations, 1948).

duction primitive in comparison with those obtaining in other industries. In additon, the predominance of small enterprises makes it difficult to transfer improvements achieved in some of the advanced countries. To facilitate a rise in efficiency, important changes in the structure of this industry must go hand in hand with the change-over to higher technology. However, the results of scientific research in one place must be adapted to the economic, social and natural conditions, as well as the level of technology of another environment before they are transferred. Pilot projects and practical demonstrations on a sufficient scale appear to be the most suitable means for such transfers. In adapting the practices of highly industrialized environments to conditions of scarcity, approaches must be developed that are less demanding on funds, time, and technical and administrative talent. For this reason, they may prove to be of interest to the highly industrialized countries as well. The building industry is so complex, and the resources involved so huge, that the required changes can be achieved only through the involvement and leadership of national governments. In this process, too, international technical and economic cooperation and exchanges are essential.

There are many phases in the life of a building, from fundamental and applied research, via the application of this research to design, production and distribution, to the ultimate use of the building. Each of these phases can benefit from scientific investigation and inquiry. Much is already being done by individual scientists, engineers and architects, by governmental, academic and professional institutions, and by production and consultant organizations. Most of them, however, lack the necessary resources, direction and common focus. Also, there is little co-ordination at the national level and practically none at the international level. A worldwide network of pilot projects, carefully researched, designed and executed, and the comparative study of data so obtained could provide a practical basis for the improvement of housing and urban development.

The usual emphasis of scientific and operational research is on conventional and traditional methods of building. Objectives are too often limited to such areas as reducing the cost and improving the quality of current building techniques. As most construction and all home building employ traditional and conventional methods, considerable economies can result from such research. A good example of this type of research is the successful application of soil as a building material using cement, asphalt or certain chemicals as stabilizing agents. In this way, an ancient method of construction is being improved through application of the current achievements in chemistry, hygiene and engineering. A systematic scientific review of traditional building methods and their improvement by means of modern technology would probably produce considerable results in a short time.

The second major area of building research and demonstration should be concentrated on the application of advances and discoveries in other branches of science and technology to construction, building design and physical planning. An example of transferring technology and organization of work from one field to another is the assembly-line production of partially or fully prefabricated structures or parts of structures. In a number of countries this method may produce the required volume and speed of housing construction or a way of overcoming a labour shortage in more conventional building. The use of plastics in construction has developed mainly as a result of the successful application of many types of synthetic materials for various purposes in other fields, such as aviation. Plastics have also invaded the housing and building field as imitations of, or substitutes for, more expensive natural materials. As a result of systematic research and development, it is now possible to create synthetic materials meeting prescribed specification both in quality, cost and weight and in aesthetic refinement. The improvement of specific qualities of certain natural materials through chemical, mechanical and thermal treatment is a similarly promising direction. Finally, the application of new tools and techniques, including those based on the computer, are important contributions to the planning and programming of housing, urban development and building.

The third major area of research in housing and urban technology is of a long-range character. It should start with a redefinition of concepts of the human habitat —from the point of view of the individual, the family and the community, and the manifold economic, social and cultural activities and functions of industrial society. The basic characteristics of structures, services and facilities required for these activities and functions should be defined, or redefined, through research. The building technology that would emerge in response to these new needs, born of new concepts, would be capable of creating a new kind of physical environment. The corresponding forms of home, community and city-region would reflect, of course, the scientific and technological advances, the new attitudes and values in economics and sociology of affluence, and the ever growing demand for education, training and culture, which have already become both necessary and attainable as a result of the emerging complexity and abundance of some of the highly industrialized countries. In the developing countries, the goals need to be focussed also on the future. Research in these fields undoubtedly merits the attention of the world's scientific community.

Research and Training

Most countries lack comprehensive research and training programmes in the technical and managerial disciplines required for urban and regional development. The need for such professional skills for all levels of planning and implementation will become more pressing as development planning gains its rightful place in governmental hierarchy as an instrument of good government of, by, and for the people. Post-graduate training for comprehensive regional planning, as well as in-service training for personnel in national, regional and local planning and development agencies, is needed in the developed and the developing countries alike. The development process is complex, and the man-made factors influencing development are diverse. As countries attain new levels of development, much research is needed to improve decision-making and implementation techniques. Research and training centres should, therefore, be established to provide the necessary scientific basis for national, regional and local policy and planning, and for the all-important efforts in civic education and citizen participation in the planning and development process.

The primary purposes of a *national research centre* would be to guide national policy formulation, the setting of national development targets, and the identification of regional goals in harmony with national policies and trends. *Regional research centres* would formulate comprehensive regional master plans, participate in their administration and execution, provide extension services and in-service training. They would also assess the efficiency and suitability of regional, local and national plans and programmes from the point of view of the region's benefit as well as its contribution to national development. In addition, however, *centres for scientific research* are needed in universities and other institutions of learning. These institutions would concern themselves with the analysis of planning for local, regional and national development; the evaluation of development practices; and, the advancement of planning and development theory. These centres would enjoy the advantage of not being under the great pressure of day-to-day operations and would thus be in a position to further the science and art of urbanism in their countries.

As previously mentioned, a major issue in the area of research and training is the lack of communication among the different disciplines involved in planning and development. These difficulties are compounded by the lack of communication between those who plan and those who represent the often contradictory objectives of special interest groups and the various groups of citizens who may support or oppose all or part of the plan. Finally, there is usually little contact between the planner and the citizen at large, who does, or should, benefit from the proposed development. A major aim of any comprehensive training and research programme in urban and regional development, therefore, should be "building bridges", establishing "communications", and reaching "common understanding" of the role and the purpose of urban and regional development planning. Another major function of such programmes would be to train, by means of suitable team research, sufficient cadres of technical and administrative personnel who comprehend the need for, and the methods

of, integrating the economic, social and environmental factors influencing planning, and even more importantly, who comprehend the political factors conditioning implementation.

Proposals for Change: Urban Explosion or Planned Growth

In spite of ever-growing debate and research, the enormity of the issues and their urgency are yet to be grasped by theorists and practitioners, let alone by political leaders or the man in the street. Thus, it is against this backdrop that we shall explore some of the strategies proposed to deal with the highly explosive urban situation. These include both proposals of merit and those which, in the eyes of the present author, should be questioned closely.

MEGALOPOLIS UNLIMITED

Nobody planned Megalopolis, it just grew and grew. Geographer Jean Gottmann has brilliantly analyzed the mammoth agglomeration of some 30 million people on the Eastern Seaboard of the United States of America. He argues that "urbanization and urbanism are two different things". He defines "urbanism" as the "distribution in space of a certain amount of material equipment", but he feels that city planners are not competent to handle urbanization. He maintains that until now "we had been involved in the organization of production, which was not too difficult, but we now have the job of finding out how to manage a way of life, and this [is] far more difficult" since it "involves a mutation of the way of life".[31] As a planner, he seems to imply that the existence of Megalopolis justifies

---

[31]Jean Gottmann, "Note on living at high densities," *Ekistics*, Vol. 18, No. 107 (Athens, October, 1964), p. 231.

its continued being. In a recent discussion he notes the following:

> New York has enjoyed an extremely successful period of growth during the past 15 years, and it is quite possible that it now is going through a recession. But there was a similar period (of recession) after the Second World War and a lot of talk then, too, that New York was finished, that the big corporations were moving westward and that the city was going down the drain. That talk ended after four or five years and in the nineteen-fifties we saw instead 150 new office towers constructed in Manhattan. It does not mean too much if a few business interests leave.[32]

He stated that New York had achieved in the last few decades a position of eminence that was not fully appreciated by the city itself. He warned, however, that it would see its place as the "capital of prestige" eroded unless it solved such problems as air pollution, water shortages and traffic congestion. In this connection he thought that the uncontrolled use of the private automobile could not be allowed to continue in the big cities. He disapproved, however, of developing smaller cities to draw the population away from the Megalopolis because "such communities did not provide adequate recreation for adults".

How does the concept of Megalopolis correspond to the concept of a workable urban environment? Is Megalopolis conducive to economic growth and social progress? Its lonely people, the seemingly uncontrollable expansion, its chaotic features, its inefficient transport—are all these desirable models? In discussing Megalopolis, Lewis Mumford has observed that the same problems of over-concentration of urban activities are to be found throughout the world, whatever the prevailing ideology. While he recognizes the possibility of reversing the present processes at work, he sees the real problem as being the present attitudes of social scientists, for, as Mumford says,

> ... sociologists and economists who base their projects for future economic and urban expansion on the forces now at work, projecting only such changes as may result from speeding up such forces, tend to arrive at a universal Megalopolis, mechanized, standardized, effectively dehumanized, as the final goal of evolution. Whether they extrapolate 1960 or anticipate 2060 their goal is actually "1984". Under the guise of objective statistical description, these social scientists are, in fact leaving out of their analysis the observable data of biology, anthropology or history that would destroy their premises or rectify their conclusions. While rejecting the scholastic doctrine of final causes, these observers have turned Megalopolis itself into a virtual final cause.[33]

Those who are inclined to accommodate the forces that have, uncontrolled and unchallenged, created Megalopolis and those who apologize for its sorry state usually wave the flag of "freedom of choice" or the people's "preference" to enjoy the opportunities that Megalopolis offers. What they neglect to point out is: what viable alternative has the celebrated man in the street to choose from? Nor do they suggest that even for the affluent society an urban environment, human in concept and function, is desirable but not yet a reality. We are still

---

[32]"Megalopolis and Beyond," Long Island University, Intercollegiate Conference, March, 1967, *The New York Times* (March 4, 1967).

[33]Lewis Mumford, *The City in History* (New York, Harcourt, Brace & World, Inc., 1961), p. 525 ff.

at the stage of neighbourhood planning. When it comes to urban development of any scale, environments are proposed that are conditioned by economics of cost rather than human benefit; most of the time technology is used to beat down nature rather than to build in harmony with it. The visual results of their free choice, maximum economics and misused technology are Disneyland-like creations and frightening backgrounds for life, vividly depicted in Hollywood's "It's a Mad, Mad, Mad, Mad World". When planners do survey and note the factors causing the emergence of Megalopolis, more often than not they fail to draw strength from this analysis for their proposals: the "rear mirror view"[34] continues to reign absolute, and "imagineering" for the present or the future remains taboo. Urbanists may be baffled by the fact that in order to overcome the traffic congestion of New York's Times Square it might be necessary to reroute the wheat shipments through the Atlantic ports, or puzzled by the fact that a housing policy that favoured suburban builders had done as much as Detroit to turn American cities into parking lots.[35]

LIMITS OF UTOPIA

The prefatory note to a special issue of *Daedalus* on the "Future Metropolis" states:

Just as any person in the late eighteenth century could not anticipate the world we live in today, so our prophecies about the future also are probably mistaken. Men in the twentieth century, however, have one advantage over their forebears: since they live in a time of change, they are conditioned to expect change, even prepare for it. Their predictions, if venturesome enough, may come somewhat closer to prophecy than those of an earlier generation.[36]

With this somewhat illusory objective in mind, new designs for the physical shape of Megalopolis are being proposed by architects and town planners in the spirit of the space age. Like their pre-war forerunners, Le Corbusier and Wright, they are fascinated with the new possibilities that science and technology offer for the physical arrangements of a well-functioning Megalopolis. The real point they are trying to prove is that "man will someday be able to live just about anywhere, and in environments of his own making, if he can alter his prejudices and learn to enjoy, rather than just put up with, the effects of science."[37]

What these "technological designs" miss are the economic, social and political instruments that society must conceive and mobilize in order to build a better city. One would also have to assume, of course, that the proposed model is compatible with the aspirations and the resources of the community for which it is designed. Also one cannot help but wonder where utopians are leading us. While any technological design is made to appear possible and places the world of plenty within the reach of all, it

[34] Marshall McLuhan, *Understanding Media: The Extensions Of Man* (New York, Signet Books, The New American Library, Inc., 1966), p. 67. McLuhan states: "If the work of the city is the remaking or translating of man into a more suitable form than his nomadic ancestors achieved, then might not our current translation of our entire lives into the spiritual form of information seem to make of the entire globe, and of the human family, a single consciousness?"

[35] Lewis Mumford, "A New Regional Plan To Arrest Megalopolis," *Architectural Record* (New York, March, 1965).

[36] L. Rodwin and K. Lynch, "The Future Metropolis," *Daedalus* (Journal of the American Academy of Arts and Sciences, Winter, 1961).

[37] W. Buehr, "New Designs for Megalopolis," *Horizon* (New York, American Heritage Publishing Co., Autumn, 1966).

also creates a problem for itself. As Martin Meyerson observes, "[eventually the major problem] . . . is not the satisfaction of wants but the creation of new wants that is challenging. . . . Planning, like utopia, depicts a desirable future state of affairs, but unlike utopia, specifies the means of achieving it."[38]

The strict dichotomy between a socioeconomic base and a technology-based environmental model, which is so generally practised in urban planning, emerges particularly clearly in the new designs for Megalopolis. This is true whether they are conceived by Buckminster Fuller's scientific mind, by the creative genius of Kenzo Tange, or by planners and builders who burrow the future cities underground, suspend them from giant masts, insert them into immense grid-like structural systems, float them in the seas or build them into the super-highway itself. As long as the dichotomy persists urban models will exhibit either an economic bias in terms of cost or a design bias in terms of monumental 'civic' architecture. Neither must be allowed to prevail.

EKISTIC ECUMENOPOLIS

"The main purpose of a human settlement is to satisfy man. Human happiness is the ultimate goal for the creation of human settlements", states the first of the "Principles of Ekistics".[39] Demetrius Iatridis, former director of the Graduate School of Ekistics, has described "ekistics" as follows:

[Ekistics is] a new discipline [which] is concerned with the future development of human settlements and seeks to effectively combine and solve the problems which have been only partially answered by the individual technical and socioeconomic sciences. It transcends the recent intradisciplinary approach to such problem solving because it focuses its attention on the settlement itself, as an entity in its own right, and draws its professional and scientific orientation from the phenomena and problems of the settlement.[40]

The principles of Ekistics are elaborated upon each year by a gathering, in Delos, of eminent figures in many fields of development, science and technology who have forcefully stated the case for environmental development within the framework of general development. They recognize the need for integration of socio-economic and environmental development and the crucial role of government in planning and regulating urban development. But the annual symposium cannot advance much more toward "ekistic synthesis". The history of similar movements shows that variations on the theme, new terminology and new names for old problems have neither clarified the problems nor have they solved them. A true interdisciplinary probe of the phenomenon "city" as an outgrowth of manmade economic and social systems is yet to come; and practical ways of translating the vision of the city for the industrial society into reality are yet to be proposed. The Athens Center of Ekistics was created to study "human settlements", and the World Society for Ekistics was established

[38]Martin Meyerson, "Utopian Traditions and the Planning of Cities," *Daedalus* (Journal of the American Academy of Arts and Sciences, Winter, 1961).

[39]*Ekistics*, Vol. 18, No. 107 (Athens, October, 1964).

[40]Demetrius Iatridis, "Social Scientists in Physical Development Planning: A Practicioner's Viewpoint," *International Social Science Journal*, Vol. XVIII, No. 4 (Paris, UNESCO, 1966).

"as an important instrument of worldwide information and education" in ekistics.[41]

The emphasis on "the settlement" may tend to perpetuate a physical bias in city building. In the past this bias has led to architectural creations of "The City", imaginative and beautiful at times, as was Le Corbusier's "Ville Radieuse", but leaving little room for development. The second principle of ekistics stresses unity "in the economic sense, in the social sense, in the political, administrative, technical and aesthetic senses" as the basis for "human settlements".[42] The need for constructively blending the mainstreams of development into a single development effort may be implied, but practical ways of actually doing it are not specifically stated.

The "City of the Future" project, initiated in 1960, is a case in point. The main focus of this project would seem to be the construction of a physical model of Ecumenopolis shaped under the impact of a world population of between 20 and 35 billion, by the geography of continents, and by a tendency of urban growth to follow linear extensions among "centers of strong ekistic development".[43] The assumption appears to be that technology and incomes will continue to progress spontaneously rather than be guided by human needs at a time when, hopefully, mankind will have moved from economic scarcity to affluence and therefore have adopted the habit of planning its economy and technology for "a human use of human beings".[44] The

planning of Pakistan's new capital, Islamabad, which led to the study of the "City of the Future", also appears to emphasize the physical side of urban development.[45] The "new city" was to "epitomize the new age, and [its] capital buildings [were] to be worthy symbols of the growing might and importance of [its] nation".[46] "GLM", the proposal of the Athens Center of Ekistics for the Great Lakes Megalopolis, again translates demographic and economic projections into patterns of physical growth.[47] In all these works, emphasis is laid on "humanizing" the city; but there is no reference to practical ways of facilitating a constructive integration of socio-economic and environmental development in order to enhance the emergence of the city-region of our time through productive interplay of the mainstreams of development.

BUILDING SOCIALIST CITIES

The planning of urban development is usually an integral part of planned national development in socialist countries. The policies and patterns followed in the Union of Soviet Socialist Republics may be used as a typical example for study and comparison.[48] In redeveloping the existing large metropolitan areas in the Soviet Union,

---

[41]"Report of Delos 3," Ekistics, Vol. 20, No. 119 (Athens, October, 1965).

[42]Ekistics, Vol. 18, No. 107 (Athens, October, 1964).

[43]Ekistics, Vol. 22, No. 128 (Athens, July, 1966).

[44]Title of the book, by the initiator of cybernetic and theories of communications and control as used in electronics, mathematics, and the computer and its many applications, Norbert Wiener, The Human Use of Human Beings—

Cybernetic and Society (New York, Doubleday Anchor Books, 1954).

[45]See various presentations in Ekistics in 1961, 1962, 1963, 1965 and 1966.

[46]"Islamabad: Preliminaries in Determining Urban Scale," Ekistics, Vol. 14, No. 83 (Athens, October, 1962). The plan for Islamabad is the work of Doxiadis Associates.

[47]"Development toward Ecumenopolis—The Great Lakes Megalopolis," Ekistics, Vol. 22, No. 128 (Athens, July, 1966).

[48]The following paragraphs draw heavily upon a paper presented by P. N. Blokhine to a United Nations Conference in Stockholm in 1961. See Planning of Metropolitan Areas and New Towns (New York, United Nations) ST/SOA/65, Sales No.: 67.IV.5.

new industries are not encouraged to settle within the area, although exceptions are made of communal service enterprises and the food and building industries. In order to obviate excessive concentration of production and people, new industries are usually located in existing, or new, medium-sized and smaller cities, which are then developed to what is regarded as the preferred size 200,000 to 250,000. Towns supporting enterprises that are interrelated with respect to their use of materials, production processes, power, transport and utility systems, are planned in accord with regional and/or sectoral, long-range development. The distribution of the population increment is guided into preferred locations through coordinated and selective industrial and agricultural development throughout the country. The national plan anticipates that over a period of twenty years industrial production will increase about 500 per cent and agricultural production about 250 per cent. This plan calls for full electrification, mechanization and automation of industry and agriculture; the introduction of new production processes, the development of new sources of energy and materials, and, a sharp rise in productivity. A balanced pattern of growth, control of undue concentrations in large cities, and an important reduction of urban/rural differences in incomes and amenities are expected to result from this policy. Regional development within the national plan is the general frame of reference for all urban development planning.

The enormity of the volume of construction and the amount of resources—physical, technical, economic and human—required to carry it out are apparent from the fact that over 800 new towns were built in the Soviet Union between the early 1920's and the early 1960's, representing an annual rate of construction of 3 million units, or over 14 units per thousand population per annum. (As will be seen from reference to Hidehiko Sazanami's chapter on housing in this book, this figure represents about the highest rate of construction anywhere.) Between 1960 and 1980 the percentage of the total population living in urban areas is expected to rise from 50 per cent to 70 per cent. Already, in 1960, about one half of this urban population inhabited cities of over 500,000. However, since it is generally held that extremely large cities (and very small settlements) are not easy to equip, and per capita costs for amenities and services are high relative to the levels of convenience and comfort to be provided, the distribution of population and the location of industrial and agricultural development are so planned as to obviate overcrowding and congestion. In agricultural areas, urban-type settlements are being established that are fully equipped with the requisite public utilities and services, and adequate educational, cultural, social and medical institutions. Living conditions in the large metropolitan areas are being improved through the intensive mechanization and automation of industry, which is expected to yield higher productivity and total wealth without necessarily increasing manpower requirements and urban expansion.

As already mentioned, the preferred size of a city is 200,000 to 250,000 inhabitants. A city of this size can offer the following: a convenient and relatively short journey between residence and workplace; relatively high standards and a sufficient variety of cultural and social amenities; and, an economically effective building organization for the city and for its maintenance.

The size and physical structure of these settlements is influenced by the communal services and facilities provided. Essentially there are three types of amenities: those designed for the daily use of the family (kindergarten and junior schools, local shops, repair facilities, markets, sports grounds and parks) are provided in a neighbourhood with a radius of 300 to 400 metres and a population of about 6,000 inhabitants; those used periodically (clubs and cinemas, cafes and restaurants, department stores and shopping centres; bank, post, telegraph, clinic and pharmacy; centres for physical recreation and parks) are provided in a residential area or district of about 900 to 1,000 metres in radius and a population of 25,000 to 35,000; and, those serving the entire city (including large theatres, concert halls and conference halls; specialized shops and large department stores; stadia and parks).

An integration of the different socio-economic functions within the city and the simultaneous establishment and development of the required structures, services and systems are basic tenets of urbanism in socialist economies. There is a gradual reversal underway from the traditional economic cost/benefit approach to a benefit/cost approach. It is anticipated that the provision of urban services and facilities will be less and less guided by strictly economic considerations and, ultimately, such services may be available "free of charge" to the user, although still a cost to the local economy.

Urban and regional development theory, policy and practice in socialist economies are based on the acceptance of a close integration of economic and environmental development and an understanding of the latter's role in shaping the physical framework for development. Consequently, environmental development in its broadest sense includes housing, communal services and facilities, as well as the requisite infrastructure for industrial and agricultural development. The provision of an environment compatible with available resources and accepted standards is considered to be a justifiable and productive investment.

It is in this context that the Soviet Union is exceptional; environmental development is carried out in advance of, or at least simultaneously with, economic development. Every economic development plan, whether national, regional or local, has its physical and social counterparts, each of which is budgeted for accordingly.

This extraordinary effort over the past four decades has already produced a wealth of practical experience of environmental and social innovations, and a variety of urban models observed in real life. It may well be that as a result of this rich experience, growing affluence, and rapid scientific and technological progress, the strong economic and physical bias in planning will change into a more flexible interdisciplinary approach to development. Intentional accommodation and adjustment, and productive interplay of the economic and social with the environmental development factors would then enhance both economic and human progress. If this were to come about, viable concepts, progressive theories and practical models of urban development for contemporary industrial society might emerge.

INTERNATIONAL COOPERATION

The many social problems that can be traced back to the growing urban crisis have caught and held the attention of the United Nations from the time of its inception.[49] But the explosive character of the

[49]Ernest Weissmann, "The Role of the United Nations in Urban Research and Planning," *Urban Affairs Annual Review*, Vol. 1; Urban Research and Policy Planning; L. F. Schnore and H. Fagin, eds. (New York, Sage Publications, 1967).

troubled situation in most parts of the world has only recently been confronted in the debates of several international bodies. The United Nations established, as a quantitative annual target for housing and urban growth in the developing countries, the construction of 10 dwellings, with the attendant urban services and facilities, for every 1,000 inhabitants. In 1965, this amounted to a total of some 24 million units—a figure expected to grow to 27 million units in 1975. However, due to slow economic and industrial growth, and especially to low levels of building technology, the present annual rate of building in most developing countries is as low as 2 or 3 dwellings for every 1,000 of their citizens. Admittedly, developing countries currently use between 14 and 25 per cent of the investment resources available for physical capital promotion for residential construction. But it is directed mainly to housing that is not of strategic importance for the economic growth of their nations, even though it may create employment and demand in the building and materials industries. Another 15 to 20 per cent of these investment resources is being absorbed by urban services and facilities. However, no measurable impact is made on the growing crisis; usually, even this heavy investment does not add more than from 2 to 3 per cent to the existing housing stock. This rate of growth is often less than the net population increase and just about one-half the rate of urbanization in the developing countries.

Against this background, the concern of governments and the international community has shifted from isolated aspects of housing or building or town planning to the causes and consequences of the urban explosion and to the concept of comprehensive environmental development as part of general development. Now, out of ear-

nest preoccupation with these problems, a body of doctrine and experience is gradually emerging; one is able to sense that it is not the lack of ideas, facts or institutions that is preventing action commensurate to the present worldwide crisis. What is missing are the resources to demonstrate convincingly the feasibility of action.[50] Nevertheless, United Nations aid, which in 1951 consisted of one project in rural housing in a Latin American country, now assists over seventy countries and territories on every continent. The trend over the last fifteen years has continued the move away from specialized aspects of housing or town planning toward comprehensive environmental development. An example is the planning for the reconstruction, in Yugoslavia, of the capital of Macedonia, Skopje—a city of 200,000 destroyed by earthquake on July 26, 1963. In October, 1965, the plans for Skopje were put on exhibition—the start of the process of formal adoption by legislative bodies in November. This was done in the presence of the International Board of Consultants, set up jointly by the United Nations and the Government of Yugoslavia, to review all planning and execution of the city's reconstruction. A variety of tested and untested techniques were applied, ranging from computer programming for emergency shelter to feasibility studies for transport and urban infrastructure. All these techniques were used simultaneously and at a sufficient scale to permit elaboration of the

---

[50]For instance, the United Nations has conceived the idea of pilot projects as a means of transferring knowledge and of adapting practical experience by showing, with the help of *external* resources, the feasibility of using *local* resources for massive and sustained action. Unfortunately, the lack of response of the affluent nations makes the creation of an adequate international fund or pool of resources and skills for this purpose a highly elusive target.

comprehensive new urban and regional plans in a comparatively short time.

The reconstruction planning for Skopje was unique in several other aspects: UNESCO (the United Nations Educational Scientific and Cultural Organization) provided aid in earth sciences; ILO (the International Labour Organization) in large-scale training for needed building skills; WHO (the World Health Organization) in the area of environmental health and sanitation; and UNICEF (the United Nations' Children's Fund) in child and family welfare. Experiences of many lands were applied. United Nations aid made for a more effective use of material, technical and financial aid from national and external sources; and the World Food Programme of the United Nations and its Food and Agriculture Organization (FAO) and the League of Red Cross Societies provided the essential extra rations and medical supplies for the rescuers and rescued and for the builders and planners during the first weeks of the emergency. In many ways, this undertaking has become a symbol of international solidarity and may well be a fore-runner of practical international cooperation in urban development in this time of crisis.

NATIONAL POLICY FOR ACTION

Japan offers a good example of a cooperative effort by a national government and the international community in developing policy for urban and regional development within the broad possibilities and practical limitations of national development. In 1960, the Government of Japan formulated its long-range development policy against the background of a sustained high rate of growth in the post-war period. The plan envisaged a doubling of national income in ten years through further industrialization.

Subsequently, the regional approach was adopted: nine geographic regions being established, each with their own development targets. The Ministry of Construction prepared a twenty-year programme for infrastructural development in the regions.[51] The stated national goals were a continuous and balanced growth of the economy; a rise in the levels of living and welfare generally; and a rational use and distribution of the country's resources and wealth in order to bring about conditions "where everyone throughout the country may equally enjoy an affluent life and every benefit that modern society can offer."[52]

In 1960, 1962 and 1964, the United Nations sponsored multi-disciplinary missions on urbanization and regional development which, in collaboration with Japanese scholars and practitioners, formulated approaches and recommendations for action in regard to a number of urban and metropolitan problems. In 1964, the United Nations team set out guides for a national policy on urbanization and regional development which are reflected in the following paragraphs.[53] A number of their recommendations have been implemented, and others are in the process of implementation. Japan's emerging national policy is an example of a combination of central planning, government guidance and regional and local initiatives directed toward the implementation of national plans through

---

[51]*Planning Bureau Blueprints of National Development Projects* (Tokyo, Ministry of Construction, 1963).

[52]*Comprehensive National Development Plan,* Economic Planning Agency (Tokyo, Prime Minister's Office, 1963).

[53]Ernest Weissmann, J. P. Thijsse and P. Ylvisaker, *Report on the 1964 Mission on Urbanization and Regional Development,* prepared for the government of Japan (New York, United Nations, 1963), No. TAO/JAP/2.

regional development in a rapidly industrializing country.

The intended effect of Japan's ten-year plan and the projected further development was to eliminate the economic and social disparities between the regions and between town and country; to close the gap between the incomes of agricultural and industrial producers; to overcome the discrepancies in the economic strength of the large as opposed to the small enterprises; and to create the required new jobs for men and women as they reached working age. In the process, the number of full-time and part-time agricultural producers was expected to decrease from 40 per cent of the country's population in 1958 to about one-third that number in 1970; industrial and power production would be increased and enjoy improved coordination nationally.

In fact, Japan has achieved an amazing post-war recovery; but urban services and utilities have now reached the breaking point in metropolitan areas. Their infrastructures and external economies would now have to be re-created, at high cost, to allow further uninhibited growth. Thus, to avoid incurring this enormous cost, a national policy has been adopted to create the requisite infrastructure and external economies through public investment in the different development regions. Central planning, regional development and local initiative are now, together, evolving a less centralized pattern of settlement, which may in time bring to the backward rural areas the benefits of industrialization and the advantages of urban ways of life.

As Japan's economic growth continues at high rates, the following kinds of approaches to developing city-regions are being stimulated simultaneously through governmental intervention. Divested of economic jargon they are:

(a) To establish through adequate public investment, industrial and commercial centres and sub-centres in the wider regions of the large, existing metropolises in order to reduce the physical congestion and other problems besetting their central cities. Such a policy would tend to relieve the pressure on the already crowded metropolitan agglomerations of the Keihin and Kansai regions and produce a more dispersed pattern of industrial and residential settlements in their wider regions, to which, for good economic and social reasons, people now migrate.

(b) To create, through heavy public investment, economic opportunities in industry, agriculture and social services in the presently underdeveloped regions in order to deflect the stream of rural migration from the already congested parts of Japan. The purpose of such policy would be, first of all, to stimulate development in areas most prepared for it (thereby relieving pressure on the highly developed regions of Kanto and Kinki) and, at the same time, create a desirable outlet for local initiative and public action.

(c) To reinforce the economies of the smaller industrial and commercial cities and to create opportunities of highly productive agricultural and industrial employment in the rural areas in order to decrease migration. For instance, industrial estates set within clusters of rural settlements may be an effective way of bringing added incomes and physical improvement to the rural areas while, at the same time, providing the necessary services and facilities to medium- and small-scale industries.

(d) To redevelop the great central cities by strengthening their economic base through increased productivity of their industries and, at the same time, improve their physical and social environments without necessarily assuming further physical expansion. In many instances, the structure of industry would have to be improved in the process by diversifying light industries; by introducing secondary industries where they do not exist, using products of heavy industry for further local processing; and, by modernizing obsolete processes in the heavy industry and mechanizing small-scale and medium-scale industries to a higher degree.

The Japanese Government is facilitating the coordination of development activities carried out by different public and private agencies in order to obtain an effective integration of physical and economic planning. A useful and practical method of initiating the process of coordination and integration is the establishment of a register of the physical implications of social and economic policies contained in the national and regional plans in the form of a *national physical plan*. This plan would indicate on maps and statistical and other graphic media, the locations, priorities and timing for:

(a) the major economic development activities, both industrial and agricultural, in Japan's nine development regions;

(b) the national power grid connecting the present and future generating plants using all sources of energy —water, hydro, thermal and nuclear;

(c) the national transport network including the major connections by road, rail and sea as a basis for closer physical integration of the islands of Japan;

(d) the system of harbours on the Inland Sea, the Japan Sea and the Pacific Ocean paying particular attention to coordinated development of harbour facilities in the different bays and opening overland connections between the harbours on both coasts of the country;

(e) the utilities systems and, in view of limited water resources, particularly the systems of water supply for human needs, for the fast growing requirements of industry, for intensive agriculture and for horticulture; and,

(f) the residential communities fully supplied with educational, health and other social and cultural facilities.

Complementing the *national physical plan*, more detailed *regional physical plans* could be established for each of Japan's nine development regions consisting of:

(a) a land-use plan designating the locations and assigning land areas to the different economic, social and cultural functions in the region; and,

(b) a register of land capabilities for agriculture, and geological and soil surveys for industrial and other urban functions in the region.

One important way of stimulating cooperation could be assistance to local and regional agencies in research and planning, and in the evaluation of development programmes. *A technical assistance programme*, adequately staffed and financed, could be jointly managed by the Economic Planning Agency and the Ministries of Construction and of Home Affairs, but would obtain full cooperation of all ministries and agencies concerned with planning and development. Under this programme, technical assistance in comprehensive regional planning, and in local and sectoral development planning could be offered to the public and private entities concerned. Furthermore, pre-investment and economic feasibility studies and social cost/benefit analyses could be carried out under this programme on behalf of regional and local bodies and other interested agencies and groups.

The increasing economic integration of Japan will be reflected in a growing administrative integration within Japan's nine regions. The precise form and timing will vary, of course, from region to region. In some cases, it may involve amalgamation of neighbouring local units, as has already taken place elsewhere, both at the prefectural and municipal levels. In other cases, integration may be expressed simply through the establishment of joint conferences and other cooperative devices. Whatever the form, coordinated planning and development across traditional boundary lines and among the several departments of the central government must be obtained. Such efforts might include:

(a) the creation of joint secretariats, at the regional level, which have the necessary resources to develop and help execute comprehensive programmes for the region;

(b) the provision of more regular and effective coordination of department activity at the highest level within the central government;

(c) the adjustment of the public finance system to provide strong incentives for, and meet requirements for, coordinated planning by individual departments, local governments and private developers; and,

(d) the establishment of research and training centres at the national and regional levels and at institutions of higher learning for both action-oriented research and research designed to develop the science and art of planning in Japan.

Conclusions

A number of general conclusions emerges from the analysis of the current urban crisis, the theories advanced to resolve it, and the ways in which workable practical action is being planned and carried out. A body of doctrine is gradually emerging. With regard to development planning, certain principles pertaining to the city's place in the region and the nation have emerged that should be embodied in any programme of development. They can be stated as follows:

(1) Environmental planning is an essential contributor to the attainment of projected economic and social goals. Economic planning and development, and to a larger degree also social development, have become definite responsibilities of governments. Therefore, environmental planning should take its rightful place in development planning along with economic and social planning, and, for purposes of their implementation, with administrative, legislative and fiscal planning.

(2) In order to facilitate integration of these mainstreams of development, the focus for development planning should be regional. The region, being a link between the nation and the local community, provides a suitable framework for the integration of, and balance among, development projects of national significance and those initiated locally.

(3) Cities are here to stay. They have been carriers of economic and social progress throughout history.

In the developed industrial society they are agents of further economic growth and places for growing population to settle, produce and enjoy the fruits of their labour. They will, therefore, grow in number and size.

With regard to the action to be proposed, careful consideration should be given to the goals to be achieved. According to the objectives, greater or lesser stress can be put upon what are, in fact, various aspects of development policy. While these aspects can also be regarded as alternatives for action, they can be treated in a complementary manner, as was illustrated above in the instance of the Japanese national development programme. Consideration should be given to the consequences of the following aspects of a development programme.

(1) In the great cities of many countries urban services and facilities have reached breaking point and their infrastructure and external economies would now have to be re-created at an exceedingly high cost to allow further unhampered growth.

(2) The benefits of industrialization and of "urban ways" can be brought to the still backward rural regions of a country through a more decentralized, national, infrastructural development policy that combines local initiatives with public investment.

(3) The establishment of industrial and commercial centres and sub-centres in the sphere of influence of existing already-crowded metropolitan regions may reduce the physical congestion and the many social problems besetting their central cities.

(4) The creation, through heavy public investment, of economic opportunities and social amenities in the presently under-developed parts of countries could deflect the stream of rural migration away from their congested regions and induce development in the areas most ready for it.

(5) The reinforcement of the economies of the smaller industrial towns and commercial centres, and the creation of employment in rural areas in the form of industrial estates and, where appropriate, in larger plants, could reduce the economic causes of migration.

(6) The redevelopment of the great central cities, through the strengthening of their economic functions with increased productivity and improved services, could add support to the improvement of their physical and social environment without necessarily resulting in further physical expansion.

Whatever the action proposed, the development programme will be better prepared and executed if due regard is given to the following:

(1) As countries advance technologically, new sources of power and new means of transportation will emerge. In the developing countries urban growth will become more rapid. But in all countries regional planning and development, combined with a judicious investment programme in infrastructure may help in guiding the growth of very large urban agglomerations into patterns providing a suitable distribution and interconnection of functions, settlements and people in city-regions.

(2) There is no universal model, optimal size or pattern for the city-region. Each situation requires its own model, which, while being

peculiar to the local situation, must also avoid too rigid a framework of development. Long-range goals must be continuously scrutinized and redefined on the basis of evaluations of economic and social achievements of current investment programmes, and in the light of results of environmental research and of scientific and technological progress.

(3) The necessary technical and administrative personnel for the planning and executing of all types of development is lacking. Consequently, there is an urgent need for planning-oriented professional training in engineering, social sciences and humanities; training at the intermediate level in different technical and social fields; and, refresher courses and orientation for those who must practice planning but who lack professional preparation. Not least is the need for education of the average citizen to enable him to participate more usefully and in a more informed way in his country's development.

There are considerable gaps in our knowledge, but there is a community of interest between the less industrialized and the more highly developed areas of the world. A concerted research effort, cooperation among nations, and exchange of information, experience and personnel may help to fill the gaps. The planning and development experience of countries at different stages of development leads one to conclude that contemporary science and technology are able to devise the requisite material basis for a vastly greater population, providing that natural, economic and human resources are developed in a planned way. Planning for such development should begin now!

No one pretends that the legitimate desire of the broad masses of the people to improve their conditions of life can be arrested or turned back, or that the rural-urban migration can be reversed. The question is, therefore, how nations can acknowledge in their planning this shift of population and use it as an instrument of progress. In all countries, there is a need to rejuvenate the cities, to return to them the clean air and elements of the natural environment, and to reduce the daily burdens of urban life. To do this we must take full advantage of the new technologies in transport and communications, in communal services and utilities, and in building. Where adequate techniques do not exist, new ones must be developed, not to build bigger and better buildings, but to initiate new departures. There must be a greater preparedness to borrow from other fields. There must be faster public transit connecting the essential functions and activity centres of the city-region, and more convenient modes of transportation. The dream of homes in their present form must be given up, and instead we must accept their contemporary counterparts in taller buildings covering less ground. Such improvements might bring efficiency, beauty and sanity to our cities. Action cannot wait, for in our time cities must progress or nations may perish.

The principal task now confronting the developing countries is to assess the appropriate share of national resources that must be spent for housing and urban development in connection with, and as a basis for, industrialization and agricultural development. In the industrially developed countries, however, the problem seems to reduce itself to a real change of pace for action. The scale of reconstruction and development must become sufficiently large to make a significant impact on their economies instead of being just another social programme under the name of urban renewal and slum clearance. Instead, urban redevelopment and reconstruction should become at this point a major basis for local and regional growth in harmony with general economic and social development in the developed countries.[54] A far-reaching change in outlook upon world problems is taking place. To an extent which might have seemed inconceivable only a short while ago, there has come increasing recognition that all people must somehow manage to live together and share the resources of the earth; that the general impoverishment of any area is a matter of concern to all areas; and, that the technical experience and knowledge acquired by society must somehow be made available to those communities that are, as yet, less advanced and less well-equipped. Indeed, it has been suggested that future historians will look back on the twentieth century not as an age of political conflicts or technical inventions, but as an age in which human society dared to think of the welfare of the whole human race as a practicable objective.[55]

This concept and the unity of purpose on which it must rely, are being tested daily through international cooperation. Humanity must forge the necessary tools for this unprecedented undertaking. For this, a change in attitudes and values concerning development is a prerequisite. For the sake of economy we are still accepting all kinds of restrictions on freedom and violations of human dignity. This attitude may have been justified in an era of scarcity when social development had to be limited to what was required to further economic goals. But can this philosophy be permitted to prevail now when a large part of the world has reached the threshold of plenty? "The masses of mankind from the poorest to the richest are preoccupied above all else with the problem of living with this great revolution that brings the promise and the prospect of the fulfilment of their hopes."[56] There is no reason why we should continue to be guided by an obsolete economic concept of national balances of payments and budgets in the midst of a paralyzing economic imbalance between the rich and the poor nations. Now, the dilemma facing the affluent nations is whether to define the levels of living they desire as a society, as well as the contribution they wish to make toward creating a viable world economy, or to continue to arm and police the world in the vain hope of isolating their lands from human suffering and inevitable social change.

The developed nations must recognize that their own social problems and physical congestion may soon reach the point of no return unless they learn how to use their ever-rising productivity for their own continued social well-being and for the attainment of an economically tranquil world. For this, however, a shift of emphasis must be accepted from a purely economic approach to development policy and planning, to a human one. The yardstick of growth could then be progress in the human condition with regard to income and social and individual welfare, and the ease, comfort and convenience of the physical environment. In fact, the creation

[54]Ernest Weissmann, *Urbanization in the World*, The National Council for Good Cities (New York, ACTION, Inc., 1961).

[55]*Preliminary Report on the World Social Situation* (New York, United Nations, 1953), Sales No. 1952. IV.11, p. 3.

[56]Walter Lippmann, *The Great Revolution*, Lecture series for International Co-operation Year (New York, United Nations, 1965).

of a desirable physical and social environment in the affluent part of the world may well become the major social issue in this century, commanding highest priority and attention by the United Nations and the governments concerned.

The annual world outlay for armaments is about 200 billion dollars in spite of the fact that existing atomic arsenals are already capable of destroying all cities on this planet not once but several hundred times. Five countries account for 85 per cent of this expenditure, and most other countries continue spending considerable shares of their revenues for armaments at the expense of development. Recent economic studies have explored the question of the possible effects of disarmament on employment. Some of these studies concluded that re-allocation to other activities of the resources now used for armaments would result in a loss of employment of about 20 per cent. According to current trends in industrial technology, including automation, this assumption may be quite correct. However, if the resources thus released were to be channelled into large-scale urban reconstruction and regional development as a basis for further economic growth, and for social improvement generally, an actual increase in employment could be obtained. Dollar for dollar spent for housing, for urban and regional development, and for the building of new towns, significantly more labour must be used than in any other industry. Urban reconstruction on a sufficiently large scale has now become the most essential element in enabling the affluent highly industrialized nations to continue progressing. Not only is it necessary to ensure economic efficiency and to improve the quality of urban life, but public action in this field, which was traditionally regarded as an unproductive expenditure necessitated by humanitarian considerations, is now recognized to be a potentially important means of obtaining and maintaining full employment. Such programmes can absorb any unemployment due to demographic growth, a decrease in employment in other sectors of production (such as agriculture or armaments), or major technological changes (such as the more general application of automation in industry and services).[57]

Some of the equipment, installations and institutions, most of the production facilities, and all of the funds and skills now used to maintain the military establishments throughout the world and carry out research, development and production of armaments, are quite suitable for the urgent job of urban reconstruction and development. Little or no conversion would be needed to make these resources useful directly.

Indeed a very small section of these tremendous *public* resources would produce a comparatively large international development fund. It is questionable as to whether nations would really feel less secure if their "over-kill" factor of more than a hundred were to be cut by 2 or 5 or even 10 per cent. Yet, resources of this magnitude would really begin to count in any United Nations sponsored war on the recognized economic and social causes of tension in a world divided into affluent and poor. As a first step in planning the strategy of this war, those metropolitan areas represented in the Centennial Study and Training Programme on Metropolitan Problems could well launch a plan to establish a comprehensive system of truly interdisciplinary centres for research and training in

---

[57]Ernest Weissmann, *Urbanization in the World*, The National Council for Good Cities (New York, ACTION, Inc., 1961).

environmental development as an aid to action in this crucial area of development. Some of the metropolitan areas already have such centres. Others may be at the point of creating them. But all those who have collaborated in this study and training programme must have recognized the concrete value that a continued international exchange of experiences, data and research would have for improving everyone's lot.

14

Governing the Metropolis: Commentary on World Opinion

Simon R. Miles

This chapter, like Chapter 8, provides the reader with a worldwide evaluation of the proposals raised by the authors in the preceding chapters. The reactions of the study groups, whether adverse or supportive, provide positive directives to those seeking improvements in the working of the governmental machinery of the metropolis.

The Functional Metropolis and Systems of Government

THE FORCES OR CONSEQUENCES OF PHYSICAL CONCENTRATION AND POLITICO-ADMINISTRATIVE CENTRALIZATION

Gorynski and Rybicki, in their essay which appears in this volume, are attempting to provide guidelines in the search for a metropolitan form that will be satisfying both as an environment and as an efficient and democratic politico-administrative system. Perhaps their main point is that although, in the past, physical, economic and even social planners have given little consideration to the implications of physical form and economic and social activities for the politico-administrative structures that have to be devised to govern them, there is no reason why this, in fact, should not be done. Thus, the main theme propounded by Gorynski and Rybicki emerges: that in developing the physical form of an existing or future metropolitan area and in providing for all the functions it is to perform, equal consideration must be given to the development of an efficient and democratic system of government, so vitally affected by the physical and functional metropolis.

This thesis is closely related to that introduced by Dupré in Chapter 11. The issue can be reduced, in essence, to one arising from a desire to maximize the efficiencies of technological development and, at the same time, retain the values of democratic control of society. Reconciliation of these two divergent interests is problematical to say the least.

At the seminar-conference Gorynski and Rybicki further analyzed the nature of this problem by elaborating upon the effect of technological developments on planning in terms of time and space. In terms of time, with the speed of development today and the resulting change, we are becoming more future-generation-oriented. In terms of space, the individual is part of a community of a larger scale. The long-term and widespread implications of decisions taken today demand a greater degree of coordination and centralization of decision-making. Furthermore, the conflict of future-oriented and present policies is felt more at the local level than elsewhere, creating a tendency to shift responsibility for resolving the conflict away to a less sensitive level of central government (less sensitive, that is, in terms of access). Given the nature of urbanization and the increasing concentration of human activity in large urban areas, there is also a tendency for what were once regarded as local services to be provided regionally and for regional services to become those of the central government. Technical services, especially, become more centralized. This centralization results in a severe loss of interest at the local level, where, as Gorynski and Rybicki maintain, only the educational, basic housekeeping and some socio-cultural activities remain.

There was general agreement that little could be done about the consequences of centralization of decision-making arising from the need for long-term planning. Indeed, the absence of this centralization and coordination is one of the great problems confronting metropolitan areas in many countries. More debatable is the question, what physical pattern of urbanization across a national territory, and urban form within any urban area, is desirable.

There was recognition that the "basic" functions, as Gorynski and Rybicki call them, that are provided over large areas have to be operated or guided from certain geographical points. The interrelated nature of these functions makes the close geographical location of their "nerve centres" advantageous in terms of coordination. Hence their concentration in a few urban areas. This concentration is enhanced by the availability of the "local" services, and the greater the concentration, the more are local services available. Thus, the economies of scale appear to increase with the degree of concentration. Unfortunately, very little research has been done on the subject of costs and benefits of concentration that can be regarded as conclusive. Most studies completed so far have focussed on the physical aspects of the costs of settlements of different sizes; few have focussed on the administrative costs. On the social costs side, work is only now being started that will offer a system for social cost accounting in social, as opposed to economic, terms.

The absence of these yardsticks forces heavy reliance upon the unquantified and more emotionally-based free expression of opinion about urbanization trends in the development of urbanization policy. For example, the existing aversion, experienced by many persons, to metropolitan growth may be largely emotional, based on an unwillingness to accept change and a desire to escape complex problems. For others, the sought-after shift from identification with a territorial group to identification with various non-territorial groups is made possible with the improved communication to be found in most metropolitan areas.

This illustration raises the question as to whether people should make the effort to comprehend a larger territorial community through their identification with various non-territorial groups. This is, however, a value-laden question. There may be nothing intrinsically wrong with "progress". It may be enjoyable for people to comprehend the non-territorial "interest community" in which they live, and it may be possible for most people, especially younger persons, to do this. However, while some functions performed by the metropolis substantiate this division of the metropolitan population by interests (as indicated, for example, by the use of social and cultural facilities), other functions, such as the provision of utilities, have to be performed for a territorial community. If the citizen is to be interested in the provision of these services to his community, then, presumably, it is necessary that he comprehend this territorial community. This may be regarded as a very laudable objective but, given the continuing growth of metropolitan areas, the value-laden question is not so easily answered. For the question also asks whether people want to be in a position where it is expected of them to comprehend a territorial community of, say, 19 million people which may emerge from the uncontrolled concentration of interest communities. It is likely that before this degree of concentration is reached, even the enthusiasts of metropolitan growth will qualify their respective positions.

The Tokyo metropolitan region already had a population of about 19 million in 1967, and if present trends continue, this is expected to increase to 33 million by 1985. Almost all study groups mentioned the continuing concentration of population as cause for alarm. The largest metropolitan areas in any one country are not necessarily growing faster than other urban centres. Paris, for example, ranks only twenty-third in France by rate of growth. But the absolute numbers involved and, more important, the continuing consolidation of the monocentric metropolis are cause for concern. Further, the general trend is for more of the growth to be comprised of tertiary sector activities, which usually prefer central locations. For example, in Tokyo one finds 10 per cent of the Japanese production and consumption, 40 per cent of economic activities and 60 per cent of the Japanese leaders in business and intellectual fields and in government services. Whether this concentration will ever cease is questionable. Athens comprised just over 10 per cent of the population of Greece in 1920 and 22 per cent in 1961. Copenhagen already contains about one quarter of the population of Denmark. A third of Argentina's population is found in the Buenos Aires metropolitan area. The study groups felt that this trend needs to be counteracted, not only in the interests of the nation and the metropolis, but, what is more, in the interest of the individual.

The continuing concentration of population and activities in metropolitan areas is largely the result of the in-migration of the rural and small town workers. This migration takes place mainly because of the disparities of wealth and access to services across the nation. Further disparities that exist within an urban area, especially in a large metropolitan area, are appreciable enough to give rise to the same effects that are taking place on a national scale.

Within the metropolitan area there is a continuing shift in the employment structure. As the metropolis grows and as the nation becomes more developed technologically, the tertiary sector increases. This shift in employment structure not only means that new skills are developed and people change jobs, but also that new jobs develop and that land-use patterns change in terms of employment structure. The general preference of the tertiary sector for a central location results in strong competition for land in the central area. The less this competition is controlled the higher the land prices rise. The inward movement of the service sector results in a displacement of secondary sector activities and of persons, rich and poor, that have been occupying these central areas. The secondary sector activities tend to move out to a point of equilibrium between efficiency and convenience of operation. This in turn tends to result in a greater separation of land uses by activity. The persons displaced from more central areas join the newly-arrived in-migrants in a hunt for a place to live. Their own outward movement tends to drive up the price of suburban land and can result in some workers living two hours away from their place of employment. While this is a much simplified description of the change and internal movement that takes place within the metropolitan area it is nevertheless representative of actuality, even in countries with a greater control over the land use than is exercised in market economies.

The rise in land values is one of the greatest problems facing many metropolitan areas. In Tokyo, land prices have risen tenfold in the last ten years. In Montreal the lack of planning control and the holding of land for speculation in suburban areas have given rise to some costly suburban sprawl of extremely poor quality. To some extent the separation of land uses by activity is desirable, and in some metropolitan areas, such as Nagoya, the problem is one of major proportions. Yet the well-planned mixing of residential and employment areas should be encouraged where possible if an attempt is to be made to reduce the length of the home-to-work trip.

Indeed, the problem is not just one of disparity in access to employment but also in access to institutions. For as was pointed out by the Madrid group in the discussions of health and welfare services, these institutions do not necessarily follow the movements of people. Unfortunately, it is often in those areas where the outward displacement of population is extremely heavy that there are inadequate intra-metropolitan transportation facilities to compensate for the physical removal of the population from these service institutions. In some metropolitan areas the situation may be beyond rescuing. In Paris, for example, two thirds of the metropolitan population is residing in the suburbs at present, and with this proportion increasing there seems to be little hope of improving the access to jobs and services in the central area.

Apart from a certain reluctance of many workers to live in the central area and the problems of satisfying the household in which there are two or more workers, the real problem in reducing the time-distance between home and work arises from the lack of social overhead capital available for investment. Both the Athens and the Tokyo study groups commented on the significance of this limited availability of social overhead capital. It was the existence of this resource in the older cities of western Europe that made them attractive as sites for early industrial development. Although the imbalance is now obvious, the situation is not as critical as that found in Tokyo, where industrialization took place without adequate social overhead investment. Tokyo now requires about four and a half times more social overhead capital invested than presently exists. Roads suffer most of all, although housing and sewers also require much higher levels of investment. Parks and railways do not fare quite so badly, but the situation is still by no means healthy. Given the variation in fiscal resources of municipalities within most metropolitan areas, intra-metropolitan disparities will exist in social overhead investments. This, coupled with the degree to which certain service institutions are centralized, will produce further disparities in the quality of service available.

With the growth of the metropolis and the accompanying trend toward the transfer of decision-making responsibility from local governments to more senior governments, the centres of decision-making become more removed from the citizen. The compromise of deconcentrated administrative offices is a partial solution, but, as has been made clear by Gorynski and Rybicki, with the physical separation of home and work the typical commuter has little interest in the affairs of either community. This is one more reason why an attempt should be made to reduce the size of the everyday "activity community" to one to which people can relate politically and socially.

NEED FOR NATIONAL FRAMEWORK

From the above there begins to emerge a mosaic of disparities on a national scale which the rural migrant seeks to avoid by going to the metropolis. Despite this rather pessimistic picture the study groups and seminar participants did not see the situation as impossible. The task ahead is to eliminate these disparities as far as possible and, at the same time, permit both a maximization of technological developments and the democratic control of the environment.

As the metropolis is a product of concentration on a national scale, any attempts to tackle the problems of the metropolis must involve the national government. National and provincial governments do not lack the means to counteract the imbalance created by the growth of one or two metropolitan areas. What is required is the will to draw up and enact necessary legislation. Yet, the lack of effective national planning was noted by most study groups. In France, ever since the breaking up of the provinces in the late nineteenth century, the government policy of political and administrative centralization has been, at the same time, a policy of "de-regionalization". The influence of this traditional attitude has been hard to break and has had its effect in retarding the development of a governmental body for the Paris region. Only with the obvious emergence of Paris as an undeniable regional fact has there been some political recognition of this. However, although the *District de la Région de Paris* was established in 1961, and even though legislative changes have been made in 1964 and 1966, the authority is still seen to represent more a regional expression of central government policy rather than an expression of the interests of the region.

It is only at the national level that many decisions regarding the location of basic functions can be made. The problems confronting national governments in regard to the development of a national policy on urbanization are often international. The effect of the establishment of the Common Market on urbanization in France has been to attract many new industrial activities to eastern France. The problem of concentration is not restricted to the policy between Paris and the provinces, but can be seen more and more in the disparity between eastern and western France.

Similar problems have affected Naples. Ever since the unification of Italy and the creation of the national capital in Rome, the South has lost industry to the North. For Italy, just as for France, the emergence of the Common Market has affected the pattern of urbanization; in this case through intensive urban development in the North. Another implication of the establishment of an international economic union, such as the Common Market, is that member countries will obviously have an interest in the healthy economic development of their trading partners. Although this was not pursued at the seminar it is a point that will probably emerge as having considerable significance for the future.

There was general agreement that the national governments must provide a national policy on urbanization that will serve as a framework within which the more detailed development of metropolitan areas can take place. Such a national policy must be realistic, however, in its proposals for deconcentration of urban growth. Deconcentration is expensive, and new towns tend to be the most expensive form of all the alternatives. Where possible, the expansion of existing towns should be encouraged if savings on social overhead capital investments are to be realized.

National plans must be comprehensive in their approach. The Naples group was critical of the Italian plan, which is only an economic plan and whose regional expression does not give sufficient recognition to metropolitan regions. The Montreal group described a newly-emerged system of regional planning in the Province of Quebec, which is already suffering from a multitude of regional bodies that each contribute to the regional plans. Also, the lack of economic or social development planning by the Canadian government makes the planning of Montreal that much more difficult. Where the federal government is responsible for certain planning activities in the metropolitan area, as with the Montreal International Airport, it seems that there is much room for improved coordination between the levels of government involved. The group estimated that approximately 20 per cent of the population of the Montreal metropolitan area now suffers unduly from aircraft noise.

There appeared to be little support for any more senior government involvement in the detailed planning and development of metropolitan areas other than is absolutely necessary for the performance of functions that senior governments alone are best suited to assume. While lack of available resources, especially skilled personnel, may make such involvement necessary, it is seen undesirable. Also, judging by the experience of Tokyo, there seems to be little to be gained from it. There, the Capital Region Development Commission, which is an organ of the Japanese Government, is charged with the responsibility of producing a master plan for the Tokyo region. The only plan it has is an assembly of plans of the local authorities and of the separate central government ministries serving the area. These are not as successfully integrated as they should be if the Commission is to justify its existence. Thus far the Commission seems to have provided little leadership to Tokyo's planners. Although the Tokyo Metropolitan Government has its own ten-year plan, it does not have jurisdiction over the entire metropolitan area. While the Capital Region Development Commission continues to function, it is unlikely that the Tokyo Metropolitan Government will be able to extend its jurisdiction, for planning or other purposes, over a larger area.

PHYSICAL AND POLITICO-ADMINISTRATIVE FORM OF THE METROPOLIS

At this juncture we should consider the structure of government for the metropolitan area. As further comments are made later in this chapter on public participation, metropolitan area finance, intergovernmental relations, and the coordination of planning and development, these subjects will be referred to now only in so far as they affect the structure of governmental machinery and the physical form of the metropolis.

There was strong support from the programme participants for a metropolitan level of government with multi-purpose responsibility. Although this form of government is not widespread, it is becoming more common. As has been recognized already, many functions have to be performed over the entire metropolitan area. The delimitation of this metropolitan area is best done by taking the commuter zone as an indication of the extent of possible jurisdiction of a metropolitan region government. Obviously this is a compromise, but some basis has to be established. This matter is discussed in greater detail in Part IV of this book. Here it will suffice to say that the commuter zone will permit the inclusion of fringe areas undergoing development. As this is likely to produce a large jurisdiction, it may well be accommodated as one regional government in a national system of regional governments. The boundaries of these governments should be quite flexible. Hopefully, with the direction of urban growth by a central or provincial government, imbalance in terms of

growth and wealth on either side of the boundary will be reduced, and there will be less need to change the boundaries.

To combat the effect of centralization of responsibility for metropolitan-wide services, the metropolitan level government should assume very few services for which it alone is responsible. The division of responsibility between this level of government and smaller units at the local level should be on a "wholesale-retail" basis rather than according to functions. In this way the metropolitan level of government would set the framework of operation for the local levels of government. This position would appear to be in agreement with the conclusions Dupré draws from John Stuart Mill in Chapter 11.

The reform of local units of government must be related to the development of the new community centres that emerge through a policy of intra-metropolitan deconcentration of central area facilities. The functional metropolitan area is attained only with the resolution of problems confronting the distribution of goods and services within the area. The Naples group saw this desired state as an "equipotential continuous system" within which all services are equally available to all points. While there has to be some modification of this as a theoretical concept, it does serve to express the desired goal. Indeed, it is this concept that underlies the 1965 plan for Paris (*Schéma directeur d'aménagement et d'urbanisme de la région de Paris*). The Paris proposal is for the development of about eight centres in what is estimated to be a metropolitan area of about 15 million by the year 2000. As each of these centres would provide services for a territorial community of about 2 million people, it is likely that seven or eight units of local government, of about 250,000 to 300,000 population each, would administer governmental services in each sub-regional community. However, the continuing existence of the 1,305 communes in the Paris region in addition to the eight *départements*, and the many *ad hoc* agencies that report directly to separate ministries of the French government, was seen by the Paris group to undermine the regional image that has to be developed for the metropolitan area. Further governmental reform must accompany the proposals for physical re-development if Paris is to achieve the sought-after sense of community and encourage the participation of citizens in the affairs of the metropolis.

A recent example of decentralization of authority in a large metropolis is provided by Tokyo, where some aspects of land-use planning and health and welfare have been developed for the twenty-three wards that make up the City of Tokyo. However, the physical relocation of the tertiary sector services located in the core of the metropolis must accompany any decentralization of authority that takes place. A comprehensive plan for the Tokyo metropolitan area is sorely needed, and it is this, plus the provision of an informal system to encourage citizen participation in the development of the metropolis, that has prompted the establishment, by Dr. Masamichi Royama, of the new Tokyo Region Comprehensive Planning Association.

Naples suffers from very similar problems. While sufficient government coordination may exist at the level of the Italian government, in the metropolitan area there are many planning bodies that lack a coordinating authority. Also, as these bodies are, for the most part, neither politically responsible nor answerable to the electorate, there is little interest in encouraging public participation in the planning process. Like many others, the Naples group favoured the introduction of a metropolitan authority, in addition to the existing local authorities, that could lend direction to planning in the region.

From the above discussion it is clear that the retention of the tier of smaller units in the two-tier system is favoured largely because of the desire to encourage public involvement in the affairs of their community. This will require that these lower-tier authorities be given sufficient responsibility to make them viable governmental bodies. They should also be given a fair measure of autonomy in spending even if the bulk of their revenue is derived from senior level grants. If the size of local units of government can be held at about 250,000 in population, it is justifiable to operate on the assumption that these units can provide effective multi-purpose government and that senior governments, including the metropolitan-wide tier, would be asked to perform only those tasks that the local governments are not well-equipped to perform. At the present time there is too strong a tendency to regard local government units as existing primarily to complement the senior governments. Understandably, until local government is reformed structurally and functionally this attitude is likely to persist.

In introducing the metropolitan-wide tier, primarily on the grounds of efficiency, one question that naturally arises is how participation is to be encouraged at this level of government. It is this question that provides the focus of the following examination of the metropolitan political process and administrative systems, based on the reactions of study groups and seminar participants to Smallwood's treatment of that subject.

The Political Process and Administrative Systems

THE SIGNIFICANCE OF CONSENSUAL GOALS FOR THE METROPOLIS

In Chapter 10, Smallwood takes as given

that, in a metropolitan area, certain parts of the three governmental functions which he discusses—service, political and redistributional—should be performed on a metropolitan-wide basis, if they are to be performed fairly as well as efficiently. Problems associated with air pollution, water supply, sewerage and transportation are no respecters of political boundaries and may require appropriate governmental authorities with jurisdictions at least metropolitan-wide in extent and in some cases larger. Given the importance of the participation of citizens in the decision-making process, efficiency is not, in itself, a sufficient basis on which to justify the regionalization of any given public function. It is more accurate to state that metropolitanization of government functions may be desirable only when technological and geographical conditions so require, when local decisions would have adverse consequences for the metropolitan region, and/or when administrative benefits are recognized and the accessibility of citizens at large is improved. Any one of these criteria may be justification enough for adopting a metropolitan approach.

Various types of metropolitan approaches include cooperative arrangements between units of local government, single-purpose districts, multi-purpose districts and metropolitan government. Each type of approach has particular advantages and disadvantages, and yet, although it is fairly easy to list the characteristics for each form in an arbitrary manner, the selection of one form over another is generally the result of existing cultural, political and social considerations rather than either efficiency *per se* or democratic control *per se*.

Since efficiency has been considered at some length in other parts of this study, the major concern here is that whatever the governmental system chosen, provision should be made for the various forms of participation necessary for the development of consensual goals pertaining to functions performed on a metropolitan-wide basis. This is just as important at this level as it is at the local, national and international levels of decision-making. It must be accepted that in some metropolitan areas it may be many years before a state is reached at which consensual goals can be established on a metropolitan-wide basis. The existing fragmentation of interests in metropolitan Manila, for example, is not only of a functional and jurisdictional nature, but is also heightened by the social settlement pattern of many "urban villages". At the present time, these urban villages are the basis of a local politician's support. The politician who wins at the polls is likely to be the one who promises most for the village in the way of jobs for the local leaders and local services such as roads, playgrounds, toilets and so on. It is this type of politics, manifested by parochialism and a help-us-now attitude, which is rife in such places as Manila and the biggest obstacle in the way of achieving "consensual goals". To the extent that a metropolis such as Manila does not have a coherent political system, these goals will not be realized on a metropolitan basis. Services will continue to be provided in a manner which reflects the greater power of some communities as opposed to others rather than the distribution of need.

If "consensual goals" are to be sought, more attention must first be given to the elements of the political process, such as interest aggregation, bargaining, compromise and accommodation and how these might take place on a metropolitan-wide level. Political parties can make a significant contribution to the processes of interest aggregation, but without a metropolitan jurisdictional base they are unlikely to pay particular attention to the development of consensual goals at this level. So long as the local authorities staunchly defend their autonomy and the national government is unwilling to impose a metropolitan authority on the metropolitan area, a metropolitan jurisdiction is unlikely to emerge. These attitudes will prevail until there is a greater awareness of the needs of the metropolitan area *per se*. The only way to bring about this awareness is through the process of political socialization. However, before discussing this and other factors that lead to a greater degree of citizen interest and participation in the affairs of the metropolitan community, and hence to the development of truly consensual goals, we will take a brief look at different types of citizen participation and at some attempts that have been made to account for variations in its form and intensity.

CITIZEN PARTICIPATION

Two major types of citizen participation can be identified. The type discussed by Smallwood is based upon citizen initiative and includes voting at elections, the use of the press as a medium for public expression, the activities of pressure groups and political parties, and the exercising of recall of political representatives. Voting in referenda might also be included in this listing, although the extent to which the referendum is used is largely dependent on government initiative. The second type, which was given more discussion by study groups and seminar participants, is that based upon governmental initiative and includes the use of advisory committees, councils, boards, public hearings, the ombudsman and consultative services.

Factors Affecting Voter Turn-Out

Accounting for varying voter turn-outs in elections was a subject of great interest to study groups and seminar participants. In light of the recent changes in local government in the London area, the voting patterns there are particularly interesting as they provide a comparison of the effect of governmental forms within the same setting, as opposed to a comparison between countries. Despite the reduction in the number of local boroughs (and hence, expansion in their average size) and the territorial expansion of the metropolitan-wide authority that took place as a result of the 1963 London Government Act, there was no significant change in voter turn-out at either level (between 30 per cent and 40 per cent). Thus, while Smallwood may be correct in his observation that elections for councils of large-scale metropolitan governments have low turn-outs, it is not necessarily true that the further expansion of these governmental jurisdictions will again lower the turn-out in future elections. Similarly, Smallwood's suggestion that small-scale units tend to have a negative influence on voter turn-out, can be neither supported nor rejected on the experiences of London before and after the structural reform that was introduced on April 1, 1965. The evidence from London does, in fact, lend support to Smallwood's conclusion that the size of the political arena is probably less significant in influencing voter turn-out than are such factors as the widespread feelings about access and identity, the stakes involved and the influence of technology. Indeed, these conclusions are also supported by the findings of the study groups working in Manila, Vancouver and Ibadan.

For example, the intensity of campaigning has been noted in several areas as a significant factor in getting out the vote. Smallwood referred to the exceptionally high turn-out in the New York City elections of 1965 as a result of John Lindsay's campaign. It is probably not too misleading to account for the intensity of this campaign by evaluating the personal stakes involved, that is, John Lindsay's political future. In London, the 1964 Greater London Council (G.L.C.) election turn-out was 44.2 per cent, as compared with 41.2 per cent for the comparable 1967 election. The 1964 campaign was particularly hard fought not simply because control of the G.L.C. was seen as a political prize but also because the new form of government to be introduced in 1965 created an unknown situation for those responsible for running the party electoral machines. Furthermore, voter turn-outs in the 1964 Borough Council elections were much higher in the suburban middle-income, generally owner-occupied, residential boroughs (40.8 per cent), than in the inner-area, lower-income, predominantly rental-occupied boroughs (27.1 per cent). In this case the effect of the degree of vested interest of the electorate is quite clear.

Although the general conclusion of the Manila group was that small political units encourage participation, this is undoubtedly based on the influence of the small-scale political base described earlier. Indeed, the additional personal advantages involved in being a politician or having political contacts in Manila probably play a strong part in accounting for the active role of political parties and interest groups in getting out what tends to be a large percentage of the electorate throughout the Manila area. In Ibadan the converse situation, whereby campaign intensity is low and seats go uncontested, results in a low turn-out.

However, the lack of a contest for seats may be due not only to the lower prestige of being a politician but also to the lesser responsibility given the level of government in question. The Vancouver group saw this latter reason as accounting for the lack of interest in local government elections not just in Vancouver, but across Canada in general. The great variation in voter turn-out from one year to another, in the Vancouver area, tends to point to an undue influence of local issues of temporal importance on the voting turn-out. For example, in 1960 and 1961 the turn-out for the three largest municipalities in the area were as follows: Vancouver (35.7 per cent and 24.5 per cent); Burnaby (26.9 per cent and 15.6 per cent); and Surrey (15.6 per cent and 27.5 per cent). For the same years, the turn-outs in Port Moody, the smallest municipality, were 37.0 per cent and 36.0 per cent. As a point of interest, it is questionable whether size of community alone is a very significant factor in influencing turn-out.

Before leaving this discussion of voter turn-outs, it should also be noted that election procedure can affect turn-out figures considerably, and it is necessary to bear such additional factors in mind when making any judgement of citizen participation based on voter turn-out figures. Most of these factors have, however, already been mentioned by Smallwood in Chapter 10. Registration may or may not be compulsory, whether or not one intends to vote. The registration may be done by government officials or the onus may be on the citizen to register himself. Similarly, voting may or may not be compulsory, The possible combinations of these variables alone make comparison difficult.

Additional factors relating to electoral procedure that may affect citizen interest in voting include: the form of the constituency (whether it is a single-member or multi-member constituency); the form of the ballot (the number of people listed on the ballot and how they are identified); the qualifications required for eligibility as a voter (apart from age, qualifications pertaining to literacy and property ownership are common); and, the timing of elections relative to one another. Each of these factors has some effect on voter turn-out, although it is not necessarily easy to determine how significant one factor is relative to another. To the extent that improvements can be made in existing conditions, these factors will be commented upon later when we turn to the ways of encouraging citizen participation in the political process.

Political Parties and Participation

Judging by the reactions of the study groups and the seminar participants, political parties are likely to make a positive contribution toward the development of consensual goals for the metropolitan community. The *raison d'être* of a political party is the furtherance of a belief in some form of policy. The more widespread the support for a party, the more likely is the establishment of consensual goals. At least this is so in theory. In practice, it is questionable whether the party structure really encourages the involvement of citizens in the development of party policy or even whether it stimulates citizen interest in public affairs and the expression of this interest through voter participation and other actions. To the extent that it does not accomplish the latter objectives, it does not really further the development of consensual goals. The widespread mistrust of the party system is based largely on the feeling that party policies, although voted for by the public, are imposed by a party bureaucracy and consequently represent the goals of very few members of society. It should be made quite clear, however, that the situation is not necessarily improved by the replacement of the party system with independent candidates. A city council of twenty independent candidates is no more likely to represent all the interests to be found in the city, whether functional or territorial, than is a council run on party lines.

What the party system does offer, however, is the mechanism to aggregate the various interests that do exist. If this mechanism is well used it can be most successful in developing consensual goals. This process involves activities both inside the party machinery in order to provide for coordinated development and also outside the party machinery in order to reach the public for purposes of receiving suggestions, for providing explanation of policy and information, and for encouraging participation of various kinds. The London group felt that the parties operating in London tend to be far more successful at internal coordination than at external relations. The strong reliance of British local government on the committee system is facilitated by the policy coordination provided by the party system. This is extremely beneficial in a large metropolitan jurisdiction where the work-load of the council is very heavy and efficiencies obtained in the decision-making process acquire added value. The Manila group put particular emphasis on the role of parties in settling internal conflicts. These conflicts, without a party system, could cause much delay in council procedure or administrative operation. The parties, and especially the opposition party, are never anxious to have internal rows become public and try to settle these disputes quickly.

The salutory role of the press is implicit here. Indeed, political parties often find it advantageous to have their own newspaper. In Ibadan, for example, both major parties have their own newspapers, which circulate widely in the metropolitan area and which contribute to the process of keeping the public informed on local political issues. It is the lack of a newspaper devoted solely to the London area that was pointed out by the London group as partial explanation for the dismal lack of knowledge that the average Londoner has of his local government. A survey carried out in June, 1966, on behalf of the G.L.C., for purposes of testing Londoners' knowledge of and reaction to the G.L.C. and the reorganized system of government, revealed that one third of those questioned did not know what the letters G.L.C. stand for; one in four did not know in which London borough he lived; and more than 80 per cent did not realize that more than one local authority receives their money. On the other hand over 70 per cent indicated that they would welcome more information on these and related matters. While the existence of parties is likely to encourage the provision of information to the public, this can not be regarded as one of the more important functions of the party machine itself, as the information is likely to be somewhat partisan in character.

Some of the other advantages of party involvement in metropolitan government are that parties provide accountability to the electorate; they provide a mechanism by which popular support for a measure can be tested quickly; they provide the co-operation necessary for the development of comprehensive programmes and at the same time allow individual members to specialize in particular fields of interest; in opposition they provide more organized debate of issues; and, they provide a broader base from which to recruit candidates for public office.

At times, the existence of the party system at the local government level can be disadvantageous. The problems of relations with senior levels of government are heightened if the parties in power at the different levels are themselves different.

The other obvious objection to parties is that the dislike of running under a party label may deter some otherwise very well qualified individuals from running for office. It might be questioned, however, to what extent these people, as independents, would be willing to settle upon any council policy if elected.

Although there are many more arguments both for and against the involvement of parties in local government politics, the major ones relating to the development of consensual goals have been given above. In balance, the involvement of parties appears to be favoured.

The Role of Interest Groups
Apart from the more formal expressions of citizen-initiated participation that voting and political party activities offer, there are also many types of interest groups that provide a system of checks and balances between community interests and governmental actions or lack of action. For these purposes interest groups are of great value. It can be argued that they provide the decision-makers with a measure of community opinion on issues, thus facilitating the development of policy. This interpretation can, however, be extremely misleading. The activities of one interest group are very often not countered by those of another group representing what may be an opposite viewpoint. This situation arises quite frequently when an opinion is expressed against a government policy. Interest groups organized to support government policy, and thus to counteract the activities of the other interest groups, are rarely received with much enthusiasm. This is largely because the populace believes that the government will make the decision in any event and is quite capable of pursuing its own policy if it so wishes. Yet, in such a situation, there is no true

reflection of the balance of views held throughout a community. This type of situation is of greatest significance if pressure is being applied on the government "behind the scenes", as is so often the case. Thus to understand citizen participation in terms of the civic efforts of "high minded citizens" working for the general welfare may be too restrictive a view. Citizen participation may well be used by vested interest groups operating behind the cloak of the general welfare.

To the extent that interest groups are designed to further specific interests of a community, whether territorial or interest in base, they are subject to the above criticism.

However, having pointed out the possible dangers inherent in interest group activity in general, it is still necessary to recognize that certain types of interest groups are of great value to society as a whole. It is important, therefore, to differentiate between the types of interest group activity.

One major differentiation can be made between the temporary and the permanent groups. A further distinction can be made by identifying those that have a territorial interest as opposed to those having a functional interest. The most effective type of interest group, measured in terms of effort exerted for results obtained, tends to be the temporary group organized to express interests pertaining to a specific territory. The example of the suburban group pressuring for improved services is extremely common in many metropolitan areas. Similarly, groups organized to push for or against renewal of an area are often very influential in the making of relevant decisions. These groups are easy to organize, given the territorial base, and often have a very visible goal.

Temporary groups organized to defend a specific functional interest also tend to have a successful record, although they may be harder to organize. An example of this type is that cited by the London group of a committee formed to fight the closure of a suburban railway line.

Some causes tend to bring territorial and functional interests together (resistance to airport expansion is a good example). Many metropolitan areas are now passing through the stage at which their original airports are ceasing to be adequate for the increasing traffic. Plans for expansion of existing airports never fail to raise the ire of local inhabitants and very often that of the entire metropolitan community. The dramatic demonstrations in 1969 in Tokyo over the expansion of Haneda airport provide an excellent example of this. The fight against the proposal for the location of a third London airport at Stansted eventually became a national issue and led to a change of government policy which, it can be fairly said, reflected the consensus of national opinion. These types of activities, which are temporary and "out-in-the-open", tend to be fairly democratic in nature. They are generally newsworthy and therefore come to the attention of the public and by being discussed in this way afford the decision-makers an opportunity to judge public opinion on the issue in question.

Of the permanent interest groups, of which there are a great number in the metropolis due to the concentration of headquarters of national institutions, there are two main types: those designed to defend interests that might in some countries be regarded as a governmental responsibility and those designed to further purely vested interests. In some instances (unions for example), there may be a mixture of the two motivations. The permanent interest group may have an organized secretariat and staff and be able to offer advice to governments on a continuing basis. Some organizations may be of a general service nature such as the Lions Club, and it may be possible for them to develop a metropolitan approach in their activities. These groups tend to obtain less publicity, which makes their good work more difficult.

For the vested interest groups, which may or may not be organized in a formal manner, lack of publicity may make life easier! The Manila study group cited the builders, realtors and the Chinese community as examples of functionally oriented groups working behind the scenes to further their separate causes, apparently with much success. It is these types of groups, more than the permanent ratepayer-type organizations, operating in the open, that require most careful observation.

The extent to which any of these interest groups makes a contribution to the metropolitan community through their individual pursuits depends almost entirely on the ability of the governmental authorities to judge the merits of their respective demands. To the extent that interest group actions result in the maldistribution of services relative to actual need, the government has failed in performing its political function of conflict-resolution and its socio-economic function of the redistribution of goods.

Government Initiated Participation

Thus far we have been discussing citizen-initiated participation. In many instances, however, the government can, and should, take the initiative in encouraging the participation of citizens in the development of policy. By inviting opinions and submissions on issues, governmental bodies can channel participatory efforts into areas where they are most needed at any one time. If this becomes an accepted practice, the public, in turn, is less likely to have to organize itself in order to protect various interests. To safeguard against individual politicians or administrators choosing to consult the public only at such times as it suits them, it is desirable to formalize these links with the public. This, however, must be done in such a way that open communication is assured, as opposed to closed consultation with what may well appear to be another arm of the government bureaucracy. Information that is not strictly confidential should be made readily available to the public. The onus should be on the government to justify the classification of information that it wishes to withhold as confidential. Too often, members of the public are told that information they are seeking is either not available or confidential, although no explanation is given as to why this is so.

As citizen participation requires a focus of interest, it is unlikely that people will identify with the metropolitan community unless there are representative symbols or focal points. This provision of the symbols is largely the responsibility of the government. The organization of common sports activities, social activities and festivals, and the erection of buildings as focal points of interest, all help to develop a metropolitan image. Again, the press can be very influential in this regard. However, without a metropolitan authority these types of activities may be not only difficult to organize, they may be unwanted.

SENIOR GOVERNMENT CONSTRAINTS ON THE METROPOLIS

It is at this point that another input, in addition to those pertaining to citizen participation, can be identified. This input derives from the internal workings of the governmental machinery. Apart from the initiative and innovation shown by political leaders and administrative officials at the local or metropolitan levels of government, the activities of senior governments in the metropolitan area are also of great significance in the overall development of the community.

It seems, however, that central and provincial governments could do more for the development of a metropolitan identity. If a metropolitan-wide authority is to be created, then, mostly likely, it will have to be a product of imposition by a senior government. Not only are senior governments reluctant to create what are often regarded as a threat to their own existence, they are often unwilling to provide for an open system of communication between local and senior governments. This is especially so if there is an intermediary provincial or state government. In Canada, for example, a municipality can negotiate directly with the Canadian government only in cases where responsibilities are assigned to the federal government. In such instances the municipality has the same rights as any private citizen. In areas of responsibility assigned to the provincial government, the municipality must deal only with the provincial government, even though it may be purely federal grants that are involved. With the growing awareness of the federal government that what goes on in the urban areas, and especially the large metropolitan areas, affects the interests of the nation, these constitutional constraints are being challenged.

This need for a much freer dialogue between levels of government applies to almost all countries. The power of the regional government of Western Nigeria to dissolve the city government of Ibadan, as it has done in 1956 and 1965, is open to question, although for many countries a lack of skilled personnel available for local government service does dictate that senior governments should have strong supervisory control over local authorities. In this instance, the initiative for establishing improved relations must be taken by the senior government involved.

PARTICIPATION IN THE INTEREST OF THE METROPOLITAN COMMUNITY

Having examined the various types of participation involved in the political process, we may now proceed to illustrate how this participation can best be encouraged to further the interests of the metropolitan community as a whole. Such a course of action essentially requires an examination of, and possible adjustment to, the process of political socialization, the size of the metropolitan community, the stakes involved (the functions performed), and the citizen's access to, and identity with, the metropolitan authorities that exist.

The Political Socialization Process

The process by which individuals learn about and internalize the values and norms of the socio-political environment is generally known as political socialization. In this transfer of social values to members of the polity, several mediating instruments (the family, school, church, community, mass media and so on) play very active roles. In a metropolitan context, the individual's perception of the metropolis as a political entity, his possible identification with this larger unit, and his participation in metropolitan public affairs may be strongly influenced by conscious changes in the use of socialization agents.

Political socialization should properly be regarded as part of the general education of the population. Among the groups, there was general agreement with Smallwood that this process has an enormous influence on the extent to which citizens will accept the new forms of government that have to be introduced in an expanding metropolian area. This implies that a greater effort should be made to get people involved in various programmes that might otherwise be regarded as catering to special interests.

Many of the citizens of metropolitan areas are born in small communities, where their particular loyalties and views of the world are shaped. The importance of imparting an understanding and awareness of urban life through the education programmes offered in rural areas has been stressed already in Chapter 8. This should apply not only to the education of children but also to the education of adults. Certainly it is in the rural areas that the challenge is greatest, for in these areas there is less opportunity to become directly involved in the affairs of the metropolitan community. Just as children in Prague receive a part of their formal education in rural areas, it may be beneficial to have school children from rural areas spend some time in city schools.

Children should be asked to make surveys of their communities in order to make them more aware of what it is that makes up a community. It was suggested at the seminar that efforts should also be made to involve them in the planning of parks, schools and facilities that they frequently use, in order that they will be encouraged to participate more in the future. While this is bound to be time-consuming for the civic officials involved, it should be recognized as an investment in the long-term education programme. Recent experiments in Philadelphia have led to the holding of some classes in most unconventional settings, such as civic buildings, firehalls and

so on. This approach can be applied just as easily in rural areas, of course.

Adults should also be organized to conduct their own community self-surveys, focussing perhaps on local problems. The identification of the problems and the formulation of solutions, leads to an involvement that is likely to give great satisfaction to the individual. Again, the time taken to organize this type of programme must be seen as an investment of a far longer-term nature than that required simply to produce some solutions for the problem at hand.

Apart from this more active involvement, there should also be a greater effort on the part of government to encourage citizen groups to organize seminars, conferences, exhibits, lectures and other programmes for citizens. The more these programmes are organized by non-governmental bodies, the more successful they will be in furthering active participation on the part of the citizen.

Earlier reference has also been made to the role of the press and, implicitly, other mass media in the political socialization process. Public officials should understand the needs of the mass media, and, at the same time, the latter should have a sense of responsibility in their handling of information on metropolitan affairs. In rural areas the role of television is of particular importance in conveying a visual image of urban life. One problem of course is that this may further contribute to increasing the pace of rural-urban migration!

Another way in which the urban poor may be able to identify with the metropolitan community is through the provision, by a metropolitan authority, of retraining programmes. This type of programme will enable many members of the urban poor to find not only greater satisfaction in their work, but also the feeling of making a contribution to society and of being part of that society.

As has already been stated, one of the key factors for the success of the political socialization process is the readiness of government officials to take the initiative and responsibility for the development of these types of programmes.

The political socialization process is probably the most important factor affecting the degree to which citizens will participate in the affairs of their community. However, the actual size and scope of the activities offered by their community as well as the access afforded to any governmental institutions are also factors of great significance.

The Size of the Metropolitan Authority and the Services It Offers

The significance of metropolitan size has been referred to several times in parts of this and other chapters. The earlier reference to local government election turnouts in the Vancouver area would seem to indicate that size alone, at any one level of government, is not a significant factor in considering citizen participation. However, as Gorynski and Rybicki make quite clear, and as reactions to their essay substantiate, the physical, economic and social functioning of the community is affected by the size of the community sub-units, and it is within these sub-units that the citizen finds it easier to identify himself, in physical terms at least.

Again, size cannot be discussed very easily without talking in terms of the functions provided by any unit of government. Although this subject is discussed in greater detail by Dupré in Chapter 11, some comment should be made on the issue here, as these functions relate to interest in participating in community affairs. It has already been indicated that local

government should be given responsibility for all the functions that it can perform without undue loss of efficiency. The services often provided at the metropolitan level, and in the opinion of the seminar participants are best provided at that level, are as follows: sewage disposal; garbage disposal; major highways and bridges; metropolitan transport; air and water pollution control; the construction of public housing; the formulation of policy and the planning for welfare and health services; the local siting of airports; fire protection; police; large space planning; major cultural and recreational activities; and, debenture borrowing. It is preferable that these services be provided by a multi-purpose authority, although the problems of achieving this have been alluded to already. At the very least there is need for coordination between local units of government providing these services in a metropolitan area where there is no metropolitan government. These coordinative devices may have to be imposed from above, but if they can be developed from local initiative they are more likely to be successful.

Some services obviously have to be shared between levels of government. The London group referred to the attempts of the Royal Commission on London Government to avoid sharing wherever possible in the interest of retaining the "health of local government". However, several services are being shared in London at the present time, notably planning, public housing and education. In fact, there seemed to be general acceptance by the seminar participants that more functions will have to be shared on a "wholesale-retail" basis if the assignment of responsibility is to be made on the basis of capability. To the extent that participatory values are enhanced by the responsibility assumed by a level of government, the above list of service functions that could be performed by a metropolitan tier of government should encourage strong citizen interest at that level.

Citizen Access at the Metropolitan Level

There is not much that can be added to what has already been said about the ways in which the citizen might be given access to governmental institutions. If participation is to be encouraged, access is obviously most important. There seems to be no ready explanation for the paradoxical observation that Smallwood derives from Almond and Verba regarding the citizens' greater voting participation at the elections of senior governments despite their feeling that they are more effective at the lower levels of government. It may well be that it is the existence of many types of interest group activity at the local level of government as opposed to the fewer groups that are found at the national level that accounts for this sense of effectiveness. This would point to the need for the development of interest groups with a metropolitan-wide interest base if access is to be provided in this form. As the problems of forming these broad-based interest groups are very real, and as the establishment of interest groups does not guarantee access, governments must be looked to in taking the initiative to establish other channels of consultation and communication in the forms described earlier in this chapter.

In summary, from the viewpoint of participatory values the multi-purpose metropolitan authority in combination with a lower tier of smaller authorities appears to be the most satisfactory solution for the provision of services in large metropolitan areas of over 2 million population. Although this figure is somewhat arbitrary in that it is derived largely from an observation of the criterion of service efficiency,

it does represent the upper limit at which a single multi-purpose body is still acceptable. This is not to deny the value of the two-tier approach in metropolitan areas of less than 2 million in population, although there is likely to be a greater loss of service efficiency with this arrangement in smaller areas.

However, while the multi-purpose metropolitan government may be regarded almost universally as the goal, it is far from being realized, or even politically realizable, in many parts of the world. The role of the metropolitan-wide *ad hoc* authority in providing a mechanism for moving toward metropolitan government in a transitional period of political difficulty was recognized by seminar participants. However, the undesirable characteristics of these bodies, which have been described in the section on utilities in Chapter 8, and pursued further by Dupré in Chapter 11, should be considered carefully before there is resort to mechanisms. An additional problem is that once there are metropolitan-wide *ad hoc* authorities, it may be difficult to justify their integration on political grounds, even though this can easily be done on the basis of efficiency.

The less formal arrangements for coordination mentioned by Smallwood were regarded favourably for purposes of transition to metropolitan government. However, to the extent that some of them are voluntary, they are extremely weak and to the extent that they are restrictive (for example, grants from senior governments for specific projects), they are detrimental to the "liberty" of local government to establish its own priorities. As these issues are taken up at greater length by Dupré in his discussion on intergovernmental relations, it is to a consideration of the reactions to his essay that we turn now.

Intergovernmental Relations and the Metropolitan Area

ATTITUDINAL ASPECTS OF INSTITUTIONAL CHANGE

Participation, efficiency and liberty are, according to Dupré, the key values that should be served by the governmental machinery developed for any metropolitan area. While none saw upholding these values as an idle aspiration, it was suggested that there are other values that should be served by the governmental machinery, the relations among its various institutions and also the relations among these and individuals and non-governmental institutions.

These additional values are: the achievement of national goals; leadership; the redistribution of wealth to give greater equality in services enjoyed; the expression, and achievement, of consensual goals for the metropolitan community; and albeit somewhat more difficult to achieve, even the development of the more abstract and intangible qualities of a metropolis such as its dignity, charisma, vitality, elasticity and its seminal force. While there may be some overlap in these values—for example, leadership may be regarded as efficiency in participation—they should be identified separately to ensure that due consideration is given to each.

If the governmental machinery (inclusive of intergovernmental relations) serves one or more of these values at the expense of others, this is only because this machinery is, in large part, the formal legislative expression of societal attitudes applied to social, cultural, political, economic, physical and technological conditions. Unfortunately, however, present-day society is often frustrated, in that the priorities given these values are very often not a reflection of present attitudes applied to present conditions, but of past attitudes applied to past conditions. This is simply a result of institutional change's being unable to keep pace with changes in societal attitudes and changes in existing conditions.

To complicate the matter further, changes in attitudes and in conditions very rarely occur at the same pace. This, too, will affect the apparent priorities in the value structure. For example, a change in existing technological conditions, as with the improvement of communications, can affect the relative significance of certain values, such as efficiency. If society does not take advantage of the technological change, then there exists what can only be regarded as a state of relative inefficiency. If efficiency as a value is to continue to be upheld, which means that societal attitudes are to remain constant, then adjustments must be made to the third variable – the governmental machinery and/or the intergovernmental relations—to take advantage of the technological change.

It is this type of situation that often confronts metropolitan communities in their fast-changing environments. The maintenance of a value structure is problematic, to say the least.

In some instances, as society evolves, societal attitudes will call not for the maintenance of values served but for a change in values. For example, social attitudes may change to the extent that a greater emphasis is put upon citizen participation in the decision-making process. Again, however, this must be given formal legislative expression in the institutional machinery, for without this change society cannot be expected to be fully satisfied. Yet it is quite understandable that since institutional change can take place only as a result of attitudinal change, changes in institutional structure cannot be expected to keep abreast of attitudinal change; it can do no more than follow, and even then somewhat erratically. It is this inability of institutional change to accurately reflect changing attitudes, and especially the inability of institutions to provide for changes that theoretically could be anticipated, that frustrates society and stifles its development.

It is against this background that we can discuss the suggestions that have been made by the study groups and the seminar participants to improve the governmental machinery and the relations among its various elements, in order to facilitate the serving of the values mentioned above.

In the developing countries, the urgency to achieve national goals ensures that sub-national goals, such as those for a metropolitan area, will be of secondary importance. The shortage of resources available for concentrating effort on more than a limited number of goals makes the realization of metropolitan goals even more remote. While it is obvious that metropolitan goals can be achieved in concert with national goals, the tremendous demands that this comprehensive planning makes upon human resources is such that many countries have been inhibited from pursuing this comprehensive approach. Thus, the Seoul study group stressed the significance of the values of efficiency, in the achievement of national goals, and leadership. This last value, leadership, which might also be regarded as efficiency of participation, is of particular significance in a country where the greater number of the population is neither educated nor properly informed of the nature of the issues confronting metropolitan areas. The replacement, in 1961, of the elected council of the Special City of Seoul by an appointed body, was recognized by the Seoul group to have achieved a great deal simply by virtue of improved internal efficiency and effective leadership. However, it is also important to note that the Seoul group thought that the time had come to introduce a two-tier system of government with elected councils at each level, as well as an improved public information system that would enable more effective citizen participation in local decision-making. This provides us with a good illustration of the need for a major adjustment in institutional structure to cope with fast-changing conditions.

In more developed countries, with a greater availability of human resources and a higher general level of education among the public as a whole, there is obviously more interest shown in giving fuller expression to the value of participation. Again, there is a concern that leadership be provided to channel this participation into the achievement of consensual goals for the metropolis as a whole. It is of interest to note, however, that if the achievement of metropolitan consensual goals is upheld as a value, an apparent threat to the liberty value is created. For it will be recalled that Dupré stresses the negative form of liberty – "the freedom from arbitrary rule". Yet to achieve metropolitan consensual goals requires the establishment of a metropolitan-wide governmental authority that will, if it is to be effective, result in the removal of many of the smaller authorities and/or their powers. This theoretically constitutes the removal of a barrier to arbitrary rule. We are once again faced with an example of a situation that requires adjustments to institutional structures if these structures are to reflect present-day attitudes toward present-day conditions.

It is not difficult to observe the increased responsiveness, and concomitant decline in arbitrariness, of senior government rule since Montesquieu's time. Nor is it difficult to justify larger units of local or metropolitan government that reflect technological change and the increased size of the community. It is likely that a strong metropolitan government will offer the citizen as much protection as he ever received from smaller units of local government. Indeed, the Boston group considered the situation in which many local authorities are free to make their own decisions that may run counter to the interest of the larger community as somewhat akin to the undesirable situation, pictured by Dupré, of interest groups operating behind a smokescreen of governmental complexity, in this case, of their own making. A similar attack upon the same issue, and one similar to that of the Boston group, was made by the groups from Budapest and Ibadan, which thought it unjust for the suburbs to enjoy a "freedom from assuming responsibility for the problems of the central city". The implications for the development of a metropolitan identity, of metropolitan consensual goals and of a machinery to give political expression to the metropolis, are obvious. All that remains, therefore, is the assurance that the citizen be able and encouraged to participate in the affairs of the metropolitan authority to the degree that he participates in local government. One might well argue that the replacement of a large number of local authorities by a few more responsible authorities would attract greater public interest. Thus, the threat to the liberty value seems to be more apparent than real.

TRENDS IN THE ALLOCATION OF FUNCTIONS
TO GOVERNMENTS

The introduction of a metropolitan-wide body is not necessarily the immediate answer to the problem of providing greater coordination for the provision of services in the metropolis, if only because of the need, in so many instances, for intermediate steps to be taken before this more desirable form of government might be established. However, the presence of such political problems does not detract from the need for a more appropriate allocation of functions between levels of government, the need for a greater focussing of interest and priority setting at the metropolitan level, and the need for a greater exploitation of other techniques for achieving cooperation and coordination between governments to enable goals to be achieved at the metropolitan, national and local levels of government. Intergovernmental relations must respond to these needs.

Investigation as to how the allocation of functions to different levels of government affects citizen interest and participation in the organization of the affairs of the community reveals that such interest and participation tends to develop in direct proportion to the responsibility assigned the level of government in question. While this concept should not be seen as a case for reducing the number of levels of government to one, it does call for the allocation of functions in a manner that will result in powerful and responsible governments commanding the interest of the public. It has already been shown that the complexity of servicing the metropolitan area is justification enough for providing organizational machinery on a metropolitan scale. It has already been acknowledged that there will be exceptions to this general rule but, by and large, the task is essentially one of allocating to a metropolitan level of government those functions and powers that will result in the metropolitan community's being serviced according to its needs.

Two major tendencies can be observed in the allocation of functions to governments: the greater involvement of central governments in a number of functional areas and the increase in the sharing of responsibility. These tend to apply throughout the world. The tendency for central governments to increase their responsibilities in a number of functional areas applies particularly in the fields of the utility services of health and welfare, and of education. Looking at this phenomenon on a national scale, it is understandable that the economic concentration that has taken place in the metropolis and which is accompanied by a depletion of economic resources, and consequently fiscal resources, in the countryside, has forced upon central governments the responsibility for financing more services on a nationwide basis to counteract this tendency toward regional disparities. It is quite natural that any central government faced with this responsibility is anxious to be assured that services are being provided efficiently and according to standards of which it approves. On these grounds centralization appears acceptable. Unfortunately, however, demographic and economic concentration can easily proceed unabated if unplanned, and the spectre of these processes being accompanied by a continuing centralization of political authority is cause for alarm.

The tendency toward a greater sharing of political and administrative responsibility can be regarded as a response to, among other things, this over-centralization. While reactions to Weissmann's proposals for a national strategy for development and Hicks' proposals for financial redistribution are discussed later in this chapter, it is appropriate to make a few comments here on the distribution of political and administrative responsibility.

Exclusive responsibility, on the part of individual levels of government, for spe-

cific functions is being superseded. A sharing of functions on the basis of a division of individual functions according to policy considerations and executive functions is preferable. The basis of this division can also be expressed in strategic/tactical terms or according to decisional/operational considerations. This system for sharing functions is in keeping with the division of labour according to competence, as proposed by Dupré. There is nothing in this concept, derived from Mill, which prevents the division of functions between governments on a decisional/operational basis as opposed to an arrangement whereby whole functions are assigned to one level as opposed to another. Such an arrangement for sharing makes it quite acceptable that policy outlines should be developed by the central government and that more detailed policy and the implementation of that policy should be largely the responsibility of lower levels of government. It is this type of principle that Gorynski and Rybicki propose in Chapter 9. This arrangement also enables central governments to establish national minimum standards for certain services, with the responsibility for meeting those standards resting with the local or metropolitan authorities. There is no reason why this decentralization of responsibility for the implementation of services should be any less efficient than if this task were to be performed by deconcentrated agencies of a central authority. However, it seems that this is only beginning to be appreciated in many countries.

Understandably, where decentralization is beginning to take place, problems are arising due to the lack of coordination of the authorities given the responsibilities as a result of decentralization. In Philadelphia, the study group observed a growing need for lateral coordination within the metropolitan area, due, in part, to this increased decentralization. Also within the United States, as elsewhere, the problem arises to assume responsibility for certain functions not adequately performed by the states and the federal departments that have to finance these local activities. This again strengthens the case for the use of a metropolitan regional body as the focal point of coordination.

The sharing of functions between governments on a strategic/tactical basis also raises some problems regarding effective implementation. The common assumption that the assignment of major policy responsibilities to an upper-tier government will "automatically" result in some vigorous execution must be questioned. A number of seminar participants described the situations in their own areas in which planning powers had been invested with metropolitan-wide authorities but whose proposals had been ignored. One particularly interesting instance of this occurred in Prague, where, it was said, despite the preparation of realistic, comprehensive plans and the absence of major governmental fragmentation within the metropolitan area, the implementation of plans has foundered through lack of coordination between agencies of government. This is due to two factors: inadequate financial resources of the planning body and its inability to recruit support for its proposals from administrative agencies whose participation in the implementation of plans is essential.

Without adequate legislative power, means of coordination and fiscal resources, planning on a comprehensive basis throughout metropolitan regions is not possible. But even then the possession of these powers may not, in itself, be adequate. There are cases in which senior governments (for instance state and provincial governments) have powers that enable

them to require the execution of plans. But these have not been exercised for lack of an ability to develop public consensus. Thus, in addition to constitutional and fiscal means by which the government can perform its duties, there must be public support for action and means by which this support can express itself to the government. This can take place far more satisfactorily at the metropolitan than at the state level. Thus, planning powers for metropolitan areas should be entrusted to a structure of government covering a constituency that is coterminous with the interests and desires of the whole metropolitan area. The essays by Gorynski and Rybicki and Smallwood point to ways in which this can be achieved.

Another aspect of sharing, which is especially relevant in the metropolitan area, is that certain services that can be provided efficiently at a metropolitan level can be offered to those sub-metropolitan units that are perhaps too small to be able to support such a service efficiently. For example, in London, some of the boroughs make use of the purchasing services of the Greater London Council. The Boston group, which recommended that a metropolitan county be established in the Boston area, envisaged that the personnel department of this metropolitan body would be used frequently by the sub-metropolitan units. In passing, it is of interest to note that the Municipality of Metropolitan Toronto makes use of the well-developed purchasing department of the lower level City of Toronto, a rare occurrence in areas with a metropolitan form of government.

If functions are to be shared, it is quite obvious that there must also be a greater sharing of the resources available to meet the necessary expenditures. This can be accomplished either by redistribution of fiscal resources through grants from senior levels of government to the lower levels of government or by offering the lower levels of government direct access to additional sources of revenue that may presently be accessible solely to senior levels of government. While comments on the latter alternative are provided in Hicks' essay and in the next section of the present chapter, some remarks can be made here on the redistribution of resources.

The common phenomenon of the rurally-oriented legislature at the national and/or state (provincial) level of government often accounts for the inequities suffered by metropolitan areas as a result of the formulae developed for the distribution of senior level conditional grants. In Philadephia it is the suburbs that are disadvantaged. For apart from the prevalence of rurally-oriented grants there are also additional grants going to the central cities, since it is very often only the central city that is able to finance the additional activities incurred as a result of the receipt of such grants. In Sydney, on the other hand, there is more concern with the diversion of funds, not only away from Sydney, but also away from the State of New South Wales. This, in itself, is significant for the metropolitan area of Sydney, as it is in fact the state government that serves as a metropolitan government "writ large" in this instance. The Sydney group, like all other groups, was prepared to recognize the need for national priorities but there was a strong plea that if national priorities are to be developed, they should be presented in a national development plan that would then form the basis for national financial programming. It was suggested that if the use of cost-benefit analysis and performance budgeting was built into the development of these programmes, more grants would find their way to urban areas. Also, on this basis, the conditional grant may appear to be somewhat more acceptable. However, more can be said about this, as it relates to priority-setting in metropolitan areas, the subject to which we now turn.

PRIORITY SETTINGS: THE RESPONSE OF SENIOR GOVERNMENTS

From the foregoing it is difficult to conclude otherwise than that the effective resolution of metropolitan problems must be carried out within the context of a national framework or policy and that this will require an intensive system of intergovernmental collaboration. However, if the efforts of national and local governments and any intermediate levels of government are to support one another in the provision of metropolitan services and in the resolution of metropolitan problems, then a focal point for this coordination is obviously required. Whatever the size of a country, in terms of territory and population, this coordination, from an administrative viewpoint, is best provided at the regional level. The fact that this may not be acceptable politically is, however, very often much more significant. The widespread support for the organization of a general-purpose metropolitan-region government cannot be ignored. In almost all cases, metropolitan systems of government should, and generally do, have more than one tier of government to sustain participatory values and at the same time provide for a more efficient assignment of functions. Where general-purpose competence has been established at a metropolitan level, there is no question that the resulting capacity to systematically set metropolitan priorities has offered important functional advantages that were previously unobtainable.

Unfortunately, even in such cases where a metropolitan level of government has been introduced, the existence of unsystematic and inappropriate restrictions from the central level of government leaves much room for improving relations between the two governments. Where the government of metropolitan areas remains locally fragmented and without an overhead tier of some sort, the local determination of metropolitan priorities is seriously impeded and the transfer of traditional, locally governmental competence to central levels is understandably accelerated. This, again, results in an undesirable degree of central government control of what are essentially metropolitan functions.

While it is extremely unlikely that the introduction of a metropolitan form of government will, or could, from a constitutional standpoint, result solely from initiative shown by local government, any imposition of a metropolitan form of government by a senior level of government should meet with the general approval of local citizens. Indeed, it may be that the senior level of government is unwilling to act until strong local support for such a form of government is demonstrated. Hence the value of strong non-governmental groups which can provide leadership in supporting activities that will lead to a more general public recognition of the need for a metropolitan focus. However, while the importance of community support for the introduction of a metropolitan form of government must not be underrated, it cannot be relied upon to bring about the desired results without the much more important existence of a positive attitude on the part of the senior levels of government.

INTERGOVERNMENTAL COORDINATION: SENIOR GOVERNMENTS

The response of senior governments to the need for greater vertical and horizontal integration still leaves much to be desired. The unwillingness of many senior governments to establish general-purpose metropolitan bodies to which funds could be given for the implementation of programmes of a high national or state (provincial) priority, has resulted in a proliferation of many special senior-government agencies in metropolitan areas. These bodies lack coordination at the metropolitan level and, more often than not, at the national level. For example, only in 1965 did the government of the United States establish the Department of Housing and Urban Development for purposes of coordinating, at the national level, the growing number of federal government programmes operating in metropolitan areas. A similar need is obvious in many countries. The Sydney study group complained that the absence of an urban affairs ministry for the state of New South Wales places an undue burden on the state cabinet as the coordinating body for the provision of services to metropolitan Sydney. The extreme difficulty of introducing metropolitan government to the Sydney region makes the proposal for the establishment of a Sydney Region Advisory Committee, attached to the State Planning Authority for New South Wales, all the more important. The major function of this committee would be to provide a focus, within the administration, of the needs of the Sydney region. In addition to the introduction of this committee, the Sydney group was particularly anxious to see the implementation of some of the suggestions made by Gorynski and Rybicki pertaining to the active involvement of citizens in public affairs. For the Sydney group, the giving of responsibility to citizens seems to be more appealing than relying upon their (more passive) participation in interest group activities.

The Philadelphia metropolitan area provides some particularly interesting illustrations of some of the conflicts that arise in attempting to establish cooperation among levels of government. Although the United States federal government has attempted to encourage local government cooperation by insisting that a regional plan be prepared if certain federal grants are to be made available to the area, the activities of the state government are sometimes in conflict with this. There are indications that the State of Pennsylvania is prepared to give special options to many of the urban local government units in the greater Philadelphia metropolitan area. In contrast, and in conflict, the State of New Jersey, with its more important metropolitan areas running into other states (among them Pennsylvania) where the dominant centres lie, is far less prepared to offer its own urban local governments the freedom to act in concert with local governments in other states. In addition, federal funds that are being channelled through the states to the urban areas are subject to certain state controls. The veto power of states in this regard can disrupt the implementation of national priorities. This again points to the need for a metropolitan level of government that could receive senior government funds, albeit according to certain conditions.

The problem confronting the conurbation known as the Randstad, in the western part of The Netherlands, is again of particular interest in this context. The physical nature of the metropolitan region—that of a necklace of urban concentrations—appears to rule out the possibility of a metropolitan tier of government for the entire Randstad region. One possibility suggested by The Netherlands study group was that the provinces should be strengthened and reduced in number thus enabling them to

be increased in size. At present they have the characteristic of being deconcentrated administrative bodies of the central government, although they appear to have little power. The effective control by the national government over local governments in The Netherlands is due mainly to the fact that about 90 per cent of local government expenditures are met through central government grants. However, this type of centralization is merely an example of what has tended to happen in those countries where local government units are generally very small. The City of Amsterdam is an obvious exception and the Rotterdam region has recently responded to the need for metropolitan cooperation through the establishment of the Rijnmond.

The Rijnmond Council has brought together twenty-three local authorities in the Rotterdam area. However, it is not regarded as a very powerful body. It is dominated by the City of Rotterdam and is weakened by its lack of powers, which are mainly of a supervisory nature. The Rijnmond's most general duties are to coordinate municipal policies in the area. With regard to planning, it does have the power to adopt one or more regional plans. In performing this function and also in issuing directions to municipal administrations regarding municipal plans for specific purposes, the Rijnmond Council replaces the role of the provincial authority. The Council is also able to issue directives to municipal authorities on other matters, such as the building of docks and industrial parks, the location of industry, housing and recreational facilities; the construction and improvement of roads, waterways, river tunnels and bridges; and action to be taken against water and air pollution. If necessary, these directives can be transformed into binding instructions. However, these powers do not seem to have been effective enough. Although the introduction of the Rijnmond Council has improved the government of the Rotterdam area, it appears that it is now time to lend strength to this body by giving it direct responsibility for the administration of certain metropolitan area functions. Nevertheless, despite the recognition of its present weaknesses, the Rijnmond experiment will have served a useful purpose if it does succeed in making quite clear the need for a metropolitan form of government and, at the same time, facilitating the introduction of such a structure.

Severel references have been made to the role of political parties in the decision-making process at the metropolitan level. In the context of intergovernmental relations, they are also of significance. A unified system of party politics common to all governmental levels will, at the very least, provide certain channels by which intergovernmental relations can take place. This will also help to focus public attention on the need to establish priorities among levels of government. However, as divided partisan control is almost as likely to exist as is undivided partisan control, it is questionable whether party politics might not underscore the differences among levels of government. While this may have to be admitted to some degree, it is likely that the advantages of coordination offered by a party at any one level, and thus the provision of a voice for that level, may make the relations among levels of government that much easier to comprehend and, therefore, to set in a framework of co-operation.

INTERGOVERNMENTAL COORDINATION: TECHNIQUES AVAILABLE TO SENIOR GOVERNMENTS

Once responsibilities for certain functions have been allocated among different levels of government, the financial considerations relating to the working relations among these levels of government acquire enormous significance. The absence of satisfactory financial arrangements is more often than not the basis of the crisis that may have to take place before a metropolitan-wide government will be introduced. For, although reference has already been made to the importance of developing an informed public and of putting pressure on the senior governments to introduce a metropolitan structure, it is a crisis of major magnitude that will likely be most effective in speeding up the introduction of metropolitan government. The insufficient resources available to local governments have resulted in their being unable to provide the services required in the metropolitan area. The alternatives, as has been noted previously, are either to open up new sources of revenue for local governments or to provide additional funds from senior governments by way of grants. The latter alternative is normally much more acceptable to the senior governments, as it enables them to maintain more control over the economy and over services provided. Although grants, whether conditional or unconditional, are often criticized, they do have several advantages. A senior government, especially the central government, is best suited to collect the more productive and dynamic sources of revenue. At the same time, on the basis of functions being allocated according to competence, the metropolitan and local governments are in a strong position to implement the services being funded by senior level grants. Also, if we put any weight on the value of the redistribution function of governments, then grants from senior levels of government must rate high on our priority list. For if we do require strong metropolitan and local governments, it is only through the grant system that redistribution can take place on a scale larger than that of the metropolitan area. Redistribution within the metropolitan area is certainly important, but it is of just as much significance to attain this on a nation-wide basis.

The problem of developing an acceptable formula for redistribution can be illustrated by reference to the conditional roads grant distributed by the Australian federal government. The government of Australia distributes the total amount of its grant for road work among the states on a formula of one third according to population, one third according to motor vehicle registration and one third according to area. Understandably, this crude formula disadvantages the small but populous states of New South Wales and Victoria. This federal grant is conditional to the extent that the federal government stipulates that not less than 40 per cent of the amount received by any one state must be spent on roads (which are not classified highways) serving sparsely populated areas. Within New South Wales virtually all of this portion of the grant goes to the rural shire councils for certain other rural road-work purposes. The remaining 60 per cent of the grant finds its way to the New South Wales Main Roads Department, where it is dealt with in accordance with state legislation, which provides that funds of the Main Roads Department must be allocated in proportions of 80 per cent to county main roads funds and 20 per cent to the County of Cumberland (the Sydney metropolitan area) main roads funds. This means that the Sydney metropolitan area, with some 62 per cent of the population of New South Wales, receives about 12 per cent of the New South Wales share of the federal government allocation.

Whether the grants should be unconditional (block) or conditional is certainly debatable, although either alternative calls for a redistribution formula. It appears that the use of block grants is confined mainly to countries in which a relatively high priority is attached to participatory values and in which local or metropolitan authorities are invested with substantial amounts of discretionary power. The high number of conditional grants relative to unconditional grants may be misleading. For example, there are only two provincial grants available to municipalities in the Metro Toronto area than can be regarded as unconditional. However, the seventy or more conditional grants are largely a product of the unsystematic approach of the Province of Ontario to intergovernmental fiscal relations.

The greater acceptance of conditional grants in developing countries has already been mentioned. The need to concentrate skilled human resources at the central level of government virtually dictates that, if these resources are to be used to the full, conditional grants be used as a tool to ensure the implementation of national policy directives. An improvement in the developing countries would be for the local or metropolitan bodies to be given more responsibility for implementing programmes financed by central government grants. The Seoul group indicated that the government of the Republic of Korea tends to implement its own programmes, rather than lend strength to the city government, by making the latter responsible for the implementation and administration of certain services. It is also of interest to note that in 1960 the senior level grants to the Special City of Seoul consisted of only 1.8 per cent of the revenues of the city government. From this finding and from the earlier remarks made regarding the lack of authority given the Special City of Seoul, one can only conclude that the absence of senior government grants, whether conditional or unconditional, does not necessarily mean freedom from arbitrary rule.

The increase in conditional as opposed to unconditional grants is a major complaint of many metropolitan areas in the more developed countries. The Sydney group indicated that in the period 1961 to 1967, specific-purpose conditional grants by the Australian federal government to the states rose by 127 per cent while general or unconditional grants rose by a mere 41 per cent. In the year 1967, conditional grants to the states were 14.9 per cent of total grants, and in 1968 the equivalent figure was 22 per cent. The fear has been expressed that the rapid increase in conditional grants is bound to erode the degree of autonomy in decision-making on the part of the states.

On the other hand, if there is to be a national plan for the general demographic, economic and social development of the country, and if national minimum standards are to be attained, then conditional grants are by far the strongest tools available to the senior governments to ensure the realization of such a plan. Although senior governments do have other powers, which they can exercise to ensure that local governments are maintaining certain standards, these tend to be constitutional powers, which they are reluctant to exercise except in the case of a dire crisis, such as a depression. In Canada, local governments are regarded as "creatures of the provinces". This provides the province with the power to dissolve the municipality, if this is necessary. In fact, the provinces content themselves with maintaining a strict watch on the borrowing of local authorities as the main means by which they are able to keep control on local government developments. In the depression years of the thirties, the provinces did have

to intervene in the financial administration of some municipalities. While it can be assumed that existing competence in metropolitan areas will generally make such drastic action on the part of senior governments unnecessary, this power of the senior governments is justified. Irresponsible borrowing on the part of local governments could bankrupt a poor country if pursued for a long period of time.

Another technique employed in creating greater coordination, especially in the provision of services, is the setting of minimum standards. One of the problems in the setting of these standards is that they may discourage the generation of programmatic innovations that should be considered as one of the particular attributes of local government. Another problem is that service levels may polarize at the level of the minimum standard, thereby distorting the order of priorities that might otherwise obtain.

Another way in which senior governments can assist metropolitan areas is by technical assistance programmes. Such programmes can be especially effective if offered by universities and similar non-governmental bodies. The dissemination of new ideas among governments in metropolitan areas takes time, and the role of educational institutions can be particularly effective in this regard. Other aspects of providing information to different levels of government relate to provision for information centres, strategically located throughout the country. The "urban observatory", which is being introduced in the United States, is a cross between the university, as an urban research resource, and the information centre.

One final device that facilitates cooperation among levels of government is the provision of direct consultation services by senior government for metropolitan area governments. The success or failure of this arrangement, however, depends upon the relations that exist between individuals involved. If successful, it is probably the most effective way of developing cooperation between levels of government. Unfortunately, it can be subjected to a sudden change in effectiveness with the change of the personalities involved. It is the development of harmonious working relations between individuals, rather than institutions and departments, that provides one of the greatest challenges in the field of intergovernmental relations.

Financing Metropolitan Government

THE SPECIAL FINANCIAL NEEDS OF THE METROPOLITAN AREA

Lady Hicks, in her essay on metropolitan finance, refers to the special characteristics of a metropolitan area—its large size, its even larger service area, and its subjection to never-ceasing in-migration—which make special demands upon governmental financial and administrative systems both within the metropolis and at a national scale. To this list of causal elements of metropolitan financial problems might be added the uncoordinated and fractionated governmental structures that exist to provide services to the metropolis. Before commenting on how governments might respond to these special needs, it is desirable to make some qualifications to these special characteristics as delimited by Lady Hicks.

In many countries, any one metropolitan area, and in many cases it is only one metropolitan area, may attract in-migrants from the entire country and, perhaps, from beyond the national borders. While it could be argued that this in-migration brings workers and wealth to the metropolis, the fact of the situation is that these migrants are generally poor and lacking in skills. They move to the metropolis to take advantage of its services and yet are rarely in a position to lend financial support to those services. The burden is felt particularly by the social services such as education, health and welfare. In Calcutta, for example, 30 per cent of the patients treated in state hospitals within Calcutta are not local residents of the city. Tremendous pressures are put upon the existing facilities: 77 per cent of families in the City of Calcutta have less than forty square feet of living space per person and, in the same city, 32 per cent of the six to ten age group have no primary education.

This growth in population without commensurate growth in financial resources to meet increased demand is nothing more than "numerical urbanization". As there is little expansion of the bases of the taxes presently available to governments in the metropolitan area, and certainly this would apply to the property tax, it is understandable that these governments are constantly pressuring senior governments to provide them with access to new sources of taxation, or to make additional grants or to relieve them of the responsibility for financing and administering certain services.

As has already been made clear in the previous section on intergovernmental relations, adjustments in financial arrangements often require, or at least are often accompanied by, adjustments in governmental structure in the metropolitan area. While the arguments for the provision of services on a metropolitan-wide basis need not be repeated here, it is necessary to raise the question once again as to whether these services should be provided by a series of special districts, by a single multi-purpose metropolitan government or by a combination of both of these approaches. The desirability of taxing the metropolitan community as a community, to bring about greater equity over the metropolitan area, tends to support the case for the multi-purpose authority as opposed to the special district. In fact, the Johannesburg study group made a point of emphasizing the undesirability of giving powers of taxation to special districts. What may occur, as happens in Athens, is that the central government will finance a number of special districts responsible for certain services that might in other parts of the country be provided by local government. In the case of Athens, the road and sewage systems are the responsibility of special districts supported by the central government, and water is provided by the private "Hellenic Water Company". The size of the budgets of these special authorities, relative to the budgets of the larger municipal authorities in greater Athens, underlines their relative importance. For example, the 1961 budget of the municipality of Athens was approximately 172 million drachmas, and that of the municipality of Piraeus was about 56 million drachmas, whereas the expenditures of the Athens Sewerage Organization and the Special Fund for Permanent Road Pavement of Athens were 89 million and 183 million drachmas respectively. The Piraeus Port Authority, another special district in the area, had revenues of 335 million drachmas that year.

Whether sufficient funds exist to finance services in the metropolitan area is, in some cases, doubtful. However, the problem of assuring adequate financial resources is not always one of lack of funds; it is due often to a lack of financial planning and management. For this reason the structure of government in metropolitan areas is of much significance. Those persons attending the seminar-conference tended to favour a two-tier system of government for the metropolitan area. A metropolitan-wide, multi-purpose body, with financial sources adequate to enable it to discharge its responsibilities, offers financial stability and viability to the sub-metropolitan units and reduces fiscal competition between these lower-tier units. On the other hand, the special district may also play a significant role in introducing a service on a metropolitan-wide basis. Given the realities of a politically fractionated metropolitan area, such as Manila, the introduction of special districts is an achievement in itself, and a big step toward the politically remote metropolitan government. Thus, the Calcutta group tended to favour a two-tier system of government, but with a number of services provided by special districts. However, the Calcutta group seemed to imply that in time these services would better be provided by the multi-purpose, metropolitan-wide body.

We turn now to an examination of the sources of funds that might best be exploited by metropolitan governments if they are to satisfy the special needs of the metropolis. We shall also consider how the management of metropolitan finance might be improved.

BORROWING VERSUS CURRENT FINANCING OF CAPITAL PROJECTS

While current financing of capital projects is seen as being highly desirable, borrowing, once utilized, is difficult to avoid. To switch from borrowing to current financing requires that during the period of transition the present community bear the burden of amortizing past debts, meeting present expenditures and financing future needs. Due to heavy debts facing most municipalities and metropolitan areas, this transition is made all the more difficult. Inflation is often seen as a justification for borrowing, but it should be remembered that such borrowing itself may feed inflation. However, if borrowing is based on the assumption that a debt may be repaid in constantly less valuable money, the risk is taken that the rate of inflation may not remain high over a long period of time.

Yet, in some instances borrowing may be unavoidable even if the required amount is anticipated well ahead of time. Long-term planning of capital outlays and the integration of these outlays with the national planning process that operates in some countries, such as Hungary, does not necessarily enable the metropolitan government to finance the very heavy capital outlays that enable it to capitalize on the economies of scale, nor does it place it in a position to make a large purchase of, say, an extensive tract of land at an opportune moment. In such circumstances it is likely that borrowing is the only way in which adequate funds can be found in the time available.

One possibility is that the metropolitan government borrow from itself by establishing a capital fund. Lady Hicks' suggestion that a consolidated loans fund is the most preferable form of such a fund was given considerable support by the Calcutta study group, although it was recognized that in Calcutta, for instance, the lack of expertise in financial management may prove a hindrance to its successful operation. The City of Johannesburg has used the consolidated loans fund scheme to move further into the capital and money markets and at the same time effectively eliminate the idea of the sinking fund and ear-marked funds. In a two-tier system of government, considerable savings to sub-metropolitan governmental units, which would pay higher rates for loans in the external market, can be offered through the establishment of a metropolitan capital fund. Similarly, senior governments can assist metropolitan and sub-metropolitan governments by establishing low-interest loan funds to reduce the cost of borrowing.

In fact, in many countries municipal authorities borrow only from senior governments. This may be so even though the municipalities can theoretically borrow on the open market. This is the case with the Calcutta Municipal Corporation, which borrows from the government of the State of West Bengal. Unless capital projects proposed are part of the state and national five-year plans, no capital funds are available to local bodies in the Calcutta area.

Whether in developing or more developed countries, metropolitan-wide bodies, by reason of their size, physical resources and better management, are in a better position to compete in the money market than are sub-metropolitan units. Although Lady Hicks does provide us with an example of the large corporation borrowing for one of its smaller neighbours, it would seem that this practice is uncommon.

ACHIEVING A DIVERSIFIED REVENUE BASE

Although some borrowing must be accepted as inevitable, there is much room for improving the sources of current revenue available for financing metropolitan area services. It is desirable that any government should have a fairly diverse tax base in order to protect itself from revenue fluctuations from any one source. Local government has traditionally had to rely quite heavily on the real property tax. Although Lady Hicks is in favour of retaining this tax as the most important source of revenue for metropolitan governments, the study groups and seminar participants gave considerable attention to improving the application of the tax and also reducing reliance upon it by improving other sources of revenue.

Opinion seems to be divided as to the most desirable basis for a real property tax. In Johannesburg, the tax is based on the capital value of the land only, and the formal impact of this tax is upon owners. The Johannesburg group noted that the main effects of shifting part of the tax away from land and onto improvements would be that small house owners would be taxed more, whereas householders on large lots and owners of semi-derelict properties would benefit. In the "downtown" areas, changes one way or another would not be very drastic. The Johannesburg group was in favour of retaining the existing basis of its property tax, although it was noted that this basis discourages the central city from the annexation of suburban areas—an action that may be in the interests of developing a metropolitan-wide government. In Calcutta, where the tax is based on the rental value of land and buildings, and the formal incidence is upon occupancy, much room was seen for improving this tax. At the present time it is included in a consolidated rate with service taxes on water supply, conservancy and lighting. The study group suggested that all of these taxes be separated and that the consolidated rate therefore be discontinued. The group also suggested that the concept of *hereditament,* which exists in Britain, should be included to enable assessment to take account of the variety of rights that are attached to certain properties. Also, the group wished to see machinery and plant assessed. In addition, assessment is very irregular and consequently very inequitable. The existence of rent controls on certain buildings further distorts the use of the rental value as the basis for assessment. If these rent controls are to continue, there should be a surcharge on the assessment of these buildings. The Calcutta group indicated its preference for a capital-value base but

mentioned that this suggestion had been rejected several times in the past in favour of the well-tried, rental-value basis.

In Istanbul, rental value is used as the basis of assessment for a land tax. In addition, the provincial government collects a tax levied on the assessed net annual rent of buildings and 25 per cent of this tax revenue is returned to municipalities, according to the derivational principle. While the Istanbul group called for tax reforms, it did not express dissatisfaction with this particular tax.

The Winnipeg group indicated its preference for a greater emphasis on the taxation of the land element and a reduction on the building element of the tax base, although no preference was expressed one way or the other for a capital-value base as opposed to a rental-value base.

Judging from the above reactions, it would seem that the system based on the capital value of the land element alone would be the most suitable for a metropolitan property tax, especially in a fast-growing metropolitan area. It is interesting to note that the Athens group suggested that such a tax should be introduced in Athens, where no property tax exists at the present time. Full-value assessments are the preference of several study groups, although the Toronto group noted that it would not necessarily be as easy to explain the merits of the full-value assessment to the local taxpayer as had been implied by Lady Hicks. The Toronto group was also wary of increasing all valuations in an area on the basis of one property sale. They noted that even with a completely *bona fide* sale, other factors must be considered. For example, "block-busting" possibilities may exist, or the parcel may have unusual value to a particular person. To illustrate this point, the group mentioned the sale of a vacant plot of land in the Toronto area that attracted four bidders at $300, $6,000, $7,000 and $20,200.

The general opinion regarding exemptions to property tax was that these be eliminated and, if necessary, replaced by a specific grant.

Discussion of the income tax at the seminar conference was limited to consideration of a low-rate income tax or payroll tax. This limitation is most likely accounted for by the fact that although the study groups were extremely interested in what Lady Hicks had to say about the access that municipalities have to the income tax in Sweden, the likelihood of local or metropolitan governments being given access to this source of taxation seemed extremely remote. On these grounds alone, it was felt that this alternative did not warrant much further consideration. In addition, constitutional problems must be anticipated if metropolitan governments are to seek a share in the income tax. This restricted outlook must be regarded as lamentable. Clearly, the considerable size of some metropolitan regional governments that will emerge, if the proposals put forward in the last part of this book are followed, is such that a share in a national income tax would be an extremely suitable form of revenue, in terms of equity, for a metropolitan government. It was also unfortunate that discussions did not compare the impact of the property tax and the income tax on the visual environment. The desperate desire to obtain increased real property investment, in addition to the abandonment of environmental values, is in large part the product of heavy reliance on the property tax for local revenues. In contrast, the heavy reliance on the income tax, as administered by the central government in Sweden, removes from local authorities the incentives to seek more development solely for the sake of increasing the municipal assessment. Environmental qualities tend to benefit from this arrangement.

The payroll tax, despite its ability to tap the concentration of high incomes earned in the metropolis, is generally regarded as having an unsophisticated base. If it were to be applied in a high-unemployment situation, as exists in Calcutta, it is likely that it would act as a disincentive for labour employment. Also, as the New York group pointed out, while a local payroll tax may well be advantageous to the central city as a means of taxing commuters, who live in areas not serviced by the city, the tax as administered by a metropolitan regional government would lose much of this advantage. However, according to the principles of horizontal equity, this would be desirable, but in terms of vertical equity, the tax is highly regressive. The Winnipeg group expressed some concern that the use of the true income tax at the local level might result in this source of taxation being over-burdened, with a corresponding mal-effect on initiative of individuals and private enterprise. Thus, although the present level of taxation in Winnipeg could be substantially increased, the group felt that this should be done through sources other than the income tax.

The levying of a local sales tax by a central city authority often results in the diversion of retail sales, and even the retail outlets themselves, to fringe areas of the metropolis. This has been the experience of New York in recent years. However, the tax presents little problem in this regard if used by a metropolitan-wide government. Like the income tax, this tax calls for a metropolitan body that reflects the concept of the socio-economic region. The sales tax is likely to be regressive, although this can be adjusted by confining its application to certain items—especially luxury goods.

Although the sales tax can be applied solely to the metropolitan region, majority opinion seems to favour its administration as a shared tax, which implies far wider application. This is partially because of the problems of collection, but also because it is in fact being used by many national and state governments at the present time. The Calcutta group supported Lady Hicks' assertion that administrative skills are lacking at the local level in India, as is probably the case for developing countries generally. This would support the case for a shared tax administered by the senior partner. Yet, if the sales tax is administered on, say, a state-wide basis, obviously the metropolis is put at a disadvantage, as the portion of the tax retained by the state is likely to be spent in such a way as to achieve a greater degree of vertical equity throughout the state. Thus, it is in the interests of the metropolis to request that a sizable percentage of the revenues be returned to local or metropolitan governments on the basis of the derivational principle.

The prevalence of street trafficking in developing countries is, as Lady Hicks has noted, a major hindrance to the collection of a sales tax at the retail level. An ingenious solution to this problem has been found by the government of Brazil. Each sales receipt represents a ticket in the national lottery. Purchasers are thus far more anxious to obtain their receipts, and revenues have increased markedly since the introduction of this system. As both the income and the sales taxes are directly related to national economic activity, the control of these taxes is best left with the national, or at least the state, government for purposes of regulating the economy. Thus, the use of these taxes by a metropoli-

tan-level government should be controlled by a set ceiling or some established relationship to national economic policies.

Of the motor vehicle tax little more was added by the study groups to the observations made by Lady Hicks. The tax is seen by metropolitan authorities as one source of revenue of which they should have a greater share. At the present time, revenues from licences, fuel and the purchase of vehicles is largely retained by senior governments. Again, this is best administered as a shared tax.

The case for user charge's being employed by local and metropolitan governments as a partial substitute for the property tax was presented by several study groups as another possible source of revenue. The New York group, arguing on the basis that the major mission of urban local governments is to allocate resources in a way which will provide goods and services necessary to support rising private standards of living, thought that the use of "private sector" pricing rules may be extremely beneficial in financing those public services. For example, transportation services, water supply, waste disposal, fire protection and some recreational services accounted for nearly 22 per cent of total current expenditure (excluding interest on debt) and nearly 55 per cent of total capital expenditure by local governments in the 38 largest American metropolitan areas in 1964-65. In some developing countries this percentage may be substantially higher. In contrast to Lady Hicks, who regards user charges mainly as benefit taxes, the New York group preferred to regard them as devices akin to resource-allocating prices, which would encourage economizing on the use of scarce resources and afford guides for public investment decisions. Although this approach also received support from the study group in the resource-scarce metropolis of Calcutta, it must be pointed out that it is easier to argue the case for pricing road services—the example elaborated upon in greater detail by the New York group—than for the other services mentioned. This is particularly so if it is implied, and presumably it is, that there could be a severe limitation on the provision of these services. The undesirability of this situation is elaborated upon by Lady Hicks.

In the City of Johannesburg, major services being provided in a new development must be financed by the developer. These services would include electric power, water and, in some cases, sewerage. The developers must also provide adequate park space, power sub-station plants, school sites and so on. These costs the developer can then include in the selling price of the developed lots or building units. Additionally, the developer must pay endowments to cover the future cost of expanding certain facilities associated with the development. Whether or not this application of user charges is direct enough to serve as a resource-allocating price mechanism is difficult to determine.

Operating costs for such services as water, electricity, sewerage and garbage disposal in Johannesburg are met by separate user charges. In Calcutta, as has already been mentioned, water, electricity and sewerage are financed out of a consolidated rate. The Calcutta group, in the interests of efficiency and fairness, would prefer to see separate user charges levied for each service, as many areas do not receive all the services in question.

In Johannesburg, there is also an arrangement that may provide an answer to the problem of taxing personal benefit or other increment that accrues to property owners from the rising values normally accompanying urban expansion. If a developer asks for a re-zoning of a piece of land, he must expect to surrender in tax up to one-half the accrued increment in land value as soon as his request is granted. If a re-zoning is carried out other than at the landowner's petition, he must pay part of his capital increment when it is in fact realized. The Johannesburg group did not indicate that there were any problems in administering this benefit tax.

While the general feeling among the study groups and seminar participants was that the above sources of current revenue should be retained as a diverse tax base for a metropolitan region authority, there was a similarly strong desire to see a greater use of senior government grants. The advantages of grants have already been discussed by Lady Hicks in her essay. For the metropolitan authority, the ability to retain responsibility for the administration of services that it would otherwise, for lack of financial resources, have to hand over to the senior government, is obviously a great advantage. Furthermore, this arrangement is desirable in that it permits a satisfactory balance between the efficient provision of a service and the offering of popular access to the administrative authority. The fact that this implies a loss of fiscal autonomy for the metropolitan or local authorities does not seem to be of such great significance as one would be led to expect from the essay by Dupré. The study groups seemed to accept that the grant-giving governments may wish to determine the standard of a service that a grant-receiving government is responsible for administering. Providing that an appropriate degree of programme determination and control is retained by the lower levels of government, it is quite acceptable for senior governments to exercise their prerogative to review the performance of local or metropolitan governments through various institutional arrangements and also to provide the mechanism for public appeal.

A second reason for favouring grants is that they provide an excellent method of achieving greater equity among metropolitan areas with different resources. This provides a strong link between metropolitan and national socio-economic policy objectives, as expressed in a comprehensive plan. In fact, the Calcutta group made a point of indicating that the achievement of national objectives should come before the achievement of equity.

Thus, both conditional and unconditional grants may be quite acceptable. The greater degree of central government participation is viewed favourably in that the central government becomes more aware of the needs of the metropolis. On the other hand, heavy financial commitment on the part of the central government is bound to be accompanied by more intensive supervision. It is largely a question of the manner in which this supervision is carried out that is crucial to the success of intergovernmental relations. The conditional grant, which is provided on a matching-grant condition, is not very popular. This type of grant can generally be utilized only by wealthier municipalities, and even these may find themselves committed to expenditures that they later discover they are unable to afford. Unconditional grants, while much favoured by the receiving authority, are criticized on the grounds that they can be easily misused and also can discourage development of local revenues. Being open to such criticism, they tend to become subject to limitations, and in effect, become conditional grants.

OBSERVING THE PRINCIPLES OF
METROPOLITAN FINANCE

What makes for a balanced revenue structure is a moot point. Presumably the structure has to provide the necessary funds and, at the same time, do this without offending the principles of metropolitan finance, as outlined by Lady Hicks. However, no one tax, or other source of revenue, can satisfy all of these principles, and while the net effect of a diverse revenue structure may achieve this, there will naturally be certain local preferences expressed as to any bias that might be manifest in the total revenue picture. It is of interest to comment on some of the observations made in this regard.

The principle of vertical equity, which involves the unequal treatment of unequals through the redistribution of income, while regarded as a highly desirable objective, could lead to too great a diversion of the financial resources of the metropolis to other parts of the country. This assumption is based on two grounds. First, there is concern that, whereas tax capacity may well be higher in the metropolis, there may also be a correspondingly higher degree of tax effort (exploitation of the tax resource) in the metropolis. Second, this erosion of the financial resource of the metropolis by the central government may impose strict limitations on the ability of the metropolis to be discriminating in the spending of locally-generated resources. This runs counter to the need for a strong degree of flexibility, on the part of the metropolis, to encourage local initiative in the improvement of environmental conditions.

These improvements can be justified according to the principle of horizontal equity, which accepts that if one metropolitan area can afford improved services, beyond the national minimum, it should be permitted to tax itself additionally for their provision. It is questionable, however, whether the sources of taxation used by the metropolis, which are mainly such regressive taxes as the property tax, should be used to pay for such improvements. The argument put forward is that the metropolis should have greater access to such non-regressive taxes as the income tax. By utilizing a progressive income tax for additional local revenue, the metropolis is furthering intra-metropolitan vertical equity. Also, not only should the metropolis have access to non-regressive taxes, but also the central government should be limited in the extent to which it can call upon the revenue potential of the metropolis. This relates to another issue associated with the principle of fiscal autonomy. The problem with local fiscal autonomy is that it cannot necessarily be reconciled with the maintenance of a strong national fiscal policy. The unfavourable differential between the contributions of the metropolis to the gross national product and what it receives in return by way of benefits appears as a basic issue of contention between local and metropolitan governments, on the one hand, and national governments on the other. It was felt by a majority of the seminar participants—perhaps somewhat selfishly, given the rural poverty in many countries —that a greater percentage of financial resources should remain within the metropolis rather than be transferred to other areas in the form of grants. The percentage of revenues generated by the metropolis that remains within the metropolis for its own use varies tremendously. In the western world it averages 15 per cent; in Poland it is about 30 per cent; while in Yugoslavia it is estimated to be about 65 per cent. Obviously, there is much scope for improving the situation in some countries if one believes that local discretion over dispersal of locally-generated funds is desirable. On the other hand, it was recognized that the needs of the rural areas could not necessarily be met by rurally-generated fiscal resources, especially if rural needs are related to the urban economy and way of life. An example of this has been provided by Philp in his essay on education (Chapter 5). The need to provide rural children with an education that will equip them for an urban way of life is bound to make some demands on outside (urban) revenues.

Some of this apparent conflict is eliminated if the metropolitan region's designs for comprehensive planning are large enough to incorporate not only the built-up urban area but also the hinterland that comprises the balance of the socio-economic region. The larger the region, the greater the degree of both horizontal and vertical equity possible in the region. The problem, although it is not insoluble, is one of disbursing funds across the metropolitan region so as not to neglect the needs of the rural parts of the region.

One last comment might be made on the criterion of economy in administering a tax. It was the opinion of the Toronto group that false economies are often practised in the interests of keeping down administrative costs. The assessment of real property is especially vulnerable to this type of practice. While an increase in expenditure on assessment procedures may not result in increased revenue, it is extremely likely to improve the equity of the tax. Frequent comments on the poor state of financial management in other metropolitan areas would tend to support this. It is to this subject of financial management that we turn now.

FINANCIAL MANAGEMENT

Lady Hicks' remarks regarding financial planning, budget preparation, accounting techniques and the use of cost-benefit studies met with general concurrence from the study groups. As can be judged from the above comments on the use of various taxes by metropolitan governments, the availability of expert management for the proper administration of these sources of revenue was one oft-repeated qualification. It is perhaps regrettable that the study groups did not extend their discussions into this general area. In part, the problem goes beyond financial administration. For example, one of the biggest problems that New York is facing now is that of finding solutions to what are termed "socio-economic equations". Both in the metropolitan area and across the country as a whole, there is a great need for a closer look at the way financial resources should be allocated to bring about desired social and environmental conditions. Given the complexity of present-day urban society, the development of satisfactory redistribution formulae can only be the product of a truly comprehensive planning approach. Unfortunately, this approach is far from being realized.

At the seminar-conference various procedures for budget preparation were discussed. One technique that avoids the presentation of inflated departmental budgets and that contributes significantly toward the vertical integration of financial administration is the preparation, by the metropolitan government staff, of an estimate of the total budget available relative to the national economic forecast. Once the executive committee has approved the total budget and departmental allocation, separate departments make out their own budgets on the basis of this allocation. The key to this process would seem to be the selection of staff to draw up the original estimates. A variation upon this arrangement for financial planning is the necessity for the metropolitan government to obtain approval from the central government for

its budget on the basis of its relationship to the national budget.

Budget preparation would seem to be a product of financial planning if it is merely a modification of a long-term financial plan. Preparation of five-year, functional, spending estimates within the framework of a twenty-year programme has to be seen in this light. Programme-performance budgeting was thought to be highly desirable as an aid for decision-making on objectives and major allocations, and for detailed planning and performance analysis of programmes. However, the limitations of its present development were recognized. One problem has been its initial introduction in an understandable way. The City of Manila first used performance budgeting in 1963, and the initial practice of preparing both performance and line-item budgets is continuing. The changeover was apparently much simpler for those agencies with more quantifiable outputs, such as public works and communications. Indeed, this changeover has not been completed for all agencies. Once again, the basic cause of the problems associated with this transition can be accounted for largely by the lack of qualified technical staff available at the time. It is unlikely, however, that any metropolitan area will be able to make transition without many teething troubles, and no doubt the City of Manila will be able to eliminate existing problems in the future. In short, the City is to be commended for having attempted the task of introducing performance budgeting in the first place.

THE PLANNING AND DEVELOPMENT OF THE METROPOLITAN ENVIRONMENT

The interrelated nature of metropolitan problems and of solutions to those problems—the main theme of this book—has necessarily resulted in some rather arbitrary organization of their discussion. This is particularly true of the discussions on the general subject of the planning and development of the metropolitan environment. What follows is a commentary on the reactions of the study programme participants to the major issues raised by Weissmann and others on metropolitan planning. However, much of what has been treated in greater detail elsewhere in this book, for example participation in planning and intergovernmental relations, has not been needlessly repeated. Hopefully, by reference to the summary on the comprehensive planning process, presented in Part IV of this book, the reader will be able to better relate all these issues to one another, including the following comments on Weissmann's essay.

There was very little questioning of what Weissmann had to say in his essay, and consequently most of the following comments are an elaboration upon his remarks.

THE EXTENT OF THE URBAN CRISIS

Although the alarming dimensions of the urban crisis, as pictured by Weissmann, appeared very real to many study groups, the study groups from Warsaw and Moscow observed that with a system of centralized planning their countries are unlikely to face many of the problems that arise solely through the absence of any planning whatsoever. Yet for most countries, whether of a planned economy type or of a market economy type, the existence of a system of national, regional and local plans does not necessarily mean that many of the problems identified by Weissmann are not being, or have not been, experienced. The very apparent disparities in standard of living that exist between rural and urban areas and also within the urban areas, obviously exist in all parts of the world. It is largely due to these disparities that the

shifting of rural populations to urban areas takes place. In part, the disparities exist because of the lack of thought given to national priorities in the past. Even with a changed pattern of settlement and human activity, they may continue to exist in the future for the very same reasons. The enormous expenditures on defence, to which Weissmann refers, is an obvious expression of this poor sense of national need.

In the developing countries, where resources are far more scarce, there seems to be a much greater awareness of the wastage of such resources as serviced urban space and funds for urban research. Yet this wastage takes place, and probably at a greater rate, in the more developed countries too. The Karachi study group strongly agreed with Weissmann's suggestion that urban research should give more attention to operational problems rather than try to explain the causes of certain problems that are now historical fact. Obviously this comment cannot be taken literally, but it does serve to maintain a correct perspective on research priorities, especially in developing countries. The Karachi group cited also the need for some basic research to prevent the further wastage of resources and some research on what national priorities should be. One of the greatest problems confronting many metropolitan areas in developing countries is the task of housing and providing jobs for rural migrants. In many instances, much money may be spent on housing migrants and in settlements removed from the metropolitan area. Yet, the lack of employment available locally for these migrants may cause many of them to return to the metropolitan area, which would result in a wastage of investment in the housing scheme. If the migrant chooses to stay he may be unemployed or under-employed, the result being a wastage of human resource. If the migrant chooses to travel back and forth between the housing scheme and the metropolis in order to retain a job, the net result is that he enjoys a lower standard of living due to greatly increased travel costs. This is an extremely simple example of the all too common lack of thought given to urban development.

Even if priorities are developed by existing levels of government, however, they must be related one to another. The lack of a satisfactory national framework, within which metropolitan planning authorities can operate, represents as great a problem as that posed by governmental fragmentation within a metropolitan area. Indeed, the Warsaw group noted that the present Polish attitude is to put more emphasis on developing a framework for the coordination of plans rather than on reducing the number of governmental units that are involved in producing and implementing the plan. Again, whereas the Corporation of the Metropolitan Area of Hanover (*Verband Grossraum Hanover*) might be regarded as the perfect example of a city-region planning authority, it lacks the necessary economic development framework that should be provided by the state and federal governments concerned.

Yet, despite this inventory of the weaknesses of existing planning, there was complete acceptance of the need to plan, which, it might be noted, has not always been the case. There was also complete agreement with Weissmann that the purpose of planning is to guide development for human progress and that the role of the city is vital to this process.

The city, however, should be seen as subject to continuing morphological change. This provides an added challenge to those charged with developing institutional frameworks within which development planning will be carried out. The

decline of the central city and the increase in population and industrial activities in the fringe areas is a common phenomenon. The central city is becoming increasingly dominated by tertiary sector activities, which, being accompanied by a decrease in central-city living, is resulting in an extended journey to work for a greater number of commuters. The implications of this for governmental structures has been referred to earlier in this chapter. The high concentration of these tertiary activities in the central city and the problems that this poses for communications is one of the basic reasons for the many suggestions that have been made for the creation of sub-centres within the metropolitan region.

Not only is the metropolis changing within itself, but it is also very often changing in relation to other urban centres both on a national and on an international scale. The Hanover group commented on the effect of the creation of the Common Market on Hanover's location relative to the centre of commercial trade for West Germany. There were also some interesting comments made with regard to the role of the metropolis as a capital city. The Buenos Aires group made reference to the consideration that has been given to moving the Argentine capital out of Buenos Aires in an attempt to divert some of the growth taking place in that metropolitan area. However, if there is anything to learn from the Pakistani experience, it is to be noted that the recent establishment of the capital of Pakistan in Islamabad does not appear to be in the least detrimental to the growth of the former capital of Karachi. Of course, what one really needs to know is whether it has affected Karachi's growth in any way whatsoever.

COMPREHENSIVE PLANNING

As Weissmann makes clear, too often there

is a tendency to equate change with progress, and too readily do we accept the changes occurring in our economic and social activities and our environment, whatever their direction, as being for the good. If human progress is to achieved, these changes must be guided by policies designed to realize society's consensual goals. Such policies have to indicate the desired direction, location, nature and pace of change. Such policies must be expressed in terms of responsibility for implementation and those institutions charged with these responsibilities must be provided with the means to perform their designated tasks. In short, these policies will be expressed as a comprehensive plan. Different institutions, whether governmental or nongovernmental, will be responsible for parts of the plan. Weissmann has illustrated why it is at the regional level that planning coordination can take place most satisfactorily. Before elaborating upon the methodology by which the regional unit is identified in the case of the metropolis, it is necessary to explain the term "comprehensive planning" as used in this book.

Although it is extremely difficult to define in precise terms and even more difficult to convey vividly, a term is required that gives expression to the on-going relationship among all systems, operating (in this instance) in the metropolis, that contribute positively or negatively to all aspects of its development. Short of inventing a completely new term, it is impossible to avoid some confusion over the meaning, or meanings, of the term selected. The term "comprehensive planning" embraces both the on-going comprehensive planning process that constitutes the forward motion of the planning and the series of comprehensive plans that mark the stages of on-going activity. Used in this sense this term came nearest to meeting the needs of the organ-izers of the Centennial Programme. However, it must be stressed that the concept enunciated here is not necessarily very similar to the interpretations previously given this term by others.

An initial and partial exception to this is the definition of a plan that is given by Weissmann. The Programme participants accepted this definition in principle but found that it did not spell out clearly enough the significance of the political and administrative structure for both the "model" and the "programme of action" to which Weissmann refers. This may be the product of generalization, and perhaps at the time of its enunciation in Stockholm, in 1961, the definition was conceived as embracing these elements. However, the first part of the definition appears to be heavily influenced by the physical manifestation of an economic and social plan, and the second part lacks reference to political action, especially the informal (as opposed to the formal, which is implied by reference to "legislative measures"). Thus, an adjustment might be made to this definition that would regard a comprehensive plan as:

a) a model of an intended future situation with respect to: i) specific economic, social, political and administrative activities; ii) their location within a geographic area; iii) the resources required; and iv) the structures, installations and landscape which are to provide the physical expression, and physical environment, for these activities; and
b) a programme of action and predetermined coordination of legislative, fiscal, administrative and political measures, formulated with a view to achieving the situation represented by the model.

This definition may require some further work regarding the enunciation of the scope of the comprehensive plan. However, as it stands, it applies equally well to the metropolitan situation as to the non-metropolitan situation.

The term "comprehensive planning process", the second element of comprehensive planning, is used to describe the process that produces and implements the plan. It should be made quite clear that comprehensive planning is a continuing process involving almost constant adjustment to a "model" as it is being developed. This adjustment is made in response to "feedback" or reactions to either proposals or subsequent developments (and lack of development). In this way the implementation of the plan is seen as being part of the process of further developing the plan through modification.

The comprehensive planning process, viewed simply, is seen to consist of the following elements: the development of an information base; goal formulation; plan formulation; plan implementation and plan evaluation. Again, however, these occur concurrently rather than sequentially and are knit together by a sixth element: the network of communication linking professionals in many disciplines, institutions, governments, voluntary organizations and individual citizens.

To be meaningful, comprehensive planning requires the integration of political, administrative, economic, social and physical planning. This is especially true when planning is directed to basic questions such as the allocation of resources, anticipation of desired population movement, location of industries and so on. It requires that there be effective coordination of planning by governmental and non-governmental institutions. It also requires that the political decision-makers have good working relationships (formal and informal) with the administration. Integrative action and an adequate knowledge base require an inter-disciplinary approach to problems involving those in the whole range of physical and social sciences. Thus, the emphasis must be on securing mutually compatible and supportive action by a wide range of organizations having different decision centres. Machinery must exist that will bring about coordination of governmental and non-governmental activities (for example, social services and the location of public and private facilities) and the various activities conducted by different levels of government (for example, education and transportation). It must also provide for the formal and informal participation by individuals, institutions, agencies and governments in the processes of defining needs and the transformation of these into policies and programmes of action.

With the above minor adjustments to Weissmann's definition of a plan, the concept of comprehensive planning as it now emerges can be applied almost anywhere, although it must be emphasized that the adaptation of the model to local circumstances will produce many variations. The important point to make is that throughout these variations, there are found the universal elements of a basic model. For the metropolitan region, as for any other region, the comprehensive planning unit has obvious implications for the development of a governmental structure. As it is desirable to illustrate the way in which all elements of the governmental and non-governmental machinery interact to service the metropolis, a summary description of the comprehensive planning process is presented in Part IV of this book. By using this as a common reference point, the reader should be able to more easily relate the many considerations, described elsewhere in the book, which have to be taken into account in order to achieve human progress according to a comprehensive plan.

THE CONCEPT OF THE METROPOLIS

Before proceeding to a discussion of the metropolitan comprehensive planning unit, it is necessary to expose what appear to be two divergent views regarding the concept of the metropolis. With this additional understanding, it will be easier to appreciate the different demands being made upon the comprehensive planning unit. One concept is that of the metropolis as the primate city or "mother city" in which can be found a variety of services unlikely to be found elsewhere in the same country. Because it is the primate city, it dominates the country's politics, economy and social life. The fact that there tends to be only one such metropolis in any one country gives special character to its problems and to the way these problems should be approached. The metropolis is commonly conceived not just as the built-up area but as a more abstract phenomenon with extensive spatial influence. The concept appears to be built of activity patterns superimposed upon one another, and whereas some influences or activities are of international significance, others are less extensive and provide a more tangible measure of the sphere of influence, for example, commuting patterns. This is the concept of metropolis as defined in more "scientific" terms by Gorynski and Rybicki in Chapter 9.

The other concept of the metropolis is that of a major urban centre, but not necessarily limited to the primate city. Again, the metropolis is seen as a service centre, but generally one servicing only a region of the country. More important, the metropolis appears to be a more tangible entity identified by a "container"—that is, it has a clear-cut boundary, for example, of a statistical unit or a political unit. This concept appears to have been developed from a general over-view of a relatively large number of metropolitan centres in any one country. For example, the Standard Metropolitan Statistical Area unit provides a definitive unit in the United States, but while this conveys an image of a physical entity, it in no way conveys an image of a great centre of power as does the image of the primate city.

The most important point to make here is that the lack of terminological distinctions and a significant basis for differentiating between areas has led to confusion regarding the comparability of metropolitan areas. In fact, any two metropolitan areas of comparable size tend to have many problems in common, although the state of technology will greatly influence the intensity of the problem. However, perhaps because these two concepts represent only the two ends of the spectrum, it is understandable that modified versions of each were held by persons from different parts of the world. With nearly all Centennial Programme participants coming from centres of well over one million population, and many of these being primate cities, there was a tendency to favour a spatial unit for the comprehensive planning of the metropolis that would be based more upon the metropolitan sphere of influence. An amalgamation of these two concepts presents us with the "metropolitan version" of the city region concept put forward by Weissmann.

THE CITY REGION CONCEPT AS THE BASIS OF THE METROPOLITAN PLANNING UNIT

The region, as Weissmann illustrates, has become more acceptable as an operative unit of development planning. This applies to both the centrally-planned economy seeking to overcome some of the difficulties inherent in detailed planning from the centre and also to a country with a market economy in which planning is mainly local and urban, and in which there is a growing awareness of the necessity to expand the spatial coverage of the "master plan" of any metropolitan area to take in the area directly influenced by, and in turn influencing, the development of the metropolitan region. In fact, both the process of devolution from above and evolution from below may be found operating in any one country. For example, Poland has passed through the phase of the regionalization of national plans to that of the expansion of city plans to the scale of the city-region.

If we accept Weissmann's assumption that urban agglomeration is basic to the developmental function, it is presumably desirable that a nation-wide structure for national development would be built upon a system of city-regions. Naturally, these regions would vary in size according to the size of their respective urban centres. However, as Weissmann makes clear, these regions are not being established solely to facilitate the production of regional plans but to establish the "ground rules for a process . . . of productive interplay among the mainstreams of development in order to produce a human condition . . . which is progressively improving as a result of the effect of economic, social and environmental progress on each other and as a result of the total contribution to human development."

Thus, while the comprehensive planning unit for the metropolis should reflect activity patterns that in turn reflect the daily lives of the bulk of the inhabitants of the metropolis, that same unit must be considered as part of a system of regional units for the national planning of the country as a whole. Thus, while the Hanover group expressed general satisfaction with the fact that the city-region of Hanover had been

given political and administrative expression with the establishment, in 1962, of the Corporation for the Metropolitan Area of Hanover, there was a pronounced dissatisfaction with the lack of regional planning by the state of Lower Saxony and the West German government, especially planning that might recognize the existence of the metropolitan region as such.

In addition to the special consideration that may be required for metropolitan areas, national policy should also encourage a viable environment in small cities and, in large countries, this will call for a concerted policy of localized concentration. Recognition of the fact that the metropolitan area is not the only region of concern and that there is a need for balanced development, emerges in different approaches to the drawing of metropolitan regional boundaries. Countries with centrally-planned economies tend to restrain the growth of the metropolitan centres and encourage the growth of smaller centres. This practice is especially well-developed in Poland, where growth centres fit a clear pattern, or hierarchy, of service centres. The Polish policy on urban growth is to achieve local concentrations in a national pattern of urban deconcentration. In this way it is hoped that the disparity that exists between urban and rural areas in terms of living standards may be reduced. The Warsaw group indicated that even with the controls available to the central government, restrictions on the rate of growth of Warsaw have met with no more than moderate success. About 15,000 migrants are added to Warsaw's population every year. Although the boundaries of the metropolitan planning area are designed to accommodate future growth, they do not reflect an anticipation of unlimited growth. Although consideration has been given, in the past, to creating one political unit of the planning area which now extends well beyond the central city, present policy favours the development of arrangements for coordination as opposed to structural reform.

In countries with market economies it is generally the fear of the political dominance of the metropolis, which could continue to grow hand in hand with the apparently never-ending economic and physical growth, that discourages national governments from giving the metropolis a distinct political expression. In instances where there has been an attempt to lend identity to the metropolis as a political unit, the authority is generally restricted to the built-up area or, if for a larger area, the powers of the metropolitan authority are often weakened by various devices. What national governments should recognize is that, whatever their policies regarding future growth, metropolitan areas require the provision of certain services on a metropolitan-wide basis. For this reason it is highly desirable that there be some form of political expression for the metropolis and, preferably, for its region. Any adjustments that are made to the political machinery should take place within the framework of the national policy devised to guide urban growth on a nation-wide basis. Hopefully, in this way any political adjustments will not run counter to economic and social measures that are built into the plan.

Before leaving the subject of the size of the metropolitan region, mention should be made of other forces that play a part in dictating its size. The political boundaries of the metropolitan authority should be as extensive as the area of urban growth and its rural environs if financial equity is regarded as a goal. The larger the area of jurisdiction, the greater the degree of equity that may be achieved. This tends to apply whether the concern is for spreading out (geographically) the impact of the property tax (which, in many instances, is the major source of revenue for financing what are, in fact, regional services), or for ensuring that the commuter pays his share toward the upkeep of central city services, or, inversely, for ensuring that the city merchant does not harbour an undue share of the profits from his trade with those in the trade area of the metropolis. Again, the need for fiscal autonomy of the metropolis calls for a large region, although this should be regarded as a more flexible demand than the others listed here.

Also calling for large metropolitan regions is the indication that public nuisances are more easily controlled by a single authority. This is probably borne out very well with reference to crime. With regard to air pollution control, the spatial scale of operation involved indicates an intergovernmental approach to the problem rather than just an expansion of the metropolitan unit. In the discussions on intergovernmental relations, however, the view was expressed that, for simplicity of dealings between governments, at the local level, an expansion of their territorial size and a consequent reduction of their numbers would go a long way to resolving the present problems arising from the sheer multiplicity of bodies involved. Also expressed was the generally-felt need for a national policy of localized concentration of urban development. This would involve a degree of deconcentration from the large cities, and obviously, such a policy should be intimately related to a policy concerning the desired size of metropolitan areas. However, it may be desirable to qualify these remarks as they apply to the metropolitan areas of developing countries. In each country a policy of greater concentration of growth in the metropolitan area may be desirable if that country is to reach the economic take-off point as soon as possible.

For the very large metropolitan region, however, it is quite conceivable that this deconcentration should take place both intra-regionally and on a nation-wide basis. For example, the deconcentration around Paris, Tokyo and London is being backed up by a policy for deconcentration (from any given metropolis) on a national scale. The point to be emphasized here is that to draw the line (for purposes of establishing the limits of the metropolitan region) between deconcentration taking place within the immediate sphere of influence of the metropolis and that taking place further afield, may be not only arbitrary but also difficult in practice unless there is a clear-cut physical, social and economic differentiation between satellite towns and new centres of growth. Certainly, however, the socio-economic region of a metropolis and its satellite towns would call for a very large spatial unit.

As the creation of satellite towns in close proximity to the existing urban area was suggested by the groups from Calcutta and Buenos Aires as possible solutions for accommodating the growth of those metropolitan areas, it may be desirable to think in terms of a lower tier of smaller units in addition to the metropolitan region units.

This last consideration was generally felt to be desirable to encourage greater participation in metropolitan affairs. This stems partly from the often-repeated need to encourage popular identity with a body politic, which itself requires a tangible territorial base (the most satisfactory being the relatively small unit of the built-up area). There is also the need to facilitate the active participation of citizens in the policy-making and programming of action (frequently found in developed countries)

and in the implementation of plans (of more interest, although not necessarily extensively followed up, in developing countries). This requirement can be met with a two-tier metropolitan government structure which permits the metropolitan-wide body to be very extensive.

Another factor is the area required for efficient administration of most services. At the Toronto seminar-conference, most discussions on the scale of operation for services favoured an efficient unit that would contain within its boundaries the area benefitting from a given service and that could also retain the flexibility of being extended beyond the present built-up area as required. There was a general feeling that the single tier of metropolitan government would prove adequate for this, although at the same time it was recognized that different services may have different areas of optimum efficiency. What was not satisfactorily answered is the need to allow for future growth. Under what circumstances should the growth take place within the present boundaries (thus permitting greater control of growth); or, under what circumstances should the boundaries move out with the growth? There was much talk of flexibility, and generally this was interpreted to reflect support for arrangements to extend the boundary of the metropolitan authority with the area of urban growth. However, as has already been noted, this type of suggestion is often not very acceptable to senior governments.

Deciding upon the criteria that should be used for the precise determination of the limits of the comprehensive planning unit proved to be a most difficult problem for all Programme participants. The commuter "catchment area" presents one of the more easily identifiable units. Apart from this, the area of optimum efficiency of services appears to be of greatest potential as a yardstick. However, although discussion took place on the utilization of cost-benefit techniques to measure the area of efficiency of operation, it was generally concluded that the technique is poorly developed to date. Indeed, until these techniques have been improved it is difficult to know to what extent their findings should be taken into consideration. On the other hand, the prospects in this field were viewed optimistically, and there was a strong plea for further research to develop these techniques to a point where a better evaluation of their merits might be made. Without wishing to jump to conclusions, it is likely that with improved cost-benefit techniques it would be far easier to allocate functions to levels of government in such a way that the spill-over effects of those functions would be internalized in the manner suggested by Dupré in Chapter 11.

Other comments made by Programme participants, on the general subject of planning, serve to underline the attitudes that have already been expressed on such subjects as intergovernmental relations and the role of citizens in the decision-making process. As such, these comments are better treated in other relevant sections of this chapter and in Part IV.

SOME PRELIMINARY CONCLUSIONS

In the concluding remarks to Chapter 8, it is noted that for the effective provision of the service function of government, a regional form of government for the metropolis is a high priority. In Part III there has been added to this a similar demand on the basis of facilitating the effective performance of the redistributive function of government and, although qualified to safeguard the value of participation, that of the political function.

It is tempting to regard a two-tier system of government, both tiers being multipurpose authorities, as being the goal to which all metropolitan area reforms should be directed. However desirable this may be, such an arrangement is far from realizable in some of the political climates that exist around the world. The question is more often one of the next step that might be taken and whether or not it will make the achievement of an end objective that much easier or more complicated. For example, it seems likely that the special district is the easier agency to introduce than the single-tier, multi-purpose agency (maybe of the central government), although the latter is more likely to lead to the creation of a two-tier, multi-purpose system. Unfortunately, it is not a simple binary choice.

The preceding chapters have also thrown much light on the allocation of responsibilities to metropolitan governments, as well as on where and how power should be used. The case for a strong metropolitan-wide governmental authority can, however, be accepted only if the power to influence the national pattern of the future urbanization of the country continues to rest with its national government.

Just how this can all be achieved, without undue conflict, is difficult to imagine. The problems are enormous. The proposals for reform must be comprehensive and ambitious, yet realistic. Part IV provides a modest comment on the comprehensive planning process that must be pursued to realize these objectives.

Part IV

The contemporary
metropolis:
Toward a system of
comprehensive
planning and management

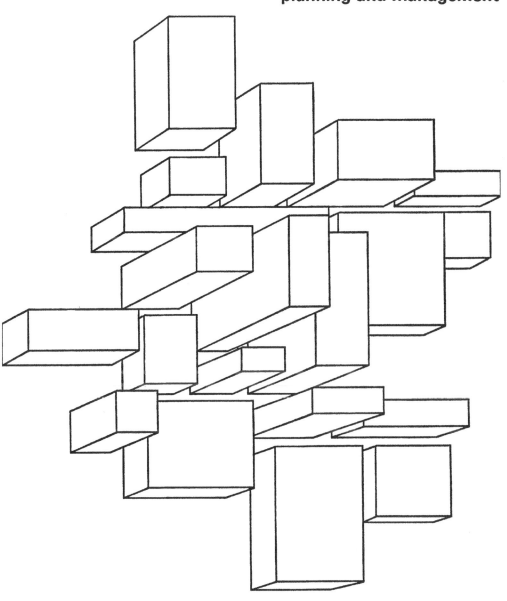

Introductory Note

The comprehensive planning process, as part of comprehensive planning, and as defined in Chapter 14, provides a suitable framework within which we can view the interrelations that exist between the issues that have been discussed in parts II and III of this book. The value of this holistic overview cannot be over-emphasized. It is vital that any decision-maker or citizen realizes the nature of this process and the role that he has to play within it. There must also be the awareness of the impact of decisions taken in one sector of activity upon others and an ability to weigh priorities within a larger frame of reference. This framework can be provided in greater or lesser detail. In this section the objective is only the provision of an outline through which the more detailed discussions presented earlier in the book can be related one to another.

It is appreciated that any attempt to describe, simply and succinctly, the entire on-going process of development of all of the elements that make up a metropolis, and the nature of the interrelations among those elements, is open to question not only at the stage of conceptualization but also at the stage of articulation. This book, like the Centennial Programme, has been more concerned with the conceptualization of the metropolis and its problems. It attempts to reflect new perceptions and thereby stimulate the extension of, and adjustment to, the concept of the operating metropolis and its interdependent parts. This process of conceptualization can be developed in depth in any one language and, as such, can be articulated. Just as conceptual thinking permits the tying together of percepts into a logical whole, so the great value of conceptualization should be the facilitation of perceptual transfer on a universal basis. The better the concepts, the easier is the transfer of percepts and ideas and the more likely are solutions to be found to problems faced.

It is at this juncture that the limitations of language as the vehicle of transfer become clear. Furthermore, these limitations are not simply those that exist between "tongues", such as Japanese and English; the constraints on communication are often very great between persons supposedly of the same tongue. Words may have different meanings to different people, and, unfortunately, it often takes a long time to discover what these words are and the nature of the difference.

Thus, the effective articulation of concepts that are expected to be understood universally is no easy task. All this is said in anticipation of some semantic confusion over the choice of the terms "comprehensive planning", "comprehensive plan" and "comprehensive planning process", each of which has been defined in Chapter 14. Each of these terms is intended to express a greater breadth of meaning than can be conveyed by the words "planning", "plan", and "planning process". Whereas "planning" is used here as something distinct from administration—and management—and for this reason has often been used with modification throughout the text, "comprehensive planning" should be understood here to embrace administration and management. A case could be made for referring to this totality as "comprehensive-planning-administration management" or some such phrase. Hopefully, the reasons why this has not been done are obvious! Thus, while there will always be different interpretations of each of these words, the only way in which we can begin to make ourselves clear is to ask the reader to accept the concepts as put forward here for the purpose of making the most of this material. The following outline of the comprehensive planning process should serve to tie together, conceptually, those perceptions made in the earlier parts of the book.

15

The Comprehensive Planning Process in the Metropolitan Region

Simon R. Miles

The comprehensive planning process, although a continuum, can be broken down into the following stages for purposes of discussion: the development of an information base; the development of a communication network; goal formulation; plan formulation; plan implementation; and, plan evaluation. It is according to this breakdown that the following discussion of the comprehensive planning process is presented.

The Development of an Information Base

The development of a satisfactory information base is essential to the planning process and, as such, can be regarded as part of the process. The present information base is extremely inadequate in most countries. This, in large part, explains the inability to plan comprehensively. The greater the information gap, the greater the need for a flexible plan to permit adjustment upon the receipt of further information. While the principle of updating and flexibility is sound, an overly flexible plan loses the basic quality of serving as a reliable guide. Also, without information, there is often a lack of awareness of actual social and economic needs.

More factual information is required on human, physical and economic resources.

Social indicators have yet to be developed to match the relatively satisfactory economic indicators. Social accounting has to be put on the priority list for further development if we are to replace the present system of accounting for social productivity through economic accounting. As Professor Bertram Gross pointed out in his plenary address to participants at the Toronto seminar, "in accordance with the high principles of national economic accounting today, building a jail is productive, training teachers is not productive. Teacher training is social consumption. . . ."

There is also the need to reconsider the breakdown of factual information. Territorial and sectoral breakdowns should recognize and utilize the comprehensive planning unit of the metropolis. Thus, it is highly desirable that statistical units coincide with such planning units. This does not mean, however, that existing statistical units should form the base for the development of metropolitan planning units. Rather, adjustments should be made to the statistical unit. Certainly, the potential of computer techniques for the collection, storage, retrieval and presentation of small-area data offers much promise for remedying this problem.

Also of importance is the need for more information, albeit of a less factual nature, on attitudes, for, as was made clear in the discussions on education and on intergovernmental relations, these play a most important part in policy-forming and the decision-making process. Again, there may be much potential in the new techniques being evolved in gaming simulation.

The development of an information base has to accompany, to some extent, the delimitation of a comprehensive planning unit for the metropolitan region. However, as the criteria for this have been well treated in Chapter 14, it will not be necessary to say anything further here.

The Development of a Communication Network

Although the need for planners to communicate with the general public is sometimes seen as a problem of information—information, for the planner, on the people's needs; and information, for the people, on the planner's proposals and the reasons for them—the problem is really more subtle and complex. Each planner needs to know what other planners are doing, and each group among the "people" needs to understand the various interests of other groups. As a matter of fact, in the formulation of goals and plans, and in the implementation and evaluation of plans, the better the quality of communication in every direction, the better the planning.

How this is to be achieved is not readily apparent. Certainly there should be opportunities for hearings when the proposals of planners are given an airing in public. There are also other ways of organizing "feedback" and to be effective, the feedback must be organized. This may be done through many voluntary organizations, labour unions and citizen organizations. Care must be taken, however, that communication channels are not overloaded. The Milanese example—a city in which forty non-governmental groups and forty governmental agencies participated in the process of examining the city's transportation problem—may not be the answer. The mass media, skillfully used, can make a most useful contribution to the comprehensive planning process. Specific reference can be made to the use of television by the Regional Plan Association in New York to conduct a metropolitan-wide survey in a project known as "Goals for the Region". This project, which utilized 5,600 citizens, encountered considerable difficulty in getting the participation of a representative cross-section of metropolitan New York (that is, in bringing in minority groups), but it did serve to highlight the potential in the media. The mass media may also be expected to create a general interest in civic affairs and establish an identity with the metropolitan area or body politic that may be of assistance in developing a better understanding of the nature of metropolitan problems. The newspapers, like television, have a potentially far greater and more useful role in this matter than they now play.

The professionals also need to develop more and better methods of communicating with one another. In addition to conferences on metropolitan issues, more might be done in the way of exchanges of visits among experts in metropolitan areas. Another proposal with great potential as an information source, and which was under consideration in the United States at the time of the conference and is now being implemented, is that of "urban observatories", which will utilize in an organized manner the resources of universities to assist governments in surveying urban problems. The universities are a major information resource, and both they and the public authorities have a responsibility for public education, the facilities for carrying out this responsibility, and the general acceptance of the public for implementing this role.

Research efforts directed toward expanding the information base should extend beyond national boundaries to draw upon international experience and a wealth of ideas that already exists. Cooperation in international research programmes provides a way in which the study of particular needs of one area could benefit from, and contribute to, the work in other areas. The obvious need is for the organization of such programmes and the development of an international communication network to service the needs of metropolitan area administrators and decision-makers.[1] The International Association for Metropolitan Research and Development (INTERMET), which is now being developed as a result of proposals emanating from this seminar-conference, will hopefully fulfil these needs.

Goal Formulation

Generally, the more extensive the information base and the communication network, the easier is the task of goal formulation. Although it is possible to observe empirically much of what is desirable, it is virtually impossible to estimate how far removed the desired condition is from the present. The same applies to giving expression to those needs, an act required as part of the process of plan formulation. The fate of the metropolis is of national concern, and metropolitan goals are closely allied with national societal goals. Society's basic goal is to improve the conditions and well-being of its members. Thus, urban society, as part of a national society, should strive to achieve economic growth that will redound to the benefit of the national society. Life in the large or the small city must be an enriching experience for all. The big question, of course, is the extent to which present needs are discounted for future satisfactions, as the latter may be achieved only after a period of concentrated economic growth. It is within this context that

[1] Many international networks do exist, ranging from the United Nations Organization and its various agencies to professional organizations. However, none of these focusses specifically on the needs of metropolitan areas or is designed to carry out multi-interest research and action programmes for the improvement of metropolitan conditions. INTERMET will attempt to meet this need.

more specific goals should be expressed for metropolitan environments. Both territorial goals and sectoral goals should be in accord with national needs at all times, and their attainment should be recognized as subservient to the attainment of national goals. Thus, while such a goal as the completion of a transportation system may be to move goods and persons, this is not an end in itself. The aim is to do this in such a way as to improve the well-being of the individual and society as a whole.

It had been proposed to the seminar discussion groups that goal formulation is a product of "interest-based planning". This process requires the planners to make available resources and information to those persons, groups and institutions charged with formulating goals. Although the planners should be involved in producing goals, they should not assume sole responsibility for this. Goals should arise out of the interplay of various factions of society, as provided for in the institutional arrangements of that society. The planners may have a significant role to play in this process, in that they attempt to reconcile competing interests in order to establish a coherent (and consensual) whole. Individual demands and societal needs have to be adjusted, one to another. It is, however, through the institutionalization of this interplay within a council or its equivalent, especially in the final stages, that results in the decision as to "who wins" from interest-based planning. Given this process, the consensual goals called for by Smallwood in Chapter 10 seem easier to achieve. However, some reservations might be expressed as to whether society's real needs would emerge from the interplay of a number of interest groups that are not necessarily representative of that society. This point appears valid in as much as, although the decision of a legislative body is theoretical-

ly representative as a decision, the proposals put to the legislative body may not reflect society's needs. Also the council may not be, in fact, representative.

The need for a national urban policy has been raised several times throughout this book. The need grows out of the recognition of the high social costs incurred in instances of unrestricted growth that do not relate to national needs and the fact that the fortunes of nations are closely related to those of their major cities. While the degree of detail incorporated into national urban policy planning might vary according to a number of factors (the size of the country, stage of technological development, administrative competence at the local level, and so on), the need for national guidance is paramount. There appears to be an obvious feeling that although metropolitan areas may be regarded as the powerhouses supplying the national grid, they are nevertheless not producing satisfactorily. There is similar dissatisfaction with regard to social conditions in the metropolis.

What, therefore, should a national policy set out to achieve in order to remedy these common complaints? In directing itself to the prime goals, the national urban policy should provide clear direction on at least the following issues: the political and administrative decision-making structure; industrial location; urban (de)concentration (matters relating to the choice of growth poles, new towns or satellite centres); the optimum size of urban centres to ensure adequate provision of cultural, social and physical requirements; manpower training (including the training of rural migrants for urban jobs); the attainment of fiscal equity among metropolitan areas and in relation to rural areas; the establishment of national minima for services; population and immigration; coordination of planning;

research on urban matters and so on. In a federal state, certain of these directives would be issued by the provincial or state governments. Within the context of the national urban policy, goals for the metropolitan region could then be determined, and, provided that a regional authority exists, much of this responsibility could be vested in such an authority.

Plan Formulation

In general terms, the answer to "What appears in the comprehensive plan for the metropolitan region?" has already been provided in the definition of a comprehensive plan given in Chapter 14. Here we shall look in greater depth at some of the points of major concern: the expression of goals in the plan; requirements of the plan in terms of space, infrastructure and resources; and the programming of action for the implementation of the plan.

THE EXPRESSION OF GOALS IN THE PLAN

Goals, while they can be stated in general terms, have to be given concrete expression in the plan. In the metropolitan region plan they must be stated in terms of regional needs. Standards have to be set up as targets and should be expressed in both quantitative and qualitative terms whenever possible. However, a strong caveat must be issued against the setting of standards that are unrealistic for the metropolitan region in question. Thus, international standards are generally ruled out—although an exception to this is certain health standards. Nation-wide standards, however, are generally acceptable because of their impact on the attainment of nation-wide balanced economic, physical and social development. Naturally, this would have to take into consideration the varying intensity of use made of services. For this reason it would be necessary to have higher standards, in metropolitan areas, for such services as transportation. The question of how the standards should be arrived at depends very much on the individual service. In part, it is a process of measuring demand (especially where choice exists, as in the case of transportation). Kain's essay on transportation has dealt at some length with the use of techniques for measuring consumer demand. In other cases, supply may have to be taken into account. As Sazanami has illustrated in Chapter 4, this is certainly the case with housing, of which there appears to be a universal shortage. It was in the discussions on housing, where the needs appear most desperate, that the discussion of standards was reduced to a discussion of socio-economic possibilities. Because of the need to provide such basic requirements as food and capital for other services, the realization of socio-economic possibilities in housing is very limited. Such a situation brings home forcefully the need to consider all social, physical, economic and cultural needs in an integrated manner rather than independently. Decisions resulting from such considerations are obviously affected by societal values.

Other considerations pertaining to standards were brought out in the discussion of utilities. Naturally, policy regarding the social aspects of utilities greatly affects the standards of services incorporated into the plan. Seminar participants generally agreed with Hanson that utilities should be provided on the basis of economic cost. However, in instances where the social responsibility of a utility is recognized, it should be compensated in its operations if it achieves certain social objectives at an economic loss.

Other problems pertaining to the setting of standards arise from the present inadequacy of the information base. This became clear in the discussions on health and welfare standards. Without adequate proof as to the effect of environmental pollutants on man, the establishment of standards for environmental conditions can be challenged as being somewhat arbitrary. Indeed, it might be simply a question of aesthetics. Unfortunately, aesthetic considerations, far from forming the sole basis on which we establish standards, tend to be set aside entirely—whatever the service being considered.

Such qualitative considerations and the societal values referred to earlier bring to the fore the question of value formation, which obviously underlies goal formulation and the setting of standards. For this reason, the key role of education is of particular interest. It should be noted that although curriculum is not discussed by Philp in Chapter 5, the impact of curriculum on metropolitan problems was well aired at later stages in the Programme.[2]

An education curriculum oriented to either the economic, cultural or social goals of society, but not to all equally, will produce a similar imbalance in the outlook and values of the children in the education system. Unfortunately, such a choice may have to be made between these value orientations in countries with limited resources. In Chapter 8 reference is made to the situation in Ibadan, where, as in many other areas represented in the Programme, there is a need to provide a uniform quality of education in rural and urban areas and, at the same time produce technicians with higher skills than are normally provided in the public education system.

There seem to be no direct answers to the questions raised by this kind of dilemma beyond the fact that this problem appears to be national rather than metropolitan in scope. To achieve a fair degree of national uniformity of standards, a higher authority (probably the central government) must assume responsibility for overall regulation as well as a role in financing the education system. Yet, there is a reluctance to see this regulation as the sole responsibility of a senior government. Within this structure, the metropolis must be permitted the flexibility to meet its special requirements (in general, the demands put upon it through the presence of minority groups of all types). Again, the potential of providing special schools with the economies of scale that operate in a metropolis should be maximized. However, just how to reconcile the generally-felt pluralistic demands of the metropolis with the need for centralized regulation and financing is a question that arises in many service fields, and any accommodation of one service should preferably take into consideration these other demands. This problem has already received much attention in several of the earlier chapters in this book, notably in Part III, where the issue of the location of power, whether financial, regulatory or policy-making, is discussed in relation to the structures that must be devised for the optimum use of that power. Later in this chapter attention is focussed on the machinery that has to be devised for the operation of the comprehensive planning process in the metropolitan region.

[2]The omission of a discussion on curriculum in Chapter 5 is due to the decision taken early in the organization of the *Centennial Study and Training Programme on Metropolitan Problems*; it was thought that too much attention might be given to this aspect at the expense of others. Whether this decision was justified is, perhaps, in retrospect, questionable.

SPATIAL REQUIREMENTS

As the concept of the metropolitan region has been treated in considerable detail in the latter part of Chapter 14, it will suffice to mention here the key dictates affecting the spatial requirements of a comprehensive metropolitan plan.

Strictly speaking, the determination of the territorial unit for planning programmes is part of the planning process. From a national perspective, it is an essential step in the allocation of planning space. One of the basic reasons why comprehensive planning rarely exists at the present time is that the allocation of planning space should be, but generally is not, a joint responsibility of all levels of government. In many countries, mainly those with market economies, the national government is unwilling to play its role because this requires recognition of, among other things, the metropolis as a political entity. Yet, if comprehensive metropolitan planning is to exist, it is vital that there be recognition of the metropolitan socio-economic unit. This unit then, must be given some political expression if it is to be a comprehensive planning unit. As has been mentioned in Chapter 14, it seems to be generally accepted that, for planning purposes, the week-day commuting distance provides the best indicator of the spatial extent of the metropolitan socio-economic unit. (This has obvious implications for the scale of the metropolitan regional governmental unit.)

Both the metropolitan area rate of growth and technological improvements in modes of transport affect the delineation of the commuter watershed. The national plan should assume prime responsibility for the general character of metropolitan growth and some responsibility for modes of technology utilized for transportation. However, the prime concern for the impact of technology in this instance, expressed by the form of transportation used for the home-to-work trip, should be with the metropolitan region plan. Certainly, some metropolitan transportation services are part of a national network. (This is obviously so with railways and to some extent with roads.) Rapid transit rail services, on the other hand, are much more obviously the concern of the metropolitan plan. This is necessary if the public is to have an opportunity to express its preferences for one type of service over another, as a choice is made not only on the basis of available technologies but also on what the area can afford. Discussion of the latter point will be treated more fully in considering infrastructural requirements later in this chapter.

All this points to the fact that in most countries (exceptions being city states), at least two levels of government have a major role to play in influencing the size of the commuter watershed and, therefore, in the delineation of the comprehensive metropolitan planning unit. It is up to the national government to play the lead role. In performing its task the national government must give recognition to the metropolis. Yet the concentration of almost all urban development in a single metropolis is, more often than not, undesirable from other points of view. World opinion seems to indicate that in such circumstances it would be beneficial if the national government were to create counter-magnets by developing certain other cities to metropolitan size and by improving the quality of their environmental services. This would also solve the political problem of the dominance of one very powerful body at the metropolitan level.

It is in the creation of counter-magnet centres for development that the application of central-place theory becomes pertinent. Theoretically, each centre selected for such development could be regarded as the central place of the city-region or planning space. The distances between the central places (or counter-magnets), like the size of the planning space, are proportionate to the size of its central place. A national pattern of centres thus emerges. Each centre, in turn, has its own system of sub-centres around it, and, if necessary, these sub-centres can perform the role of planning-space centres of a smaller scale.

Thus, the relationship between a national system of city-regions and the national responsibility for allocation of planning space is clear. The Polish experience in this regard has been referred to in Chapter 14, and perhaps it is in that country where this thinking has been most developed to date.

INFRASTRUCTURAL REQUIREMENTS

A seemingly universal problem confronting major metropolitan areas is the inadequate investment in infrastructure, a problem heightened by the equally obvious disparities that continue among areas within the metropolis. This problem is best viewed in terms of infrastructural investment per capita. For example, despite economies of scale, areas of high-density population (daylight or night-time) tend to suffer from an inadequate per capita infrastructural investment. One such manifestation of this is the high-density slum area with its very much over-used facilities. It is unfortunate, however, that the association between high population density (in terms of land area) and poor living conditions, has become so widely accepted. High density living conditions *per se* need not result in poor living conditions. The distinction has to be made between population density in terms of land area alone and population density in terms of each or all of the different elements of the urban infrastructure (the more important of which

are power facilities, drainage facilities, fixed transportation facilities and, included with this definition, buildings) plus internal and external living space. However, because this distinction is generally not made, any inadequacy of infrastructural investment has to be inferred, and somewhat inaccurately to say the least, from the ratio of population to unit area of land. Because of this association of high density with poor living conditions, any improvement in living conditions is often confined to the negative approach of deconcentration of population, that is, moving people out to rural areas or to new towns.

While these new towns may provide adequate facilities, this approach has to be seen more as one of diluting the ills of the metropolis than tackling the problem at its source by improving infrastructural investments.

This should not be interpreted to mean that new towns are ruled unacceptable; they are not. It simply means that in many countries the new investment in infrastructure is not always concentrated where it is needed. There should be an attempt to improve living conditions while maintaining present densities, not merely to keep the central areas "alive", but also to counteract the extremely high costs of deconcentration. This is particularly so in developing countries, where such costs tend to be prohibitive. Deconcentration of population also requires the concomitant deconcentration of jobs. This again is extremely difficult to achieve, especially in developing countries, precisely because of the high infrastructural investment required to make such a move attractive to, and productive for, industry and commerce. Furthermore, in some of the bad examples of deconcentration efforts, the costs have probably been felt more by the individuals involved than by the tax-paying public at large.

Thus, in the developing countries especially, but perhaps the same applies to the more developed countries, it is usually preferable to invest more in the existing centres. Where so many services are being over-used at the present time (for example, the Tokyo transportation services are estimated to be operating at 300 per cent of planned capacity, at times), the problem becomes a question of investment priorities.

What should our priorities be? There was general agreement among those participating in the Centennial Programme that the need for basic shelter takes top priority. This need, as with the basic infrastructural requirements with which it is related, is much more marked in the developing countries. The approach to the housing problem requires consideration not only of the need to capitalize on existing services and to maximize productivity of land, but also of the minimum needs of the individual and his ability to pay for these needs. The approach that tends to satisfy the former requirements is to build upwards rather than outwards. This approach, however, has given rise to problems in some areas where high-rise housing has been found unacceptable. Instances have arisen where tenants would not move into dwellings "in the sky". This provides very good evidence of the need for a stronger information base—in this case information on popular attitudes, as referred to earlier in this chapter. Another basic approach tends to meet the latter requirements of the individual rather more than those of economic productivity of land and services. This involves the provision of the minimum necessary amount of private (internal) living space and a compensating maximum amount of public (external) space, which is communal. However, either approach requires large-scale development if schemes are to be comprehensive, and this, in time, requires either the nationalization of land or the exercising, by the municipal authorities, of powers of expropriation. This is a subject to which we shall return.

The Japanese approach to the improvement of the social overhead investment per capita ratio is of interest. The theory maintained is that conditions can best be improved in the metropolis if land uses are separated. This separation of functions obviously demands a very good communication network between land-use zones. Thus, investment in metropolitan transportation is seen as having high priority. Furthermore, this acceptance of intra-metropolitan concentration of activities is accompanied on a national scale with the building up of counter-magnets, as was referred to earlier in this chapter. The Japanese approach is to develop each counter-magnet in the context of its larger metropolitan region. For example, Nagoya might be regarded as the primary counter-magnet for Tokyo and Osaka. In this instance, therefore, Nagoya is being developed not only as a counter-magnet but also as a regional centre. The desired aim is to concentrate important service activities of the Nagoya metropolitan region (known as the Chubu region) within the built-up area of Nagoya. Infrastructural development is therefore designed to further this aim. Thus, for the larger region to prosper, Nagoya must prosper, and all the efforts of the region are concentrated in Nagoya. This in turn helps to develop Nagoya as a counter-magnet to Tokyo.

However, the heavy capital investment required for development of infrastructure calls for very careful programme budgeting. Whereas urban areas in such countries as the United States may be in a position to plan for obsolescence, for the underdeveloped countries this approach represents an obvious waste of resources. The real interest is in supporting heavy initial investments in infrastructure that will last a relatively longer period of time. Naturally, it is the funding of this heavy initial investment in, say, housing that proves to be the greatest problem. Judging by the observations made by the Study Programme participants, the building industry may well be efficient, but nowhere is there evidence of an actual reduction of costs, and increase in standards, in the production of dwelling units.

Before leaving the subject of infrastructural requirements, and at the risk of being repetitive, it is of interest to observe that the foregoing has an impact on the structure of metropolitan government. The enormous costs involved in making any investment in infrastructure, plus the area-wide nature of so many services, often call for the acceptance of service responsibilities by a metropolitan-wide governmental body. This could be either a single-purpose or multi-purpose body. However, the only justification offered for single-purpose bodies is that they may be politically more acceptable and may be a step toward the formation of multi-purpose bodies. Contradicting this view, however, is that experience tends to indicate that the existence of single-purpose bodies generally makes reform more difficult to achieve at a later date.

Another argument in favour of the multi-purpose body is that it facilitates the establishment of priorities between services. It has already been stressed that all urban services should be regarded only as means to achieving societal goals and that an excellent service is not an end in itself. A multi-purpose body can help retain this perspective, which tends to get lost with a single-purpose arrangement. It is this latter

situation that may result in a transportation or water authority's attempting to dictate future direction of development. Again, there is always the need to consider the entire system of a service rather than part of a system on its own. Transportation provides an excellent example. As Kain points out in Chapter 2, there is a tendency for too much attention to be given to the main haul of a transportation system (for example, the rail car portion of a rapid-transit system) and not enough to the feeder and distribution elements (generally motor vehicles) at either end of the main haul. This short-sightedness naturally negates the utility of, in this case, the main haul of the system. In summary, therefore, the coordinated provision of any one service and the integration of service priorities in terms of planning and financing to meet the desired goals of society are facilitated through a multi-purpose body as opposed to single-purpose bodies.

RESOURCE REQUIREMENTS

A major element in the formulation of plans is to ensure the adequate supply of resources necessary for the development and implementation of the plan. These resources fall into the general categories of physical, human and financial. However, the treatment they receive here is not all-embracing. With regard to physical resources, consideration is limited to land, the most important resource. With regard to financial resources, the discussion is again limited, as this has been well treated in chapters 12 and 14. Further, the following comments are also limited to the discussion of these resources as they pertain to the development of the metropolis.

Land as a Physical Resource
Discussion of land as a natural resource is excluded here. Also excluded, therefore,

is discussion of the natural products of land: food, wood and minerals, as used for construction materials or power. The planning of the production of these resources is part of the comprehensive national plan within the context of which one should view the comprehensive planning of the metropolis.

As a resource, land is, with few exceptions, relatively static. Land cannot be produced or expanded easily. Exceptions to the rule obviously do exist; the best examples are the production of land through the creation of polders, in The Netherlands, and the loss of land through coastal and riverine erosion experienced in all parts of the world. However, despite this relatively static supply, the demand for land in metropolitan areas increases as economies expand.

Thus, the problem facing those devising a metropolitan area land-use policy is to increase the productivity of existing land. Essentially, this involves determining the optimum use of the land and providing conditions that make that usage possible. Determining optimum use presents many problems, both for the market economy and for the planned economy. Theoretically, in a market economy under conditions of static supply, any increase in demand for land by any one sector of the economy should result in an overall change of land values and a considerable shuffling of land uses. Although the land-use pattern is not that sensitive to changes in land values, and although the market has generally had to operate within the confines of some controls, considerable wastage of buildings and infrastructure has occurred because of the volatile nature of the land market in areas where competition for land is strong.

In a planned economy, optimum use may be easier to achieve in that there exists a far more effective system of control of land use through the nationalization of much or all of the land. Furthermore, this is backed up by a system that can ensure development according to the land-use plan. Without a market mechanism, however, it is virtually impossible to know whether the land-use plan is optimal in economic terms, although considerable advances have been made in the use of computers to simulate optimal conditions. Depending upon its level of sophistication, the use of cost-benefit analysis could replace the "market" as an indicator, and mention should be made here, therefore, of the techniques being developed in Poland in this field. The replanning of Skopje, in Yugoslavia, provided the testing ground for the "threshold" optimization technique developed by the Polish planners working on that United Nations mission. The technique involves cost-benefit studies of different land-use patterns. It is understood, however, that the technique suffers the normal problems in that it is underdeveloped in the social accounting aspects. However, as this technique is now being applied in Warsaw, it should be possible to refine it as experience is gained.

The determination of the optimum use of land in a metropolitan area and the metropolitan land-use plan that emerges from this process, must take into consideration the same factors, whatever the system utilized. These are the physical factors of availability of land according to location and size of land parcels—which will obviously be subject to technological changes with the requirements of expansion; the economic factors of the existing uses and services in the metropolitan area and also the costs of providing additional services; and the social factors that will affect such decisions as the broad patterns of land use, the relationship of one use to another and how much land should be given over to any one use. Also to be taken into consideration may be certain existing constraints, such as the use of land for revenue purposes. Apparent conflicts that might arise between, say, the existing transportation system, pointing to a linear development, and the socially desirable pattern of small communities, can only be reconciled in terms of the goals set.

Once the allocation of land use is decided upon, it is necessary to ensure availability of land and provide a means by which land may be acquired. The Toronto seminar discussions on housing generally concluded that nationalization (or at least intermittent expropriation) of land is necessary if large tracts are to be assembled, thus enabling large-scale and economical housing developments to take place. As can be imagined, the same could be said of other land uses. However, while recognizing that nationalization of land solves many problems, the seminar participants readily accepted that this measure is not essential for assuring the satisfaction of demand for all land uses. In general, the needs of the housing market account for the greatest amount of expropriation or nationalization by governments. The references, in Chapter 8, to the Peruvian Land Bank well illustrate the need for powers of expropriation. Such powers are essential to all levels of government and should be incorporated into planning legislation. With a properly developed comprehensive plan, there should be no conflict between governments as a result of several levels of government exercising powers of expropriation in a metropolis.

Once measures have been devised to ensure the availability of land, it is necessary to ensure that the optimum potential is realized. (Obviously, this will have been assumed in the original calculations of allocation of optimum land use and must be followed up.) Beyond the normal system of financial priorities, this is seen to call for the maximizing of present technologies and the sponsoring of innovation through the encouragement of research. Quantitative improvements are called for in facilitating the more intensive use of buildings (especially schools) and the land they occupy (as with the development of multi-storey schools and factories). Qualitative improvements in the environment should also be regarded as part of the process of realizing the optimum potential use of land.

There was considerable debate in the seminar sessions on the advantages of mixed land uses as opposed to separated uses. The Japanese case for separation of land uses has been put forward earlier. The case for mixed use is very strong, though, especially on the grounds that it brings greater interest into the city—but also because it is absolutely necessary in the smaller self-contained community within the (polycentric) metropolitan region. For the future, of course, the rapid technological changes that are occurring today may well have considerable effects on the form of the city. For example, the development of new communications networks and correspondence systems for mass education could do away with educational institutions in their traditional form.

The impact of legislation should also be mentioned here. The availability of condominium legislation in many countries has permitted higher-density development of freehold dwellings, which seem to remain more popular than those in leasehold. Those representatives from countries without condominium legislation generally felt that it would be desirable and that it would assist in increasing the productivity of land in cities. Indeed, this is understood to be a major reason for introducing condominium legislation into Puerto Rico, where land is scarce and individuals wish to own their dwellings.

If there is to be an optimization of land use in metropolitan areas, the techniques for determining use require that decisions pertaining to the whole metropolitan area be made from one office. Naturally, this points to land-use planning as a function of a metropolitan region authority. The general feeling among the Toronto seminar participants was that while such an authority should coordinate economic-development planning and physical planning, it should also provide separate departments for these two activities in order to ensure that physical planning is not dominated by economic-development planning. Below the level of the metropolitan region there should exist local planning bodies responsible for the detailed action plans. More will be said on this in a later section on programming for action.

Human Resources

Just as there is a need to maximize the potential of land as a resource, there is also the similar need for a far better utilization of human resources. Seminar discussions ranged from unemployment and underemployment of present skills to the underemployment of potential skills. To reduce the underemployment of potential skills there exists a vital need to retrain and re-educate those from all levels of society as technological advancements reduce the utility of their present-day skills. All these aspects of maximizing the potential of human resources affect all economies and technologies although, as has been observed earlier, the acute strictures imposed by limited financial resources have affected developing countries far more than developed countries in tackling this problem. Despite the present-day facility for international interchange of technological in-

novations, only the more developed countries are in a position to capitalize upon every technological advancement. Yet, to do so requires constant retraining of their human resources, and whereas their need is not as great, in absolute terms, as is that of the developing countries, this retraining is equally as necessary in terms of making maximum use of the technologies available.

Perhaps it is necessary to distinguish between updating the skills of an individual in his own field and retraining him for another career. Whereas the former process must be accepted as necessary, there is much scope for reducing the requirements of the latter. Essentially, this latter retraining process is aimed at redressing a present imbalance between skills required and skills available. It is an imbalance that the comprehensive p'an should be able to eliminate in large part. The sectoral and territorial breakdown of economic data, to which earlier reference was made, should provide the required information on the quantitative and qualitative characteristics of the desired work force at any point of time within the scope of the plan. Understandably, planning for future employment requirements, as incorporated into the plan, is a national responsibilty. Reference has been made, in Chapter 8, to the desire of local governments to have senior governments assume the greater share of education costs where this is not already the case. However, whereas senior governments should bear the responsibility for anticipating demand for skills, setting general guidelines and maintaining overall control, the metropolitan region and local authorities should be responsibile for retraining immigrants and rural-urban migrants. The local authority is more likely to be sensitive to the potential of the individual and able to react to his needs in the context of the local needs of the metropolis. There are obvious exceptions to this, as, for example, with Negroes in the southern United States and Indians in Latin America.

One last point should be mentioned before leaving the subject of the development of human resources in metropolitan areas. Many persons living in the metropolis grew up in the rural areas, and this will continue to be the case for some time to come. It is in the rural areas that the central government has to ensure that a child's education will equip him for an urban life if he so wishes it. Thus, it is in these areas that there exists a real challenge to the central government. For it is only through providing this type of education that much of the later retraining can be avoided. The implications of these demands of the government are more pertinent at the national level where the emergence of a labour force has to be balanced with employment requirements in the various sectors of the economy.

Financial Resources

There is little more that can be added to what has already been said in chapters 12 and 14 on the availability of financial resources, their development and management and their relative merits and demerits. It has already been stressed that the problem of assuring adequate financial resources is not always one of a lack of funds, but it is often a matter of inadequate financial planning and management. This applies just as much to the more developed countries as it does to the developing countries.

Another major problem associated with the development of an equitable tax base is the size of the tax jurisdiction. As has been made clear in the earlier treatment of the principles of vertical and horizontal equity, the use of a large metropolitan regional tax jurisdiction greatly facilitates the achievement of these desired goals.

The other points of major significance, that have been mentioned and that bear heavily upon the success or failure of the comprehensive metropolitan plan, relate to the composition of the tax base and its impact upon the economic, social and physical development, not only of the metropolitan community but of the country as a whole. A general feeling that emerged from the writings and discussions associated with the Study Programme was that there should be greater use of the more progressive taxes, such as the income tax, and that these should be collected and redistributed by the senior governments. However, if this type of measure is to be implemented it should be complemented with a greater degree of freedom of discretion in spending on the part of the grant-receiving metropolitan authorities. However, details on the problems involved and the various solutions that might be tried in tackling them are provided in chapters 12 and 14.

THE COMPREHENSIVE PLANNING MACHINERY AND THE COORDINATED ACTION PROGRAMME

Thus far, we have examined both the process by which societal goals are given expression and the inputs of infrastructure and resources required to make possible the realization of those goals. It is now necessary to turn to the problem of developing a comprehensive planning machinery that will permit the coordinated programming of action for the realization of the plan.

This machinery has both formal and informal structural elements. It is the mechanism by which individuals, groups and institutional bodies communicate with, and relate to, one another to bring about the plan. It is crucial to societal development, and in many ways it is an expression of society. The machinery that does not facilitate individual and institutional communication in the planning process will lead to frustration; similarly, the absence of an ordered relationship will produce confusion, a non-productive plan and obstruction to a society's development. The precise form of the comprehensive planning machinery will obviously vary according to the society it is designed to serve.

Within this context we can proceed to an examination of the attributes and characteristics of comprehensive planning machinery as it provides for the coordination of all levels of government activity. Additionally, the integration of non-government activities is a function of this machinery. Many of these desirable attributes have emerged in earlier parts of this book. Here their interrelationships and contributions to coordinated action are considered.

Perhaps the most desirable characteristic of the comprehensive planning machinery is that it should recognize the oneness of the socio-economic metropolitan region. While the concept of this region varies according to local conditions, and there may well be disagreement locally on its precise delimitation, a rough approximation of the metropolitan region is certainly attainable. Concrete expression of the metropolitan identity must exist if the public is to perceive metropolitan problems as a metropolitan responsibilty. Similarly, the administration must be structured in such a way that it can tackle such metropolitan problems on a metropolitan-wide basis.

As can be gathered from a reading of Part III of this book, there are many ways in which this metropolitan identity can be achieved. Politically, it is highly desirable that a metropolitan-wide legislative body exist to decide upon metropolitan-wide issues. According to the size of the metropolis, this can be either a one-tier or two-tier structure. A three-tier structure is more likely to defeat the objective of coordination in all but a very large metropolis, where it may be justified on the basis of permitting greater popular participation. The choice of the number of tiers, the size of the metropolitan tier and the functions it should be assigned should be guided by the need to make the metropolitan tier the most important governmental unit in fact and in the minds of the public. Additionally, greater popular identification with the metropolis can be achieved through direct election of metropolitan councillors. It is likely to be less confusing for the electorate if, in the instance of the two-tier system, the local and metropolitan legislatures utilize electoral districts with coterminous boundaries where possible.

A general-purpose metropolitan body appears to be favoured over the alternative of a number of special agencies. Only with a general-purpose body will consensual goals be enunciated, and the study Programme participants seemed to be in general agreement with Smallwood that the expression of such goals may be regarded as a key to achieving identity for the metropolis. Special agencies, particularly prevalent in the utilities and services fields, but also formed in regulatory activities, often acquire considerable powers, and yet they are normally not politically responsible. They can form an impediment to the establishment of a general-purpose metropolitan form of government by providing an alternative method of performing functions over a wide area. Control becomes difficult, and their functions are not coordinated with related functions of other government organizations. At best, they can be regarded as a good substitute for a more democratic and responsible metropolitan approach, often owing their existence to the difficulty of creating a general-purpose body. Their existence is often justified on the (dubious) claim that they are technically efficient and free from political influence. The proliferation of this type of formal agency is generally regarded as confusing to the public and militating against effective participation.

The general-purpose body offers the advantages of greater comprehensibility; however, it should also be adaptable to the many demands made of it. As was indicated earlier in this chapter, the seminar participants general'y agreed that comprehensive planning is interest-based and that the reconciliation of conflicting metropolitan-wide interests should take place in the metropolitan legislature. The formal structure providing access to this arena of reconciliation should be able to accommodate the occasional desire for expression on the part of many pressure groups and institutions that are not part of the formal decision-making machinery. While the formal machinery might be designed to accommodate direct participation, through councillors, committees and so on, it cannot be expected to cater to all, and especially to temporary interests.

The precise size of sub-metropolitan governments is difficult to discuss meaningfully, as this factor is so dependent upon local conditions. However, in any two-tier system, the lower-tier units should all be from about 100,000 to 250,000 in population judging by existing examples. These figures are intended to represent a balance between the dictates of service efficiency and participation, a balance that was sought throughout the discussions on the politico-administrative structuring of the

metropolitan machinery and its relations with local and senior levels of government. It is of interest to note, however, that participatory values were generally given priority consideration by the seminar participants. This might be seen as an understab'e reaction to the pressures for an efficient administration that will be able to operate at the new scale of urban living made possible by technology. If technological advances are not to be the sole dictates of a form of government, participatory values must be retained.

This issue came to the fore in the seminar discussions of the essay by Gorynski and Rybicki. As can be appreciated from reading Chapter 9, Gorynski and Rybicki do not accept that the politico-administrative structure of the metropolis should have to accommodate itself to physical form as dictated by technological change. Instead, the authors adopt a novel and refreshing approach in seeking a form for the future metropolis, in terms of physical and politico-administrative structure, that would permit a maximum of public participation without loss to the efficient functioning of the metropolis. The authors' further elaboration upon this theme and their comments presented at the Toronto seminar on the effect of the forces of space and time on the structure of government and the decision-making process have already been presented in Chapter 14. Gorynski and Rybicki are prepared to accept a trend toward more centralization for plan formulation but they feel that as far as implementation is concerned, this should be decentralized where possible. To this end, the authors presented an indication of the assignment of plan-formulation, plan-implementation and plan-supervision (evaluation of service management) functions, giving due regard to time and scale. This is shown in modified form in Figure

Figure 15-1
The Allocation of Responsibilities for Elements of the Comprehensive Planning Process

Space ╲ Time	current (one to two year plans)	medium (five to ten year plans)	perspective (fifteen to fifty year plans)
local	R F S ⎡ L F S I ⎤		
regional		C F S ⎡ R F S I ⎤ L F I	
national			⎡ C F S I ⎤ R F S I L I

Key: L local government F plan formulation
      R regional government S supervision
      C central government I implementation

    ⎡_____⎤ unquestionable allocation of responsibility

Figure 15-2
The Changing Allocation of Planning Responsibilities as Influenced by the Forces of Space and Time

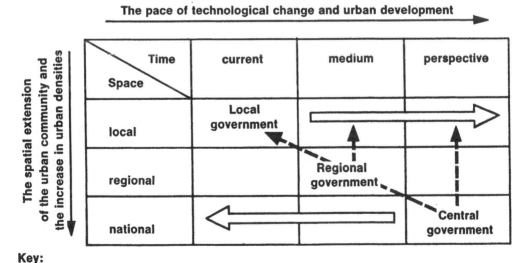

The pace of technological change and urban development →

The spatial extension of the urban community and the increase in urban densities ↓

Space ╲ Time	current	medium	perspective
local	Local government		
regional		Regional government	
national			Central government

Key:

 Direction of expansion of plan formulation and implementation functions

▬▬▬► Expansion of supervision function

15-1. Only those functions that can be easily assigned are indicated in Figure 15-1. The empty boxes are areas in which responsibilities are open to question, although, as Figure 15-2 indicates, the influence of the increased pace of development and also that of increased urbanization and higher densities have resulted in an expansion of planning functions of all levels of government.

The expansion of powers implies a greater sharing of responsibilities. However, if functions are to be shared, and the general conclusions to be drawn from Part III of this book are that there should be more sharing, then there should be a strong focus upon the metropolitan level. Senior governments can most effectively set out policy and review local programmes while lower levels of government can most effectively implement programmes. For some functions, as mentioned in Chapter 14, the metropolitan authority may be given almost complete responsibility.

The assignment of responsibilities indicated in Figure 15-1 and implied in Figure 15-2 calls for strong, and therefore large, regional (metropolitan) government units capable of carrying out planning functions efficiently. Any one regional (metropolitan) government is then broken down into well-defined and clearly-shaped local government units sensitive to, and able to accommodate, democratic participation. The metropolis can meet these demands, but whereas the traditional metropolis is a monocentric metropolis, Gorynski and Rybicki indicate that a polycentric metropolis is more suitable for this type of development in the future. The diagrams in Figure 15-3 indicate that the po'ycentric metropolis can more easily combine physical structure with a politico-administrative structure to give balanced communities throughout its lower tier.

Given the functions that have to be provided in the metropolis, it is likely that each of these sub-metropolitan centres will take on a character of its own. For example, it is particularly desirable that one such centre retain the unifying characteristics of the central city of the monocentric metropolis. This pattern of related centres would be repeated on a national scale if the nation-wide development of metropolitan growth points, as recommended earlier in this chapter, is pursued. It is also of more than academic interest to note that, as with monocentric metropolises, the specialization of functions between polycentric metropolises might repeat the pattern exhibited within any one metropolis. This is certainly the conclusion to be drawn from the classification of metropolitan areas by function, as presented by Jones and Forstall in Part I of this book. Thus, there emerges a strong relationship between the metropolis as a functional unit and the nation as a functional unit. This tends to strengthen the politico-administrative interrelations or at least it strengthens the case for the improvement of these relations if they are weak.

Thus far, comment on the comprehensive planning machinery has been directed to those attributes that facilitate cooperation; namely, the identification of the scope of a function, the assignment of functional responsibility, the willing involvement of individuals and institutions, and the development of a close association between physical structure and the politico-administrative structure of the metropolis and the nation.

It is with the discussions of national, metropolitan and local plans that we come to those elements of the comprehensive planning machinery that give expression to priority-setting and the coordination of action in terms of space and time. Natural-

Figure 15-3

The Reconciliation of the Physical Form and the Politico-Administrative Structure of the Metropolis

a) The monocentric metropolis

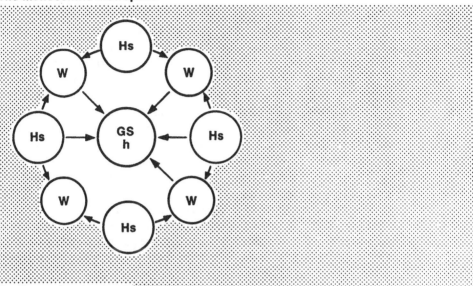

b) The linear development

c) The urbanized (polycentric) region

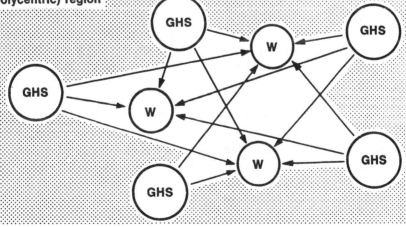

Key: G   government   H,h   housing   S,s   services

   W   work (basic production)   ░░░░   open space

ly, without cooperation it will not be possible for all levels of government to take part in the formulation of long-term, socio-economic policy-making, and neither will it be possible to achieve the consistency of decisions taken at various levels, as this implies exchange of information and co-ordination of activities. These are two vital elements of comprehensive planning.

Most Study Programme participants were generally agreed that priority-setting should be focused at the metropolitan level. As has already been stated, the metropolitan regions are regarded as the growth points in national development. Problems adversely affecting metropolitan development will inhibit sound national development. Thus, national governments should reassess their priorities regarding expenditure of national resources if the figures given in Chapter 14 regarding local discretion over spending of locally-generated financial resources are to be changed.

Although priority-setting should focus on the needs of the metropolis, the general framework for the metropolitan plan (which will give expression to those needs) can be developed only within the context of the national plan. The national plan must be a physical and economic development plan designed to hasten the social development of society as a whole. The plan must express clear policies on growth areas and sectors. For more detailed information on the content of such a plan the reader is referred to Chapter 13. Essentially, these plans must be open-ended, or flexible. They must be reviewed at established intervals. Time factors in the plan should be adhered to as they will be used by metropolitan and local planners as policy guides. Means of implementation must be assured beforehand. The national plan must also be flexible enough to deal with changes in the economy that may be brought about by external influences such as war or the international migration of labour. The metropolitan regional plan will consist of both a general outline plan and a detailed master plan. Again to avoid repetition, these are described, along with the detailed action plans for sub-metropolitan units, in Weissmann's essay.

The financial arrangements that exist to strengthen coordination of plans have already been elaborated upon. Legislative measures were not debated at great length in the Toronto seminar. The general feeling was that legislation should always be flexible enough to permit change and experimentation. The contribution of legislation was somehow seen as far less significant than participation in bringing about cooperation among governments and between the governmental and non-governmental sectors. However, in order that this statement is not misinterpreted, it should be said that most Study Programme participants felt that the lack of comprehensive planning is in large part a result of the lack of supportive legislation. At the same time, however, just as this legislation will not appear without public pressure and support, it will also be of little avail unless there is an interest in action and a spirit of co-operation.

While everything that has been said on participation applies equally to the individual, to institutions and to economic enterprises, perhaps it should be stressed here that comprehensive planning is not necessarily centralized planning but, rather, integrated planning. It does not call for any greater ownership of enterprises by the governmental sector, although, in the interests of society, it may well call for greater control of the plans of privately-owned economic enterprises.

An interesting example of a middle-of-the-road approach, (between national centralized planning and city planning under a free enterprise system) is that of Yugoslavia. There the approach is that of market planning, in which all economic enterprises are free to develop their own development plans, dispose of their profits, elect their own management, and behave like any enterprise in a market economy. The government intervenes in the market enomomy through such planning devices as credit policy, funds allocated for the development of underdeveloped regions, and so on, although few funds are made available for future development plans. The Yugoslav government does not impose its plans, but rather relies on the willingness of individuals, institutions and enterprises to accept such plans and implement them. It is with this example in mind that we turn to the implementation of plans.

Plan Implementation

While implementation is identifiable as a part of the comprehensive planning process separate from plan formulation, there is a close two-way relationship between these two elements. The action programme of the plan indicates clearly the timing of implementation and the spatial scale of any one operation. Also, administrative planning is needed to assign responsibilities for plan implementation and to develop the increased capability needed to that end. As the implementation of the plan progresses, it is evaluated in terms of the desired effect. On the basis of observations, changes are made to the plan model and to the programme of action. Providing that the plan is a clear and all-embracing guide at any given point of time, there should be few problems with implementation.

However, there are two major weaknesses in any plan. First, the plan cannot anticipate a major change on the international scene that will affect the economic or political future of the country in question. Similarly, it cannot anticipate any revolutionary technological change in any one field of endeavour. All it can do is provide a means by which adjustments can readily be made at the level affected by the change, thus absorbing the impact of a mal-effect or capitalizing on a good effect as it relates to plans at other levels. Second, the plan cannot guarantee successful working relations among individuals, institutions and enterprises. All it can do is provide for the most satisfactory machinery to facilitate these relations.

It is in the market economies that the greatest challenge to the implementation of the plan is experienced. Here, the government sector has to develop a framework within which the operations of private enterprise are known to be both conducive to the common good and, at the same time, rewarding to the private sector. In this instance, the responsibilities must be shared not only between levels of government but also between the governmental and non-governmental sectors. The attempts by the French government have probably been among the most successful, as compared with many other market-economy countries, although the integration here has been achieved largely through governmental centralization, while the participation of the non-governmental sector tends to be limited mainly to that of the larger "corporate machines". It has been emphasized earlier in this chapter that lower levels of government should be given major responsibility for plan implementation, provided that there is provision for supervision by senior governments. Where the private sector plays a part, the question as to which level, or levels, of government it should work with will depend on the scale of the project in question. For example, the day-to-day operations of a privately-owned utility involve close liaison with local authorities, whereas expansion plans may be of main concern to the metropolitan-wide government, if there is one. Furthermore, the maintenance of national standards of service or the need for any international dealings (for example, with power supply or telephone linkage) require that decisions be made by the central government. This rather obvious illustration is given only to indicate that essentially the need for contact with different levels of government is not affected by whether or not the ownership of the utility, or any other service, is governmental or non-governmental.

In both the market economy and the planned economy, the detailed action plans should be formulated and implemented at the local level. The responsibility of the senior governments is to ensure that the long-range and medium-range plans will materialize over the given period of time. The key to this strategy, and the key to local government strategy in regard to private enterprise, is to identify appropriate levers of development, which, once "pulled" or implemented, will initiate further development by lower levels of government and/or private enterprise. Obversely, refusal to pull the levers, which is to maintain the "status quo", is just as effective a control device. The number of development levers decreases with the extension of the time horizon and the space horizon of the plan. There are few levers available to the senior government; for example, port location (or the lending of relative emphasis to different locations); airport location (the location of a third major airport in South East England is a national rather than local issue); railway and major highway location; provision of power supply (which is especially significant in those countries that do not have adequate power coverage for all parts of the country); and, the selection of development areas (both new towns and regional development projects). The fulfilment of those functions for which the central government is responsible should also be seen with this strategic role in mind. For example, the location, and even precise siting, of national defence facilities and major government offices will have considerable effect upon the activities of lower levels of government and private enterprise.

Key levers available to the metropolitan or regional governments are: the provision or location of such major transportation links as expressways and rapid rail transit, including the number and location of access points; the provision of major sewerage, water and, in some cases, power facilities; the location of such major institutions as hospitals and universities; the location of regional parks; and, the location of major shopping centers.

The key levers available to local government and designed to stimulate private development are the provision or location of certain health and welfare institutions, schools, local streets, local parks, sewers, water supply and, in some cases, power supply. It was agreed that housing should not be used as a lever. This list is by no means complete, as different strategies of development have to be adopted according to local conditions. The point to be made is that by utilizing these levers of development, flexibility is still offered to private enterprise and lower levels of government in pursuing their respective roles. Also while these key levers have been described in terms of their impact on spatial development, the timing of implementation is obviously just as important.

The other type of lever open to governments is the provision of financial support or incentives for development. As this has been discussed in greater detail in Part III of this book, and by Hicks in particular, it will suffice at this juncture merely to underline once again its significance as the major means by which senior governments can assist in the improvement of conditions in the metropolitan areas and elsewhere. Again, there is no need for more than a brief second mention of the fact that technical assistance offered by senior governments, while not a lever, is also a useful instrument in assisting in the implementation of plans. The scarcity of expertise, especially in developing countries, makes valid the concentration of specialists and consultants. This relates back to the development of an adequate communications network. The consultative capacity of universities could be improved, along with the development of a system for interchange of information between universities and other institutions in metropolitan areas.

Voluntary groups and individuals may also wish to participate in the implementation of the plan. Voluntary groups tend to be most active in the promotion of the social welfare and cultural activities of a metropolis. While they represent a response to a felt need, it may also be said that their presence is indicative of the unresponsive attitude of the public sector toward the services that tend to be regarded as luxuries rather than necessities. On the other hand, once voluntary agencies are established it often becomes difficult to coordinate their services without heavy financial support on the part of the government. This is unattractive to the public sector as such an arrangement is generally far less efficient than a unified governmental service. Another problem in relying on voluntary agencies for essential services is that they are extremely vulnerable to a reduction of funds, the supply of which is a measure of the economic health of the local community.

Mention should be made of the productivity of the participation of individuals in self-help community development programmes, especially in a labour-intensive economy. Of greater importance is the fact that such participation tends to result in a sense of community pride and can be regarded as socially productive. In cities, as opposed to rural areas, work programmes of this type tend to be small in scale and require intensive supervision by involved and devoted individuals. It is in project development and evaluaton, as opposed to the provision of on-going programmes, that the voluntary agencies, universities and informal volunteer groups can play a most constructive role.

Evaluation and Modification of the Plan

Plan evaluation and modification is an essential part of the comprehensive planning process that acquires greater significance with time. Given a good comprehensive plan incorporating all of the elements enumerated throughout this chapter and Chapter 14, plan evaluation and modification should serve to keep the plan updated. This applies to the long-, medium-, and short-term plans. Thus, at any one time a current plan should exist. Unfortunately, what tends to happen at the present time is that new plans are developed at set intervals of time, with a view to attaining what may be new goals, without sufficient attention being given to the degree of success of measures utilized in the previous plan period. If, on the other hand, the evaluation of the plan accompanies its implementation and undesirable results or implementation techniques are observed, they can be tackled immediately.

While this may appear an enormous task, its feasibility and success are largely dependent on the existence of a feedback process whereby public reaction to the plan may be brought to the attention of all those involved in plan formulation and modification. As has already been mentioned, feedback can be disorganized, and even when organized it is difficult to judge whether a truly comprehensive reaction has been obtained. Thus, it is of added interest here to note the attempt to simplify this process and our understanding of it through the application of systems analysis and also through the use of gaming simulation. The object of these approaches is to forecast the results or effects of plan implementation prior to its implementation in fact. Although those participating in the Study Programme realized that these techniques are undeveloped at present, several participants felt that they offer much potential that is still to be explored. This is yet another field in which further research is needed and in which an international study programme could prove most fruitful.

Index

For Product Safety Concerns and Information please contact our EU
representative GPSR@taylorandfrancis.com Taylor & Francis Verlag GmbH,
Kaufingerstraße 24, 80331 München, Germany

Printed and bound by CPI Group (UK) Ltd, Croydon, CR0 4YY
01/05/2025
01858584-0001